RWBB   ROGER WILLIAMS COLLEGE LIBRARY
TH437 .P33 1984
Parker, Albert D
Planning and estimating heavy constructi

3 1931 00062 7967

DATE DUE

APR 8 1991		
MAR 1995		
DEC 12 1995		
DEC 11 2001		
JAN 02 2007		
GAYLORD		

D1064330

Planning and Estimating Heavy Construction

ALBERT D. PARKER, P.E.
DONALD S. BARRIE, P.E.
ROBERT M. SNYDER, P.E.

McGraw-Hill Book Company

New York   St. Louis   San Francisco   Auckland
Bogotá   Hamburg   Johannesburg   London   Madrid   Mexico
Montreal   New Delhi   Panama   Paris   São Paulo
Singapore   Sydney   Tokyo   Toronto

ROGER WILLIAMS COLLEGE LIBRARY

TH
437
P33
1984

4-23-85

Library of Congress Cataloging in Publication Data

Parker, Albert D.
  Planning and estimating heavy construction.

  "This is a revision and expansion of the two
Parker books, Planning & estimating underground
construction and Planning & estimating dam construction"—

  Includes index.
  1. Building—Estimates. 2. Underground
construction—Estimates. 3. Dams—Estimates.
I. Barrie, Donald S. II. Snyder, Robert M. (Robert
Michaels) III. Title.
TH437.P33 1984   624   82-20796

ISBN 0-07-048489-9

Portions of this book were previously published under the
titles *Planning and Estimating Underground Construction* and
Planning and Estimating Dam Construction, both by Albert D. Parker.

Copyright © 1984 by McGraw-Hill, Inc. All rights reserved.
Printed in the United States of America. Except as permitted
under the United States Copyright Act of 1976, no part of this publication
may be reproduced or distributed in any form or by any means,
or stored in a data base or retrieval system, without the
prior written permission of the publisher.

1 2 3 4 5 6 7 8 9 DOCDOC 8 9 9 2 1 0 9 8 7 6 5 4 3

ISBN 0-07-048489-9

The editors for this book were Joan Zseleczky and Susan Thomas,
the designer was Jules Perlmutter, and the production
supervisor was Teresa F. Leaden. It was set in Baskerville
by Bi-Comp, Incorporated.
Printed and bound by R. R. Donnelley & Sons Company.

Contents

Preface

Since the first publication of Al Parker's twin books on planning and estimating underground and dam construction, much of the basic demand for traditional massive heavy construction projects has shifted to the developing countries. Here in the United States, major heavy construction activities now include projects for rapid transit, wastewater facilities, heavy marine facilities, pipelines, and the renovation and repair of highways and bridges. Traditional construction activities such as the building of dams, tunnels, and powerhouses associated with major hydro and flood-control projects are becoming less and less prevalent. However, in Australia, Canada, and China and in the developing countries of South America and Africa, water and power projects are being carried out on an unprecedented scale. Japan has developed a completely new approach to soft-ground tunneling which is beginning to be appreciated around the world. This book is intended to update the material included in the Parker books to show the present state of the art in dam and tunnel construction. Because of their increasing importance, coverage has been broadened here to include applications of heavy construction techniques in areas other than dams and tunnels and to recognize the significant contribution of other countries in the technological advancement of the industry through innovative research and development.

Heavy construction planning and estimating requires a comprehensive review and evaluation of the plans and specifications, along with a thorough understanding of the conditions to be encountered at the jobsite. An understanding of the production capabilities as well as the advantages and disadvantages of a wide range of construction equipment, construction systems, and alternative methods is essential if the work plan, estimate, and schedule for a particular project are to prove both practical and economically sound.

No other branch of the construction industry offers a challenge comparable to that facing the heavy construction estimator and his associates. Within a short period of time, they must prepare a complete plan for implementing a major project and develop a competitive bid price for a unique situation subject to all of the variable impacts of nature, personnel, inspectors, regulations, and political constraints. Heavy construction estimating requires experience, judgment, and skill, since there are so many different construction systems, types of construction equipment, and construction methods which can be used in planning and bidding a project. Because of the opportunity for individual innovation on the part of the estimators and the financial consequences of these choices, heavy construction planning and estimating is one of the most challenging and potentially rewarding careers in the construction industry.

Part One of this book, including Chapters 1 through 3, provides an introduction to the field, discusses heavy construction production plant capabilities, and points to the ever increasing importance of quality control, governmental regulation, and safety requirements. Part Two, including Chapters 4 through 9, contains an updated version of fundamental information originally set forth in Parker's *Planning and Estimating Dam Construction*. Part Three, including chapters 10 through 13, similarly updates the state of the art, incorporating information first presented in *Planning and Estimating Underground Construction* while also presenting new material on soft-ground tunneling. Part Four, including Chapters 14 through 16, is completely new (with exception of the material on powerhouses) and concentrates on such heavy construction applications as pavements, pipelines, rail networks, powerhouses, stations, bridges, and other massive concrete or steel structures. This section also discusses applications of heavy construction techniques for piling, dredging, and other marine construction. Part Five, including Chapters 17 through 20, substantially revises and updates example estimates for a tunnel and a dam originally included in the Parker books. It also contains chapters on heavy-construction estimating procedures and joint-venture and bid-preparation guidelines.

The revising authors wish to acknowledge their debt to Al Parker for his efforts in the preparation of the original twin volumes, and to express their appreciation to his widow, Helen Parker, who has proved most cooperative and helpful in the preparation of this book.

Donald S. Barrie

Robert M. Snyder

DISCLAIMER

In recent years women have begun to take increasing interest in the field of heavy construction. Unfortunately, because the field was occupied solely by men throughout so much of its history, the language of heavy construction has grown up around the assumption that its practitioners would always be men. We welcome women to the field and ask the indulgence of any women who may read this book if we sacrifice accuracy to linguistic convenience by using such terms as *radio man* when we mean *radio operator*. To change all such terms would result in many an awkward and laughable phrase.

ONE

Overview of Heavy Construction

Chapter 1 Introduction to Heavy Construction

The purpose of this book is to describe the study and computations that are necessary in planning a heavy construction program and in estimating the cost of the project. In order to plan and schedule the construction, the engineer must be fully familiar with the many different types of heavy construction plant and equipment and with the different construction methods. Before estimating the construction cost the engineer must first select the equipment and methods to be used on the particular project or any portion of the project where heavy construction methods and equipment are to be utilized. It is for this reason that equipment and construction methods applicable to dam construction, tunnel construction, and other heavy construction projects are described before estimate preparation is introduced.

Population growth and rapid development of resources in emerging nations have resulted in growth of construction projects throughout the world to provide for water storage, land irrigation, power generation, and flood control. Transportation needs have generated numerous requirements for heavy construction techniques in the building of vehicular tunnels, bridges, canals, causeways, highways, airports, and rail networks. The demand for energy has produced massive pipeline projects such as the Alaska oil pipeline, which challenged the state of the art in heavy construction, numerous gas and coal-slurry pipelines, and other major mining applications. Massive concrete and steel structures associated with fossil and nuclear power plants as well as applications in transportation projects continue to provide a growing demand for heavy construction expertise throughout the world.

In the United States pumped storage projects, rapid transit projects, the development of smaller hydroelectric sites and power plants, and the maintenance and renovation of the existing dams, tunnels, bridges, highways, and other facilities will continue to provide a stable demand for heavy construction knowledge in the foreseeable future.

PURPOSE OF AN ESTIMATE

The primary purpose of an estimate is to furnish a logical cost and time basis for a bid for a heavy construction project, with a bid price that will allow the contractor to make a reasonable profit. Heavy construction is a competitive field, and it is necessary for the contractor's organization to prepare an accurate estimate of the cost of doing the work. If the estimated cost is too low and the contractor is the successful bidder, he will lose money on the project. Conversely, if the estimated cost is much higher than the actual cost, the contractor will have very little chance of being the low bidder and hence will not get the job. Therefore, the estimate must be based on the use of the most economical construction system, on performing work with construction plant and equipment units that have been selected to operate efficiently as component parts of the construction system, on proper work scheduling, and on accurate costing of all work items. The contractor places a great amount of confidence in the skill and integrity of his estimator, since proper bidding is essential to his success. Estimating is less important outside the construction industry; the contractor is the only manufacturer who must price his product before it is produced.

If the contractor is the low bidder on the construction project, the estimate has several secondary uses. All the adjustments made before bidding are carried back into the body of the estimate, and it is then published as a "budget" estimate. This budget estimate provides the job management with a list of the plant and equipment required to construct the job, is used as a guide in determining the methods of constructing the work, and furnishes the basis for job control, since the estimated cost can be compared with actual cost and estimated production can be compared with actual production. This comparison points out the cost items that are critical and indicates when work falls behind the schedule and should therefore receive attention. Cost-accounting procedures and cost accounts are established in accordance with the estimate's format and the cost items and divisions used in the estimate. If there are changes in the work or in job conditions, or changes that affect job completion which are sufficiently extensive to warrant presenting a claim to the owner, the budget estimate is used as a basis for preparation of the claim.

If the contractor is not the successful bidder, he should compare his estimate with the bid submitted by the successful bidder. This comparison, with field inspections of the low bidder's work progress, may indicate improvements in construction planning that can be incorporated into future estimates.

HEAVY-CONSTRUCTION ESTIMATING METHOD

The heavy-construction estimating method is used for dams, tunnels, powerhouses, and similar projects. This estimating method treats each construction project as a separate entity, and the construction-cost computations are based on the use of the construction equipment units, the plant facilities, the labor rates, the labor productivity, the working conditions, the work schedule, the work sequence, the subcontract prices, and the permanent material and equipment prices that apply to each project.

This estimating method requires that the estimator take an overview of the entire project. The specification work constraints, the type of project, its location, any special requirement such as stream diversion, the topography, the work scheduling, and the work sequence are all studied and evaluated. A construction system is selected on the basis of these factors as well as considerations of economy and the required completion date. A construction schedule is then prepared indicating the time available for each construction operation. Selection of construction equipment of appropriate type, size, and capacity completes the construction system. The work required for each work item is planned so that the construction equipment is used to maximum capacity. The number of men, hours of equipment operation, quantities of consumable supplies and repair parts, the subcontracted work, and the required permanent materials and equipment are listed for each work item. The listings are then costed using the labor rates, the supply cost, the price of repair parts, the subcontract prices, and the materials and equipment quotes that apply to each project. The individual work-item costs are then summed to arrive at the total cost for direct labor, consumable supplies, subcontract work, and permanent materials and equipment.

Plant and equipment depreciation chargeable to the project is estimated by preparing a detailed list of all required construction equipment units and plant facilities. Acquisition plus erection cost, less resale value at the time of project completion, represents the depreciation for each unit.

Indirect expenses for supervisory costs and labor, bond premiums, taxes, and insurance are also estimated. These costs vary with work volume, time required for construction, and the amount of the contract. Escalation is estimated by determining the labor and material expenditures that will be made in each yearly time period, estimating the percentage that the labor and material cost will increase during each period, and then computing the total increases.

The total estimated cost for the project is the sum of the costs of labor, consumable supplies, subcontract work, permanent material and equipment, equipment depreciation, indirect expense, and escalation.

UNIT-COST ESTIMATING METHOD

The unit-cost estimating method is used to estimate the cost of all but heavy construction projects and requires much less effort than the heavy-construction estimating method. Work quantities are established for each bid item and then multiplied by the applicable unit cost. The sum of these extensions represents the prime contractor's on-site construction cost. To this figure are added the payments to subcontractors and the cost of procuring permanent equipment, materials, and facilities, to arrive at the total estimated cost. This procedure varies in minor ways among contractors. Some contractors use unit costs that include all on-site costs. Other contractors use unit costs that exclude equipment depreciation and/or indirect expense, and these costs are then computed separately and added to the extended total.

The unit costs used in an estimate are based upon man-hours that have been developed by each contractor from his experience in similar work. These costs can be readily adjusted for changes in labor productivity or in labor rates. However, there is no accurate method of adjusting these unit costs to reflect the use of different types of construction equipment.

In preparing an estimate using the unit-cost method, the major task is taking off the quantities of work. Estimating skill is required to make proper use of, and adjustments to, the unit costs. Computer manufacturers have developed programs for taking off quantities, for storing and using unit man-hour production rates and material costs, and for making and totaling the cost extensions. The computer greatly lightens the estimating work load for contractors who use the unit-cost estimating system.

COMPARISON OF ESTIMATING METHODS AND THEIR APPLICATIONS

A heavy-construction estimate is prepared using the plant and equipment, the labor rates, and the labor productivity applicable to the specific project. A unit-cost estimate is prepared using work performance on similar projects. Adjustments must be made for changes in labor rates and for labor productivity, but there is no way to make accurate adjustments for changes in equipment use. Because of the savings in estimating effort, contractors use the unit-cost method for construction projects where unit costs do not vary from project to project as a result of changes in sizes or types of construction equipment. However, the heavy-construction method must be used to produce estimates for projects where the costs do vary, as in tunnel, dam, and powerhouse projects.

Projects of these types are constructed most economically using large amounts of construction equipment. Furthermore, different types of equipment are required to construct similar projects, since the physical characteristics at the job locations will always be different. On such projects, construction cost for the major work items is primarily composed of the cost of operating, maintaining, and depreciating the construction equipment. Changes in equipment can result in large differences in the unit costs of the major work items and in the total construction cost.

Heavy-construction estimating can be simplified by using the unit-cost method for estimating the cost of many minor work items that are constructed by hand labor, such as the construction of built-in-place wood forms, the placement of reinforcing steel, hand excavation, electrical work, plumbing, and architectural work. Unit costs for these items will vary among projects only because of differences in labor rates and labor productivity.

The unit-cost estimating method can be used to prepare road-construction cost estimates since contractors in this field maintain their own fleets of construction equipment which are used on each of their road projects at standard depreciation rates. Since all projects use the same basic equipment and are charged the same depreciation rates, unit costs will not vary among projects.

The unit-cost method is also suitable for estimating the cost of industrial, commercial, and building construction. For such projects, the prime contractor's on-site labor and construction equipment cost is a smaller portion of the total construction cost; the greater portion of the construction cost is for permanent materials, equipment, facilities, and subcontracted specialty work. The prime contractor's construction equipment will be basically the same on all jobs, and in any case represents such a small portion of his on-site cost that a change in equipment types will not result in significant changes to unit costs. However, the use of man-hour units in developing extended labor costs has become prevalent in view of the wide variation in hourly labor rates at different project locations. Because of both varying wage rates and inflationary increases over the years, contractors' records kept in man-hour units are much more useful in preparing unit-cost estimates than records which are kept in dollars.

COMPUTER ESTIMATES

Computer preparation of estimates can eliminate many tedious calculations and is an excellent way of preparing estimates, provided the computer's limitations are recognized. The computer does exactly as it is instructed; it cannot be used as a substitute for estimating ability. In computer jargon *GIGO* still controls—that is, garbage into the computer, garbage out of the computer. A great amount of planning ability is needed to instruct the computer on all the necessary details involved in preparing a heavy-construction estimate. The overall planning discussed earlier must still be done. The construction system must be determined, plant and equipment components selected, production rates and hourly operating and maintenance costs established, crew sizes determined, and supply consumption estimated.

After this planning is completed, data can be fed into the computer, and it will produce a typed estimate. This estimate should be checked to determine if the computer was properly programmed. An example of inaccurate programming is seen when overbreak concrete is programmed to reduce instead of increase the pay quantity unit cost. Proper programming requires both an experienced estimator to establish the instructions for the computer and a good programmer to see that these instructions are properly translated into computer language.

The program used should include equipment hours; plant hours; man-hours; fuel, oil, and gas consumption; and repair parts consumption. The best programs output typed estimates in the same basic form that is used for calculator estimates. Such programs have been developed by certain consultants and contractors but are not generally available from computer manufacturers. The manufacturers' programs normally use the unit-cost method and have its same limitations. Most of the inaccurate heavy-construction computer estimates have been produced by personnel who understood computer programming but who were not experienced in heavy-construction estimating. The computer cannot supply estimating knowledge; it supplies only rapid computations based on adequate instructions.

Progressive heavy construction firms have generally adopted computer estimating, since it can greatly reduce the amount of manual work done in the estimating department. Computers were first introduced into many construction companies by using them to prepare payrolls, maintain equipment records, record costs, and make detailed schedules from precedence diagrams. They then were introduced into the estimating department to make quantity takeoffs of regular structures, to compute cycle times for haul trucks and scrapers, and for other recurring applications. Invariably as the estimating personnel become acquainted with the computer, they recognize the advantages of using it to prepare the complete estimate and they develop many innovative applications. Innovative uses of microcomputers are now being developed by progressive estimators.

A good first project with the computer for an estimating department is to use it to prepare the budget estimate for a project that was successfully bid with a longhand estimate. The estimate data can be the computer input; an estimate will be produced that will furnish a large amount of detail helpful for job control. Some estimators who have both computer and calculator estimating experience think this is the right application for computer estimates. They believe that a bid estimate can be turned out more rapidly by the longhand method, but that if the bid is successful, a computer estimate should be run so that the details of the estimate will be more readily available.

CALCULATOR ESTIMATES

Calculator or longhand estimates have been used in the heavy construction industry since its infancy. If they are prepared in the right format, they provide great flexibility. Estimate costing of part of the project can be performed while the remaining work is being planned. At any stage of estimate preparation, it is relatively simple to make changes to completed work when new work concepts are conceived. With a computer, planning is often complete before the estimate is costed. With calculator estimates, there will be more chance of arithmetical errors than with computer estimates; however, programming errors will not occur. Arithmetical errors can be reduced by using an estimating format that provides checks for all major arithmetical computations.

Since calculator estimates more readily illustrate estimating procedures, this type of estimate is used for illustrative purposes in Chaps. 18 and 19.

CONSTRUCTION-SYSTEM SELECTION

A construction system comprises the construction time, manpower, construction equipment, construction plant, and supporting facilities as an integrated unit for performing all the work required to complete a project. Successful heavy-construction estimating requires the selection of construction systems that can produce the work in the most economical manner.

Because of its high productivity and despite its high purchase cost, construction equipment is more economical to use on heavy construction projects than hand labor, given today's high labor cost. As a result, the cost connected with operation and depreciation of the construction equipment generally forms the largest portion of the total construction cost. Different equipment units will differ in their operating, maintenance, and depreciation costs and in their rates of production. Construction cost is lowest when construction is performed with equipment having the most favorable relationship between cost and productivity. However, the construction equipment that can be used on any heavy construction project is limited to the types that will function together as an integrated construction system. Thus the construction system must be determined prior to the selection of construction equipment.

Main Construction System

The main construction system is the coordinated arrangement of the subsystems into one complete operating unit. The system must have sufficient capacity so that crash scheduling will not be required. It must be capable of constructing the major items of work with maximum efficiency, and yet with enough flexibility to supply the minor demands placed on the system during the construction of the minor work items.

The main construction system is used as the control of the composition and production capacity of each subsystem so that a balanced operation will result without any activity gaps. Main-system review will indicate whether any equipment supplied for one subsystem can also be utilized in another subsystem. For example, excavating equipment supplied to the excavation subsystem can often be utilized later in the concrete subsystem for either aggregate-pit or rock-quarry excavation. This review will also reveal whether all equipment supplied to the subsystem is in balance with the work to be accomplished. Often large expenditures for equipment are planned for a minor amount of work. Systems review may show that this work should either be done with rented equipment or be subcontracted.

Subsystems

The specification constraints, construction schedule, work sequence, and climate and topography of the site must be studied and evaluated, since they control the selection of construction subsystems to be used to construct the project. That is, the subsystems selected must meet the work constraints and comply with the work conditions, and also must be economical and capable of functioning efficiently together as components of the main system.

The number of subsystems required depends on the type of project. For a concrete dam, an excavating and a concreting subsystem are minimum requirements. Subsystem selection is a complex problem because so many different types of work are involved in dam construction. Work may include site clearing, tunneling, river diversion, common excavation, rock excavation, drilling and grouting, concrete construction, rock quarrying, rock embankments, earth-fill embankments, powerhouse construction, permanent material installations, and equipment erection. The types of work that must be performed for the construction of any heavy construction project vary with the purpose of the project, the type of facility, the site characteristics, and other factors. Because the work required for heavy construction projects may differ so widely, contractors have developed a large number of subsystems. Each can be used economically for some projects but will not necessarily be efficient in the construction of projects with different working conditions.

Limitations on Construction System Development

Work constraints limit the contractor's development of new construction systems. Most of these work constraints are delineated in the specifications or established by labor contracts and are therefore totally binding upon the contractor. Many specifications are so out of date that they contain work constraints that do not apply to modern construction practices and are, in fact, a great hindrance to the development of better construction systems. The systems that have developed for applications shown in this book are the most suitable to use with present specification and labor constraints. These are continually becoming more expensive and complex as additional work constraints are added to the specifications or included in labor contracts. Most system improvements to date have been the result of contractors' efforts in developing new systems, or of their modification of systems to include new work concepts developed by other industries. Japanese and Swedish companies have been especially innovative in developing new and improved heavy construction systems.

SCHEDULING

The construction schedule is utilized as the coordinating tool in the systems approach to heavy-construction estimating. It shows the time available for constructing each phase of the project. By using the schedule in conjunction with work quantities, the desired production rates and work sequence can be established, and from this, the required construction equipment and facilities determined. The construction schedule is also used for *resource leveling*—that is, the scheduling of the construction of noncritical work items at time periods when the construction equipment is not used to full capacity on critical items. Without resource leveling, additional plant and equipment may be required for these noncritical work items. To make the construction schedule as useful as possible as a coordinating tool, it should be prepared in a form that illustrates the flow and sequence of all operations and yet is simple enough to be readily understandable.

Advantages of Network Diagramming

The bar chart and several types of network diagramming have been used to draft schedules for heavy construction projects. In the past, the bar chart has been widely used because it is simple and readily understandable. Recently it has been modernized by some contractors to show work sequence, with triangles indicating the original schedule, the revised schedule, and the actual schedule. Bar-chart schedules do not show work flow and the interrelation among activities, nor do they highlight essential work. Because network diagramming provides methods for illustrating these points, it has largely replaced the bar chart in the heavy construction industry.

Network diagramming enables the construction planners to study the effect that the performance and completion of one operation will have on others. Use of a network diagram to determine the critical path in dam construction, for example, will show mobilization of plant and equipment, plant and equipment erection, construction of the diversion facilities, dam excavation, dam placement, and the placement of gates and other mechanical items. Even if the critical path is known, the diagram is helpful in the determination of the minimum construction time, since it indicates whether there can be any overlapping of the activities that are part of the critical path. It also indicates when the construction of noncritical work items can be scheduled, thereby facilitating resource leveling. Furthermore, network diagramming can assist in scheduling the receipt of permanent materials and equipment so that on-site construction will not be delayed. Another advantage of the logic diagram is that it can be readily revised as construction is performed, reflecting any change in work performance or in job planning.

Comparison of Network Diagramming Methods

The principal types of network diagramming presently in use are the *program evaluation and review technique* (PERT) and the *critical path method* (CPM). With the PERT system, arrows are labeled with events and connect to numbered nodes so that each event is described by the number in the node at the start of the event and the number in the node at the end of the event. PERT can use three estimated times for each activity: the optimistic time, the pessimistic time, and the most likely time. The CPM method using an arrow diagram is very similar, the difference being that the arrows are labeled with activities and only one estimated time is used. In order to properly show work sequence both methods of scheduling require the utilization of dummy activities. The preparation of these arrow diagrams is so complex that errors frequently creep into any complicated network, usually because of the absence of some required dummy activities.

Both of these methods of diagramming are so complicated that special knowledge is required for their preparation; therefore, this function is often performed by specialists or consultants who may or may not have had construction experience and whose only job contact is from occasional visits. Many consultants specialize in the preparation of PERT or CPM arrow diagrams for contractors who must submit schedules of this type to the owner, but who do not want to be bothered with their preparation. Since these construction schedules have not been prepared by the job-construction personnel, the construction forces may classify them as consultant-prepared, theoretical diagrams, ignore them, and not keep them up to date. When this happens, their main usefulness is destroyed. Another disadvantage of a consultant-prepared schedule is that it is often so complex, with so many activities and dummy activities, that only an arrow-diagramming specialist can comprehend its ramifications.

The dummy activities required by arrow diagramming are often confusing to job personnel, who prefer a schedule that shows only required activities. Furthermore, a change or modification in these diagrams often requires additional arrows, necessitating a renumbering of the arrow nodes; thus, an activity may not retain the same number for the life of the project. To summarize, arrow-diagram schedules are often ponderous to use and very difficult to follow in determining the overall job direction and construction-system requirements. For those interested in PERT or CPM arrow diagramming, there is considerable published data on these methods.[1.1]

Precedence Diagramming

The CPM method which uses precedence diagramming is a simplified form of network diagramming. It is quite similar to a flow chart and does not require the insertion of dummy activities. It has all the advantages of PERT and arrow-diagram CPM, and none of the disadvantages.[1.2] It is not as stylized so the diagramming can be done in many ways. Activities are placed in circles or squares and given an activity number which does not have to be changed during the life of the project. The squares can also carry a cost-code number, earliest starting dates, latest starting dates, time required to complete the activities, and the calendar date of the start of completion of any operation, as illustrated in Fig. 1.1a.

The squares can be plotted on a time scale and connected with arrows showing the sequence of operations. These arrow connections are illustrated by Fig. 1.1b, which shows that foundation concrete cannot be placed until all

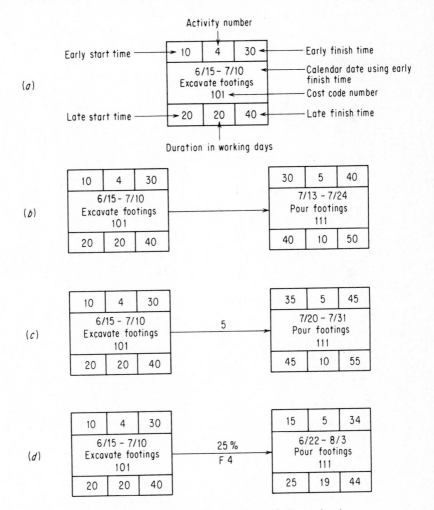

Time unit is working days based on a 5-day workweek

Figure 1.1 Precedence diagramming.

the foundations are excavated. Restrictions and constraints occurring between activities can be shown. For example, Fig. 1.1c indicates that concrete cannot be placed in the footings until 5 days after all the excavation is completed. Figure 1.1d indicates that foundation pouring can start after 25 percent of the foundation has been excavated but cannot be completed until 4 days after all footings have been excavated. With these simple designations, a complete diagram can be constructed. Many other diagramming and designation methods are used for precedence diagramming, depending on the owner's requirements or the contractor's practice. Many proprietary programs are now available for both mainframe and microcomputer application utilizing precedence diagrams. Useful overall references include Stanford University Publications.[1.3,1.4]

ESTIMATING ABILITY

To plan a heavy-construction estimate and develop the most efficient construction system, the estimator must have practical construction knowledge, including familiarity with the types of equipment and facilities required to accomplish the construction, the crew sizes required for each operation, their established wage rates, and the time required to complete each construction phase. For a dam estimate he must understand construction-plant layout, aggregate manufacturing, refrigeration, mixing-plant operation, form layout, concrete-placement methods, dam-foundation excavation requirements, methods of diversion, economical dirt-moving procedures, quarry operations, embankment compaction, equipment installation, labor regulations, equipment maintenance, camp construction, logistic support, overhead cost, insurance coverage and rates, safety practices, subcontract administration, bond rates, escalation, cash requirements, and accounting procedures.

It is difficult for an estimator to visualize all the problems that will be encountered in construction unless he has had construction field experience. This experience can be supplemented by job inspections, the perusal of past job records, the study of job reports published in trade magazines, and the reading of construction books. Construction experience is invaluable in determining labor production for operations where machines are not the basic production units but production is the result of labor working with hand tools. The amount of work that construction labor will produce with hand-held tools is the hardest part of an estimate to prepare, since it is dependent on the efficiency of the laboring force and the planning, knowledge, and leadership of the supervisory force. Productivity of this laboring force is not mathematically determinable but must be based on the judgment and experience of the estimator.

In contrast to this, the labor cost of operations where production is accomplished by utilizing equipment and plants can be computed mathematically.

These types of operations are machine excavation, truck haulage, embankment construction, aggregate production, mixing and cooling plant production, cableway or crane operation, etc. The number of men required for each unit of equipment and for each plant is dependent on their size and type. The time required for constructing any facility is dependent on equipment and plant production rates, which are primarily a matter of determining the production capacities of machines. Therefore, labor-cost estimating for these types of operations is based on equipment production rates and varies only in a secondary manner with the ability and efficiency of the workmen.

The time available for the preparation of a competitively bid estimate is so limited that it is imperative that the estimator have the ability to schedule his time so he can spend as much time as possible on the major work items and only a minimum amount of time on the items whose total cost will be relatively minor.

ESTIMATING DATA

Heavy-construction estimating is a complex subject and requires such a wide range of construction knowledge that estimators should maintain reference files of estimating data. Catalog information and other data secured from equipment manufacturers will provide information for the selection of the types of construction equipment and of construction plant. Files should be maintained on the available plant and equipment, its production capacities, and on repair parts and supply consumption. If any complicated facility, such as an aggregate plant, refrigeration plant, or mix plant, or if any specialized machinery, such as a tunnel-boring machine or a cableway, is required, the manufacturer's representatives will provide design assistance and assist in plant selection so that the proper facilities are provided to accomplish the work requirements. Manufacturers will also supply quotations on equipment and construction plants to enable the estimator to properly price these facilities.

Copies of the latest labor agreements are important to the estimator as they furnish information on craft jurisdiction, working rules, working hours, wage rates, and fringe benefits. In order to determine the payroll burden, files should be kept current on insurance quotes and on state and federal payroll taxes.

A file of costs from construction projects and a file of completed estimates will provide information on construction methods and equipment selection and will assist in determining the equipment's consumption of fuel, oil, and grease; maintenance cost; labor productivity; and the like. These files furnish invaluable assistance as a check against estimated unit costs. If differences occur, they should be analyzed to determine if they are a result of estimating errors, of the use of different construction systems, of changes in labor rates and labor productivity, or of differences in efficiency of the supervisory personnel used on the project.

A list of interested subcontractors and material suppliers should be maintained. These should be reliable contractors who have a good record of work performance. The estimator can then depend on these firms to quote on permanent materials and subcontracts, thus relieving him of this responsibility.

A file of articles from trade magazines describing the construction of major projects may furnish solutions to many of the problems the estimator encounters.

PLANT AND SAFETY REQUIREMENTS

An understanding of plant and safety requirements is necessary for all types of heavy-construction estimating. Chapter 2 discusses the layout and details of aggregate plants, gravel processing plants, batch plants, asphalt plants, and ready-mix plants. Chapter 3 summarizes inspection, testing, and safety and health requirements which are becoming increasingly important on most construction projects.

DAMS, QUARRIES, AND EMBANKMENTS

Dam estimating is so complex that it is necessary to develop the construction background before it can be explained. In Chaps. 4 through 9, background information on construction scheduling, construction-system and equipment selection, construction planning, and cost estimating is presented. This consists of defining the types of work required for dam construction, explaining how each type is performed, describing the available construction equipment and plant facilities, explaining how their work capacity can be computed, and explaining the controls for construction-system selection. All information presented in the first nine chapters is a prerequisite to understanding the information presented in Chap. 19, which explains how the construction planning of a dam located within the United States is performed and which illustrates preparation of an estimate of its construction cost.

TUNNELS AND UNDERGROUND STRUCTURES

A thorough understanding of underground work is necessary before preparation of a tunnel estimate. Chapters 10 through 13 present background information on methods and equipment utilized in the construction of conventional hard-rock tunnels, in soft-ground and mixed-face tunnels, in shafts and underground chambers, and in concrete linings. Chapter 18 presents a sample tunnel estimate for a fictitious Sierra tunnel driven by conventional methods.

OTHER HEAVY CONSTRUCTION APPLICATIONS

Additional information is presented in Chaps. 14 through 16 covering other heavy construction applications, including piling, caissons, dredging and marine construction; concrete and steel structures; and highways, airports, canals, pipelines, and rail networks.

ESTIMATING CONSIDERATIONS

Chapters 17 through 20 develop a comprehensive set of heavy-construction estimating procedures, present sample estimates for both a tunnel and a dam project, and discuss joint venturing and preparation of the actual bid.

SUMMARY

The descriptions and information included in this book are very general in order to furnish an overall concept of the procedures and methods used in planning, scheduling, and estimating heavy construction projects. Emphasis is on the contractor's approach to heavy construction rather than on the design features. A more detailed discussion on individual applications can generally be found by consulting the references listed at the end of each chapter as well as in other more specialized publications.

REFERENCES

1.1 Gabrial N. Stilian, *PERT: A New Management and Control Technique,* American Management Association, New York, 1962.

1.2 John W. Fondahl, *A Non-Computer Approach to the Critical Path Method for the Construction Industry,* Technical Report No. 9, Stanford University Dept. of Civil Engineering, The Construction Institute, Stanford, California, 1961.

1.3 John W. Fondahl, *Methods for Extending the Range of Non-Computer Critical Path Applications,* Technical Report No. 47, Stanford University, Dept. of Civil Engineering, The Construction Institute, Stanford, California, 1964.

1.4 John W. Fondahl, *Short Interval Planning Using SIPCPM,* Technical Report No. 270, Stanford University, Dept. of Civil Engineering, The Construction Institute, Stanford, California, 1983.

Chapter 2 Heavy Construction Plant Requirements

This chapter discusses *general plant,* which services a job as a whole, rather than *specific plant,* such as tunnel shields, which are required only for particular types of construction. Specific plant is described in other chapters which pertain to various types of construction activities.

Most equipment can be provided with enclosures, heaters, and air conditioners to provide operator comfort in severe climates. All equipment must be provided with appropriate safety devices, such as roll-over protection, seat belts, fire extinguishers, lights for night operation, and backup alarms for the protection of operators and nearby workmen.

MOTOR GRADERS

The motor grader is the universal tool for finishing surfaces to grade, for light spreading, for excavating roadside ditches, for preparing road-mix surfacing, and for general roadway maintenance. It is also useful for plowing snow, though a snow-thrower attachment will be needed if snow depth is great or if the snow must be moved beyond the end of the blade.

A motor grader should be a part of every scraper spread and a part of every hauling fleet to keep haul roads in good shape and free from tire-damaging rocks. Good haul-road maintenance will ensure higher production because it will permit faster haul speeds and reduce equipment maintenance costs, particularly tire costs.

Motor graders come in a range of weights from about 25,000 to about 60,000 lb and with engines ranging from about 125 to about 250 hp.

BOTTOM-DUMP AND END-DUMP TRAILERS

Bottom-dump and end-dump trailers are designed to be used with the same two-wheeled tractors that are used with scrapers, or with four-wheeled tractors.

Bottom-dump trailers are available in a capacity range of 50 to 150 tons. They are usually loaded with belt loaders or wheel loaders, or from loading hoppers. Loading with shovels, draglines, clamshells, front-end loaders, or similar equipment which drops loads into the hauling unit is very hard on the gates and gate-operating mechanism. Bottom-dump trailers deposit their loads in windrows in the disposal area. These windrows must in most cases be spread with bulldozers or motor graders. Bottom-dumps are also sometimes used for other purposes, such as hauling gravel to a screening plant, where they can dump directly into a receiving hopper.

End-dump trailers have many of the characteristics of end-dump off-highway trucks. They are usually loaded with the types of equipment not suitable for loading bottom-dumps. Because of the articulation between the tractor and the trailer, they have smaller turning circles than off-highway trucks. For this reason they are often used for haulage from large-diameter tunnels or from underground chamber excavations.

BELT LOADERS AND WHEEL EXCAVATORS

A belt loader is towed by a tractor taking a continuous shallow cut. Thus it is useful where it is desirable to avoid mixing soil strata or where it is desirable to harvest only the aerated (or moistened) surface soils from a borrow pit. Subsequent cuts will be made after the new surface has been aerated or moistened. Although a belt loader can load almost any type of hauling equipment, it is most commonly used with bottom-dump trailers. Smaller self-propelled units are sometimes used for recovering sand or gravel from stockpiles.

Another type of belt loader is usually referred to as a *rough loader*. This machine is stationary and is fed by bulldozers, often working in shot rock or other difficult materials that require very sturdy construction of the loader. These machines are most commonly used with end-dump off-highway trucks.

A wheel excavator is usually designed for very high capacities, often in the range of 100 tons per minute. The excavating buckets are mounted on the rim of a large wheel, which in turn is mounted on a boom that can be raised, lowered, and slewed as it progresses through the cut. The bottoms of the buckets are sometimes made of a mesh of steel chains which better expel sticky material onto the conveyor belt that delivers the excavated materials from the wheel to the hauling equipment. Because the wheel excavator works best at a depth approximating the wheel diameter, all the materials excavated at one pass are well blended together. Delivery is usually to bottom-dump trailers, with an arrangement so that the flow can be switched from one haul unit to another without stopping the belt. At Oroville Dam, California, a wheel excavator charged a long conveyor system at ground level which ultimately delivered the material (gravel spoils left by earlier gold dredging) to storage hoppers, which in turn loaded standard-gauge railcars. These cars were dumped by a rotary car dumper near the dam, and another conveyor delivered the material to rubber-tired hauling units on the dam.

It is apparent that wheel loaders are very expensive machines to purchase, but they have low operating costs per unit of excavation. They are suitable only for very large projects.

A *trenching machine* is similar to a wheel loader, but is designed to excavate relatively narrow trenches and to deliver the spoils to a windrow alongside the excavation, or to dump trucks moving along with the trencher. It also differs from a wheel excavator in that it is designed to do its work below the level of the ground on which the machine moves instead of above that level, as is the case for a wheel loader.

SCRAPERS

The more common form of scraper is the one in which the front of the machine is supported by a two-wheeled, rubber-tired tractor. The less common form is one with wheels at both the rear and front of the machine, which is towed by a four-wheeled, rubber-tired tractor or by a crawler tractor. Only the more common type will be discussed here.

Scrapers are usually purchased as complete units of scraper plus tractor or two scrapers in tandem plus a tractor. In some single-unit machines the scraper is provided with an engine and drivetrain to supply additional power. Some of the smaller single units are provided with equipment for self-loading. Scrapers vary in capacity from about 10 to about 50 yd^3 or more for tandem units.

The scraper performs by slicing off a shallow stratum of earth as it is pulled through the cut. This is a particularly useful feature where it is desirable to prevent mixing a thin stratum of soil with the underlying soils. After the scraper is completely loaded, the bowl is raised and the unit starts for the disposal area. Most scrapers have sufficient power to travel in the range of 25 to 35 mph (40 to 55 km/h) on a firm and level haul road. Scrapers are not commonly used for hauls longer than about 1 mile (1.6 km), partly because of problems with heat buildup in the tires. Scrapers are always fitted with retarders to reduce dependency on brakes on downgrades. The bowl can be dropped for additional braking in an emergency situation.

Scrapers without self-loading features may be self-loaded if they have sufficient power. This will involve shallow cuts and long loading times. Alternately, scrapers can be push-loaded, using a rubber-tired or crawler tractor. In this case the number of scrapers should be so calculated that neither the scrapers nor the pusher is idle for much of the time.

Scrapers are sometimes used for loading rock which has been ripped or blasted. This is, of course, very hard duty, but the decreased machine life and increased maintenance cost may, on careful study, be found to produce lower excavation costs than would other methods. Part of this saving may arise from the fact that scrapers deposit their loads in relatively thin and uniform layers which may not need to be spread by other machines.

Scraper performance depends upon the available horsepower and torque and upon grade and rolling resistance. Although the estimator should be able to calculate the effects of these several factors, he will, in practice, use the tables and performance curves calculated by the manufacturer for each machine model. Use of this data will speed the process and will reduce errors in the calculations.

POWER SHOVELS AND RELATED MACHINES

Power shovels may be engine-powered or electrically powered. The electric machines have lower operating costs and lower maintenance costs. Electric shovels are usually in the 6-yd^3 bucket size or larger. The largest diesel shovels are in the range of 15-yd^3 bucket size.

Shovels used in heavy construction are almost invariably crawler-mounted. Very large shovels, used mostly in open-pit mines, sit on a steel "tub" when digging and "walk" on pontoons when traveling. The common range of sizes for construction use is from 1 to 20 yd^3.

There are so-called "combination machines" which can be converted for shovel, dragline, clamshell, crane, or backhoe use, but they are primarily designed either for shovel or backhoe use or for crane, dragline, or clamshell use. Machines designed primarily for shovel or backhoe service have short crawlers to allow plenty of room for the bucket to complete its excavation arc. Machines designed primarily for crane, dragline, or clamshell work do not have this limitation and have long crawlers for stability, or they may be truck-mounted.

Shovels are used to excavate materials above the plane on which the machine is located, though they can work slightly below that plane. Shovels dig in an arc away from the machine and are most efficient working a face height which fills the bucket in one pass. The factors which affect shovel production rates are the height of the face, the angle of swing to dump the load, the type of material being excavated, operator efficiency, and job and management conditions. Job and management conditions are good if the flow of the excavation is solid and is kept clean, if hauling vehicles are always promptly and properly spotted, if maintenance is good, and if the rock is shot to sizes that the shovel can readily handle when it is working in a rock cut. The optimum height of the face is usually about equal to the height of the dipper-stick trunnion.

Backhoes are used to excavate materials which are below the plane on which the machine is located, though they can work slightly above that plane. Backhoes dig in an arc towards the machine, and the depth to which they can excavate is a function of the boom and handle lengths. Backhoes most often handle buckets which range in size from $1/4$ to 6 yd³, but far bigger units have been built for special requirements, such as underwater excavation. Buckets are available in a wide range of widths for general excavation or for trench excavation. The factors which determine backhoe production rates are similar to those that determine shovel production rates.

Draglines are used to excavate materials which are below the plane on which the machine is located. Draglines perform by pulling a bucket towards the machine. Draglines used in heavy construction commonly use buckets in a range of from $1/2$ to 20 yd³, but far larger machines are used in open-pit mining. The factors which determine dragline production rates are similar to those that determine shovel production rates.

Clamshell excavators are used to excavate materials which are below the plane on which the machine is located, and are sometimes used for loading materials from stockpiles on the surface or for unloading materials from railroad cars or barges. Bucket sizes usually range from $1/2$ to 8 yd³. Bucket weights vary, depending upon whether they are heavily built for excavation of tough materials or whether they are lightly built for rehandling loose materials. Variations of the clamshell principle are orange-peel buckets and rock grapples.

Safety with this class of equipment, as with all other classes, requires good maintenance, frequent inspections, and sensible working practices. The safe lifting capacity should not be exceeded. Work should be laid out to avoid danger to men and equipment working near or with the machines. No work should be permitted near power lines. Electric shovels and draglines must be properly grounded and equipped so that failure of the grounding system will automatically de-energize electrical circuits.

Production rates for power shovels and draglines are best estimated by use of the tables prepared by the Crane and Shovel Association. These are reprinted in the latest edition of the *Caterpillar Performance Handbook*.[2.1] An example of the use of these tables for a power shovel is given here.

Example

   (a) Size: 6 yd³.
   (b) Material: well-blasted rock.
   (c) Job conditions: good.
   (d) Management conditions: excellent.
   (e) Operator efficiency: normal.
   (f) Face height: 80 percent of optimum.
   (g) Angle of swing: 120°.
   (h) The correction factor for job and management conditions, selected from the table on job efficiency, is 0.78.

(i) The 6-yd^3 bucket will, on the average, contain 4.2 yd^3 of loose material based on a fill factor of 0.7 (from the table on bucket-fill factors).

(j) The correction factor for height of face and angle of swing is 0.86 (from the table on applicable correction factors).

(k) This shovel should be able to excavate 575 yd^3, bank measure, per 60-minute hour under ideal conditions and if the bucket is completely filled on each pass (from the table on estimated hourly production).

(l) It is assumed that the shovel will in practice operate 50 minutes of every hour. The actual production will then be equal to the tabulated hourly production **(k)** times the job efficiency factor **(h)**, times the swing and depth factor **(j)**, times the bucket-fill factor **(i)**, times the percentage of the 60-minute hour which will actually be used **(l)**, as follows:

$$\text{Actual hourly production} = \underset{\textbf{(h)}}{0.78} \times \underset{\textbf{(i)}}{0.70} \times \underset{\textbf{(j)}}{0.86} \times \underset{\textbf{(k)}}{575} \times \underset{\textbf{(l)}}{0.83}$$

$$= 224 \text{ yd}^3 \text{ bank measure per hour}$$

MISCELLANEOUS EQUIPMENT

Electric welders may be diesel- or gasoline-powered or may be all-electric. The diesel- or gasoline-powered units are usually mounted on two-wheeled trailers or on service trucks, and are useful in areas away from electric power. The all-electric models may be motor generator sets or may be rectifier units. The electric types cost less to purchase than the engine-driven types and have lower operating and maintenance costs.

Water pumps may be engine-driven, electrically driven, or air-operated. Air-operated pumps are typically centrifugal units. Engine- or electricity-driven pumps may be reciprocating or rotary-type positive displacement units, diaphragm type, or centrifugal. Diaphragm pumps are often designed to function well even if the water to be pumped carries with it a considerable volume of trash. Centrifugal pumps may be used where there is a suction lift, but are more efficient when submerged. Pumps which operate with the impellors below the water surface may be driven by a vertical shaft from an engine or motor located above the water surface. These are frequently referred to as *turbine-type* pumps. Pumps referred to as *submersibles* are electric-powered, with a watertight motor directly coupled to the impellor. Air-lift or jet pumps are occasionally used. Air-lift pumps use compressed air introduced at the bottom of the well to induce upward flow. Jet pumps use water jets directed upward from the bottom of the well to induce upward flow.

Wellpoint systems are used for dewatering pervious soils. A wellpoint consists of a vertical tube with a screened intake and check valve. The wellpoints are connected to a header which in turn is connected to a vacuum pump. Since the lift is a suction lift, wellpoints are not effective at depths greater than 20 ft (6 m). If used for greater depths, two or more stages must be installed as the excavation progresses.

AIR PLANT AND DRILLS

Air plant consists of air compressors, air-operated equipment, and the distribution system which brings the air to the tool. Air plant and equipment used in tunneling is described elsewhere.

Air compressors are generally classified as *stationary* or *portable*. The stationary machines are almost always driven by electric motors; the portables are driven by diesel or gasoline engines. Older compressors and stationary compressors are usually two- or three-stage reciprocating machines; the newer portables are usually rotary machines. Most construction applications require 100 lb/in² at the tool; so most compressors are adjusted to deliver air to the system at about 105 to 110 lb/in². Portable compressors are available which will handle from 65 to 1,800 ft³/min; stationary machines are available in the range of 500 to 1,500 ft³/min. Portable compressors are generally air-cooled; the stationary machines are generally water-cooled. Compressors operated in urban environments require muffling or sound insulation for the engine and for the air intake.

The required capacity for a stationary air plant will be determined by the air consumption of the tools, by the maximum number of units which will be drawing air at any one time, and by line losses due to friction or leaks. In calculating air requirements it must be realized that air consumption increases with altitude. After the requirements have been tabulated, the size of pipes can be calculated. The greater the volume of air which is stored in receivers, the lower will be the compressor capacity needed to meet peak demands. Adequate capacity should be installed so that at least one unit can be idle for maintenance without inconvenience to the work.

The distribution system should provide blowoffs for water condensation in the lines and should include dryers if the system is to be operated during freezing weather. Pipe sizes can be calculated from any handbook on fluid flows; probably the best for this purpose is the one published by the Compressed Air and Gas Institute.

One of the biggest users of compressed air on construction projects is the rock drill. For production drilling of blastholes the most common tool is the track-mounted drill. It can drill in any direction and will handle drill steels in 20-ft lengths and bits up to 6 inches in diameter. Used with a down-the-hole hammer, holes up to 9 inches in diameter are feasible. These machines usually have sufficient tractive power to tow their own portable compressor. They are often fitted with winches to enable them to travel on steep slopes. Hydraulic drills mounted on tracks can be powered by diesel engines, thus being entirely independent of air or electric supply lines. Large-hole drills are most often diesel-powered.

Drills for use underground are described in the chapters which pertain to rock tunneling and shaft sinking. Drills for foundation excavation and quarry work are described in their respective chapters.

The most common form of rock drill, whether air or hydraulic, contains a piston which imparts a rapid series of blows to the striker, which in turn strikes the drill steel and causes the bit to strike and cut the rock. It also causes rotation of the drill steel and thus the bit. Some drills use independent rotation. The drill steel is hollow. Air or water passes through the drill steel and holes in the bit, cooling the bit and conveying the cuttings to the surface. The bit may be integral with the drill steel, but is more often attached to the steel and is thus readily replaceable. Bits for soft rock may be steel; bits for hard rock are steel with carbide inserts. The steel is purchased in lengths which are appropriate for the length of the drill feed, and additional lengths are added as needed. If the rock is rich in silicates, which are dangerous to human health, the cuttings must be captured in a dust collector. Wet drilling will reduce the dust hazard and is mandatory for underground use. Drillers and men working near drills should be required to wear goggles for eye protection and ear plugs or ear covers for hearing protection.

For large-diameter or deep holes the striking mechanism is contained in a down-the-hole drill, which eliminates the attenuation of the striking force which would otherwise occur in the drill steel. Down-the-hole drills are operated by compressed air which is conveyed through the hollow drill steel.

Hand-held air tools include jackhammers, compactors, paving breakers, clay spades, torque wrenches, saws, chipping hammers, concrete vibrators, etc. Diamond drills, grouting equipment, small pumps, and winches are often air-operated. A great volume of compressed air may be required for cleaning rock and concrete surfaces and for operating air-water jets or sandblast equipment to prepare construction joints in concrete.

CONCRETE BATCH AND MIX PLANTS

If concrete is to be produced at jobsite, a plant will be needed to measure the ingredients and to mix the concrete. The plant may produce mixed concrete or may measure (*batch*) the ingredients and then deliver the batch to mixer trucks for mixing en route to the placement site. Central mix plants may discharge to agitator cars, mixer trucks, or other specialty pieces of concrete hauling equipment.

Modern batch plants are usually controlled by a small computer that can batch any required quantity of any of several mix designs which are stored in its memory. Each mix design will consist of different proportions by weight of one or more sizes of coarse aggregate, fine aggregate, cement, pozzolan, water, ice (if required), and admixtures such as a water-reducing agent. Most specifications require all ingredients except admixtures to be batched by weight; some specifications permit water to be batched by volume. Some specifications permit cumulative weighing of the various sizes of aggregates and of cement and pozzolan; some specifications require separate weigh hoppers and scales for each ingredient. All scales should be tested weekly with test weights to ensure accuracy. Most specifications require automatic measurement of moisture content in the fine aggregate and automatic adjustment of water batching to compensate for that absorbed moisture.

Batch plants may be portable or may be the stand-up type. The portable plants are suitable for small projects. Most portables use cumulative weighing of coarse and fine aggregates and of cement and pozzolan. They are typically low profile and have small storage bins, with the aggregates fed to the weigh hopper by a built-in conveyor. The stand-up plants are suitable for large projects. They have large storage bins and gravity-feed to the weigh hoppers. Most specifications require that coarse aggregates be rescreened and washed immediately before going into the storage bins. The rescreen is usually set on top of the plant where the charging conveyor terminates. The charging conveyor is arranged so that fine aggregate is neither rescreened nor washed. Both portable and stand-up plants are usually fitted with tilting mixers. If the maximum aggregate size is 1½ in (4 cm) or smaller, turbine-type mixers are occasionally installed for central mixing.

Supplemental silos for cement and pozzolan are usually necessary to ensure a constant supply. Transfers from the supplemental storage to the bins or silos in the plant are usually handled pneumatically, but may be handled by a combination of screw conveyors and bucket elevators. Supplemental storage for aggregates may be in bins or silos, or they may be stored in piles on the ground. Recovery from the piles for small plants will usually be by use of a front-end loader. Larger plants will probably use a conveyor in a reclaiming tunnel under the storage piles, with clam gates to admit the desired sizes of aggregates.

If the plant is to operate in cold weather, it may be necessary to heat the mix water, and perhaps the aggregates, to comply with the requirements of the specifications. If the plant is to operate in hot weather or if the specifications require low-temperature concrete, it may be necessary to chill the water, to cool the coarse aggregates, or to substitute ice for most of the water in the mix. It is unusual to provide cooling for fine aggregate, cement, or pozzolan. Both the heating requirements and the cooling requirements can be reduced by insulating the bins and silos. Cooling requirements can be reduced by using reflective paints on the bins and silos and by sprinkling and shading the aggregate storage piles. The fine-aggregate storage piles should not be sprinkled, because this will reduce the quantity of chilled water or ice which can be added to the mix. Sprinkling the fine aggregate will also cause trouble in batching because of wide variations in moisture content as the excess moisture drains to the bottom of the bin and thus into the weigh hopper.

All specifications require timers to ensure proper mixing time and interlocks to ensure that batching cannot start until all weigh hoppers are completely empty and that no ingredient is double-batched or omitted. Most specifications require that the plant operator and the inspector be able to observe the action within the mixer drums. Some plant layouts are such that this requirement can best be met by installing a television camera and monitor. Bins should be fitted with low-level and high-level detectors, with indicators provided at the control panel. Most specifications require that space be provided for a batch plant inspector and for facilities to sample and test the ingredients and the concrete.

Transit-mix trucks are available in sizes ranging to 8 yd^3. Ratings and requirements for transit-mix trucks are established by Standard Specifications for Ready-mixed Concrete, ASTM C94. The capacity ratings are higher if the trucks are used for transporting premixed concrete than if they are used as transit mixers. Concrete may also be hauled in "bathtub" trucks, rail-mounted agitator cars, or buckets. A cableway job may use transfer cars to move the concrete from the batch plant for transfer to the bucket carried by the cableways.

SCREENING, CRUSHING, AND SAND PLANTS

Screening or crushing and screening plants are required to process gravels, sands, or crushed rock in graded sizes. Typical end products include aggregates for portland cement or asphaltic concrete, railroad ballast, base course for roadways, filter material for dams, and gravel or crushed rock surfacing.

The feed to such plants may be sands and gravels, or it may be broken rock from quarries, from excavation required for other work, or occasionally from talus slopes. If the feed is expected to contain boulders or rock fragments larger than the plant can handle, the delivery should be over a grizzly, which will eliminate the oversize. Because deliveries arrive load by load, while these plants function best at a uniform rate of feed, it is prudent to provide a surge pile at the receiving plant, or perhaps immediately downstream of the primary crusher in a crushing plant. Material in the surge pile or bins should be recoverable without additional handling except conveyance on a belt. This may not be practical in a small portable plant, where the surge pile is probably composed of material that has not passed over the grizzly and is recovered when needed by a front-end loader. It is thus in dead storage instead of live storage, as would be the case in a larger plant.

Screening Plants

A screening plant may incorporate no crushers if the feed contains only a small percentage of material larger than the specified product and if the gradation of the feed supplies the approximate distribution of grain sizes. If this is not the case, it will usually be necessary to provide at least one crusher in order to avoid wasted effort in supplying the plant and in disposing of the surplus materials in some size classifications. The screening section of any plant will consist of one or more vibrating screens, each with several decks of screen cloth. The coarsest screen will be at the top of each deck and will discharge the oversize to a crusher or, if no crusher is provided, to waste. The output of the crusher, if used, will be returned to the screening cycle. The other decks will discharge to conveyors which deliver the products to stockpile. The pan under the screen may deliver the fines to waste, to storage, or to a sand plant. Screening may be performed wet or dry. Wet screening is preferable, since it provides a cleaner product, but in some arid locations it may not be feasible. Rotating screens are sometimes used instead of vibrating screens. These are built with the coarsest screen at the inlet end and the finest screen at the far end. Rotating screens are more effective than vibrating screens for scrubbing undesirable coatings from gravels. If washing on the screen is not sufficiently effective in removing deleterious materials, a log washer may be required. The gravels are scrubbed under water by an arrangement of paddles which move the product to the upper end of an inclined trough. Waste is removed with the water from the lower end. The capacity of any screening plant is a function of the screen area. Each deck will deliver a percentage of undersize, but if the screen fabric is intact it will deliver no oversize. Since the largest dimensions of the material control the screening action, thin flat pieces will be delivered with the larger-sized fractions instead of the smaller-sized fractions of the plant output.

Crushing Plants

A crushing plant may consist of a single unit, or it may consist of a primary crusher, a secondary crusher, and perhaps a tertiary crusher operating in series. Each crusher returns its output to a screen, and the oversize is recirculated. Properly sized materials are conveyed to stockpiles.

There are several types of crushers, each suitable for some stages of crushing and for some type of rock. Each type of crusher is available in a wide range of sizes and can be adjusted to vary the size of the output. The principal types of crushers are described below.

Jaw Crushers

Jaw crushers are very common machines. The crushing action is developed by movement of a movable plate toward a fixed plate. The opening at the top is wider than the opening at the bottom, and the size of the rock fragments is progressively reduced as the feed works its way downward. Jaw crushers are most often used as primary crushers. Very large jaw crushers are built for this purpose. They work well with all but the softest rocks and produce a well-graded range of sizes.

Gyratory Crushers

These machines, which are widely used, effect their crushing action by the eccentric rotation of a hardened conical steel column within a conical steel bowl. Reduction proceeds as the feed moves downward. Gyratory crushers are most commonly used as primary units, but may also be used as secondary units. Very large gyratory crushers are available for use as primaries. Gyratory crushers work well with all but the softest rocks and produce a well-graded range of sizes.

Cone Crushers

Cone crushers work on the same principle as the gyratory crusher, but the movable element, the cone, is inverted. They are used where a uniform-sized product is required, generally as secondary or tertiary units. They work well with all but the softest rocks and are preferred for rock which tends to break into "flats."

Roll Crushers

These units consist of two rolls on horizontal shafts. The rolls may be smooth or may be ribbed. One roll is mounted so that it can be moved to vary the opening between the two rolls. Roll crushers are used for secondary or tertiary crushing. Roll crushers are not useful for some types of rock, especially those that tend to break with conchoidal fractures, because of the large proportion of "flats" which result. A cubical shape for crushed rock is most desirable for almost all purposes.

Impact Breakers

Impact breakers are effective for secondary or tertiary use with the softer rocks. The rock is broken by a rapidly turning rotor with vanes that throw the rock against a hardened steel plate. A grate at the bottom limits the size of rock fragments which can be discharged.

Autogenous Impact Breakers

These are used as secondary or tertiary units for soft rocks. The rock is broken by the action of two rapidly turning rotors with vanes that throw two streams of rock fragments so that they collide within the breaker chamber, breaking one rock fragment against another.

Hammer Mills

Hammer mills are another form of impact breaker. A hammer mill consists of a number of hammers which can swing through limited arcs as the rotor to which they are attached turns within the crushing chamber. The rock is crushed between the hammers and the breaking plates which are attached to the chamber walls. Openings at the bottom determine the maximum size of the output.

Sand Plants

A sand plant may clean and size natural sands, or may produce sand from quarry rock or from required excavation, or may use both types of feed.

The sand may be cleaned and sized in a screw classifier or in a sand classifier. The *screw classifier* consists of a spiral screw operating in an inclined trough of flowing water. The unclassified sand is fed into the lower end, and the screw conveys the sand to the upper end, where it is recovered. Silts and clays are discharged over an adjustable weir at the lower end. The feed rate, rate of rotation of the screw, velocity of water flow, and angle of inclination of the trough can be varied as necessary so that the classified sand will satisfy the requirements of the specifications.

The *sand classifier* operates by introducing the feed into a tank in which there is a flow of water from end to end. The coarser particles settle near the feed end, and the finer particles settle near the end where the overflow weir is located. Clays and silts are carried off in the overflow. A series of valves in the bottom of the tank permit the various sand fractions to be recovered. The several fractions are recombined as necessary to meet the specified sand gradation.

If sand is to be manufactured from rock or from gravels the first step is crushing to a proper feed size, usually ½ in (1 cm) or less. The most common crushers used for this purpose are rolls and cone crushers, the latter of a type called a *short-cone crusher*. After crushing, the material is fed to a rod mill, or less frequently to a ball mill. Either mill consists of a cylindrical steel shell which is rotated about its horizontal axis. The shell is fitted with renewable liner plates and charged with steel rods or with cast-iron balls. The feed passes through a central opening at one end of the mill. As the mill rotates, the rods or balls are lifted and then dropped to produce a tumbling action that reduces the feed to sand, which is discharged through a central opening at the discharge end of the mill. A ball mill is used if a very fine sand is required. The discharge from the mill, together with the natural sand if a blended product is to be produced, goes to a sand classifier or to a screw classifier.

STORAGE FACILITIES

All sands, gravels, and crushed rock must be properly handled to avoid segregation, so that the product as withdrawn from storage maintains the gradation that it had when going into storage. One cause of degradation is breakage, particularly in the larger particle sizes. The other cause is segregation. As materials spill off the end of a conveyor belt, the light fractions tend to fall vertically, but the coarser fractions tend to roll to the far side of the pile. This tendency will be reversed in the case of sand being discharged in a strong breeze. Both degradation and segregation can be controlled by causing the conveyor discharge to pass through a rock ladder.

TRUCKS

Trucks designed to be used for hauling the spoils of excavation on heavy construction projects will probably be too wide, too high, and too heavy for operation on public highways, and are referred to as *off-highway vehicles.*

The most common off-highway vehicle is the end-dump truck. Off-highway end-dumps range in capacity from about 20 to about 120 tons, though larger vehicles are used occasionally, particularly in open-pit mining. Most are purchased with rock bodies, specially reinforced to resist shocks of being loaded by shovels or draglines. The engine exhaust passes through ducts in the body so as to prevent freezing of wet materials in cold climates. Off-highway end-dumps are usually capable of speeds of 40 to 45 mph (65 to 75 km/h) on firm and level haul roads.

Truck size must be consistent with excavator size. The truck bed should well exceed the dimensions of the bucket if the truck is to be loaded by a shovel, and should be low enough to clear the tipped bucket if the truck is to be loaded by a front-end loader.

Aside from the matter of target dimensions and loading heights, the estimator must search for an economic balance between excavator size and truck size. The larger units usually require a greater capital investment, but they have lower unit operating costs, particularly for labor. Small trucks may travel faster, and if one unit breaks down a lesser percentage of hauling capacity is lost than would be the case with a smaller spread of larger trucks. A common rule of thumb is that the excavator should be able to fill the hauling vehicle in four or five passes.

Manufacturers' handbooks provide data and curves from which to estimate travel speeds for any combination of load, grade, and rolling resistance.

On-highway types of trucks used in heavy construction may include end-dumps and bottom-dumps. On-highway vehicles must be used in hauling through urban areas and may be the most economical choice for long hauls elsewhere, generally because of the high speeds which can be attained. Tire problems are less severe, and haul roads and bridges can be designed for the lower gross vehicle weights.

Special off-highway vehicles are sometimes developed for special uses. Examples are a 7,000-gal tanker trailer or a low-bed trailer for hauling three 4-yd^3 concrete buckets, either of which can be coupled to a two-wheeled, rubber-tired tractor.

Most other trucks used in heavy construction are standard vehicles, such as mixer trucks for concrete, flatbeds, fuel trucks, service vehicles, man-haul buses, etc.

CRANES

Cranes are certainly the most ubiquitous type of construction plant, for it is a rare job that does not need crane service, if not directly, at least for servicing other classes of construction plant. Most jobs need mobile cranes, crawler-mounted, rail-mounted, or truck-mounted, but some jobs can be served by nonmobile cranes, such as guy derricks or stiffleg derricks, or by floating cranes.

Cranes may have lattice-type booms, in which case boom length is varied by inserting or removing boom sections, or may have telescoping booms, in which case boom length can be changed at will. Telescoping booms are common only on truck cranes.

Until recent years all cranes were "cable" machines, using wire ropes for load lines and for boom hoist lines, with shaft and gear arrangements for swinging. Recent years have seen the development of the hydraulic crane, where hydraulic cylinders are used for the boom hoist and for extending or retracting the telescoping boom. Other functions are performed by hydraulic motors. Hydraulic cranes are common only in the truck-crane configuration, and then not in the larger machines.

Crawler Cranes

Crawler cranes will not be selected if great mobility is required, since they move slowly and since the cost of undercarriage maintenance becomes very high if much time is spent in walking. Many crawler cranes are purchased as combination machines. This configuration, with the necessary accessories, permits use of the machine as a lifting crane, as a clamshell excavator, as a dragline, as a backhoe, or as a shovel. Other attachments can be purchased so that the machine can be used for pile driving or for large-hole drilling. The nominal or rated capacities of cranes are based upon conditions which are not often applicable to actual requirements encountered on the job. It is therefore necessary when selecting a crane to ascertain that it will safely handle the required loads with the length of boom which will be actually needed, at the necessary radius, and on the ground surface which will actually be encountered. Some crawler cranes are fitted with tracks which can be moved outward to increase transverse stability. Some very large machines are fitted with rings or with rolling counterweights to increase stability in any orientation.

Truck Cranes

Truck cranes are superior to crawler cranes if job requirements dictate that they move around a lot. Truck cranes travel faster, and undercarriage maintenance will not be a problem. Although truck cranes can be adapted to dragline work and other uses, they perform more satisfactorily as lifting cranes or as clamshell excavators. Truck cranes incorporate outriggers which can be extended to increase stability when working. Truck cranes are rated under conditions similar to those used for crawler cranes, and the same factors must be considered when selecting a machine for a particular job. Truck cranes may be "cable" machines with drum hoists and fixed-length booms, or they may be hydraulic machines with telescoping booms and hydraulic boom hoists, with only the load line being wire rope.

Rail-Mounted Cranes

Rail-mounted cranes are similar to crawler or truck cranes, except that they are mounted on railcars. Other rail-mounted cranes, called *gantries,* have the crane atop a frame, permitting trains or trucks to pass freely under the gantry crane, which is often an advantage. The gantry also increases the vertical reach. Gantry cranes, and less frequently whirlies, are used on trestles, for placing concrete, and for other crane duties in the construction of dams and powerhouses.

Floating Cranes

Floating cranes may be specially constructed machines or may consist of a crawler or truck crane chained down to a barge. Sometimes the upper works of a standard crane are mounted on a pedestal on a barge. Specially constructed floating cranes are most commonly built as derricks, with no swing. Orientation is changed by swinging the barge.

Derricks

The most common types of nonmobile lifting machines are guy derricks and stiffleg derricks, although the upper works of a crawler or truck crane may occasionally be mounted on a pedestal to serve a particular need. Guy derricks and stifflegs usually have low line speeds and low swing speeds, but often have great lifting capacities. Both use a boom, supported by boom hoist lines from a vertical mast. The guy derrick has a swivel at the top of the mast, from which guy lines radiate to support the mast. This arrangement will permit full 360° rotation of the boom. The stiffleg derrick employs structural legs rising from concrete anchor blocks. This arrangement limits the boom swing, but avoids the necessity for guy lines.

Tower Cranes

Tower cranes may be fixed or mobile. A vertical tower supports a horizontal boom, on which travels a carriage, and on the opposite end, a counterweight. The load line descends from the carriage. Thus, by swinging the boom and tramming the carriage, an entire work circle can be covered. If the machine is rail-mounted, the entire machine, with load, can travel to extend the coverage. Some tower cranes include a climbing feature, whereby additional tower sections can be added as the work becomes higher, without dismantling the machine. Really tall towers must be tied off to the completed work to ensure complete stability.

Operating Safety

All cranes and derricks should be operated and maintained with safety as the most important consideration. Operating safety includes ensuring a safe working surface, staying within the safe lifting capacity, and keeping clear of power lines. Maintenance safety includes frequent and regular line and sheave inspection, brake maintenance, and frequent examination of the structure, looking particularly for bent members, incipient weld failures, and corrosion around connections. Boom stops, limit switches, and load-indicating devices should be regularly inspected and tested. All people working around cranes should know and use the proper hand signals, but only one person should be authorized to direct the crane operator, except in an emergency. In assessing safe load limits it is essential that the tables or diagrams for the machine be consulted. The nominal rating for any machine equals the safe load only for minimum boom length, minimum load radius, and perhaps for only one orientation of the load relative to the position of the crane carrier.

COMPACTION EQUIPMENT

The construction of embankments and of roadways requires compaction of the soils, rock, or courses of gravel or crushed stone to achieve the engineering qualities upon which designs are predicated. The equipment necessary to compact these materials may be self-propelled, towed, or manually propelled.

Self-propelled compaction equipment includes flat rollers, sheepsfoot rollers, segmented rollers, grid rollers, and rubber-tired rollers. All except the last may incorporate vibrators to provide dynamic forces to supplement the static weight of the roller. The smaller machines, suitable for small or obstructed areas, may be manually guided. Occasionally, proper distribution of hauling equipment routes, or tractors alone, may provide all required compaction.

Towed compaction equipment is in all respects similar to the self-propelled equipment except for the need of a separate prime mover. The tractors used for this purpose supplement by their weight and vibration the compactive effort of the machines which are being towed.

Manually propelled compaction equipment may be operated by gasoline engines or by compressed air and may effect their work by impact or by weight and vibration. All have a flat compacting surface.

The applications of the above types of compaction equipment are shown below:

Type of Compactor	Used on:
Flat rollers	Sands, gravel, and clay mixtures (and for paving)
Sheepsfoot and segmental	Clay and soils rich in clay
Grid	Rock
Rubber-tired	All soil types and rock

Because proper moisture content is critical to achieving maximum soil density, the spread of compaction equipment will probably include a water truck for adding moisture or disc plows and harrow for loosening the soil to promote aeration and reduce the moisture content.

Compaction of deep in situ sands may be accomplished by use of vibrating probes. Compaction of deep in situ silts and clays may be accomplished by construction of sand drains or paper drains. Compaction of rock fills may be accomplished by dumping the rock from substantial heights with the use of monitors to wash sand and rock fragments into the interstices between the larger stones.

TRACTORS

Tractors are the great prime movers of the construction industry. Both crawler types and rubber-tired types are widely used for bulldozing, ripping and pulling, or pushing other equipment.

Rubber-tired tractors are principally used for pulling or pushing other equipment, or for service where frequent moves are necessary. The four-wheeled version can also be used for bulldozing. The two-wheeled version is used only for pulling such equipment as scrapers, end-dumps, or bottom-dump wagons. Rubber-tired tractors may have engines with from 150 to 600 hp. Smaller farm-type tractors are frequently used for light work and are often fitted with trench hoes, bulldozer blades, augers, and other attachments.

Crawler tractors have better traction and lower ground pressure than equivalent rubber-tired machines, and of course do not have the problem of tire-cutting when working in shot rock, but working in sand causes fast wear on the crawlers. The range of engine sizes ranges from 60 to 700 hp, but some of the larger machines are available in side-by-side or tandem units under the control of one operator. These double units may approach an engine capacity of 1,000 hp.

As prime movers, crawler or rubber-tired tractors may be used to pull scrapers, end-dumps, bottom-dumps, compactors, or other equipment, or may be used to push-load scrapers. Crawler tractors are also used for ripping rock, and either type may be used for bulldozing. The equipment manufacturers provide data which will assist in selecting the tractor with sufficient tractive power and speeds for the intended purpose.

Ripping rock by the use of crawler tractors fitted with rippers is frequently a cheaper method of breaking up rock than is drilling and blasting. It is sometimes worthwhile to combine ripping with light drilling and blasting. Rock which might otherwise be drilled and blasted may be ripped in cases where the noise and the shock waves induced by blasting are not acceptable.

The most common criterion of the rippability of rock is the velocity of sound waves in the rock. These velocities are ascertained by the use of seismic-measuring equipment, generally performed by specialists in this work. The *Caterpillar Performance Handbook* contains charts of recommended limits of rippability for various sizes of tractors.

Tractors used as push-loaders for scraper spreads are often fitted with rippers so that they can utilize time which would otherwise be lost to loosen the soil and thus reduce scraper-loading time.

Bulldozers are used for moving earth over short distances and for leveling and spreading windrows or piles of soil or rock which have been delivered by hauling equipment. They are also used for land clearing, usually with a rake-type clearing blade which allows the earth to fall free as the machine moves ahead.

Bulldozer blades may be angled or may be fixed. The angle type is most useful for pioneering work on steep side slopes and is available only for crawler-type tractors. The angle can be altered, the blade can be turned to discharge to the right or to the left, and the blade can be tilted back and forth. None of these adjustments can be made while the machine is moving, but the blade can be raised and lowered at any time. The angle dozer requires a heavy steel arch to support the fulcrum about which the blade is pivoted. This additional weight causes faster wear on the front idlers of the crawlers.

The fixed bulldozer blades may be straight or bowl-shaped and can be fitted to either crawler or rubber-tired tractors. The blades can be tilted back and forth, but are not otherwise adjustable. Hydraulic power is used to raise or lower the blade. Some blades are specially reinforced where they will come in contact with the push block on a scraper, and may in addition have hard rubber cushions over the reinforcement. These special blades are usually narrower than standard blades, to avoid danger of cutting the scraper tires.

Tractor manufacturers provide handbooks which can be used to estimate the production rates for various combinations of tractors and bulldozer blades.

There are a number of attachments which will increase the range of duties which tractors can perform. Examples are winches, snowplows, brush and tree cutters, V blades for land clearing, "stingers" for splitting trees and stumps, root plows which cut off roots below the ground surface, and side booms for pipeline work. One interesting application is the use of a heavy chain, drawn by two tractors, for clearing land.

FRONT-END LOADERS

There are two types of front-end loaders: crawler and rubber-tired.

Crawler-type front-end loaders range in bucket capacity from 1 to 5 yd³. They are used where not much travel is required, where low ground pressure is an advantage, and where the ground surface would be hard on rubber tires because of cutting by sharp rock fragments or because of high temperatures, as in handling hot slag or cinders.

The rubber-tired-type front-end loaders range in bucket capacity from 1 to 20 yd³. The larger size machines are articulated. These machines are more useful than crawler-type front-end loaders or shovels when the work involves frequent moves around the jobsite. Tire life in shot rock can be improved by using protective tire chains, or by using steel shoe tire arrangements, which also improve flotation if working on swampy ground.

Buckets for front-end loaders are selected for the type of materials to be excavated and for the conditions of loading vehicles. Buckets for loading shot rock are heavily constructed, are fitted with teeth, and may have a V-type cutting edge to improve penetration into the working face. The heavier construction uses up some of the lifting capacity. Smaller buckets are used for heavy materials, and larger buckets are used for light materials such as coal or volcanic cinders. The limit on bucket size is the lifting capacity of the loader. Side-dump buckets are available for some models. These are particularly valuable for working in tunnels, where working space is limited.

There are a number of attachments available for front-end loaders. One is a clam gate which enables the loader to grasp hard-to-handle materials, such as the spoils of building demolition. Other attachments replace the bucket, and are used for handling brush and logs or for converting the loader for use as a fork-lift. Loaders can also serve as tractors and can be fitted with rippers for light-duty service.

REFERENCES

2.1 *Caterpillar Performance Handbook,* 12th ed., Caterpillar Tractor Co. (Peoria, Illinois), 1981.

Chapter 3 Quality Control, Government Regulations, and Safety

Heavy construction projects are increasingly affected by quality assurance and control requirements, government regulations, and worker safety and health considerations. Regulatory agencies in the United States have become major controlling factors which the heavy construction estimator must try to understand; and the establishment of the Occupational Safety and Health Administration (OSHA) has brought strict national standards of worker safety. The problems these regulations impose are compounded by the intensively competitive nature of the heavy construction industry, where short-run expediency in potential cost-cutting areas often seems necessary for business survival. In today's construction environment in the United States and other fully developed countries, the estimator who does not try to properly assess the impact of quality, regulatory, and employee safety requirements upon the overall cost of the project will be equally as negligent as his counterpart who has an inadequate understanding of the capacities and production capabilities of alternative choices for production methods and equipment. And regulatory factors will become increasingly important throughout the world as the developing countries evolve, and consideration of the effects of the heavy construction project upon the environment, the community, and the individual worker in those areas approaches the current significance of such consideration in the United States.

QUALITY ASSURANCE AND CONTROL*

Traditionally the resident engineer is charged with the overall responsibility for inspection on a heavy construction project. Inspection is designed to ensure that the project is in fact constructed in accordance with the plans and specifications and in compliance with tolerances and requirements which are specified or mandated by applicable law. In most states, for example, a division of dam safety or other similar organization is charged with the responsibility of approving plans and specifications of dams exceeding a specified height or storage capacity and with the continuing responsibility for periodic safety inspections and reporting requirements for existing dams and for those under construction.

The massive failure of the Teton Dam on the Snake River in Idaho in 1976 initiated a major emphasis upon regulatory requirements for quality assurance against failure on both existing dams and new projects throughout the United States and the world. These more stringent requirements have generally been implemented under the traditional organization framework in most heavy construction applications. However, in the nuclear power industry a "zero defects" program mandated by the Nuclear Regulatory Commission must bear a major share of the responsibility for massive overruns and for the practical abandonment of fixed-price contracting on nuclear plants in the United States.

The heavy construction industry must continue to recognize that economics is crucial to survival and that the economies of the quality of design and the quality of the finished facility must be preserved, taking full account of the safety of the environment, the community, and the individual worker.

Economics of Quality Assurance

Quality assurance must begin in the design phase. Figure 3.1 illustrates the relationship between the cost and the value of quality. In theory the optimum level of quality occurs when the marginal cost of each unit of added quality equals the benefit derived from such additional refinements. This point is illustrated in Fig. 3.1 where the slopes of the two curves are equal. Above this point the cost of additional quality exceeds its benefit. Below this point the benefit of added quality exceeds its cost.

Quality construction, of course, must be a function of the particular application and risks associated with failure. The quality requirements for the design and construction for a major dam whose failure would have a catastrophic effect would differ considerably from those for a tailings pond dam in a remote location whose failure would have a negligible effect upon people or property.

* This section is based in part upon a chapter from Donald S. Barrie and Boyd C. Paulson, Jr., *Professional Construction Management*, McGraw-Hill, New York, 1978.

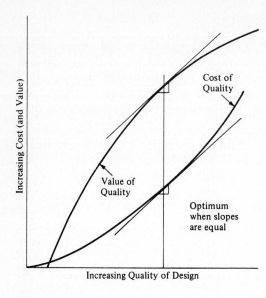

Figure 3.1 Economics of quality of design. (Adapted from Elwood G. Kirkpatrick, *Quality Control for Managers and Engineers,* Wiley, New York, 1970, p. 8.)

The cost of achieving quality requirements during the construction phase is proportional to the cost of skilled labor, materials, equipment methods, and supervision utilized, as well as to the cost of monitoring and inspecting the work to verify the output quality and to correct or repair defective work. Figure 3.2 illustrates the economics of quality and conformance. Optimum quality conformance costs represent the minimum of the sum of quality control costs and their associated construction costs.

Figure 3.2 Economics of quality of conformance. (Adapted from Elwood G. Kirkpatrick, *Quality Control for Managers and Engineers,* Wiley, New York, 1970, p. 10.)

On major heavy construction projects throughout the world where the need both for the avoidance of a catastrophic failure and for maximum feasible economy are recognized, a favorite method of considering both quality assurance and economy is achieved through the appointment of an independent consulting board of recognized experts of several disciplines. This board reviews concepts during the design phase for both quality assurance and economy and continues to follow the project through the construction phase, making recommendations regarding the cost-quality trade-off in the event of changed physical conditions, proposed design changes, or other newly discovered factors which impact upon the overall quality-cost trade-off.

Tasks, Functions, and Responsibilities

Early in any project, tasks and functions associated with quality control must be identified and responsibility for their accomplishment must be determined. Figure 3.3 illustrates the responsibilities for a number of quality control functions. Owner, engineer, and contractor all have duties and responsibilities associated with quality control.

Figure 3.4 illustrates a flow diagram showing quality-control aspects of concrete construction.

Statistical Approach

Methodology for most modern quality-assurance programs in heavy construction requires a basic understanding of probability and statistics. Statistical control can be divided into two major categories. The first deals with quality characteristics which can be measured in a quantitative manner. The second deals with qualitative observation or attributes. Control charts and sampling procedures can be designed to help achieve optimum quality control at an economical cost. Figure 3.5 shows a typical example illustrating the portrayal of mean values, standard deviation, and distribution.

GOVERNMENT REGULATIONS*

Federal regulations in the United States have expanded significantly during the past decade, presenting many new demands upon both the estimator and the project management in assessing the cost of compliance with controls, approval requirements, funding constraints, and numerous other agency requirements applicable to both public and private projects. Similar regulatory requirements are prevalent in almost every country, with varying degrees of

---

* This section is based in part upon a chapter by Franklin T. Matthias, in Donald S. Barrie, ed., *Directions in Managing Construction*, Wiley, New York, 1981.

LINEAR RESPONSIBILITY CHART

• KEY •

●	PRIMARY RESPONSIBILITY
▲	JOINT RESPONSIBILITY
■	APPROVAL RESPONSIBILITY
○	MUST BE CONSULTED
△	MAY BE CONSULTED
□	AUDITS OR REVIEWS

	OWNER				ENGINEER				CONSTRUCTOR							
	Manager of New Construction	Chief Engineer	Board of Review	Site Representative	Project Manager	Project Engineer	Responsible Design Engineer	Materials Engineer	Project Superintendent	Construction Superintendent	Site Q.C. Supervisor	Site Q.A. Supervisor	Home Office Q.A. Chief	Purchasing Agent	Startup Engineer	Regulating Agency
Select quality objectives	▲				▲				▲							▲
Define activities affecting quality	■	△	□		●	△	△	△	○							□
Specify quality standards	■	△	□		■	●	△	△	○							□
Prepare quality control manual	■	△	□		■	○		△	○				●			□
Prepare quality control procedures		□			■					○	●		■			□
Prepare construction method procedures				□		□			■	●	○	□				□
Prepare welding and NDT procedures		□	□		■			●								□
Establish design criteria	■	●	□													□
Perform design	■	■	□		■	■	●									□
Define vendor quality control requirements	■				■	●	○	○							△	□
Prepare procurement documents	■	△	□		○									□	●	
Evaluate vendor quality capability	■	△	□		■	○			○						●	□
Inspect off site manufacturing		□	□			□	△						●		○	□
Control distribution of plans and specifications				□		○			○		●					□
Specify sampling plans		□	□			□	△		○				●			□
Train and qualify craftsmen				□					○	●	○	□				□
Train and qualify inspection personnel			□	□							●	□				□
Direct construction operations									○	●						
Inspect work in progress			□	□		□	△		△	●	□					□
Accept work in progress			□	□	△	□	△			●	□					□
Stop work in progress	△		□	□	○		△		○	●						□
Inspect materials upon receipt				□			△	△		●	□					□
Monitor and evaluate quality trends			□	□	□					●						□
Maintain file of quality control documents			□	□						●						□
Determine disposition of nonconforming items	■	△	□	△	●	△	△	△	○		○	○				□
Investigate failures		○	□	□	○	○	△	△	△	△	●	△	△			□
Release systems and components to operations				□						□	●				■	□
Conduct flushing and cleaning operations				■											●	□
Conduct preoperational testing			□	■											●	□
Accept completed plant as to quality	●	○	△													■

Figure 3.3 Linear responsibility chart (LRC) for quality control organization. (From "System for Control of Construction Quality," *Journal of the Construction Division*, ASCE, March 1972, p. 31.)

complexities, all of which must be given consideration by the estimator in the development of the schedule and the estimate of cost.

Major Regulatory Agencies

Federal regulatory agencies in the United States which have major impact upon the construction industry include the *Environmental Protection Agency*

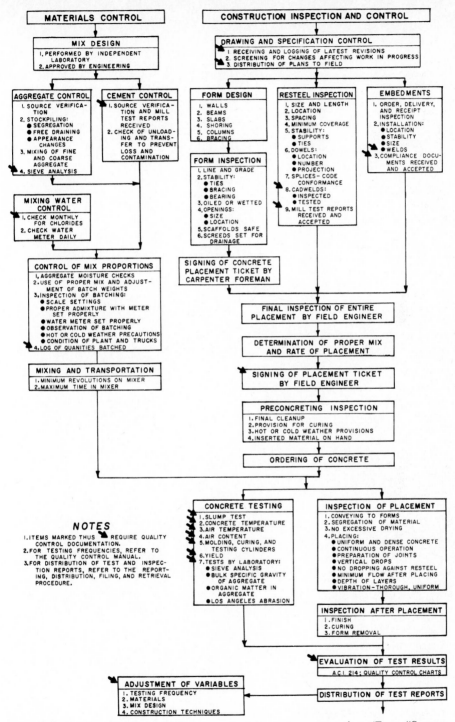

MATERIALS CONTROL

MIX DESIGN
1. PERFORMED BY INDEPENDENT LABORATORY
2. APPROVED BY ENGINEERING

AGGREGATE CONTROL
1. SOURCE VERIFICATION
2. STOCKPILING:
   ● SEGREGATION
   ● FREE DRAINING
   ● APPEARANCE CHANGES
3. MIXING OF FINE AND COARSE AGGREGATE
4. SIEVE ANALYSIS

CEMENT CONTROL
1. SOURCE VERIFICATION AND MILL TEST REPORTS RECEIVED
2. CHECK OF UNLOADING AND TRANSFER TO PREVENT LOSS AND CONTAMINATION

MIXING WATER CONTROL
1. CHECK MONTHLY FOR CHLORIDES
2. CHECK WATER METER DAILY

CONTROL OF MIX PROPORTIONS
1. AGGREGATE MOISTURE CHECKS
2. USE OF PROPER MIX AND ADJUSTMENT OF BATCH WEIGHTS
3. INSPECTION OF BATCHING:
   ● SCALE SETTINGS
   ● PROPER ADMIXTURE WITH METER SET PROPERLY
   ● WATER METER SET PROPERLY
   ● OBSERVATION OF BATCHING
   ● HOT OR COLD WEATHER PRECAUTIONS
   ● CONDITION OF PLANT AND TRUCKS
4. LOG OF QUANTIES BATCHED

MIXING AND TRANSPORTATION
1. MINIMUM REVOLUTIONS ON MIXER
2. MAXIMUM TIME IN MIXER

NOTES
1. ITEMS MARKED THUS ➤ REQUIRE QUALITY CONTROL DOCUMENTATION.
2. FOR TESTING FREQUENCIES, REFER TO THE QUALITY CONTROL MANUAL.
3. FOR DISTRIBUTION OF TEST AND INSPECTION REPORTS, REFER TO THE REPORTING, DISTRIBUTION, FILING, AND RETRIEVAL PROCEDURE.

CONSTRUCTION INSPECTION AND CONTROL

DRAWING AND SPECIFICATION CONTROL
1. RECEIVING AND LOGGING OF LATEST REVISIONS
2. SCREENING FOR CHANGES AFFECTING WORK IN PROGRESS
3. DISTRIBUTION OF PLANS TO FIELD

FORM DESIGN
1. WALLS
2. BEAMS
3. SLABS
4. SHORING
5. COLUMNS
6. BRACING

FORM INSPECTION
1. LINE AND GRADE
2. STABILITY:
   ● TIES
   ● BRACING
   ● BEARING
3. OILED OR WETTED
4. OPENINGS:
   ● SIZE
   ● LOCATION
5. SCAFFOLDS SAFE
6. SCREEDS SET FOR DRAINAGE

RESTEEL INSPECTION
1. SIZE AND LENGTH
2. LOCATION
3. SPACING
4. MINIMUM COVERAGE
5. STABILITY:
   ● SUPPORTS
   ● TIES
6. DOWELS:
   ● LOCATION
   ● NUMBER
   ● PROJECTION
7. SPLICES – CODE CONFORMANCE
8. CADWELDS:
   ● INSPECTED
   ● TESTED
9. MILL TEST REPORTS RECEIVED AND ACCEPTED

EMBEDMENTS
1. ORDER, DELIVERY, AND RECEIPT INSPECTION
2. INSTALLATION:
   ● LOCATION
   ● STABILITY
   ● SIZE
   ● WELDS
3. COMPLIANCE DOCUMENTS RECEIVED AND ACCEPTED

SIGNING OF CONCRETE PLACEMENT TICKET BY CARPENTER FOREMAN

FINAL INSPECTION OF ENTIRE PLACEMENT BY FIELD ENGINEER

DETERMINATION OF PROPER MIX AND RATE OF PLACEMENT

SIGNING OF PLACEMENT TICKET BY FIELD ENGINEER

PRECONCRETING INSPECTION
1. FINAL CLEANUP
2. PROVISION FOR CURING
3. HOT OR COLD WEATHER PROVISIONS
4. INSERTED MATERIAL ON HAND

ORDERING OF CONCRETE

CONCRETE TESTING
1. SLUMP TEST
2. CONCRETE TEMPERATURE
3. AIR TEMPERATURE
4. AIR CONTENT
5. MOLDING, CURING, AND TESTING CYLINDERS
6. YIELD
7. TESTS BY LABORATORY:
   ● SIEVE ANALYSIS
   ● BULK SPECIFIC GRAVITY OF AGGREGATE
   ● ORGANIC MATTER IN AGGREGATE
   ● LOS ANGELES ABRASION

INSPECTION OF PLACEMENT
1. CONVEYING TO FORMS
2. SEGREGATION OF MATERIAL
3. NO EXCESSIVE DRYING
4. PLACING:
   ● UNIFORM AND DENSE CONCRETE
   ● CONTINUOUS OPERATION
   ● PREPARATION OF JOINTS
   ● VERTICAL DROPS
   ● NO DROPPING AGAINST RESTEEL
   ● MINIMUM FLOW AFTER PLACING
   ● DEPTH OF LAYERS
   ● VIBRATION – THOROUGH, UNIFORM

INSPECTION AFTER PLACEMENT
1. FINISH
2. CURING
3. FORM REMOVAL

EVALUATION OF TEST RESULTS
A.C.I. 214; QUALITY CONTROL CHARTS

ADJUSTMENT OF VARIABLES
1. TESTING FREQUENCY
2. MATERIALS
3. MIX DESIGN
4. CONSTRUCTION TECHNIQUES

DISTRIBUTION OF TEST REPORTS

Figure 3.4 Flow diagram, quality control of concrete construction. (From "System for Control of Construction Quality," *Journal of the Construction Division*, ASCE, March 1972, p. 33.)

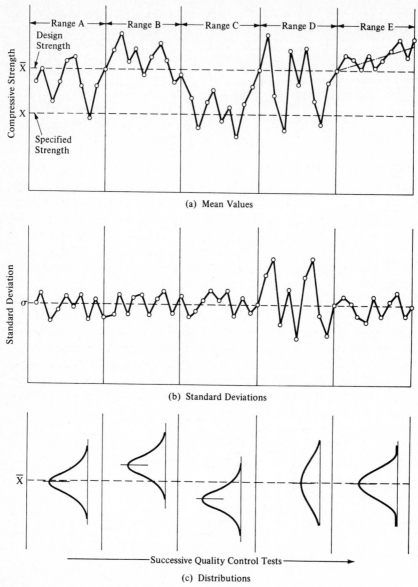

Figure 3.5 Control charts for mean and standard deviation. (From Donald S. Barrie and Boyd C. Paulson, Jr., *Professional Construction Management*, McGraw-Hill, New York, 1978, p. 323.)

(EPA), the *Nuclear Regulatory Commission* (NRC), the *Occupational Safety and Health Administration* (OSHA), the *Department of Energy* (DOE), the *Department of Transportation* (DOT), the *Corps of Engineers* (COE), the *Department of Commerce* (DOC), the *Office of Federal Procurement Policy* (OFPP), the *Federal Office of Contract Compliance* (FOCC), and the *Bureau of Reclamation* (BR). The Corps of Engineers and the Bureau of Reclamation each also fund specific projects through congressional appropriation. Table 3.1 lists the principal regulatory and funding agencies which affect construction in the United States.

Regulatory Cost and Delays

Federal regulations have increased almost exponentially in the past decade. In 1936, 2,411 pages of regulations were printed in the *Federal Register,* in 1970 there were 20,036 pages, and in 1976 there were 60,221 pages. In 1979 information developed from a state governors' conference stated that the annual regulatory cost to the nation was $600 billion. Figure 3.6 shows that the incremental annual cost to 48 companies of compliance with regulations published by six regulatory agencies was in excess of $2.6 billion.

Regulatory actions which are most prominent on the heavy construction scene are oriented to environmental protection, public and occupational safety and health, minority employment, utilization of minority contractors, and the control of banking and the money supply. Reporting requirements can present a substantial cost burden throughout the construction process; regulatory requirements and approvals may also be the source of costly delays. A study entitled "Cost of Delays in Construction"[3.1] rated the principal

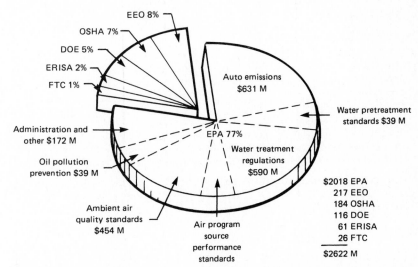

Figure 3.6 Incremental costs to 48 companies of regulations published by six regulatory agencies. (*Civil Engineering*, ASCE, September 1979.)

TABLE 3.1 **Principle Regulatory and Funding Agencies Significant to Major Construction Projects**

Entity	Initials	Regulatory Agency	Funding Agency
Environmental Protection Agency	EPA	√	√
Nuclear Regulatory Commission	NRC	√	
Department of Labor	DOL	√	
Occupational Safety and Health Administration	OSHA	√	
Employment Standards Administration	ESA	√	
Equal Employment Opportunity Commission	EEOC	√	
Federal Energy Regulatory Commission	FERC	√	
Department of Transportation	DOT		
Federal Highway Administration	FHA*	√	√
Urban Mass Transit Administration	UMTA*	√	√
Federal Aviation Administration	FAA*	√	√
Department of Energy	DOE*	√	√
Department of Agriculture	DOA		
Rural Electrification Administration	REA*	√	√
Soil Conservation Service	SCS†	√	√
Department of Commerce	DOC		
Economic Development Administration	EDA*‡	√	√
Corps of Engineers	COES¶	√	√
Bureau of Reclamation	BR¶		√
Office of Management and Budget	OMB	√	
Office of Federal Procurement Policy	OFPP	√	
Federal Office of Contract Compliance	FOCC	√	
National Aeronautical and Space Administration	NASA	√	√
Department of State	DOS		
Agency for International Development	AID*	√	√

* Regulation through financing controls

† Projects normally small and handled in house

‡ Former regulation of Minority Business Enterprise policy

§ Jurisdiction under, in, on, or over navigable waters

¶ Projects funded by Congress—engineering and construction normally in house; construction by contract

SOURCE: From Table 15-1 by Franklin T. Matthias in Donald S. Barrie, ed., *Directions in Managing Construction*, Wiley, New York, 1981.

reasons for delays in construction and made an evaluation of the effects of these delays. Two groups answered the question, "Have construction delays due to governmental regulation occurred on the projects with which your organization is involved?" They answered as follows:

500 Engineering Firms Listed in *Engineering News-Record*
Substantially	56%
Quite a bit	35%
Hardly at all	9%

532 Consulting Engineering Firms in Ohio
Substantially	47%
Not at all	13%

Answers to other questions in the *ENR* 500 survey indicated project delays of 36.6 months, with 47.2 percent of project cost attributed to the regulatory delays. The Ohio engineers' answers indicated delays of 29.4 months, with 27.7 percent of the project cost attributable to the delays.

On the other hand, after many initial delays a major oil pipeline in Alaska was completed within 13 months of start of construction.[3.2] An enlightened and vigorous management effort identified and involved 42 agencies. Management approval efforts were diverted to the major interests, jurisdiction, and degree of influence of each, and the success of the campaign is evident in the completion schedule.

Mitigating Regulatory Problems

Anticipating the costs and effects of regulatory requirements upon the project schedule is a difficult and challenging task for the heavy construction estimator. Engineers and construction management can ease the burden of meeting regulatory requirements by developing joint programs throughout the project to achieve timely compliance. Mitigating adverse affects of regulatory constraints will include the identification both of the specific requirements and of the individual agencies and even individuals who will have the major role in the enforcement of the requirements. The preparation of a program for dealing with the requirements must be accompanied by the preparation of a program for dealing with the individual regulatory representative, i.e., for establishing mutual understanding and respect so that the inevitable problems can be resolved in as cooperative and friendly an atmosphere as possible.

SAFETY AND HEALTH

The establishment of OSHA has led to increased awareness of worker safety requirements on all domestic heavy construction projects. Not only are companies liable for fines for unsafe practices, but individual superintendents and managers are also subject to citations and penalties. However, safety in heavy construction is also sound business practice. A study prepared by Stanford University for the Business Roundtable[3.1] showed that the actual cost of accidents is about double the cost of benefits paid through insurance claims. The estimator should outline the framework for a comprehensive jobsite safety program and provide the proper resources in the estimate for its successful implementation.

TYPICAL EMPLOYEE SAFETY BOOKLET

A booklet of safety principles is often given to each employee at the time of hire as a part of an overall jobsite safety program. The following excerpts are typical entries of such a handbook.

Personal Protection

1. Hard hats shall be worn at all times on this project.
2. Eye protection of the proper type shall be worn when exposed to any hazards and at all times while working inside existing buildings and/or adjacent to other employees.
3. Safety belts, properly tied off, shall be worn when scaffolding or other safe work platforms are not available.
4. Adequate and proper footwear shall be worn at all times on the project.
5. Respirators will be furnished for your protection when you may be exposed to hazardous fumes or dust.

Safety equipment is designed to protect you from injury. Take care of it, turn in damaged equipment for repair, and remind your fellow employees to do the same.

Housekeeping

Orderliness, as housekeeping is sometimes called, is the foundation of our safety program. Clean and orderly work areas are absolutely necessary if we are to have a safe job.

1. Immediate work areas, especially those at heights on scaffold or platforms and in similar confined spaces, shall be free of debris.
2. Space used for passage such as walkways and stairs and areas around ladders must be kept clean. Debris is not only dangerous; it makes your job more difficult.
3. General scrap should be disposed of as soon as possible. Remove or bend over nails when stripping forms.

Manual Material Handling and Storage

All material shall be properly stacked and secured to prevent sliding, falling, or collapse. Aisles, stairs, and passageways must be kept clear to provide for the safe movement of employees and equipment and to provide access in emergencies.

1. Use proper lifting techniques when manually handling materials:

   - Get down close to the load.
   - Keep your back straight.
   - Lift gradually, using your legs. Do not jerk.
   - Get help for bulky or heavy loads.
   - Shift your feet when turning. Do not twist waist.

2. Material stored inside structures under construction shall not be placed within 6 ft (2 m) of any hoistway or other inside floor opening, nor within 10 ft (3 m) of any outer wall which does not extend above the top of the material stored. Stored materials shall not block any exit, fire-fighting equipment, panel board, etc.
3. Pipe, conduit, and bar stock shall be stored in racks or stacked and blocked to prevent movement.
4. The quantity of materials stored on scaffolds, platforms, or walkways shall not exceed that required for one (1) day's operation.
5. Materials shall never be thrown or dropped from a distance of more than 5 ft (1.5 m). The drop area shall be barricaded to protect personnel from being struck by falling materials. Trash chutes are required when dropping material from heights above 5 ft (1.5 m).
6. When stripping forms or uncrating materials, bend or pull out protruding nails.

Hand Tools

The proper tool in the proper condition, used at the proper time, can do much to prevent injury.

1. Every tool has a purpose. Use it for that and only that. Don't improvise.
2. Keep edges sharp, striking faces smooth, struck ends dressed, and handles tight.
3. Use the proper strength tool for each job.
4. Tools have a proper place. Keep them there when not in use.

Electrical Equipment

All temporary electrical equipment used on the jobsite shall be listed by an approved testing laboratory (Underwriters Laboratories, Inc., or Factory Mutual Laboratories) for the specific application. All temporary electrical installations shall conform to the National Electric Code.

1. All electrical tools and equipment shall be grounded. All 120-V single-phase, 15- and 20-amp nonpermanent circuits shall either be protected by ground fault circuit interrupters (GFCI) or an assured grounding program requiring regular inspection and color coding of all cords, tools, and outlets.
2. Damaged or defective electrical tools shall be returned immediately for repair.
3. Electricians are the only employees authorized to repair electrical equipment. Tampering with tools or equipment may result in an employee's discharge.
4. When it is necessary to work on energized lines and equipment, rubber gloves, blankets, mats, and other protective equipment shall be used.
5. Temporary electrical cords shall be protected or elevated. They shall be kept clear of walkways and other locations where they may be exposed to damage or may create tripping hazards.
6. Splices in electrical cords shall retain the mechanical and dielectric strength of the original cable.
7. Energized wiring in junction boxes, circuit breaker panels, and similar places shall be covered at all times. Hazardous areas shall be barricaded and appropriate warning signs posted.
8. Temporary lighting shall have guards over the bulbs. Broken and burned-out lamps shall be replaced immediately.

Hand and Portable Power Tools

1. Inspect your tools daily to ensure that they are in proper working order. Damaged or defective tools shall be returned for repair.
2. Power saws, grinders, and other power tools shall have proper guards in place at all times.
3. Power tools shall be hoisted or lowered by a hand line, never by the cord or hose. When using many tools, or working near others using such tools, you shall use appropriate personal protective equipment. If you have questions about the protective equipment or safety rules, ask your supervisor.
4. An approved safety check valve shall be installed at the manifold outlet of each supply line for hand-held pneumatic tools.
5. All pneumatic hose connections shall be fastened securely, by clips or chain.
6. Safety clips or retainers shall be installed on all pneumatic tools to prevent the accidental ejection of the tool from the barrel.
7. Only those trained and licensed may use powder-actuated "guns" or stud tools.

Excavations and Trenches

1. Trenches 5 ft (1.5 m) or deeper shall be shored or sloped back to the angle of repose. Any excavation in unstable soil may require shoring or sloping. Each excavation shall be inspected daily by the responsible superintendent.
2. Where vehicles or equipment operate near excavations or trenches, the sides of the excavation shall be shored or braced as necessary to withstand the extra load and vibration. Also, stop logs or other barricades shall be installed at the edges of such excavation.

A convenient and safe means shall be provided for workmen to enter and leave the excavated area. This shall consist of a standard stairway, ladder, or ramp securely fastened in place, suitably guarded or protected in locations where men are working.

Ladders

1. Job-made ladders shall be constructed to conform with the established standards.
2. Industrial-type, heavy-duty ladders shall be used on this job. Metal ladders shall *not be used* in any area where an electrical shock hazard exists.
3. Broken or damaged ladders shall not be used. Repair or destroy them immediately. Ladders to be repaired shall be tagged, "Do not use."
4. All straight ladders shall be tied off or secured at the top. Do not splice together short ladders to make a longer ladder.
5. The base of the ladder shall be set back a safe distance from the vertical—approximately one-fourth of the working length of the ladder. Ladders · shall not be placed against movable objects.
6. Ladders used for access to a floor or platform shall extend at least 3 ft (1 m) above the landing surface. The areas around the top and base of ladders shall be free of tripping hazards such as loose materials, trash, cords, hoses, etc.
7. You shall face the ladder at all times when ascending or descending. Be sure that your shoes are free of mud, grease, or other substance which could cause a slip or fall.

Scaffolding

Each scaffold shall be inspected and approved by responsible supervisory personnel prior to initial use and after alteration or moving.

1. There is no such thing as a temporary scaffold. All scaffolding shall be erected and maintained to conform with established standards.
2. Guardrails, midrails, and toeboards shall be installed on all open sides of scaffolds more than 7½ ft (2 m) in height.

3. Scaffold planks shall be at least 2-x10-in full-thickness lumber, scaffold grade or equivalent.
4. Scaffold planks shall be cleated and shall extend over the end supports at least 6 in (15 cm)—but not more than 12 in (30 cm).
5. All scaffolds shall be fully planked. Scaffold planks shall be visually inspected before each use. Damaged planks shall be destroyed.
6. Access ladders shall be provided for each scaffold. Climbing off the end frames is prohibited unless their design incorporates an approved ladder.
7. Adequate mudsills or other rigid footing, capable of withstanding the maximum intended load, shall be provided.
8. Scaffolds shall be tied on to the building or structure at intervals which do not exceed 30 ft (9 m) horizontally and 26 ft (8 m) vertically. Doubled #12 wire or equivalent shall be used to secure.
9. Where persons are required to work or pass under a scaffold, a screen of 10-gauge, ½ in wire mesh is required between the toeboard and the guardrail.
10. Overhead protection is required if employees working on scaffolds are exposed to overhead hazards. Such protection shall be a 2-in plank or the equivalent.

Power Cranes and Shovels

1. Cranes shall be located on firm, level footing. When needed, substantial timbering (cribbing) or mats shall be used for leveling and for safety, distributing the weight of the crane on the underlying material.
2. Tag lines shall be used to control loads that have a tendency to swing.
3. Lifting capacity and operating speed of a crane or other equipment specified by the manufacturers shall be conspicuously posted and never exceeded.
4. Shock-absorbing-type "boom stops" shall be provided on crane booms.
5. Do not stand or walk under a suspended load.
6. Riding on a load or headache ball is prohibited.
7. Blades, buckets and shovels on earth-moving equipment shall always be lowered to the ground when equipment is parked or unattended.
8. All boom equipment shall carry a sign reading, "All equipment shall be positioned, equipped, or protected so that no part shall be capable of coming within 10 ft (3 m) of high voltage lines."
9. Motor truck cranes shall have electrical insulation applied to the booms and insulated links for the hooks.
10. Crane hand signals shall be used and understood. An illustration of them shall be posted at the jobsite.
11. The operator shall inspect his crane daily and have all safety defects corrected prior to use.

Welding and Burning Operations

1. Fire resistant blankets shall be used while burning or welding over or near any combustible materials.
2. Authorization for welding or burning shall be obtained from the supervisor of the involved area.
3. You shall be sure that suitable fire-extinguishing equipment is available in your work area.
4. You are responsible for maintaining your burning or welding equipment in safe operating condition.
5. When burning or welding, you shall wear approved eye protection, with suitable filter lenses.
6. Keep all welding leads and burning hoses off to sides of floors, walkways, and stairways. You are responsible for seeing that your equipment complies with safe practices at all times.
7. Never weld or burn on barrels, tanks, piping, or other systems which may have contained either combustible or unknown products without first obtaining approval from your industrial relations representative or other responsible authority.

Welding

1. If your eyes are exposed to flying objects from chipping slag or other welding-cleaning activity, you shall wear approved eye protection under your hood.
2. When you arc-weld near other workmen, they shall be protected from the arc rays by noncombustible screens or shall wear adequate eye protection.

Burning

1. Do not use matches to light torches. Spark igniters shall be used.
2. A special wrench is required to operate the acetylene cylinder valve: the wrench shall be kept in position on the valve at all times.
3. All oxygen and acetylene bottles shall be secured in an upright position at all times.
4. Appropriate gas bottle carriers shall be used while lifting or lowering gas bottles to various elevations.
5. Protective caps shall remain on cylinders when cylinders are not in use or are being transported.

Ventilation and Protection

1. Welding, burning, and heating performed in confined spaces may require general mechanical or local exhaust ventilation to reduce the concentrations of smoke and fumes to acceptable levels. Your safety supervisor shall be consulted prior to starting these operations.

2. If adequate ventilation cannot be provided, employees shall be provided with, and required to use, air-supplied breathing apparatus.

3. When welding, cutting, or heating metals containing or coated with toxic metals, such as zinc, lead, cadmium, or chromium-bearing metals, in the open air, you shall wear filter-type respirators.

General Principles in Review

1. Falls.   Falls result in by far the most serious accidents. Tie off ladders, use handrails and life belts. Be extra cautious at heights.

2. Caught Between.   Getting caught or pinched between two objects generally means someone wasn't alert, for this hazard is usually quite obvious if one just looks. Always leave yourself a way out.

3. Struck By.   Falling or flying objects result in many injuries. Wear your personal protective equipment. Keep out from under or in front of equipment, materials, or operations that may cause this type of injury.

4. Strains.   These nearly always involve improper pushing, pulling, or lifting.

All four of these types of injuries can be avoided if you use normal care and common sense in doing your job.

Fire Protection and Prevention

In a fire, assure the safety of all personnel, be sure it has been reported, and then use the appropriate fire-fighting equipment until help arrives.

1. Familiarize yourself with the location of all fire-fighting equipment in your work area.

2. Tampering with fire-fighting equipment is grounds for discharge.

3. Learn the classifications of fires:

  • **Class A:** ordinary combustible materials such as wood, coal, or paper; use a water or liquid-type fire extinguisher.
  • **Class B:** flammable petroleum products or other flammable liquids; use a dry powder or carbon dioxide fire extinguisher.
  • **Class C:** fires in or near energized electrical equipment where, because use of water or liquid fire extinguisher would be hazardous, a "nonconducting" type such as carbon dioxide must be used.

4. Only approved solvents should be used for cleaning and degreasing. The use of gasoline and similar flammable products for this purpose is prohibited.

5. Combustible and volatile liquids must be handled only in approved, properly labeled safety cans.

6. When you must weld or burn near combustible materials, move them, cover them with fire-resistant fabric, or wet them down. Fire watch must be posted while burning or welding near combustible materials.

7. Do not attempt any work involving a source of ignition near a pit, sewer, drain, manhole, trench, or enclosed space where flammable gases may be present.

8. Keep the work area neat. An orderly jobsite reduces the fire and accident hazard.

9. The use of open fires is prohibited.

USE OF EXPLOSIVES

Good practice in using explosives is summarized in a list of do's and don'ts adopted by the Institute of Makers of Explosives, February 1, 1964. The list is reproduced here, and an explanation or clarification of some of the rules is given after each grouping.*

Do's and Don'ts

The term *explosives,* as used herein, includes any or all of the following: dynamite, black blasting powder, blasting caps, electric blasting caps, and detonating cord. The term *electric blasting cap,* as used herein, includes both instantaneous electric blasting caps and all types of delay electric blasting caps. The term *primer,* as used herein, means a cartridge of explosives in combination with a blasting cap or an electric blasting cap.

When Transporting Explosives.

Do obey all federal, state, and local laws and regulations.

Do see that any vehicle used to transport explosives is in proper working condition and equipped with tight wooden or nonsparking metal floors with sides and ends high enough to prevent the explosives from falling out. In an open-bodied truck the load should be covered with a waterproof and fire-resistant tarpaulin, and the explosives should not be allowed to contact any source of heat such as an exhaust pipe. Wiring should be fully insulated so as to prevent short circuiting, and at least two fire extinguishers should be carried. The trucks should be plainly marked so as to give adequate warning to the public of the nature of the cargo.

Don't permit metal, except approved metal truck bodies, to contact cases of explosives. Metal, flammable, or corrosive substances should not be transported with explosives.

Don't allow smoking or unauthorized or unnecessary persons in the vehicle.

---

* Comments after each grouping are based upon information set forth in Gary Hemphill, *Blasting Operations,* McGraw-Hill, New York, 1981, a valuable source for safety considerations involving explosives.

Do load and unload explosives carefully. Never throw explosives from the truck.

Do see that other explosives, including detonating cord, are separated from blasting caps and/or electric blasting caps where it is permitted to transport them in the same vehicle.

Don't drive trucks containing explosives through cities, towns, or villages or park them near such places as restaurants, garages, and service stations, unless it cannot be avoided.

Do request that explosive deliveries be made at the magazine or in some other location well removed from populated areas.

Don't fight fires after they have come in contact with explosives. Remove all personnel to a safe location and guard the area against intruders.

The foregoing incorporate the Department of Transportation (DOT) regulations and general rules of safety concerning the transportation of explosives. The intent is to avoid fire and shock, either of which could cause a detonation of the load.

When Storing Explosives.

Do store explosives in accordance with federal, state or local laws and regulations.

Do store explosives only in a magazine which is clean, dry, well ventilated, reasonably cool, properly located, substantially constructed, bullet- and fire-resistant, and securely locked.

Don't store blasting caps or electric blasting caps in the same box, container, or magazine with other explosives.

Don't store explosives, fuse, or fuse lighters in a wet or damp place, or near oil, gasoline, cleaning solution, or solvents, or near radiators, steam pipes, exhaust pipes, stoves, or other sources of heat.

Don't store any sparking metal or sparking metal tools in an explosive magazine.

Don't smoke or have matches, or any source of fire or flame, in or near an explosive magazine.

Don't allow leaves, grass, brush, or debris to accumulate within 25 ft (8 m) of an explosives magazine.

Don't shoot into explosives or allow the discharge of firearms in the vicinity of an explosives magazine.

Do consult the manufacturer if nitroglycerin from deteriorated explosives has leaked onto the floor of a magazine. The floor should be desensitized by washing thoroughly with an agent approved for that purpose.

Do locate explosives in the most isolated places available. They should be separated from each other, and from inhabited buildings, highways, and railroads, by distances not less than those recommended in the *American Table of Distances*.

The storing of explosives is regulated by various agencies; however, they all use the *American Table of Distances*[3.3] as their guide. Again it is obvious that the rules are to avoid contact of the explosives and fire and shock (concussion).

When Using Explosives.

Don't use sparking metal tools to open kegs or wooden cases of explosives. Metallic slitters may be used for opening fiberboard cases, provided that the metallic slitter does not come in contact with the metallic fasteners of the case.

Don't smoke or have matches or any source of fire or flame within 100 ft (30 m) of an area in which explosives are being handled or used.

Don't place explosives where they may be exposed to flame, excessive heat, sparks, or impact.

Appendices.

Do replace or close the cover of explosives cases or packages after using.

Don't carry explosives in the pockets of your clothing or elsewhere on your person.

Don't insert anything but fuse in the open end of a blasting cap.

Don't strike, tamper with, or attempt to remove or investigate the contents of a blasting cap or an electric blasting cap, or try to pull the wires out of an electric blasting cap.

Don't allow children or unauthorized or unnecessary persons to be present where explosives are being handled or used.

Don't handle, use, or be near explosives during the approach or progress of an electrical storm. All persons should retire to a place of safety.

Don't use explosives or accessory equipment that is obviously deteriorated or damaged.

Don't attempt to reclaim or to use fuse, blasting caps, electric blasting caps, or any explosives that have been water-soaked, even if they have dried out. Consult the manufacturer.

These recommendations involving the use of explosives are quite self-explanatory. It should be noted that the blasting caps must be handled with even greater care than explosives.

When Preparing the Primer.

Don't make up primers in a magazine, or near excessive quantities of explosives, or in excess of immediate needs.

Don't force a blasting cap or an electric blasting cap into dynamite. Insert the cap into a hole made in the dynamite with a punch suitable for the purpose.

Do make up primers in accordance with proven and established methods. Make sure that the cap shell is completely encased in the dynamite or booster and so secured that in loading no tension will be placed on the wires or fuse at the point of entry into the cap. When side-priming a heavy wall or heavy-weight cartridge, wrap adhesive tape around the hole punched in the cartridge so that the cap cannot come out.

Primers are quite hazardous to handle because the sensitivity of the blasting cap is present with all the force of the explosive.

When Drilling and Loading.

Do comply with applicable federal, state, and local regulations relative to drilling and loading.

Do carefully examine the surface or face before drilling to determine the possible presence of unfired explosives. Never drill into explosives.

Do check the borehole carefully with a wooden tamping pole or measuring tape to determine its condition before loading.

Do recognize the possibility of static electrical hazards from pneumatic loading and take adequate precautionary measures. If any doubt exists, consult your explosives supplier.

Don't stack surplus explosives near working areas during loading.

Do cut from the spool the line of detonating cord extending into a borehole before loading the remainder of the charge.

Don't load a borehole with explosives after springing (enlarging the hole with explosives) or upon completion of drilling without making certain that it is cool and that it does not contain any hot metal or burning or smoldering material. Temperatures in excess of 150°F (65°C) are dangerous.

Don't spring a borehole near another hole loaded with explosives.

Don't force explosives into a borehole or through an obstruction in a borehole. Any such practice is particularly hazardous in dry holes and when the charge is primed.

Don't slit, drop, deform, or abuse the primer.

Don't drop a large size, heavy cartridge directly on the primer.

Do avoid placing any unnecessary part of the body over the borehole during loading.

Don't load any borehole near electric power lines unless the firing line, including the electric blasting cap wires, is so short that it cannot reach the power wires.

Don't connect blasting caps or electric blasting caps to detonating cord except by methods recommended by the manufacturer.

As with all the do's and don'ts, these tips on drilling and loading are advice on how to prevent accidental detonation of the explosives.

When Tamping.

Don't tamp dynamite that has been removed from the cartridge.

Don't tamp with metallic devices of any kind, including the metal end of the loading poles. Use wooden tamping tools with no exposed metal parts, except nonsparking metal connectors for jointed poles. Avoid violent tamping. Never tamp with the primer.

Do confine the explosives in the borehole with sand, earth, clay, or other suitable incombustible stemming material.

Don't kink or injure fuse or electric blasting cap wires when tamping.

You must be even more careful when tamping the primer charge; you are better off not to tamp it at all.

When Shooting Electrically.

Don't uncoil the wires or use electric blasting caps during dust storms or go near any other sources of large charges of static electricity.

Don't uncoil the wires or use electric blasting caps in the vicinity of radio-frequency transmitters, except at safe distances. Consult the manufacturer or the Institute of Makers of Explosives pamphlet, *Radio Frequency Hazards.*

Do keep the firing circuit completely insulated from the ground or other conductors such as bare wires, rails, pipes, or other paths of stray currents.

Don't have electric wires or cables of any kind near electric blasting caps or other explosives except at the time and for the purpose of firing the blast.

Do test all electric blasting caps, either singularly or when connected in a series circuit, using only a blasting galvanometer specifically designed for the purpose.

Don't use in the same circuit electric blasting caps made by more than one manufacturer or electric blasting caps of different style or function, even if made by the same manufacturer, unless such use is approved by the manufacturer.

Don't attempt to fire a single electric blasting cap or circuit of electric blasting caps with less than the minimum current specified by the manufacturer.

Do be sure that all wire ends to be connected are bright and clean.

Do keep the electric cap wires or leading wires disconnected from the power source and short-circuited until ready to fire.

When Shooting with Fuse.

Do handle fuse carefully to avoid damaging the covering. In cold weather warm slightly before using to avoid cracking the waterproofing.

Don't use short fuse. Never use less than 2 ft (0.6 m). Know the burning speed of the fuse and make sure you have time to reach a place of safety after lighting.

Don't cut fuse until you are ready to insert it into a blasting cap. Cut off an inch or two to ensure a dry end. Cut fuse squarely across with a clean sharp blade. Seat the fuse lightly against the cap charge and avoid twisting after it is in place.

Don't crimp blasting caps by any means except a cap crimper designed for the purpose. Make certain that the cap is securely crimped to the fuse.

Do light fuse with a fuse lighter designed for the purpose. If a match is used the fuse should be slit at the end and the match head held in the slit against the powder core. Then scratch the match head with an abrasive surface to light the fuse.

Don't light fuse until sufficient stemming has been placed over the explosive to prevent sparks or flying matchheads from coming into contact with the explosive.

Don't hold explosives in the hands when lighting fuse.

In Underground Work.

Do use permissible explosives only in the manner specified by the United States Bureau of Mines.

Don't take excessive quantities of explosives into a mine at any one time.

Don't use black blasting powder or pellet powder with permissible explosives or other dynamite in the same borehole in a coal mine.

Before and after Firing.

Don't fire a blast without a positive signal from the one in charge who has made certain that all surplus explosives are in a safe place, that all persons and vehicles are at a safe distance or under sufficient cover, and that adequate warning has been given.

Don't return to the area of any blast until the smoke and fumes from the blast have been dissipated.

Don't attempt to investigate a misfire too soon. Follow recognized rules and regulations, or if no rules or regulations are in effect, wait at least 1 hour.

Don't drill, bore, or pick out a charge of explosives that has misfired. Misfires should be handled only by, or under the direction of, a competent and experienced person.

When Disposing of Explosives.

Don't abandon any explosives.

Do dispose of or destroy explosives in strict accordance with approved methods. Consult the manufacturer or follow the Institute of Makers of Explosives pamphlet on destroying explosives.

Don't leave explosives, empty cartridges, boxes, liners, or other materials used in the packing of explosives lying around where children or unauthorized persons or livestock can get at them.

Don't allow any wood, paper, or any other materials employed in packing explosives to be burned in a stove, a fireplace, or other confined places, or to be used for any purpose. Such materials should be destroyed by burning at an isolated location out-of-doors, and no person should be nearer than 100 ft (30 m) after the burning has started.

The do's and don'ts contain a considerable amount of good advice in a particularly condensed list. It is important to study and understand them. There is no excuse for not having the knowledge, because a copy of the do's and don'ts generally comes in all cases of explosives and boxes of blasting caps.

ADDITIONAL INFORMATION

Additional information on constructive safety can be found in a series of checklists published in *Construction Methods and Equipment*[3.4] and in the *California Administrative Code*.[3.5]

REFERENCES

3.1 Technical Report #260, Stanford University, Department of Civil Engineering, Palo Alto, CA.

3.2 *Blasters' Handbook,* E. I. duPont de Nemours & Co. (Wilmington, Delaware), 1980.

3.3 *American Table of Distances,* The Institute of Makers of Explosives, November 5, 1971 (from Reference 3.2).

3.4 "Safety and Health on Worksites," *Construction Methods and Equipment,* nos. 1–14, June 1972– June 1973.

3.5 Title 8, Industrial Relations, part I, Department of Industrial Relations, chap. 4, Division of Industrial Safety, *California Administrative Code.*

Part

TWO

Dams, Quarries, and Embankments

Chapter 4 Diversion, Cofferdams, and Unwatering

The method of stream diversion must be determined prior to selection of the construction system since it controls when the construction equipment and plant can be erected, the method and sequence of dam excavation, and the method and sequence of material placement in the dam. Since it is part of the critical path of a dam's construction schedule, it controls the time required for construction. Selection of the diversion method is therefore one of the first tasks that should be undertaken in planning dam construction.

The method of stream diversion must be one that will not endanger life or cause excessive property damage when the diversion facilities are subject to maximum flood flows. If the method chosen restricts the flow of the water past the damsite, it must either prevent overtopping of the diversion facilities, or it must permit controlled overtopping. For economic reasons it must be the simplest method that will divert the water past the damsite without delaying construction, with a minimum exposure to overtopping. To determine whether diversion facilities should be designed for controlled overtopping or with sufficient capacity to prevent overtopping, the overtopping cost (taking into account how frequently overtopping could occur) should be compared with the cost of providing increased diversion-facility capacity. Based on this study, the determination of whether the diversion facilities should be designed with sufficient capacity to handle flood flows with a frequency of 1 year in 5, 1 year in 10, 1 year in 20, and so on, is simply a matter of construction economics. It is usually more economical to risk occasional overtopping than to provide sufficient diversion-facility capacity to handle all flood flows.

Overtopping may result in increases in the direct cost, indirect cost, plant and equipment cost, and escalation cost of the project. The direct cost will increase if the diversion facilities are damaged as a result of the overtopping; if flood waters must be removed from the construction areas; if debris must be removed from the construction area; or if formwork, scaffolds, bridges, or other temporary or permanent works must be repaired or replaced. If overtopping results in an extension of the construction period, the indirect cost will increase since most indirect costs are proportional to the time required for construction. Plant and equipment cost increases will be the result of damage to the construction equipment, or of construction delays which, by increasing the time required on the job, reduce the salvage value of the equipment. Construction delays can also result in escalation cost increases since higher wage rates and higher supply prices may come into effect. The extent of the delay and the increased cost due to overtopping is dependent upon the design of the diversion facilities and the severity of the flood flow.

Dam construction contracts vary with respect to the division of responsibility for the selection and design of diversion facilities. On some jobs this is done by the owner; other contracts allow the contractor to select and design the diversion facilities; and in some cases the job of designing and selecting is a joint effort by both parties. Similarly, depending on the contract documents, the cost exposure to overtopping may be entirely the owner's responsibility, the responsibility of the contractor, or the responsibility may be divided. Insurance covering the cost of overtopping can be obtained either by the owner or by the contractor. If there is a force majeure clause in the contract,* the contractor can recover from the owner the cost of any overtopping of the diversion facilities by unprecedented floods. In this case, partial relief may also be available under the Disaster Relief Act from the national government for either the contractor or owner.

Even if the diversion method is selected and designed by the owner, the contractor's estimator must completely review the diversion system so that he can plan and estimate the cost of its construction, its effect on the other work, the time required for its construction, the chances of overtopping, and the cost to the contractor if overtopping occurs. When the contractor is to select and design the diversion method, which is a large factor in the total construction cost, the low bidder is often the contractor who has selected and designed the most efficient method with the minimum exposure to overtopping. If it is a joint effort, the contractor must thoroughly review the owner's diversion method in order to properly select and design his portion of the facilities and determine his exposure to overtopping expense, as well as to determine the effect of the diversion method on the construction time and cost of the total project.

The remainder of this chapter discusses the streamflow characteristics that should be compiled for diversion planning, methods of diversion and the

---

* This clause provides relief for the contractor for conditions over which he has no control.

types of damsites for which they are suitable, cofferdam construction, methods of restricting subsurface water inflows, and foundation unwatering.

STREAMFLOW CHARACTERISTICS

Before attempting to select the diversion method, the volume of runoff that will pass by the damsite and the anticipated volume of flood flows with varying frequencies should be determined and plotted. The basis for this study is the historical record of streamflow at or near the damsite, and the longer the period for which records have been kept, the more accurate will be the results. Graphs should be prepared showing the volume and the calendar months when flows could occur for floods rated as 5-, 10-, 15-, 20-, 50-, and 100-year floods. (This rating, of course, does not mean that these floods will occur in exactly this frequency, but is rather a measure of exposure.) Graphs should also be prepared showing the maximum daily flow, the average daily flow, and the minimum daily flow.

If streamflow records are not available or have been kept for only a short period of time, the design of the diversion facilities must be based on the area of the watershed and the amount of anticipated precipitation, correlated with the streamflow records that are available. Though this is a logical computation, it cannot replace streamflow records, and insufficient data can result in errors in determining the diversion facilities and dam storage capacities. Dams have been built where, because of the inadequacy of the records which determined the design, there has never been sufficient runoff to fill their reservoirs. In other cases, principally in mountainous, arid locations, flash floods resulting from the combination of a heavy snowpack and a warm spring rain can often result in peak flows far in excess of those indicated by the often meager available records.

For more detailed information on how to compute and tabulate streamflow, books specializing in this field should be reviewed.

HYDRAULIC-MODEL STUDIES

The best method of stream diversion to use at any damsite is dependent on the streamflow characteristics, the water velocities that will occur during diversion, the type of foundation material at the damsite, the type of dam that is to be built, the topography at the damsite, the methods and equipment with which the dam material will be placed, the sequence of placement of the materials in the dam, the time required for construction of the dam, and the cost that would result if overtopping of the diversion facilities occurred.

If large flows and high velocities are involved, diversion-design calculations should be substantiated by hydraulic-model studies. Such studies have been used to verify and improve diversion methods planned for many dams, including those on the lower Columbia River and the Guri Dam in Venezuela. These studies assist in developing solutions for inadequacies in the diversion plan. Specific points that may be checked by hydraulic-model studies are the water velocities that will occur along the cofferdam sides or during cofferdam closure, and the size of rocks that will be required to prevent cofferdam erosion and effect closure. The hydraulic-model studies performed for Guri Dam were used for this purpose as well as to determine the best diversion method. The study for McNary Dam on the Columbia River indicated that the best method of cofferdam closure was the gradual raising of the closure dike with concrete tetrahedrons. For the Itaipu project, which is located on the border of Brazil and Paraguay, six models of the diversion structure were tested at Brazil's University of Paraná hydraulics laboratory before final choice of the second-phase diversion scheme was made. Diversion of a river as large as the Paraná, with an average flow of 316,000 ft³/s (8960 m³/s) had never before been attempted. The chosen diversion scheme was designed for a maximum flood of 1,070,000 ft³/s (30,000 m³/s) compared to a recorded peak flow of 1,003,000 ft³/s (28,400 m³/s) recorded in March 1929.[4.1]

DIVERSION THROUGH THE DAM

Diversion through the dam is the most economical method when dams are to be constructed on large rivers where flood flows are of such volume that the construction of diversion tunnels, diversion flumes, or diversion channels is a major undertaking. Irrespective of stream size, this diversion method has wide application at sites where the water channel is wide enough to permit the dam to be built inside a cofferdam and still retain a sufficiently wide river or diversion channel to handle the flood flows. After the first-stage cofferdam is complete, the dam within is completed to sufficient height to handle flood flows; the cofferdam is then removed to permit diversion of the stream through the sluiceways, and a second-stage cofferdam is constructed to permit another section of the dam to be completed in the dry. Two or more stages may be utilized, dependent upon the overall diversion requirements. Dams that have been constructed with this type of diversion include those built on the lower Columbia River, the lower Snake River, and the St. Lawrence River.

When this diversion method is used for a typical concrete dam, diversion is handled in three or more stages. First, a cofferdam enclosing a portion of the water channel is constructed at the damsite. Then the cofferdammed area is unwatered, the enclosed dam foundation area excavated, and the foundation surface prepared to receive concrete. Concrete is poured in the enclosed dam monoliths until they reach an elevation that will be above the impounded water surface to be encountered during second-stage diversion. Typical construction during first-stage diversion is shown in Fig. 4.1.

Figure 4.1 First-stage diversion at Amistad Dam, Texas. (C. S. Johnson Division of Koehring Co.)

If the water is to pass over low blocks during second-stage diversion, alternate blocks are left low; this alternate low- and high-block concrete placement is required to furnish clearance for the transverse forms and to position a high block on each side of a low block to support the bulkhead which restricts the water flow while concrete is being raised in the low block. The number of blocks to leave low, and the elevation of these blocks with respect to the upstream pool can be computed using the broad-crested weir formula:[4.2]

$$Q = 3.087 \, L \, (h + h_v)^{3/2} \, C$$

where Q = maximum theoretical discharge, ft³/s
         L = width of block, ft
         h = depth of water on block, ft
         h_v = velocity head of approach, ft
         C = compensation factor for end restrictions and other losses. For this application, its value is approximately 0.85.

If during second-stage diversion, the water is to pass through diversion conduits instead of over low dam blocks, these conduits, and provisions for future closing of the conduits against a head of water, must be incorporated into the dam. Figure 4.2 shows the conduit technique in use during the construction of Guri Dam in Venezuela. The river was exceptionally large for handling in this manner.

Figure 4.2 Second-stage diversion through diversion conduits, Guri Dam, Venezuela.

The number, size, and elevation of the diversion conduits can be computed using the orifice-flow formula. Since this formula involves a factor that varies with the shape of the conduit entrance and would require many sketches to explain, it is beyond the scope of this book. The interested reader is referred to hydraulic textbooks that cover this matter in detail.

While concrete is being placed inside the first-stage cofferdam, the dam foundation excavation and preparation can be performed on the dam abutments so that these areas will be ready for concrete placement when the conversion is being made from first- to second-stage diversion.

After the dam concrete in the first-stage cofferdammed areas has been raised above the rock foundation, the stream-dividing leg of the second-stage cofferdam is erected on top of any of the upstream and downstream concrete aprons and connected to the dam concrete. This portion of the second-stage cofferdam is erected at this time and located so that upon final completion of the second-stage cofferdam, the dam concrete that was placed during first-stage diversion will extend underneath the second-stage cofferdam and will connect to the concrete placed during second-stage diversion.

Second-stage diversion starts at the beginning of the first low-water period that occurs after the concrete within the first-stage cofferdam has reached an elevation that will be safe from unplanned overtopping. The first-stage cofferdam is removed, then the dividing leg of the second-stage cofferdam is connected to the opposite banks of the river by upstream and downstream cofferdam legs. This second-stage cofferdam will enclose the unexcavated portion of the river channel and when it is completed, water will be impounded upstream of the cofferdam until it reaches an elevation that will furnish sufficient head to pass the water through conduits in the dam or over low blocks. The second-stage cofferdammed area is unwatered; the enclosed dam-foundation area is excavated and prepared for concrete placement; and concrete placement is started. During construction of the second-stage cofferdam, dam-concrete placement can continue in the high blocks in the first-stage diversion area and in the blocks in both dam abutments.

After the concrete in the dam monoliths, enclosed by the second-stage diversion, reaches the height of the upstream leg of the second-stage cofferdam and any required grouting is completed, third-stage diversion can commence.

Third-stage diversion begins with the removal of all cofferdams, allowing the water to pond against the dam and to flow either over low blocks in the dam or through diversion openings constructed for this purpose. If the low-block system is used for third-stage diversion, concrete is poured in these blocks by cofferdamming off one low block with an upstream bulkhead and then placing concrete in this block until the level of the concrete is above the upstream water surface. The bulkhead is moved from block to block, and as each block is raised, the water surface behind the dam will also gradually rise until both the concrete and the upstream water surface reach an elevation that will permit the riverflow to be handled by the permanent river outlets, or, if there are no outlets or the outlets are of insufficient capacity, then until the water flows over the permanent spillway. When this elevation is reached, the bulkhead will no longer be required and concrete can be placed in a normal manner. If the diversion-conduit method is used for third-stage diversion, the conduits are usually left open until the dam is topped out. The openings must then be closed at the upstream ends with gates capable of resisting the head of water in the reservoir. The diversion conduits are then plugged with concrete, the concrete is cooled, and contact grouting is performed.

If high-pressure gates are used to close the openings, a safeguard to ensure closure of the openings is to install stop-log slots ahead of the gates so that stop logs can be dropped in these slots if the gates become inoperative. The time between bid advertising and bid submittal is usually too short to allow the estimator to make detailed designs of these closure gates, so he must secure this information from manufacturers. Closure-gate design is covered in detail in other books.[4.2]

Unless special provisions are made for overtopping, diversion through the dam has limited application for earth- and rock-fill dams, as the flow of water over the unprotected dam embankment will cause rapid erosion. Earth- and rock-fill dams have been constructed featuring diversion schemes that utilized routing of flood waters through a nominal-sized tunnel or conduit by impounding water against the upstream face. Rock-fill dams have also been designed to be overtopped during the flood peak, and "flow-through" diversion schemes in the absence of an impervious core have also been successful for individual applications. These schemes generally utilize some form of downstream slope reinforcement to prevent loss of rock at the toe.

On the Ord River project in Western Australia, embankment placing was completed in two stages. After completion of the first stage, the embankment was protected by armored stone and steel toe reinforcement to permit flood flows to pass over it. Actual flow during the planned overtopping exceeded 30,000 ft³/s (8,500 m³/s). Government surveys made before and after the flood passage indicated negligible movement and no loss of embankment materials.[4.3]

On the other hand, three dams (Brindle Drift, Lesapi, and Xonxa) were constructed in southern Africa making use of downstream zones of reinforced rock fills to control major construction flood. Relatively minor failures occurred during overtopping of the Brindle Drift and Lesapi dams, and at Xonxa a major failure occurred. Utilizing lessons learned after the above failures, the Googong Dam constructed near Canberra, Australia, successfully withstood overtopping by a major flood for a period of about 24 hours in which the maximum depth of water over the rock fill was about 10 ft (3 m).[4.4]

With earth- and rock-fill dams, diversion through the dam has its greatest application at damsites where, during the low-flow season, the water in the stream will decline to a trickle or to nothing. Then a portion of the dam is omitted to provide a channel for flood-flow passage, and the remainder of the embankment is placed to a height that will be safe from overtopping when the channel-flow passage is plugged. This height can be computed by evaluating the storage capacity of the reservoir upstream from the dam, the elevation of water-release openings, and the elevation of the spillway crest. After placement of this part of the embankment and at the beginning of a low-water season, the water is piped or flumed through the dam, the dam foundation is excavated in the former channel, and fill is placed in the embankment opening until the embankment reaches the bottom of the pipe or flume level. The pipe or flume is removed, and fill placement in the embankment opening must then proceed at a rate that will prevent the water in the dam reservoir from reaching the top of the fill. When this section of the dam embankment is brought up to the height of the remaining embankment, fill is placed across the entire dam until it is topped out.

Diversion through the dam can be used to advantage in the construction of combination dams containing earth or rock fill and concrete spillway and/or powerhouse sections. First-stage construction consists of cofferdamming the concrete section, placing concrete in this section, and providing facilities in the concrete section for the passage of water during second-stage diversion. During second-stage diversion, the water is passed through the concrete section, the earth- or rock-fill section is cofferdammed, and the embankment is completed. During the third-stage diversion, the concrete section is completed. Many combination dams are constructed with the concrete section located on one abutment, which simplifies the construction of the first-stage cofferdam. Examples of this are the Wells and Wanapum dams constructed on the Columbia River.

When stream diversion is through the dam, the type of cofferdam that is used varies with the volume of water, the area available for cofferdam construction, the water velocities adjacent to the cofferdam, and the water velocities which will occur. When the velocities are moderate despite the decreased waterway, earth- or rock-fill cofferdams can be safely used. If the river channel is narrow and space is at a premium, or if high velocities occur adjacent to the dividing leg of the first-stage cofferdam, the best method of constructing the dividing leg of the cofferdam is with steel-sheet-pile cells or with timber cribs. If the upstream and downstream shore connecting legs of the cofferdams are exposed to high water velocity near the dividing leg, the sheet-pile cells or cribs are carried back a short distance on each leg. If high velocities occur when a cofferdam is being closed, then closure must be made with rocks large enough to resist the transporting action of the water.

Other closure methods may be required if velocities are such that closure cannot be made with large rocks. Cofferdams have been closed by raising their surface with precast tetrahedrons. Cofferdams used when high water velocities are encountered across the total area have been constructed by building timber cribs on shore with the bottoms tailored to fit the streambed. These are then launched, floated into place, and sunk in position. This type of cofferdam was used to construct Bonneville Dam on the lower Columbia River. In other cases, erosion-resisting abutments are constructed on each side of the closure opening; temporary cofferdam closure to stop the flow through the opening is made with a trussed bulkhead floated into place between these abutments, allowing the cofferdam to be completed behind the bulkhead. This system was used in closing the first-stage cofferdam at Long Sault Dam on the St. Lawrence River. When site conditions present definite problems, unique closure methods have been devised, such as closing the channel by blasting in the river banks or by casting a concrete bulkhead on the bank and moving it into position by an explosive charge.

Construction of the dividing leg of the second-stage cofferdam does not present any difficulties, as it usually is constructed in the dry within the cofferdammed area during first-stage diversion. The cofferdam is located on top of any downstream apron concrete and must be tied into the upstream and downstream sides of the dam concrete. Due to space restrictions and because part of the cofferdam is located on the concrete downstream apron, it is usually of timber-crib or sheet-pile-cell construction. The downstream section of this cofferdam may be subject to fast-flowing water as the water is discharged from the diversion openings or over low blocks in the dam, which may result in a negative force on the cofferdam, causing it to slide toward the water. If it is built where these conditions will occur, the cofferdam must be designed to resist this action.

DIVERSION THROUGH A SPILLWAY[4,5]

After 10 years of effort, Chinese engineers utilizing 30,000 workers on the Gezhouba hydroelectric project successfully completed closure of the main channel of the Yangtze River in early 1981. The upstream main channel cofferdam featured a 20-m-deep grout curtain and double concrete core walls keyed into bedrock. The huge river was successfully diverted into a 27-gated spillway completed in 1980. The design capacity of the diversion works is 2,360,000 ft³/s (66,800 m³/s), and initial flood flows of this magnitude were successfully handled in mid-1981. Construction of the second and larger powerhouse (1,750 MW) is now underway and scheduled for completion in 1986. Total installed generating capacity of the completed project will be 2,715 MW. Generation of initial power from first-stage construction began after the successful diversion in mid-1981.

Possibly one of the most massive projects in history could be the proposed Three Gorges project on the Yangtze River. The Gezhouba project will act as a regulating dam for the massive project located 25 miles (40 km) upstream. The huge flood-control power/navigation/irrigation scheme contemplates an 8,500-ft (2600-m) -long, 650-ft (200-m) -high gravity dam, a 25,000 MW powerhouse, and a five-step navigation lock.

DIVERSION THROUGH A TUNNEL

This method of diversion is used at damsites located in narrow, steep-walled canyons where space is not available for other types of diversion. It is also used at damsites when the diversion tunnel can be utilized as part of the permanent facilities. Dams that have diversion tunnels converted to permanent use are Hungry Horse Dam, where the diversion tunnel became part of the glory-hole spillway; Hoover Dam, where some of the diversion tunnels were used to convey water to the powerhouse and the others were used as part of the spillway; Oroville Dam, where diversion tunnels were used to convey water to the powerhouse; and Glen Canyon Dam, where diversion tunnels were converted to spillway use.

This diversion method requires the construction of a diversion tunnel or tunnels in the dam abutments to convey the water past the damsite. Whether the diversion tunnels are lined with concrete or left with bare rock surfaces depends on the characteristics of the rock. If the tunnel must be supported with steel sets, the tunnel is usually lined in the supported section to protect the supports from erosion action. The upstream portal of the diversion tunnel must be located a sufficient distance upstream of the damsite so that the upstream cofferdam can be positioned between the tunnel portal and the upstream edge of the dam excavation. To prevent damage to the dam foundation, the diversion tunnel should be located so that there will be a minimum distance of 100 ft (30 m) of solid rock between the bottom of the dam and the closest part of the diversion tunnel or tunnels. The designed capacity of a diversion tunnel is a function of the waterway area and the head of water on the tunnel entrance; this head determines the velocity of flow in the tunnel. In designing the system, the cross-sectional area of the tunnels must be economically balanced with the height of the upstream water, which in turn determines the height of the upstream cofferdam. For the upstream cofferdam, a minimum freeboard height of 5 ft (1.5 m) should be allowed.

A preliminary determination of the diameter of a diversion tunnel and the height of the upstream cofferdam required to handle a given stream flow can be made utilizing Fig. 4.3. This chart applies to tunnels that are submerged at their inlet and have a free flow from their outlet. If a water velocity is assumed, for example 25 ft³/s (0.7 m³/s), the diameter of the diversion tunnel, the velocity head, and the friction loss per 100 ft (30 m) of tunnel can be taken from the chart. Then the difference in feet between the elevation of water impounded by the upstream cofferdam and a theoretical point that is 0.8 of the tunnel diameter above the invert at the tunnel's downstream portal is equal to the total head. The total head consists of the friction loss in the tunnel, the velocity head (both of which were secured from Fig. 4.3), and the entrance head. The entrance head for a square-tunnel opening is equal to one-half the velocity head, and for a bell-mouth tunnel opening it is equal to one-fourth the velocity head. If lower upstream cofferdam heights are desired, the above-described use of the chart can be reversed.

The upstream cofferdam can be constructed as a small earth- or rock-fill dam with an impervious core, or the fill can be faced with an impervious upstream blanket when there is ample space available for its construction. If the main dam is either earth or rock fill, its design often incorporates the upstream cofferdam as part of the permanent dam. When earth- or rock-fill cofferdams are used, stream closure velocities should be checked to determine if final closure of the cofferdam can be accomplished with earth or rock. When cofferdam space is at a premium or when water velocities during cofferdam closure are excessive, either a timber-crib cofferdam or a steel-sheet-pile cofferdam is used.

A downstream cofferdam can be of simpler construction since it will be subjected only to small tailwater heads and since its closure is generally made after the upstream cofferdam is completed, it can be of earth or rock fill, timber cribs, or steel-sheet-pile cells. Factors that influence the type of construction are the space available, whether it will be exposed to water being discharged from a diversion tunnel, and the tailwater elevation. When downstream cofferdams are exposed to the discharge from diversion tunnels, they must either be concrete or have their surfaces protected from the velocity of the discharged water, and they must be of sufficient height to provide protection against the hydraulic jump that occurs when the water is discharged from the tunnel.

When the dam has reached an elevation safe from overtopping from the

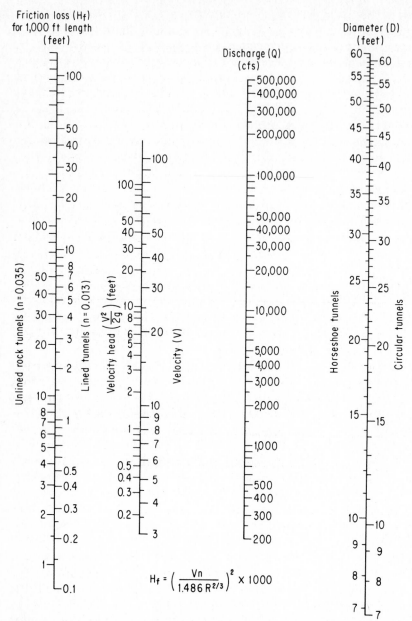

Figure 4.3 Friction loss in full-flowing tunnels. (Data taken from C. V. Davis, ed., *Handbook of Applied Hydraulics*, 2d ed., McGraw-Hill, New York, 1952.)

maximum anticipated flood, the upstream cofferdam is removed or leveled, and the diversion tunnels are closed. The dam elevation that will be safe from overtopping can be computed by evaluating the capacity of the reservoir behind the dam, the amount of water that can be handled by the permanent water outlets incorporated in the dam design, the ability to bypass flood flows through powerhouse facilities if they are incorporated in the construction, the elevation of the crest of the permanent spillway, the capacity of diversion conduits that may be placed in the dam for second-stage diversion, and, for concrete dams, the flood flows which can be passed over low blocks without damage to construction facilities or downstream installations. Whether the upstream cofferdam must be removed or leveled depends upon specification requirements.

Closure of the diversion tunnels is accomplished by using temporary portal gates or stop logs to shut off the water flow through the tunnel until permanent concrete tunnel plugs can be placed. The concrete plugs are located directly under the dam so that the plug will form part of the impervious curtain in the dam foundation. The procedure for placing this concrete plug depends on two particular job conditions: the depth of water that will be impounded in the reservoir during the closure operation and the necessity to maintain a minimum streamflow through the diversion tunnel until the water impounded by the dam reaches an elevation that permits water to pass through the permanent stream-release facilities.

If several diversion tunnels are constructed, they often have different upstream invert elevations in order to simplify the closure operation. The tunnel with the lowest invert elevation often can handle all the low-water flow, leaving the other tunnels dry. This permits the placement of the concrete plugs in these higher tunnels in the dry without the need of any temporary closure gates.

When it is not necessary to maintain any water flow through the last (or only) tunnel, temporary closure can be made with precast-concrete stop logs placed in slots constructed at the portal before first-stage diversion is started. The stop logs must be designed to resist the total head of water that will be impounded by the dam during the time between temporary closure and completion of the permanent concrete plug.

Plugging of the diversion tunnels becomes more complicated when it is necessary to maintain a minimum streamflow during the time required for the water in the reservoir to reach an elevation that will allow the minimum flow to pass through permanent water-release facilities in the dam. One of the simplest methods of accomplishing this was used by Kaiser Engineers at Detroit Dam, Oregon. To maintain the required streamflow, a small tunnel was excavated which branched off from the main diversion tunnel a short distance from the upstream portal, and which daylighted adjacent to the upstream portal of the main diversion tunnel. The portal for this small tunnel was equipped with a remotely operated high-head gate that was capable of resisting the water pressure. Because of the small size of the tunnel opening, this gate was not too expensive or difficult to install. Before diversion, the main diversion-tunnel closure was made, concrete stop logs, designed to resist the anticipated head, were dropped in the main tunnel opening. The minimum streamflow was then passed through the small branch tunnel. The concrete plug was placed in the dry by passing the minimum flow through a 36-in-diameter pipe which extended through the area where the concrete tunnel plug was to be placed. When the upstream water reached an elevation that permitted the required streamflow to pass through the permanent river-discharge outlet, the gate was closed over the bypass tunnel, and the 36-in-diameter pipe was filled with concrete, which completed the tunnel plug placement.

When temporary diversion conduits are located in the dam to release the required streamflow, they must be closed in the same manner as that used to close diversion conduits in cases where the entire stream is diverted through the dam. High-head gates are required, and upon closure the diversion openings are plugged with concrete. This concrete is often placed by prepack methods to reduce shrinkage.

When concrete dams are constructed at sites where there are great variations in flood flow, economics may make it desirable to size the diversion tunnels and build the upstream cofferdam to handle less than the maximum flood flow. The diversion facilities may only have sufficient capacity to handle floods that might occur in 5 years, but incorporated in the overall design should be provisions that will allow greater floods to be passed through the damsite without extensive damage to any facility. These provisions may consist of spillways located on the upstream and downstream cofferdams to permit overtopping, and scheduling of work so that the dam excavation and the dam concrete can be completed to streambed height during the low-water season. In lieu of cofferdam spillways, the cofferdam crest and its downstream slopes may be protected with paving or gunite so that its surface will not be eroded when the cofferdam is overtopped. Construction equipment must also be located free of the area subject to floods or be of a type that can be readily removed from the area when flooding is anticipated. The diversion facilities for the construction of Detroit Dam were designed in this manner. During construction, the upstream cofferdam was overtopped five times. This cofferdam was built of timber cribs that were not damaged during the overtopping, and when a minor portion of the downstream cofferdam was washed out, its replacement cost was negligible.

To hold diversion cost to a minimum, when earth- or rock-fill dams are constructed at sites where a tunnel is used for diversion, the tunnel may be sized to handle only the low-water flow with a low cofferdam. But, in conjunction with the storage capacity in the reservoir, it is sized to protect the dam from overtopping after the dam embankment has been raised sufficiently to allow ponding of an upstream reservoir and the creation of head on the tunnel. When this type of diversion is used for earth-fill dams, diversion is handled in four stages. First-stage diversion commences at the start of the low-water season and consists of diverting the low flow through the diversion tunnel, excavating the dam's foundation below stream level, and raising the dam embankment to streambed elevation. Second-stage diversion occurs during the high-water season when the flood waters are passed over this completed section of the embankment. Since the dam embankment was not raised above the streambed elevation, the erosion of this embankment will be slight. During this period of second-stage diversion, placement of the dam embankment is stopped, but excavation for the dam's foundation on the abutments above the water level can continue. Third-stage diversion begins at the start of the second low-water season when the water is again passed through the diversion tunnel. During this low-water season, the previously placed embankment is cleaned off, and the embankment is raised to an elevation that will develop enough upstream pool capacity and enough head on the diversion tunnel to store and pass flood flows and thus keep the embankment safe from overtopping. This allows the remainder of the dam to be completed irrespective of high- or low-water flow. Fourth-stage diversion is the closure of the diversion tunnel with stop logs or gates and the placement of the permanent concrete plug. Trinity Dam, California, was constructed using this method of stream diversion.

The tunnel diversion method can be modified in many ways to fit particular site conditions. This was done for the construction of Akosombo Dam, Ghana, Africa.[4,6] This was a rock-fill dam located at a site that was covered by a deep pool of water created by a downstream channel obstruction. The foundation rock was overlain by a layer of deposited sand. The large variation of river flow, from a minimum of 800 ft³/s (23 m³/s) to a maximum of 500,000 ft³/s (14,000 m³/s) made the passage of the flood flows through the dam the most economical method of diversion. A 30-ft-diameter diversion tunnel was constructed to handle normal flows. Before the placement of the cofferdams was begun, the sand was dredged from both the dam and cofferdam foundation areas. Dredging for the downstream cofferdam foundation area extended to 217 ft (66 m) below water level. The cofferdams were constructed below water by dumping rock fill. The cofferdams were sealed by first placing a transition zone and then an impervious blanket on the outer side of each cofferdam. Above water level, rock was placed in the cofferdam by end-dumping. After the cofferdams were closed, the enclosed area was unwatered and the dam was constructed. The cofferdams were overtopped by flood flows before they were completed, with comparatively little damage resulting from this overtopping.

Diversion-tunnel Construction

Because of the small amount of excavation required, a tunnel used for diversion should be constructed with a minimum of equipment. Rubber-tired excavation equipment and jackleg drills are often used. Tunnel construction is covered in part three of this book.

DIVERSION THROUGH AN ABUTMENT

First-stage diversion through a channel or flume located on one of the dam's abutments is used at damsites that do not have wide enough streambeds to permit diversion through the dam but do have sufficient space on one abutment for a channel or flume. The topography of the abutment must permit the channel or flume to be located at an elevation low enough that an economically sized cofferdam will furnish sufficient upstream impounded water head on the channel or flume to allow the passage of maximum flood flows. The dam must also be of a configuration that allows the use of this method of diversion without delaying construction and that will, during second-stage diversion, permit the passage of flood flows through openings in the dam, over low blocks in the dam, or over a permanent spillway.

The economical height for the first-stage cofferdam can be determined by comparing costs using this diversion method with those of other diversion methods. At Guri Dam in Venezuela, diversion was accomplished with the diversion-channel method, and a large amount of material had to be excavated from one abutment to construct the diversion channel. The diversion channel was long, with dikes constructed along one edge. Ten sheet-pile cells were used for erosion protection at the diversion-channel entrance. The channel location was so high on the abutment that a 145-ft (44-m) -high cofferdam was required to raise the upstream water to an elevation that permitted it to flow through the channel. These diversion facilities are illustrated in Fig. 4.4. Even with the extreme height of the upstream cofferdam, this was the most economical type of diversion to use at this narrow damsite, where flood flows of 600,000 ft^3/s (17,000 m^3/s) were anticipated.

The height to which concrete must be poured in the dam during first-stage diversion is controlled by the method of handling the water during second-stage diversion. If water is to be passed over low concrete blocks, these blocks only need to be poured above the downstream tailwater before starting second-stage diversion. If the water is to be handled by diversion conduits, through permanent openings in the dam, or over the dam's permanent spillway, most of the concrete in the dam must be poured before the start of second-stage diversion. When these last two methods of handling the water are used, often a second-stage cofferdam must be erected across the entrance of the diversion channel or flume.

When second-stage diversion is started, the first-stage upstream cofferdam must be removed. If this is a high cofferdam, the work may be quite extensive, as in the removal of the first-stage cofferdam at Guri Dam (Fig. 4.5).

After the removal of the first-stage cofferdam and, if required, the con-

Figure 4.4   First-stage diversion with a channel, Guri Dam, Venezuela.

Figure 4.5 Removal of the first-stage cofferdam, Guri Dam, Venezuela.

struction of a second-stage cofferdam across the entrance to the diversion channel or flume, the remainder of the diversion work is similar to that discussed under diversion through the dam. If the dam is all earth or rock fill, second-stage water can only be handled over the dam spillway.

Depending on space available and closure velocities, the cofferdam can be of earth and rock fill, steel-sheet-pile construction, or timber-crib construction. The first-stage diversion cofferdam constructed at Guri Dam was an earth- and rock-fill cofferdam, and the cofferdam was easily closed during a low-water flow of approximately 7,000 ft³/s (198 m³/s).

The channel or flume method of diversion can be used to great advantage in the construction of concrete dams with earth sections, when the diversion channel can be located through the earth-embankment section of the dam. The concrete portion of the dam and the portion of the sections that do not conflict with the diversion channel are constructed in first-stage diversion. Then at the start of a low-water season, the water can be diverted through the concrete portion of the dam and the earth embankment can be placed across the diversion channel. The final diversion stage is the completion of the concrete section of the dam.

In the past, such as during construction of Canyon Ferry Dam in Montana, diversion flumes were often used. Because of the rapid rise in timber and timber-framing costs, diversion channels have replaced flumes, but small wooden flumes are still used for diverting small streams past damsites.

DIVERSION THROUGH A
DIVERSION-CONTROL STRUCTURE

The chosen diversion program for the Itaipu project on the Paraná River in South America was a three-phase program, incorporating a major diversion channel, combination concrete arch and embankment cofferdams, and portions of the finished permanent dams and powerhouse. Figure 4.6 shows the second phase, in which the river is diverted into a channel through a permanent diversion structure partially completed to el. 472 ft (144 m). The diversion structure will ultimately be completed to el. 738 ft (225 m) and will house the permanent power intakes. Figure 4.7 shows the second-phase diversion, and Fig. 4.8 the final phase.

DIVERSION THROUGH A
PIPE CONDUIT OR A FLUME

The pipe-conduit method is a single-stage diversion procedure widely used in the construction of earth- or rock-fill dams on small streams. A ditch is excavated through one abutment at, or slightly below, the final foundation excava-

Figure 4.6 Second-phase diversion, Itaipu project, Paraná River, South America. (International Engineering Co., Inc.)

tion elevation. A concrete conduit is then constructed in this ditch. At the upstream end of the conduit either a permanent gate structure or an intake structure is erected, so that the diversion conduit will also serve as a permanent river outlet. If a permanent gate is not installed on the upstream end of the conduit, a permanent valve or provision for such a valve is constructed at the midpoint of the dam. After completion of the conduit, the water is diverted through it by cofferdams. When dam construction is completed, the diversion conduit is converted to a permanent river outlet and water storage is started in the reservoir.

The pipe-conduit method has very limited application in concrete construction, but when it can be used, it is one of the simplest diversion procedures. It is applicable when small concrete dams are constructed on small streams or on streams where the flow can be rigidly controlled by upstream facilities. The streamflow must not exceed the amount that can be carried through the damsite by the conduit or a small flume. This diversion method can also be used to handle the low-water flow when the dam foundation can be excavated and concrete placed to streambed elevation in one season. Then the flood flows can be passed over low blocks in the dam.

In concrete-dam construction when the pipe or flume is carried through the dam at streambed elevation, the section that spans the dam must be supported so that the excavation for the dam's foundation can be completed alongside and beneath it. Small upstream and downstream cofferdams, usu-

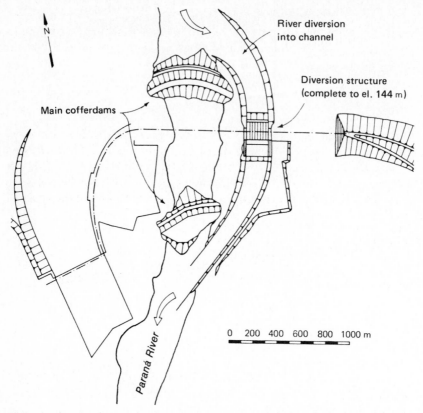

Figure 4.7 Second-phase diversion, Itaipu project, Paraná River, South America. (From "The Bi-National Itaipu Hydropower Project," *Water Power and Dam Construction,* October 1977.)

ally of earth-fill construction, are required to divert the water into the pipe or flume. After concrete has been placed in the dam to the bottom of the flume or conduit, an opening is left so that the conduit or flume can remain in place until the dam is completed. Upon completion of the dam, the conduit is removed and the opening left for its passageway through the dam is filled with concrete. In some cases where pipe conduits are used, instead of blocking out for the conduit, the pipe is embedded in the concrete and used as part of the permanent stream-regulating facilities.

Figure 4.9 shows low-water flow being handled by use of a flume during the construction of the afterbay dam on the American River Project. At this dam, one-half the foundation excavation and concrete placement to streambed level was accomplished during the first low-water season, and the other half was done in the following low-water season. During the first low-water season, diversion was accomplished with a large steel pipe. During the second low-water season, the water was carried by the flume (shown in the photograph), which was supported on a concrete block that had been poured during the previous low-water season.

COFFERDAMS

The number and size of cofferdams required to accomplish diversion vary with the method of diversion, the size of the river, the type of dam, and the site conditions. The available space to construct the cofferdam, the height of the impounded water, and the maximum water velocities that will be encountered during diversion and during cofferdam closure will control the type of cofferdam.

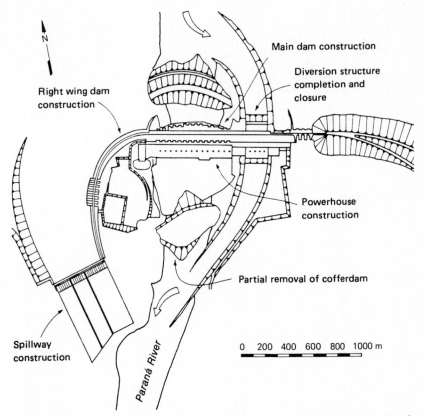

Figure 4.8 Final-phase diversion, Itaipu project, Paraná River, South America. (From "The Bi-National Itaipu Hydropower Project," *Water Power and Dam Construction,* October 1977.)

Figure 4.9 Diversion with a flume, American River project, California. (American River Constructors.)

Earth- or Rock-Fill Cofferdams

Earth- or rock-fill cofferdams are the most economical types to use if space is available for their construction, if the cofferdam foundation is impervious, if suitable fill material is available, and if water velocities are not too great. They can be placed on pervious foundations if a water-cutoff wall has been constructed through the pervious material. They can be constructed to any height if they are properly designed. For example, the first-stage earth- and rock-fill cofferdam at the Guri damsite in Venezuela was approximately 145 ft (44 m) high. The greater the height, the more the attention that must be given to cofferdam design. The passage of water through these cofferdams is prevented either by an impervious core or by use of an upstream blanket. These water barriers must be well-designed to prevent leakage, and the cofferdam must have sufficient weight downstream of the impervious core to resist the static head of the impounded water. Velocities that will be encountered in final closure must be computed; if they become critical, final closure of the cofferdam must be made with large rocks, rock necklaces, gabions, tetrahedrons, tetrapods, or other precast-concrete shapes. *Rock necklaces* are several large rocks that have been drilled and strung on a cable. *Tetrahedrons* and *tetrapods* are names for concrete castings that have projecting arms that interlock with each other. Earth- and rock-fill cofferdam design is similar to earth-

or rock-fill dam design, and design data can be found in various books on this subject.[4.7] Rock particle sizes suitable for final closures at different water velocities are given in other publications.[4.8]

Shot-in-Place Cofferdams

Cofferdams have been constructed in narrow gorges by blasting the canyon walls to form a river barrier when difficult access, high stream velocities, and year-round flows prohibited other methods of cofferdam construction. This construction procedure necessitates placement of all rock in one large blast. The upstream face can then be sealed with a blanket of impervious material. This type of cofferdam was used at Mayfield Dam on the Cowlitz River in Washington, but after a long delay in attempting to seal the upstream face, the contractor was forced to resort to an expensive program of chemical grouting before the structure would serve its purpose. However, reports on the use of this method in Russia indicate that it can be quite successful.

Timber-Crib Cofferdams

In the past this type of cofferdam was used when there was limited space for cofferdam construction, when the cofferdam sides were exposed to eroding water velocities, or when the cofferdam closure was made against high-water velocities. Because of the large increases in the cost of timber and timber-framing labor, steel-sheet-pile cell cofferdams are replacing timber-crib cofferdams. The latter are now used only for special purposes, such as when cofferdam closure requires that cribs be floated into place and then sunk, or to prevent damage to the concrete when cofferdams are erected on concrete aprons.

A timber-crib cofferdam is framed from large timbers, usually 12-in square, in a crib pattern. The timbers are drift-bolted or bolted together, and the enclosed areas are filled with rock. Water tightness is provided either with a double layer of vertical planking across the face of the cofferdam or by placing impervious fill in some of the cofferdam pockets. If the cofferdam is to be floated into position and sunk, enough pockets must be floored so that sinking can be accomplished by filling the pockets with rock.

A timber-crib cofferdam is stable when the resultant of the water pressure against the cofferdam and the weight of the cofferdam passes through the middle third of the cofferdam. To accomplish this, the cofferdam's width must equal its height. The cofferdam can be stable with a width less than its height if inside the cofferdammed area an earth fill is placed against the cofferdam. A timber-crib cofferdam can be overtopped without damage if the top is paved with large rocks, sheathed with lumber, or paved with concrete or asphalt. When timber cribs must be removed on job completion, they often are constructed with holes drilled in the crib corners to simplify the future placement of explosives.

When timber-crib cofferdams are to be floated into position and sunk, the streambed on which the cofferdam will rest should be sounded and the foundation profile determined. The bottom of the cribs can then be tailored to match the streambed. Based on the Bonneville Dam cofferdams, the amount of timber used in the construction of a timber crib is approximately 11 percent of the total crib volume. Both constructed-in-place and floated-into-position types of timber-crib cofferdams were used for the construction of Bonneville Dam as described in the reference article.[4.9]

Steel-Sheet-Pile Cofferdams

Steel-sheet-pile cofferdams are now used instead of timber-crib cofferdams when space is limited, where high water velocities occur along the cofferdam sides, and where cofferdam closure must be made against a high differential head causing high velocity during cofferdam closure. Besides having a cost advantage over timber cribs, they can give a more positive water cutoff, since the piles can be driven through overburden material to rock. Steel-sheet piles are used in circular cell cofferdams, double-diaphragm-wall cofferdams, and as a single-diaphragm wall reinforced with a rock fill.

Cellular cofferdams are used when large, deep cofferdams are required, since these units are individually stable and do not require bracing. The use of individual cells has an advantage in that each cell can be completely filled after erection to allow progressive construction. Such cells must be designed to prevent overturning, sliding, and rupturing of the piles in the interlocks. Design formulas for cell construction can be secured from manufacturers' literature.[4.10] A rule of thumb for sheet-pile cell design is that the diameter must equal the height. Cell diameter can be reduced by placing an earth berm inside the cofferdammed area against the cells. When large-diameter cells are required, overstress in the pile interlocks can be avoided by the use of high-strength piles or a cloverleaf cell. Cloverleaf cells are divided into four smaller cells and provide resistance to overturning and sliding. Pile interlock stress is reduced since it is a function of the maximum radius of the component cells. The number of piles in each of the various-sized cells and of the piles in the connecting arcs between cells, the dimensions of cells, and the volume of fill required per foot of cell can be secured from sheet-pile catalogs.[4.11] When cells are placed on bare rock, the bottom of the cells should be sealed with sacked concrete or other impervious material to cut off water inflow, to prevent the water flowing past the cell from removing the cell fill, and to prevent cofferdam leakage when the cofferdammed area is unwatered. Sealing of the cell bottom is done after the cells are driven by cleaning out the cell and then having divers place sacked or tremie concrete against the piles on the water side of the cell. To reduce damage from overtopping, the top of the cell can be paved with large rocks, concrete, or asphalt. If overtopping is anticipated, damage can be reduced by installing gates or weirs in the cofferdam to allow flooding of the cofferdammed area before overtopping occurs. Cellular cofferdams on the Columbia River have been completely submerged during flood flows with only minor damage resulting.

The major construction problems encountered in sheet-pile cell construction are threading of the piles in the cell and maintaining the position of the piles until they are driven. This problem is very critical when the cells are located in water on bare rock. Piles can be accurately threaded around the cell and held in the proper position by using a cell template. After all piles are placed, they can be progressively driven two piles at a time a distance of approximately 4 ft (1 m). This procedure is repeated until all piles are driven the required distance. To speed up pile placement, piles can be threaded in pairs on shore and then placed in the cell. Cell fill can be placed by end-dumping trucks from the last cell, by using clamshells that place fill from piles dumped by trucks on previously completed cells, or secured from barges. The fill can also be placed with dredges. Water is used for compaction purposes during the filling operations. Drainage holes should be placed on the interior side of the cells so that hydrostatic pressure, which would increase the stress on the pile interlocks, will not be developed in the cell. Also cells should be free-draining so that they can better withstand overturning and sliding. This type of cofferdam has been used for diversion purposes during the construction of a large number of dams.

Double-diaphragm cells can be used instead of circular cells when water velocity during driving is not a problem. This type of cofferdam consists of parallel multiple arcs connected by cross walls at each arc intersection. The advantage of double-diaphragm cells over individual cells is that fewer piles are required. The disadvantage of this type of cofferdam is that in order to prevent distortion of the cross walls the fill must be placed along the whole cofferdam with the fill height in each cell approximately the same as in the others. Because use is limited to locations where low water velocities occur and because the filling method is more difficult, double-diaphragm cells are seldom used for diversion purposes.

When a cofferdam is to withstand only low heads, a single-diaphragm wall of sheet piles supported on both sides by rock fill can be used for cofferdamming purposes. Since the piles transmit all loads to the rock, there will be no tension on the interlock, resulting in more leakage than with circular cells. This type of sheet-pile cofferdam was used at Rocky Reach Dam in Washington.

Precast-Concrete Cofferdams

A novel method developed in 1930 on the Saguenay River at Chute à Caron, Quebec, consisted of building a precast-concrete dam, standing it on end at the side of the main channel, and tipping it over at the proper time into the main channel. Essentially the method involved the following major features: (1) a fixed pier, relatively massive, to carry the greater part of the weight and thrust; (2) a small pier, carrying part of the weight, to be blasted away; and (3) a cylindrical rolling face on the fixed pier designed and placed so that the precast dam would fall into an accurately predetermined position. The river channel at the point of final closure was about 110 ft (34 m) wide and 28 ft (9 m) deep, with a water velocity of about 20 ft³/s (0.6 m³/s).[4.12]

Floating Cofferdams

A reinforced-concrete floating cofferdam weighing 5,200 tons (4.7 million kg) enabled successful reconstruction of the outmoded spillway during the raising, by 10 ft (3 m), of the Chief Joseph Dam, Bridgeport, Washington. The unique floating cofferdam enabled spillway reconstruction to be completed without curtailing power production or impeding river flow. After reconstruction was complete,[4.13] the cofferdams were used as anchors for a debris boom.

Concrete Training Walls

The use of concrete training walls may be economical in training the water that is discharged from downstream tunnel portals so that the downstream cofferdam will not be eroded. (This was done at Detroit Dam, Oregon.) They also may be used in the dual capacity of training the water and cofferdamming. A third use is to control the discharge from diversion conduits so it will be carried past construction areas, such as those required for powerhouse construction. Concrete training walls can also be used as the dividing walls of the first-stage cofferdam at narrow damsites when diversion is through the dam and space is at a premium.

When concrete cofferdams or training walls are used, they should be constructed with holes for explosives cast in the concrete to facilitate their removal.

SUBSURFACE WATER CONTROL

If the cofferdam is erected on pervious overburden materials, large amounts of water may flow through this material into the cofferdammed area. Because such inflows have created problems at some damsites, the permeability of the local materials should be studied to determine whether the excavated area will be subject to water inflows. Sometimes river gravels are so well graded and compacted that inflows under cofferdams will be of only minor consequence. This was the case at Detroit Dam, Oregon, where excavation was carried downward through 80 ft (24 m) of gravel to expose suitable foundation rock. This excavation was adjacent to the upstream cofferdam, which was placed on top of the gravel; but since the gravel was tight, there were no major water inflows through the gravel beds. At other damsites where there were shallow layers of gravel underneath the cofferdam, large water inflows into the excavated area created pumping problems and also resulted in erosion of the sides of the excavation. If it is determined that inflow either under the cofferdam or through the abutments will be encountered, provision for handling it should be made prior to the start of excavation. Water inflows can be reduced by predraining the area, by increasing the path of percolation of the water, or by constructing cutoff walls.

Predraining

Most often, groundwater inflows are prevented by the installation of wellpoints or wells to predrain the excavated area. If excavation is to be done through cohesionless material ranging from silt to gravel, the wellpoint system will not only predrain the material but will also stabilize the slopes of the embankment by keeping them dry. Wells may be constructed by predrilling or jetting, depending on soil conditions. Wellpoints are from 10 to 15 ft (3 to 4.5 m) long and are driven directly into place or are installed in sand-filled wells spaced in regular intervals around the perimeter of the excavation. Each wellpoint is connected to a header, and water is removed from the header by one or more pumps, depending on the size of the excavated area. Additional groups of wellpoints, headers, and pumps, known as *wellpoint stages,* are installed as the excavation is deepened. In highly permeable soil, the use of perforated cased wells packed with gravel and drained with vertical pumps, or a combination of these wells and a wellpoint system, may be more economical than wellpoints alone. Layout of a predraining system employing either wellpoints or wells is a highly specialized task, usually requiring the services of a specialist in the field of dewatering. A typical wellpoint installation is shown in Fig. 4.10.

Increasing the Path of Water Percolation

The volume of water inflowing through the sides of the excavation can be reduced by increasing the path of its percolation. This can be done by placing a blanket of impervious material on the bottom and sides of the river channel, starting at the upstream edge of the cofferdam and continuing upstream as far as necessary. Blankets are often used to increase the path of water percolation under and around concrete dams but are seldom used for diversion purposes, since they do not provide a positive solution. Furthermore, the reduction of water inflow into a cofferdammed area may require the placement of a large quantity of material. Therefore, this method of reducing the

Figure 4.10   Wellpoints installed at Trenton Dam, Nebraska. (John W. Stang Corp.)

water inflow is generally not resorted to unless unanticipated inflows are encountered after a cofferdammed area has been unwatered.

Cutoff Walls

The third method of controlling groundwater inflow is by the construction of cutoff walls. Cutoff walls can be constructed of impervious material by the Cronese, or bentonite slurry-trench, method or by the Vibroflotation method, or the existing overburden material can be formed into an impervious wall by grout injections or by freezing. Another method of constructing a cutoff wall is by driving sheet piles through the materials until they contact the rock surface.

The *bentonite slurry-trench method* of producing an impervious cutoff wall requires that a trench be excavated to bedrock. The material on each side of the trench is retained in position by stabilizing the excavation with a heavy slurry of bentonite, which is placed in the trench as it is excavated. When the trench excavation has progressed sufficiently so that backfilling operations will not interfere with excavation, the bentonite slurry is progressively displaced with impervious fill material, forming an impervious cutoff. This type of cutoff wall was used to enclose the excavation area during the construction of Wanapum Dam, Washington. In this case, portions of the slurry trenches were 90 ft (27 m) deep.

For the construction of the second powerhouse at Bonneville Dam on the Columbia River, the Portland District of the U. S. Army Corps of Engineers designed a 2-ft (61-cm) -thick concrete slurry wall for control of groundwater into the excavation. Similar walls have been constructed to depths of over 200 ft (61 m) through a variety of soil conditions. Other applications of the slurry wall featuring an ingenious system of circular elements and intermediate panels were used for the Wolf Creek Dam in Kentucky and for the Manicougan Dam in Northern Quebec, Canada.[4.14]

Loose sand and gravel material can be compacted into a cutoff wall using a patented process called *Vibroflotation*. This process is based on the use of a long, slender tube (called a Vibrofloat) containing an electrically driven eccentric weight to furnish the vibrating action and equipped with a water jet. The Vibrofloat is swung from a crane and inserted into the loose material at regular intervals. The water jet aids in the sinking action, and some preconsolidation takes place as the Vibrofloat descends. When it has reached the desired depth, the water jet is turned off. As the Vibrofloat is slowly retracted, the soil is compacted to approximately 70 percent density, and additional sand must be added. This method has not been generally used for cofferdam cutoff walls, but it has been used to compact material under permanent earth-fill embankments. For example, it was used in the construction of the cutoff wall under the earth-fill abutments at Priest Rapids Dam in Washington. Also, the barrages (diversion dams) constructed for the Pakistan irrigation system made extensive use of this method of water cutoff.

When the rock at the damsite is of such porous nature that water can readily flow through it, water inflow into excavated areas can be reduced by establishing a grout curtain in the rock. The preferred grouting procedure is to complete the grouting before excavation is started. If grouting is done after the excavation has been started, the head differential between the water surface and the bottom of the excavation may create sufficient water velocity through the rock to wash away the grout before it can solidify. However, if grouting after excavation must be performed, admixtures can be used to accelerate the setting of the grout, or the excavated area can be flooded prior to grouting to eliminate the head differential. It is difficult to form a cutoff wall with cement grout in most overburden material because it is hard to control the flow of cement grout. Recent improvements in chemical grouting have made its use feasible in overburden materials that cannot be grouted with cement. In one instance, a porous layer of sand and gravel covered canyon-floor bedrock at the site of a new high dam; 330,000 ft³ (9,400 m³) of cement and chemical grout was injected to form two upstream curtain walls that minimized water flow in the alluvium. Dam excavation then proceeded in the dry.[4.15]

A frozen impervious wall of overburden materials can be formed by excavating wells on a close pattern in the material and then freezing it by circulating cold brine through piping installed in these wells. In order to form an efficient frozen barrier without too great an expenditure for wells and cold-brine pumping, the flow of the groundwater past these wells should not exceed 2 ft/h (0.6 m/h). If the groundwater flow exceeds this velocity, sufficient closely spaced wells must be excavated to form a wide zone of freezing. The history of the problems encountered with this freezing method during the construction of High Gorge Dam in the state of Washington provides useful information on this type of cutoff.[4.16]

Steel-sheet-pile cutoff walls are seldom used by contractors because of the expense involved in their construction and because this method does not put enough tension on the pile interlocks to cut off all the water inflow.

UNWATERING

Cofferdam unwatering requires two types of operations. Initial unwatering is required when the cofferdammed area contains impounded water after cofferdam closure, when the cofferdammed area is flooded, or when cofferdams are overtopped. After the initial unwatering is completed, continuous maintenance unwatering is required to keep the area free from water.

The capacity of pumps required for the initial unwatering is dependent on the volume of water impounded by the cofferdam, the amount of cofferdam leakage, the water inflow through overburden or rock, and the time allowed for unwatering. Since time is always of critical importance to a contractor, the plant is usually sized to accomplish unwatering in the shortest practical time, often in less than one week. This initial unwatering requires more pumping capacity than is required for maintenance unwatering. Since initial unwatering pumps are only required for a short period of time, large-capacity vertical-head pumps are usually rented for this purpose. Installation of pumps of this type is shown in Fig. 4.11. Most such pumps are installed adjacent to the downstream cofferdam with discharge pipes running through or over the top of the cofferdam, which is the location that will result in the lowest pumping head. When the pipes are placed on top of the cofferdam and when it is not necessary to measure the water discharge, the pumping head can be reduced by extending the pipes so that they discharge below the downstream water surface.

The amount of work required for maintenance pumping depends on the type of system installed. A wellpoint system requires no additional maintenance pumping. When vertical or horizontal pumps are used for unwatering, they are installed over sumps excavated in the cofferdammed area. These sumps must be deepened when the excavation reaches the same depth as the sumps; and as the sumps are deepened, the pumps will also have to be lowered to reduce the suction head. Also, ditches must be excavated and maintained in the excavated area to lead water to these sumps. In addition, many small portable pumps are required to pump out small excavated areas located below the general rock surface. The main unwatering pumps are usually vertical pumps since this type of pump can be easily raised or lowered. A short

Figure 4.11 Unwatering pumps, Oviache Dam pumping project. (John W. Stang Corp.)

section of hose connecting the pump to the discharge line will provide flexibility when the pumps are lowered. The size and number of pumps required is dependent on cofferdam leakage and the amount of water that will pass under the cofferdams and through the banks of the excavation. If a cofferdam encloses a large area or if extensive areas of adjacent hillsides drain into the cofferdammed area, rainstorms or melting snow may at times contribute very large volumes of water. Estimating the quantity of water that must be handled is one of the most difficult problems to solve in dam-construction planning, because it is hard to anticipate the conditions that will be encountered. For example, at Wanapum Dam very little pumping was anticipated, since a slurry cutoff trench was constructed around the excavated area to reduce water inflows. Nonetheless, it was necessary to install an extremely large maintenance pumping plant to remove artesian water that entered the cofferdammed area through the holes drilled in the rock during the excavation of the dam's foundation.

When high upstream cofferdams are located adjacent to the upstream edge of the dam excavation, the combination of cofferdam height plus excavation depth may result in such a high static head that large volumes of water may enter the excavated area through the material that forms the cofferdam's foundation. This occurred at Guri Dam, Venezuela, where the water elevation on the upstream side of the cofferdam was 175 ft (53 m) above the bottom of the excavation. To handle this inflow of water, a low secondary cofferdam was erected between the future concrete dam location and the main cofferdam. The water that was collected behind this secondary cofferdam was conveyed through the excavated area by a large-diameter pipe that was left in place when the dam concrete was poured. This pipe drained the area upstream of the dam until just before the upstream cofferdam was removed. The flow in the pipe was then blocked off, and the pipe was filled with concrete.

Maintenance pumping should always be closely reviewed, as the history of most dam construction is that maintenance requires a much larger and more complex water-handling system than was anticipated when construction was originally planned. A careful analysis should always be made of every possible condition that could increase pumping-capacity requirements or add to the duration of the pumping period.

REFERENCES

4.1 "The Bi-National Itaipu Hydroelectric Project," *Water Power & Dam Construction*, October 1977.

4.2 Calvin Victor Davis, ed., *Handbook of Applied Hydraulics*, 2d ed., McGraw-Hill, New York, 1952.

4.3 Lewis L. Oviard and Joseph L. Jordan, "Rockfill Quarry Experience, Ord River, Australia," *Journal of the Construction Division*, ASCE, March 1980, p. 39.

4.4 Phillip J. N. Pells, "Reinforced Rockfill for Construction Flood Control," *Journal of the Construction Division, ASCE*, March 1978, p. 85.

4.5 "Closing the Gap for Gezhouba," *Engineering News-Record*, August 20, 1981, p. 28.

4.6 "Highest Underwater Cofferdams Keep Akosombo Dry," *Engineering News-Record*, June 20, 1963, p. 64.

4.7 Subcommittee of Small Water Storage Projects of the Water Resources Committee of the National Resources Committee, *Low Dams, a Manual of Design for Small Water Projects*.

4.8 John A. Havers and Frank W. Stubbs, Jr., eds., *Handbook of Heavy Construction*, 2d ed., McGraw-Hill, New York, 1971, pp. 155–160.

4.9 C. I. Grim, "Cofferdams in Swift Water for Bonneville Dam," *Engineering News-Record*, September 5, 1935, p. 315.

4.10 *Steel Sheet Piling*, U. S. Steel Company (Pittsburgh, Pennsylvania).

4.11 *Steel Sheet Piling Catalog 433*, Bethlehem Steel Company (Bethlehem, Pennsylvania).

4.12 C. P. Dunn, "Blasting a Precast Dam into Place," *Civil Engineering*, June 1981, p. 16.

4.13 "Floating Cofferdam Takes Prize," *Civil Engineering*, June 1981, p. 16.

4.14 Safdar A. Gill, "Applications of Slurry Walls in Civil Engineering," *Journal of the Construction Division*, ASCE, June 1980.

4.15 *Cement and Chemical Grouting-203*, Intrusion-Prepakt, Inc. (Cleveland, Ohio).

4.16 "New Technique Used for High Gorge Ice Barrier," *Engineering News-Record*, January 22, 1959, p. 28.

Chapter 5 Foundation Excavation and Preparation

This chapter explains how the excavation and preparation of a dam foundation are planned, scheduled, and accomplished. It describes the construction equipment that is used to perform this work and discusses the selection of that equipment.

Foundation excavation and preparation are larger factors in successful dam construction than is indicated by their share of the total cost. If they are not well planned and scheduled, they may delay the construction of the project, causing increases in job overhead and in labor and material escalation, and a decrease in equipment salvage proportional to the additional time that the equipment will be required on the job. This work requires special consideration because the differences in specification requirements and in type of inspection will cause construction cost to vary greatly among jobs. A change of interpretation involving one short sentence of the specifications may result in thousands of dollars of cost. Since foundation excavation and preparation must be completed before the placement of the structures can start, it is important that they be done by the most efficient system possible, and that they be scheduled to reduce interference with other construction activities.

SPECIFICATION REQUIREMENTS

Depending on the specification requirements, type of structure, and physical characteristics of the foundation materials at the site, the work required for foundation excavation and preparation may be quite extensive. The complexity of this work is suggested by Fig. 5.1, a flow diagram of the work that may

97

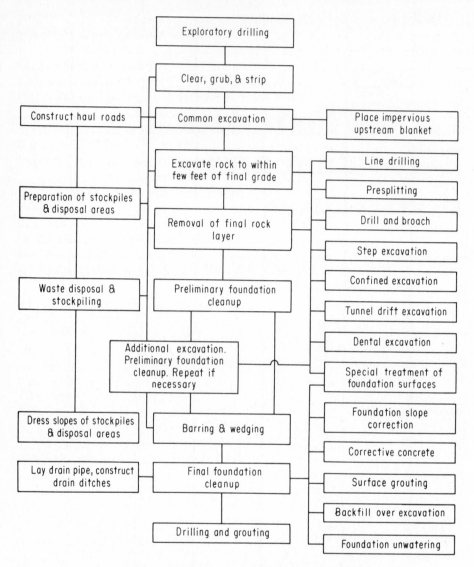

Extent of work varies with specification requirements
and type of dam foundation

Figure 5.1   Work-flow diagram for the excavation and preparation of a dam foundation.

be required to excavate and prepare the foundation for a dam. The number of these operations that are required at each damsite varies with the specification requirements and the type of dam. The work necessary for a dam project can only be determined by a thorough study of the plans and specifications. The work-flow diagram can, however, be used as a checklist for such a study.

The foundations for different types of dams have to be excavated and prepared in different ways. For an earth-fill dam the excavation under the impervious core and filter blankets must be carried to sound rock. The foundation preparation of this area may require the performance of many of the items of work shown on the work-flow diagram. The remainder of the dam foundation area need only be cleared and stripped, with the exception that any soft, unsuitable common material must be removed.

For a rock-fill dam with an impervious core, the excavation under the impervious core and filter zones must be carried to sound rock. The common material must be removed from the remainder of the foundation area, but foundation preparation under this area is seldom required.

For a concrete dam, the entire foundation area must be excavated to sound rock. The foundation preparation of the entire dam foundation is extensive in scope and often includes many of the items shown on the work-flow diagram. Powerhouse foundation excavation and preparation are similar to those required for a concrete dam. Often the powerhouse location is such that a large amount of presplitting is required for vertical and near-vertical surfaces.

Only part of the work required by the plans and specifications for foundation excavation and preparation is listed as unit-price bid items in the bidding schedule. When the contractor prepares his bid, the cost of performing any work listed in the bidding schedule can be placed directly against this item. The cost of performing work that is not listed in the bidding schedule must be combined with the cost of performing listed work in order to arrive at a combined cost, which is placed against the listed bid quantities. When work covered by bid items overruns in quantity, the contractor will receive adequate compensation because payment will be received for actual quantities of work performed. When work that is not covered by bid items overruns, the contractor will not receive compensation because the bidding schedule will not reflect the change in work quantities. This procedure of combining two or more items of work into one bid item often works to the disadvantage of the contractor.

The type of foundation excavation and preparation work required by the specifications but not separately listed in the bid schedule is expensive to perform. Moreover, the extent of this work may be difficult to determine from the specifications, since it is often defined as work necessary to meet the requirements of the owner's engineer. This makes it difficult for the contractor to determine what the final cost will be for these unlisted items when he submits his bid. If the owner's engineer considers the performance of unnecessary work essential, it is an unfortunate situation for both the contractor and the taxpayers or owner. In order to recover the cost of this work, the contractor must go through the trouble and expense of requesting a change order or filing a claim. When the change order is paid, the cost of this unnecessary or overrestrictive work is then paid for by the taxpayers or owner. If a claim is rejected, expensive arbitration or litigation often results.

Restrictive work procedures that cannot be anticipated by the contractor may be enforced by the owner's engineer when insufficient foundation exploration has been performed prior to bidding. In order to compensate for this inadequate foundation exploration, the owner's engineer may make the contractor explore the rock at the damsite while performing excavation, by ordering the contractor to excavate the rock in shallow layers and to perform preliminary rock cleanup on each successive rock surface exposed. By examination of these cleaned-up surfaces, the engineer can determine when the excavation has reached a depth that will provide a suitable rock surface for the dam's foundation. This procedure will hold pay quantities to a minimum; however, if excavation by this multiple shallow-layer removal procedure is not called for in the specifications as a job requirement, but is later required by the engineer, it will increase the contractor's excavation cost without furnishing him adequate compensation.

Excavation in multiple and shallow layers increases the contractor's anticipated excavation cost in several ways: the drill holes are smaller in diameter and more closely spaced; there are more restrictions on the use of explosives; loading and hauling efficiency declines when shallow layers of rock are removed and excavation is repeated many times across the same area. If the square feet of preliminary rock cleanup is paid for as a separate item, the contractor will be reimbursed each time he performs this work. If it is not a separate item and if he has not included the cost of this repetitive preliminary rock cleanup under other items of work, the contractor will incur additional cost without any compensating revenue.

The foregoing discussion may have presented a pessimistic approach to dam foundation excavation and preparation, but each job is different and whether the contractor will be adequately compensated for this work depends on the job specifications, the number of bid items, the competency of the inspectors, the particular job conditions, and the competency of the contractor's estimator. Many jobs have been constructed without any conflicts between the contractor and the owner's engineers during the performance of foundation excavation and cleanup. However, on jobs where change orders have been requested, a great number of these change orders have been for additional work required to excavate and prepare the dam foundation.

WORK DESCRIPTIONS
AND EQUIPMENT APPLICATION

Descriptions of the work that may be required to excavate and prepare the foundation for a dam or powerhouse are given on the following pages. These descriptions appear in approximately the same order as on the work-flow diagram (Fig. 5.1). Although these procedures are described as separate operations, in practice many of them can be done simultaneously. The construction equipment required to perform each phase of the work is described in detail, to provide assistance in the selection of proper equipment. Finally, there is an explanation of how the required number of each type of equipment unit is established.

Exploratory Drilling

When exploratory work at the damsite is not sufficient to determine the characteristics of the dam foundation material, the specifications may require the contractor to do exploratory drilling so that the owner's engineer will have more information on foundation conditions. When this is required, bid items are usually included in the bidding schedule to reimburse the contractor for this work. Since this work is of a special nature, it is usually subcontracted to specialists. If the dam contractor wishes to do the work using his own resources, a diamond-drilling setup is required.

Clearing, Grubbing, and Stripping

This step involves the removal of trees or other types of vegetation, the removal of all stumps or roots, and the removal of all topsoil containing organic material. Whether the topsoil must be removed before common excavation can commence is dependent upon the extent of the vegetation and whether the common excavated material is to be used for backfill or fill in the permanent construction. Specifications may require that the contractor cut, pile, remove, and sell all merchantable timber. This costs the contractor more than he can recover on the sale of timber, but there is often no other solution to the problem. In some areas where merchantable timber is of substantial value, logging contractors can often handle this phase of the work far more economically than can the dam contractor.

If the area to be cleared is small and sparsely timbered, this work can be done with chain saws and standard excavation equipment. If there is a large timbered area to be cleared, then special equipment is required for clearing and grubbing. This specialty item is often subcontracted to clearing contractors who own this type of equipment and who can rapidly mobilize it and move on the job. If the prime contractor wishes to do this work with his own forces, economical cost can seldom be obtained without the use of tractors equipped with clearing blades, clearing cabs, and root rippers. Tractors equipped with logging winches may be required to snake logs down from steep slopes. Rubber-tired front-end loaders equipped with log tongs are needed for loading logs. On highways, logging trucks and trailers are required for hauling logs to sawmills. Bulldozers and tractors equipped with brush rakes are used for general cleanup. Often a final cleanup must be performed after the reservoir is filled or during filling. Small boats utilizing a log boom can sweep the surface of all floating debris for removal from a convenient cove or beach.

Any stripping that is required can be accomplished by the same equipment used on the common excavation.

Construction of Haul Roads

Haul roads are required to provide access to the excavation areas and from the excavation areas to the waste disposal and stockpiling areas. This is a nonpay item of cost, so the contractor must spread the cost of road construction to the pay items of excavation.

Haul-road construction is a priority work item, since the various required accesses must be provided before the excavation of the dam foundation can be started. Road construction must continue throughout the excavation process, since access roads have to be relocated as excavation progresses and haul roads to the disposal areas changed as excavation advances and disposal areas are filled.

Bulldozers and crawler-type percussion drills are used for pioneering the roads. The roads can then be widened by drilling and shooting and side-casting the excavated material over the bank. Specifications concerning the disposal of material from access-road construction should be closely checked. In some areas the excavated material cannot be side-casted but must be loaded into trucks and hauled to fills or to designated disposal areas.

Preparation of Stockpile and Disposal Areas

It may be necessary to clear, grub, and provide drainage for stockpile or disposal areas. As previously mentioned, clearing and grubbing are often subcontracted, since these are specialty work items. If disposal areas are located over natural drainage channels, considerable expense may be incurred in installing culvert pipe or in providing other methods of maintaining the natural drainage. Selected rock pavement may be required under stockpile areas to prevent the stockpiled material from being contaminated by the local material. This work can be performed by the same equipment selected for access-road construction or for the foundation excavation for the dam and powerhouse.

Common Excavation

Common and rock excavation are the basic work items for foundation excavation. *Common excavation* is the removal of all loose, small-size, unsuitable material which overlies the dam foundation. Specifications define the difference between common and rock excavation, and the definitions vary with the contracting agency. In some cases, common and rock excavation are lumped together in the bid schedule and can be removed together and combined as *unclassified excavation.*

The construction methods and equipment used for performing common excavation vary with site conditions and the thickness and extent of the deposit. Thin layers of common material can be removed from steep slopes by using large, open-bottomed, crescent-shaped drag scrapers. A typical crescent scraper setup consists of an anchor on top of the abutment and a tractor, equipped with a logging winch, at the bottom. Cables are reeved from the logging winch to the anchor and scraper so that the scraper can be pulled up and down the slope. The anchor on top of the abutment may be a fixed anchor, a tractor, or an adjustable bridle attached to two fixed anchors. Another method often used to remove thin layers of common material is by high-pressure water jets discharged from hydraulic monitors. When these are used, it must be permissible to discharge the muddy water into the natural drainage channel or to clarify it in settling ponds. To provide flexibility, the monitors can be mounted on crawler tractors. A typical hydraulic monitor is shown in Fig. 9.5. When thin layers of common material occur on abutments with flat slopes, then bulldozers can strip and push the common excavation into piles. These piles can be loaded into trucks with front-end loaders.

Scrapers are used to remove deeper deposits of common material from damsites where abutment slopes are not excessive and haul distances are not great. The most economical method of loading common excavation is by scraper, since the scraper participates in the loading operation. The most economical method of hauling common material is by bottom-dump truck, as this type of haulage vehicle has the best truck weight to payload ratio. Scrapers are economical to use on short hauls when the savings on loading cost are greater than the increased haul cost. Bottom-dump haulage is most economical on long hauls when the savings in haul cost are greater than the increased loading cost. When bottom-dumps are used on long hauls, they can be top-loaded with belt loaders, front-end loaders, or shovels. Haul trucks are loaded by portable belts when thick deposits of common material are located on sidehill slopes. With this type of deposit, maximum production can be secured by using bulldozers to shove material onto the belt loader. Heavy-duty belt loaders can be secured in 48-, 54-, and 60-in widths. Production depends on the number of feeding bulldozers, length of push, type of material, width of belt, and capacity of hauling units. Figure 5.2 shows bottom-dump trucks being loaded by a belt loader.

Front-end loaders are used when conditions are not perfectly suited for belt loaders. Loading trucks with shovels is a standard excavation method when small amounts of materials are involved and when shovels that have been purchased for rock excavation are available. Shovels are also used when hard-digging material is removed from confined areas.

Draglines are used on common excavation when the excavated material is removed from below the level of the loading equipment or when material is

Figure 5.2   Loading bottom-dump trucks with a belt loader. (C. S. Johnson Division of Koehring Co.)

excavated below a water surface. Small hydraulic backhoes can be used to good advantage in removing pockets of common material occurring in the rock formation.

When large quantities of submerged common materials cover the dam foundation, dredges for the removal of this material may be economical. Dredges were used on Akosombo Dam in Ghana, Africa, to remove thick beds of submerged sand from the dam foundation in areas that were slightly more than 200 ft (61 m) below the water surface.

Rock Excavation to within a Few Feet of Final Grade

Rock excavation is the major work item for foundation excavation. It consists of the removal of all weathered and unsuitable rock material to within a few feet of final grade. Sound rock may have to be removed to provide a stepped surface on the dam abutments or to eliminate abrupt irregularities in the rock surface. When nearly vertical sidewalls must be maintained, line drilling, presplitting, or drilling and broaching may be required. When a narrow zone of unsuitable rock passes through the dam foundation, the rock must be removed with equipment of limited maneuverability. These circumstances define *confined excavation*.

At most damsites, rock excavation is a drill, shoot, load, and haul operation, but at some locations, ripping of the rock may be practical. If the rock appears rippable, it should be checked with a refraction seismograph, which checks the time it takes for seismic waves to travel through the material. Seismic velocity tests are performed by some equipment suppliers or by consultants who specialize in this field. Almost all rocks are rippable when the seismic velocity in feet per second is 6,500 (2,000 m/s) or less, and some sedimentary rocks with seismic velocities of 8,500 ft/s (2,600 m/s) are rippable. When hauls are short, the ripped rock can be loaded and hauled with scrapers pulled by crawler tractors. For longer hauls, scrapers pulled by rubber-tired tractors are employed. When this type of unit is used, the requirement that the scrapers be push-loaded with crawler tractors must be rigidly enforced. Any wheel spinning of the scraper power unit will cause extreme tire wear and skyrocket the costs. Belt loaders are also used for loading this material into trucks. Another method is to use bulldozers to push the ripped material into piles and then load it into trucks with front-end loaders or shovels. Further information on rock ripping is available in manufacturers' publications.

If drilling and shooting are required, the rock is seldom deep enough to warrant using drills larger than percussion drills mounted on track-type drills (see Fig. 5.3). This type of drill is used for holes from 2¾ to 6½ inches in diameter and for depths up to 50 ft (15 m). Some major recent improvements include the development of crawler-mounted drill jumbos, extendable booms, and hydraulic drills.

Figure 5.3 Medium-sized drifter drills mounted on air-powered tracks. (Joy Manufacturing Co.)

Drilling in rock for foundation excavation requires small diameter holes at a close spacing, so that good fragmentation and bottom breakage will occur. The harder and more massive the rock, the closer the drill hole spacing must be and the greater the amount of explosives needed to fragment the rock. In rock of normal breaking characteristics, when average fragmentation is desired, the spacing and burden of small-diameter blast holes are related to the depth of rock, as long as the proper powder factor can be maintained. It is desirable to do subgrade drilling of from 1 to 3 ft (0.3 to 1 m), depending on the hole spacing; however, this may be controlled by the specifications. Table 5.1 gives recommended hole spacings for holes 3½ in and smaller in diameter for various depths of rock:

TABLE 5.1 Hole Spacing for Various Depths of Rock

Rock Depth (ft)	Hole Spacing (ft)
2½ to 3	2½ × 2½
3½ to 5	3 × 3
5 to 8	3½ × 3½ to 4 × 4
8 to 12	4 × 4 to 4½ × 4½
12 to 18	5 × 5
18 to 24	5½ × 5½
24 to 30	6 × 6 to 8 × 8

SOURCE: *Blasters Handbook,* Canadian Industries, Ltd. (Montreal), 1959.

The powder factor that can be obtained for different diameter holes on various hole spacings can be determined using Tables 5.2 and 5.3.

Densities of typical dynamites may range from 0.8 to 1.6 g/cc, while densities of blasting agents may vary from about 0.5 to 1.7 g/cc. Ammonium nitrate fuel oil (AN/FO) placed by gravity will run about 0.80 and 0.98 g/cc in density when pneumatically loaded. See Chap. 9 for additional information covering massive rock excavation and quarries.

Three kinds of explosives are generally utilized for blasting operations in foundation excavation and for other operations adjacent to the dam: nitroglycerin and ammonium dynamites; ammonium nitrate fuel oil mixtures (AN/FO); and high-density slurries. Choice of the most economical type of explosive for foundation rock excavation depends upon the nature of the operation, that is, the size of the excavation, the type and hardness of the rock, and the wetness of the location.

After the rock has been drilled and shot, it is loaded by shovels or front-end loaders into rear-dump haulage units. Because of their lack of maneuverability, the large-sized rear-dump units are seldom used on dam foundation rock excavation; instead, trucks of the 35- or 50-ton class or smaller are used. If the rock can be loaded with a front-end loader, this method is preferred to shovel

TABLE **5.2** **Hole Burden**

Hole Spacing (ft)	Yd³ Rock per Lin Ft of Hole
2½ × 2½	0.23
3 × 3	0.33
3½ × 3½	0.45
4 × 4	0.59
4½ × 4½	0.75
5 × 5	0.93
5½ × 5½	1.12
6 × 6	1.33
8 × 8	2.37

SOURCE: *Blasters Handbook,* E. I. duPont de Nemours & Co. (Wilmington, Delaware) 1980.

loading because of savings in capital and operating cost. Only one operator is needed on a front-end loader, while small shovels require an operator and an oiler and large shovels require an operator and two oilers. Another advantage in the use of the front-end loader is that its purchase price is much less than that of a shovel of similar production capacity. Truck capacity and the size of loader or shovel should be keyed together so that three or four loaded buckets will fill the truck to capacity.

Compressed air for the drills can be piped from a stationary compressor plant or secured from portable compressors. Compressed air from electrically driven, stationary compressors is cheaper than air from portable compressors, since electric motors cost less to purchase and operate than do diesel engines. Portable compressors are used prior to the erection of the stationary compressed-air plant and when their maneuverability warrants their usage.

Rock excavation should be planned so that the face of the excavation has sufficient height for efficient loading. This also results in more efficient utilization of drill holes and explosives. Shovel loading is maintained at maximum

TABLE **5.3** **Pounds of Explosive per Lineal Foot of Hole**

Density, g/cc	0.80	1.00	1.15	1.30
Hole diameter, in				
2½	1.70	2.13	2.45	2.77
3	2.45	3.06	3.52	3.98
3½	3.34	4.17	4.80	5.42
4	4.36	5.45	6.26	7.08
5	6.81	8.51	9.79	11.07
5½	8.24	10.30	11.84	13.39
6	9.81	12.26	14.10	15.93

SOURCE: *Blasters Handbook,* E. I. duPont de Nemours & Co. (Wilmington, Delaware) 1980.

efficiency when the bank of shot rock is the same height as the vertical distance from dipper shaft (dipper-stick pivot shaft) to ground level. This distance varies with size and make of shovel.

If thick layers of rock must be removed, the rock can be drilled with large drills, explosives can be bulk-loaded, and large-sized shovels and trucks can be used in the excavation and hauling operations. Some rock-fill dams are designed so that the rock fill is secured from large rock cuts required for the development of either a spillway or a power-intake channel. The procedure used to remove rock from massive cuts is similar to the procedure used in a rock quarry, as discussed in Chap. 9.

Waste Disposal and Stockpiling

This forms a part of all excavation work items since all excavated material must be directly placed in embankments, utilized as backfill, used as a source of aggregate material, stockpiled for future use, or placed in waste-disposal areas. The disposition depends on the characteristics of the excavated materials, the type of dam, and the specification requirements. Equipment for this type of work is the haulage equipment used in the excavation process and graders, water trucks, or similar road maintenance equipment. At the disposal or stockpile locations, bulldozers are required to spread the dumped material.

Line Drilling

This is the drilling of regularly spaced, small-diameter holes along the desired vertical, or nearly vertical, rock surface of the area to be excavated. Holes are usually 2 to 3 inches in diameter, spaced apart from two to four times the hole diameter. The drilling operation is performed with crawler-type drills. After the holes are drilled, the enclosed rock is drilled and shot. Line drill holes are not loaded, and the adjacent blast holes have less explosive than the remainder of the shot. By proper location of the drill holes and by proper use of explosives, the rock is broken to the plane of weakness formed by the line-drilled holes, leaving part of the circumference of the holes visible. Because of the large amount of drilling resulting from the close hole spacing, line drilling has declined since the development of presplitting.

Presplitting and Cushion Blasting

Presplitting is a technique that extends line drilling a step farther. Holes are usually 2 to 4 inches in diameter at a spacing equal to one-half the burden from the shot. These holes are set off first, and the explosion cracks the rock along the line of the drill holes. When the remainder of the rock is later drilled, shot, and excavated, this presplit rock surface forms the side walls of

TABLE **5.4** **Presplitting Hole Spacing and Loading**

Hole Diameter (in)	Hole Spacing (ft)	Explosive Charge (lb/lin ft of hole)
1½ to 1¾	1 to 1½	0.08 to 0.25
2 to 2½	1½ to 2	0.08 to 0.25
3 to 3½	1½ to 3	0.13 to 0.50
4	2 to 4	0.25 to 0.75

SOURCE: *Blasters Handbook*, E. I. duPont de Nemours & Co. (Wilmington, Delaware) 1980.

the excavated areas. Recommended hole spacing and explosive charges for different-sized holes are given in Table 5.4.

Cushion blasting is similar to presplitting, except that the main mass is removed first. Drilling is similar to presplit practice and can be performed either prior to primary blasting or just before removing the final material. Cushion blast holes are loaded with light, well-distributed charges, completely stemmed, and fired after the main excavation is removed. The larger the drill holes, the more cushioning effect is realized. Recommended hole spacing and explosive charges for different sized holes are given in Table 5.5.

If controlled blasting is not a separate pay item in the bidding schedule, the contractor must include the cost of this work with the rock excavation. See Chap. 9 for a discussion of use of explosives for foundation and controlled blasting operations.

Drilling and Broaching

This is a technique that will also produce a smooth, excavated rock surface. Final trimming of the sides of the excavation is accomplished by drilling holes along the desired excavation surface and breaking the rock along the plane of the holes, using expansion by mechanical means. Since the advent of presplitting, drilling and broaching are seldom used, although the Japanese are pioneering some advances in this area.

TABLE **5.5** **Cushion Blasting Hole Spacing and Loading**

Hole Diameter (in)	Average Hole Spacing (ft)	Average Burden (ft)	Explosive Charge (16/lin ft of hole)
2–2½	3	4	0.08–0.25
3–3½	4	5	0.13–0.50
4–4½	5	6	0.25–0.75
5–5½	6	7	0.75–1.00
6–6½	7	9	1.00–1.50

SOURCE: *Blasters Handbook*, E. I. duPont de Nemours & Co. (Wilmington, Delaware), 1980.

Confined Excavation

At many damsites, a zone of unsound or fractured rock may cross the damsite at any angle. This zone is often called a *gut section*. The zone of unsound material may be narrow, with vertical side walls forming restricted areas for equipment operation. When equipment operating space is restricted, excavation cost increases and is often set up in the bidding schedule as a separate bid item called *confined excavation*.

If there is any reason to anticipate that work of this nature will be required, it should be listed as an item in the bidding schedule. Otherwise, when this work is necessary, the contractor must submit a change order to receive proper compensation.

Step Excavation

Step excavation may sometimes be required for the foundations of concrete dams. This necessitates that the sloping rock surfaces of the dam abutments be excavated in horizontal and vertical stepped surfaces. This is always a nonpay item and is considered a part of rock excavation.

Special Treatment of Foundation Surface

Special excavation procedures and surface treatment may be needed when unstable types of rock are used for the dam foundation. Special treatment is often requested when the dam foundation is formed by shales that will air-slake upon exposure. Specifications often require that the foundation be protected by leaving the last foot of material in place until 24 hours before concrete placement. Or they may require that the foundation surface be covered with gunite or sprayed with bitumen as soon as it is exposed.

When the rock in the dam foundation consists of stratified rock layers, specifications may require that these rock layers be tied together by closely spaced vertical anchor bolts. Specifications may prohibit blasting when excavation is being performed for other types of foundations. In this case, rock excavation must be done by mechanical methods.

Removal of Final Layer of Rock Excavation

Specifications often require that the final few feet of rock be removed by controlled drilling and blasting, using small holes drilled with close spacing. Subdrilling is not permitted. To secure adequate breakage, the hole spacing should not exceed the depth of rock to be excavated. A large amount of this drilling is done with jackhammers. Barring and wedging is required to level the foundation and to remove any loose rock. Small backhoes are useful in this work to eliminate some of the hand labor. The thin, excavated rock layer is pushed into piles by a bulldozer and loaded in rear-dump trucks by either a front-end loader or a shovel. Final rock removal is shoveled or placed by hand into skips which are picked up by cranes and dumped into trucks. Figure 5.4 shows barring and wedging of the rock foundation of a concrete dam. A rock skip is shown in the background.

Dental Excavation

Dental excavation is the removal with hand-held tools of small particles of soft rock or other unsound materials that extend into the dam foundation. If the unsound rock occurs in large pockets, small backhoes can be utilized for its removal. It may be necessary to excavate shallow shafts when the unsound materials extend in depth. To avoid requests for change orders, this work should be included as a separate item in the bid schedule.

Tunnel Drift Excavation and Backfill

When unsound rock extends into the abutments, specifications often require that this rock be removed by following it with tunnel drifts. Tunnel drifts may also be driven for use as grouting or drainage galleries when extensive grouting or drainage is necessary for stabilization of the rock in the dam abutments. Drifts can often be driven with small, air-track drills or jacklegs, using a front-end loader to load and haul the excavated material out of the tunnel. This item of work is usually paid for under appropriate bid items.

Figure 5.4 Barring and wedging of the rock foundation for a concrete dam.

Foundation-slope Correction

Foundation-slope correction may be needed in the preparation of the foundation surfaces for the impervious core of earth- and rock-fill dams. When it is a requirement, any rock surface that has a downstream slope in reference to the axis of the impervious core must be changed to slope upstream. This work is included under the pay item of rock excavation. It generates small pay quantities while requiring relatively great expenditures for labor, since it must be done with hand-held equipment. Prior to bid submittal, it is difficult to estimate the amount of work that will be required, since the extent can only be determined when the foundation surface has been exposed. When this work is more extensive than was estimated, the contractor can only recover his additional cost by requesting a change order.

Barring and Wedging

Barring and wedging involve the removal by bars, picks, and wedges of the last layer of rock on the foundation surface. The material usually removed is loose rock or drummy rock that has been loosened by other excavation procedures. Backhoes can be used in this operation, and a great amount of hand labor is required. The contractor is often paid for this work under the rock-excavation item.

High-Pressure Jetting

Foundation cleanup can often be facilitated by use of air and water jets, particularly in the impervious core section. High-pressure air and water cutting equipment can often be utilized in core sections that require foundation cleanup comparable to that required for a concrete dam.

Preliminary Foundation Cleanup

The owner's geologist and inspectors often request that the foundation be given a preliminary cleanup so that they can examine it and determine if it is suitable for the dam foundation. After this has been done and the engineers have examined the rock surface, they may find that it is not suitable for the dam foundation. Then another layer of rock must be removed, and the engineers will again request a preliminary cleanup of the rock surface. This procedure may be repeated many times until a suitable foundation surface is established. If bid items and pay quantities for this work are not included in the bid schedule, the inspector's extensive use of this procedure will be a large item of cost to the contractor.

Preliminary foundation cleanup requires the removal of all loose material and all water from the foundation surface by air and water jets. In some areas removal of the water may necessitate construction of small cofferdams, ditches, and sumps with the installation of small pumps. Some of this material can be removed with backhoes, but the majority of the work must be accomplished by hand labor.

Final Foundation Cleanup

Final foundation cleanup is required when a suitable foundation surface has been established. The work requirements are the same as for preliminary foundation cleanup, except that final cleanup must be completed more thoroughly. This should be a separate pay item in the bidding schedule; otherwise the contractor must include this cost under rock excavation.

Surface Grouting

When the foundation surface under earth-fill dams contains cracks or small fissures, it may be required that these be filled with hand-placed cement grout. The specifications should be checked to determine whether this is a requirement and whether it is a separate pay item.

Corrective Concrete

Corrective concrete may be required to treat the foundation for the impervious core of earth- or rock-fill dams. The impervious core cannot be effectively placed and compacted underneath overhanging rock ledges, which form when unsound material is removed from the foundation surface. To provide a suitable foundation surface, specifications may require that these spaces be filled with concrete. If this work is scheduled before the mixing plant is ready for operation, it may be necessary to provide a small, portable mixing plant to furnish the concrete sometimes called *dental concrete*. This is usually a separate bid item.

Backfill Overexcavation

If the contractor overexcavates, this overexcavation may have to be backfilled to grade. Depending on the type of dam structure and specification requirements, selected backfill or lean concrete may be used. Often in small areas refilled with backfill, compaction can only be accomplished with hand compactors. The cost of this work is the responsibility of the contractor.

Unwatering of the Foundation Surface

The dam foundation surface must be free of water before placement of the dam material can commence. This work may be quite extensive in the bottom of river channels and may require the construction of special cutoff cofferdams, drainage ditches, small sumps, and localized pumping. Job supervision often describes the work as "being performed with tablespoons and mops."

Drainage Ditches and Drain Pipe

This job may require the construction of drainage ditches backfilled with gravel, ditches containing drainage pipe and backfilled with gravel, or the installation of half-round tile drains. When required, they are usually paid for under separate bid items. Ditch excavation is done with hand-held jackhammers and small backhoes; a considerable amount of hand labor is also required.

Placement of Impervious Blanket on Sides of Channel Upstream from the Dam

It may be necessary to place an impervious blanket on the sides of the channel upstream from the dam to cut off any water flow through the abutments when dams are constructed on abutments containing pervious material. If this blanket is required, bid items and estimated quantities are included in the bid schedule.

Dressing the Slopes of the Stockpiles or Disposal Areas

Dressing the slopes of stockpiles or disposal areas is often required and can be performed with bulldozers. Sometimes specifications call for mulching and seeding of dressed waste disposal surfaces. In this case, equipment for broadcasting seed is needed. Cost of this work must often be included in the bid items for common and rock excavation.

Drilling and Grouting

If a concrete dam is to be constructed, only a shallow grout curtain is placed prior to covering the rock surface with concrete. This curtain consists of holes approximately 25 ft (8 m) in depth filled with low-pressure grout. The main grout curtain is formed after the foundation is covered with concrete. (This procedure is explained in detail in Chap. 6.) For earth- or rock-fill embankment foundations, the complete grout curtain is installed before any fill is placed. Lengths of holes and the required grouting pressure vary with local rock conditions.

Drilling and grouting are normally covered by separate bid items in the bidding schedule. Since the technique is a specialty work item, it is often subcontracted. If the contractor wishes to do the work with his own work force, he must provide diamond drills and a grouting setup. The equipment for drill-hole grouting is illustrated by Fig. 5.5.

Figure 5.5   Grouting equipment. (Gardner-Denver Co.)

THE EXCAVATION-CONSTRUCTION SYSTEM

The planning and scheduling of the excavation and preparation of the dam foundation controls the selection of the number of operating construction equipment units required. Planning is started by making a takeoff of the work quantities for each section of the dam foundation that is scheduled as a separate operation. Access roads and haul roads should be laid out, and haul distances and grades established. The excavation sequence should be established by scheduling the work so that a minimum amount will form part of the critical path for the dam construction and so that excavation equipment requirements can be minimized without delaying project completion. These data are then used to select the types and number of equipment units.

QUANTITY TAKEOFFS

Quantity takeoffs are made to check the bid quantities, to list all the work essential for foundation excavation and preparation, to compute the quantities of work that must be done during each phase of the excavation, and to check how reimbursement will be received for each item of work. Dam excavation may be scheduled so that portions of the dam foundation are completed and dam placement is started in these portions before the remaining excavation is completed, or all foundation excavation may be completed before any dam placement is started. The sequence of work performance should be established before starting the quantity takeoffs for foundation excavation so that quantities can be tabulated in accordance with the planned performance of the work. If a powerhouse is included with the dam contract, these excavation quantities should be listed separately.

Bid item descriptions and bid quantities shown on the bidding schedule cannot be used as the scope of the work. Work scope must be determined from the specifications, since many items of work may be called for by the specifications but will not be listed as bid items. As an aid in determining the extent of the foundation work for any project, the specifications can be reviewed and compared to the work-flow diagram presented at the beginning of this chapter.

Quantity takeoffs should properly designate whether the quantities are bank or loose yardage. *Bank yardage* is the amount of material in its natural state. *Loose yardage* is the volume which the material will occupy after it has been excavated. *Swell* is the increase in yardage from in-place to loose yardage. The loading equipment loads bank yardage. The transportation equipment hauls loose yardage, and the capacity of this equipment must be computed using loose yardage and weights of the loose material. Payment will be in bank yardage. For estimating purposes, all yardage should be converted to bank yardage, which will prevent confusion and simplify computations.

Materials vary so much that the only method of accurately determining the relation of bank yardage and weight to loose yardage and weight is by actual measurement. An approximation of these data for certain materials is listed in Table 5.6.

TABLE 5.6   Comparison of Bank Yardage to Loose Yardage

	Bank Yd³ (lb/yd³)	Swell (%)	Swell Factor	Loose Yd³ (lb/yd³)
Clay, dry	2,300	25	0.80	1,840
Clay, wet	3,000	33	0.75	2,250
Earth, dry	2,800	25	0.80	2,240
Earth, wet	3,370	25	0.80	2,700
Earth, sand, gravel	3,100	18	0.85	2,640
Granite	4,500	50–80	0.67–0.56	3,000–2,540
Gravel, dry	3,250	12	0.89	2,900
Gravel, wet	3,600	14	0.88	3,200
Limestone	4,200	67–75	0.60–0.57	2,620–2,760
Loam	2,700	20	0.83	2,240
Sand, dry	3,250	12	0.89	2,900
Sand, wet	3,600	14	0.88	3,200
Sandstone	4,140	40–60	0.72–0.63	2,980–2,610
Shale, soft	3,000	33	0.75	2,250
Slate	4,590–4,860	30	0.77	3,530–3,740
Trap rock	5,075	50	0.67	3,400

SOURCE: *Production and Cost Estimating*, Terex Corp. (Hudson, Ohio), 1981.

The quantity takeoff should list the thickness of both common and rock excavation so that the most economical use of excavation procedures can be developed. The common excavation takeoff should tabulate separately the quantities of silt, sand, clay, boulders, etc., that will be encountered. The takeoff of rock excavation should show the various types of rock that will be encountered so that drilling speeds and explosive consumption can be properly estimated. The capacities of the disposal areas should be checked to prove their adequacy. Dams have been put out for bids when the specified disposal areas were inadequate. Depending on the job specifications, the cost of providing more disposal capacity could be the contractor's responsibility. If additional disposal areas are required, the cost involved in their development and the cost of any additional haul should be determinable.

LAYOUT OF CONSTRUCTION ROADS

Construction roads are required for access from the contractor's installations to the excavation area and from the excavation area to the stockpiles or to the waste-disposal areas. Depending on the particular site conditions, the cost of providing the roads may be a major cost to the contractor. It is necessary to prepare a road layout to decide the amount of work required for their construction. It is important not only to locate the original haul roads required but to determine the relocation that may be necessary for completion of the excavation.

The other reason for preparing a road layout is that it is needed to estimate the turnaround time for scrapers or trucks used for hauling excavated materials. Haul-road layout is the most important factor in determining truck-turnaround time, as most of the errors in this computation can be attributed to the fact that actual haul-road grades are quite different from those originally estimated. In comparison, errors due to assigning the wrong truck speeds to specified grades are usually of a minor nature.

SCHEDULING

The excavation of the dam foundation should be scheduled to minimize the excavation time that is part of the project's critical path, to keep the amount of excavation equipment that must be purchased to a minimum, and to perform excavation in a manner best suited to site conditions.

The less the amount of excavation time that is part of the critical path time, the less the time required for the project's completion. Reducing the overall construction time reduces the overhead cost and amount of escalation, for both costs vary in direct proportion to construction time. Similarly, reduction in overall construction time diminishes plant and equipment write-off, since the shorter the period that equipment is on the project, the greater its resale value and the smaller the amount of equipment depreciation that must be charged to the job.

The critical path time for foundation excavation and preparation will be reduced if the work is done while other critical work items are being performed. Among the factors influencing the critical-path construction time are considerations of whether excavation can be done concurrently with the construction of the diversion facilities, whether all the dam foundation must be excavated before dam placement can commence, whether one section of the dam foundation can be excavated and dam placement done in this section while the remainder of the dam foundation is excavated, and whether powerhouse excavation can be delayed until the dam foundation is excavated.

Specifications may require that the dam placement start at the lowest point in the dam foundation and that the total length of the dam be placed continuously, with a minimum difference in elevation between any two points. If so, the dam foundation must be completely excavated before dam placement commences; therefore, the schedule cannot readily be arranged to shorten the critical path.

If the damsite is in a steep canyon and if tunnels are used for river diversion, the dam abutments can often be excavated from the top of the abutment down to stream level while the diversion tunnels are being constructed. The stream must be diverted before the remainder of the dam foundation can be excavated. Under these conditions, only the amount of excavation that is below stream level becomes part of the critical path.

Some damsites are wide enough to permit one side of the dam to be excavated before excavation is started on the remainder. When specifications permit, dam placement can be done in this side section while the remainder of the dam is excavated. With this procedure, the excavation of the first section of the dam foundation is the only part of the excavation that is part of the critical path, and the remainder of the excavation is a noncritical item.

When the river channel is wide and diversion can be made through the dam, the contracting agency may divide the construction work into two or more contracts, with each diversion stage let as a separate contract. Then the excavation sequence for the entire project is taken out of the control of the contractor. The excavation for each contract becomes part of the critical path of each contract, and the total time required for the dam construction will be extended.

There are too many ways to schedule excavation to reduce the critical path time to discuss them here. The point is that excavation is such an important item in job planning that considerable effort should be spent on arriving at the best possible solution.

When the excavation work is scheduled, it is necessary to correlate this work with other similar project work so that the excavating equipment can be used on as much work as possible; this will reduce the amount of excavating equipment required for the total project and the amount of equipment write-off. It may be possible to schedule the work so that the same equipment can be used for foundation excavation, in the aggregate pit or quarry, for construction of earth- or rock-fill embankments, for construction of diversion facilities, or for road construction. Proper job scheduling will increase the quantities of work done by a piece of equipment and thus decrease the unit cost of equipment write-off. An equipment-use schedule should be prepared so that equipment usage can be planned for the total project. This scheduling of equipment use is called *resource leveling* by technicians.

Work should be scheduled to fit site conditions so that excavation can be done for a minimum of cost. At steep damsites, the preferred method of performing excavation is to start work at the top of each abutment and excavate downward until the bottom of the dam foundation is reached. This method of excavation prevents material from sliding or being shoved into completed areas during execution of the work, and the excavation crew does not have to return to any completed areas.

When the abutments have an abrupt slope, when the excavated material can be shoved over the edge and there is space on the canyon floor to receive it, shelf excavation can be started at the top of each abutment. This shelf is lowered as each excavation round is drilled, shot, and shoved over the bank. The equipment remains on this shelf until the shelf is excavated to the canyon floor. After the excavated material falls on the canyon floor, it can be loaded into trucks and hauled to disposal areas. Excavation cannot be done below streambed level until the stream has been diverted.

If the abutments are too steep for haul-road construction and if there is insufficient space on the canyon floor to receive excavated material, abutment excavation must wait until diversion has been completed. After diversion, the excavated material can fall into the cofferdammed area, where it can be loaded into trucks and hauled to disposal areas. When abutments are flatter, excavation can still be started at the top of each abutment, but roads must be constructed to furnish access to the excavation areas and to provide haul roads for the transportation of the excavated material to the disposal areas. On flat damsites, the absence of grades and the ease of haul-road construction allow foundation excavation to be performed by any method.

SELECTION OF EQUIPMENT

Before the final selection of excavation equipment is made, the use of this equipment in the concrete subsystem, embankment subsystem, etc., should be reviewed so that the type that is capable of performing as many tasks as possible can be selected. This will keep expenditures for plant and equipment to a minimum and will result in a low project write-off for the equipment. On the other hand, excavation equipment must be selected that will perform the excavation economically. The preferred equipment and methods used for each item of foundation excavation and preparation were systematically described in the first part of this chapter.

To achieve low direct cost, the equipment must be selected to fit the depth of the excavation and the type of haul roads. Shallow common excavation requires different equipment from that suitable for the excavation of deep, massive flat layers. Small drills are used on shallow rock layers, but large drills can be utilized when rock excavation must be done in depth. Length and steepness of haul roads influence the selection of haul units.

When the equipment can only be utilized on foundation excavation, the yardage of common and rock excavation will control equipment selection. When there are only small quantities of common excavation but large quantities of rock to be excavated, the purchase of special equipment for the small amount of common excavation cannot be justified. One solution to this problem is to excavate common and rock as one operation, if the specifications or job planning do not require their separation. When this is possible, the same bid price should be submitted for common and rock excavation so that separate quantity surveys for each class of excavation will not be required. If common and rock excavation must be removed separately, rock-loading and haulage equipment can often be used to excavate both materials. Use of rock excavation equipment on the common excavation may slightly increase the direct cost of common excavation, but the total cost of common excavation will be low, because the amount of equipment write-off chargeable to common excavation will be held to a minimum.

When foundation excavation is the main use of the excavation equipment and there are large quantities of common but only small quantities of rock, then expenditures cannot be justified for the purchase of rock excavation equipment such as drills, shovels, etc.; instead, it should be rented. Often rock excavation can be subcontracted to contractors who specialize in this work and therefore own this type of equipment.

At flat sites, it may be desirable to excavate one abutment, start dam placement in this abutment, and, as dam placement continues, excavate the remainder of the dam. The quantity of material to be excavated in the first abutment then controls equipment selection. The remainder of the dam foundation can be excavated with less emphasis on the time required for completion, which may result in more work being performed with the equipment, thus reducing write-off.

After the types of excavation equipment have been selected, the number of required units of each type of equipment can be determined by establishing what the production rates will be for each type of unit and then dividing the scheduled rate of production by these production rates to establish the number of equipment units that must be in operation. This number must be increased by one-third to arrive at the number of units to be purchased. This one-third factor is required because only 70 to 80 percent of the equipment will be available for operation at one time. The remainder will be out of service, being repaired, waiting to be repaired, or being serviced. To this equipment fleet must be added the required servicing equipment and facilities, such as service trucks, tire trucks, repair shops, and tire shops; road maintenance equipment such as water wagons and graders; and pickups and man-haul trucks for transportation of supervisory personnel and workmen. This last group must be treated as project equipment since the same service equipment and facilities may be suitable to use for all the project work and need not be duplicated in every construction subsystem.

The number of rock drills required for excavating rock for the dam foundation is readily computed. First, the scheduled quantity of bank cubic yards of rock to be excavated per day is multiplied by the lineal feet of drill hole required to break one cubic yard of rock. The number of rock drills that must be in operation is then determined by dividing this product by the lineal feet of drill hole produced by one drill in one day. Of these three quantities, the one that is most difficult to determine accurately is the lineal feet of hole produced by a drill in a working day. The best method of determining this is to make drilling tests on the rock; the next best method is to base production on past drilling records for similar rock. The third choice is to send rock samples to drill manufacturers so that they can run laboratory drilling tests on the rock. The last and least satisfactory solution is for the estimator to base drill production on his judgment and published data.

Drill manufacturers publish penetration rates for their drills, based on drilling a medium-hardness granite 100 percent of the time. Therefore, penetration rates have to be reduced to allow for the nondrilling time required for moving drills, changing steel, and freeing bits, and for time wasted by the drilling personnel. The amount that penetration rates should be reduced varies with the hardness of the rock. The harder the rock, the longer it will take to drill one hole and the greater the amount of available time spent on drill penetration. The lineal feet of hole produced per drill per shift will vary between 70 percent of the penetration rate in hard rock and 40 percent in soft rock.

The following figures are presented to illustrate how rock type influences drill penetration rates, hourly drilling rates, and bit life. The drill penetration rate in medium-hard Sierra granites might be approximately 12 in per minute; a drill will produce 45 lin ft of hole per hour; and a bit with 10 resharpenings (the maximum) will drill 260 lin ft of hole. In extremely hard gabbro, the drill penetration rate could be 4.5 in per minute, drill production will be approximately 15 lin ft of hole per hour, and bit life will be approximately 120 lin ft of hole with 10 bit resharpenings. For soft argillite, the drill penetration rate may be approximately 4.5 ft per minute, the drill production will be approximately 100 lin ft of hole per hour, and the bit life approximately 600 lin ft of hole with 10 bit resharpenings. These drill speeds and bit lives are based on those of percussion drills using carbide insert bits. Utilization of hydraulic drills can increase the drill penetration rate in hard rock substantially, sometimes from 50 to 100 percent. Bit life can also often be increased.

The volume of compressed air required for excavation can be found by multiplying the number of operating drills by the air consumption. To allow sufficient air for leaks, hole blowing, and drill inefficiency, 125 ft³/min of compressed air should be supplied to a jackhammer, 550 to 600 ft³/min to a crawler-mounted 4½-in-bore-diameter drifter drill, 800 to 900 ft³/min to a crawler-mounted 5-in-bore-diameter drifter drill, and 1,000 to 1,200 ft³/min to a crawler-mounted 5½-in-bore-diameter drifter drill.

If a shovel is used for loading material, the preferable method of estimating its production is to evaluate performance records on similar work. Lacking this information, shovel production rates can be secured from tables in various publications.[5.1,5.2] Production figures given in these publications are for perfect working conditions and for 100 percent work efficiency. These production figures must be reduced for the anticipated equipment efficiency, the types of material, the height of bank, the angle of shovel swing, and the efficiency of the operator. Table 5.7 gives estimated shovel production figures suitable for foundation excavation with 70 percent efficiency, 90° boom swing for truck loading, and, since efficient bank heights are seldom encountered in work of this type, an assumed bank height of 40 percent of optimum. If operations are such that the bank height will always be at optimum, these figures can be increased by 25 percent. Optimum bank height for any shovel for maximum loading capacity is level with the dipper-stick pivot shaft.

A rule of thumb for determining the production of an articulated front-end loader is that loading capacity will equal that of a shovel whose bucket size is from one-half to two-thirds the bucket size of the front-end loader. The cycle time of a front-end loader is greater than a shovel cycle, since the loader must accomplish with travel and turning that which is done by the swing of the shovel boom.

To illustrate the production rates that have been achieved with front-end loaders, consider the 10-yd³ front-end loader that was used during the construction of a freeway south of Portland, Oregon. When it loaded blasted rock

TABLE 5.7 Shovel Production in Bank Yards per Hour for Dam Foundation Excavation; 70% Efficiency, Bank Height 40% of Optimum, 90° Swing

| Bucket Size (yd³) | Sand and Gravel | Common Earth | Clay | | Rock | |
			Hard	Wet	Well Broken	Poorly Broken
½	50	45	35	25	30	20
1	110	90	70	50	55	40
2½	210	170	140	110	135	80
3½	280	260	220	160	170	120
4½	350	310	260	190	240	160

Figure 5.6   Ten-yd³ front-end loader. (Caterpillar Tractor Co.)

into rear-dump trucks, it reached a production high of 465 bank yd³/h. When it loaded fines and small rocks, production increased to 560 bank yd³/h.[5.3] Figure 5.6 shows a 10-yd³ loader of this type.

The production achieved with belt loaders varies with the ease with which the material can be bulldozed onto the loader, the pushing distance, the grade on which the bulldozers operate, the number of bulldozers, and the availability of trucks. At Portage Mountain Dam 60-in-belt loaders could load continuously because they discharged onto conveyor belts. Loader production in the sand and gravel moraine was 3,000 tons per hour.[5.4] Converted to bank yardage this was 1,850 yd³/h. With easily pushed material, 115-ton bottom-dump trucks have been loaded at a rate of 1,800 bank yd³/h by a 60-in-belt loader fed by four D-9 bulldozers.[5.5] Any change from these optimum conditions will reduce this loading capacity. Loading capacity can always be computed with bulldozer production, which is the controlling factor. Bulldozer capacity is given in manufacturers' literature.[5.1] Again, this production must be adjusted for job efficiency and other operating conditions. After production computations are completed, the result should be adjusted as judgment and experience dictate.

Backhoes are often useful in both excavating and loading foundation material. Figure 5.7 shows a typical hydraulic backhoe.

The number of operating trucks or scrapers required to balance loading equipment capacity is determined in the following manner. The cycle time of this equipment is computed and then divided into a 50-minute hour to establish the number of cycles per hour per unit. The cycles per hour are then multiplied by the unit capacity to establish the hourly production per equip-

Figure 5.7 Typical hydraulic backhoe. (Bucyrus-Erie Co.)

ment unit. This can then be divided into the scheduled hourly production rate to establish the number of required equipment units. A check of these figures can be secured from production charts published by equipment manufacturers.[5.1] Again, these charts must be adjusted for job efficiency and other factors.

In order to compute a turnaround cycle, it is necessary to know the swell and the weight of the loose material so that the allowable payload can be determined. The turnaround time is a combination of waiting time at the loadout point, loading time, haul time, waiting time at the dump, dumping time, and return time. Truck loading time is controlled by the loading capacity of the loading equipment. Scraper loading time varies with the type of pusher and type of material to be loaded. The haul and return time is computed from haul distances, grades, type of roads, weight of the empty and loaded vehicle, and vehicle operating characteristics. Particulars and examples of production factors, selecting equipment, developing cycle durations, and assembling the estimate are all discussed in the *Handbook of Heavy Construction*.[5.6]

The following discussion illustrates how a turnaround cycle can be computed for a 35-ton rear-dump truck hauling broken granite. Haul roads are assumed to be well maintained and covered with gravel surfacing. The first 3,000 ft of haul will be down a 10 percent grade, and the remainder will be level. Truck specifications pertinent to this haul are payload capacity of 35 tons and heaped cubic yard capacity, on 3 to 1 slope, of 27 yd^3. The granite weighs 4,500 lb/yd^3 in the bank; therefore, if a 60 percent swell factor is used, 1 bank yd^3 will be the equivalent of 1.6 broken yd^3, and each broken cubic yard will weigh 2,700 lb, or 1.35 tons. Assuming that the average truck load will be 25-yd^3 loose measurement, then this broken yardage will weigh 34 tons, which does not exceed the truck's payload weight capacity. This establishes the truck's payload at 15.6 bank yd³. Trucks will be outfitted with dynamic braking so that a loaded truck can maintain a speed of 10 mph down the 10 percent grade; for the level hauls, top truck speed will be 30 mph. The empty truck will maintain an average speed of 25 mph on the level and when it climbs the 10 percent grade, its speed will drop to 15 mph.

Truck Cycle	Min	
Load:		
Wait	0.5	
Load	4.0	
Pull out	0.2	
Subtotal		4.7
Haul:		
3,000 ft down a 10% grade at 10 mph	3.4	
2,000 ft level at 25 mph	0.9	
Subtotal		4.3
Dump:		
Wait	0.5	
Dump	1.0	
Pull out	0.2	
Subtotal		1.7
Return:		
2,000 ft level at 25 mph	0.9	
3,000 ft up a 10% grade at 15 mph	2.3	
Subtotal		3.2
Total		13.9 min
Number of cycles in a 50-min hour		3.6
Bank yd hauled per hour at 15.6 yd³ per truckload		59 yd³

When all the foregoing tabulations and computations have been completed, the list of plant and facilities required for this operation can be compiled.

REFERENCES

5.1 *Caterpillar Performance Handbook*, 12th ed., Caterpillar Tractor Co. (Peoria, Illinois), 1981.

5.2 P & H *Excavation Production Calculator*, Harnischfeger Corp. (Milwaukee, Wisconsin), 1956.

5.3 "10-Yard Wheel Loader/Truck Team Speeds, Oregon Rock Cut," *Roads & Streets*, September, 1969.

5.4 Irvine Low, "Portage Mountain Dam Conveyor System," *Journal of Construction Division*, ASCE, September 1967.

5.5 "Modified Equipment Moves 45000 Yards Per Shift," *Roads & Streets*, April 1969, p. 48.

5.6 John A. Havers and Frank W. Stubbs, eds., *Handbook of Heavy Construction*, 2d ed., McGraw-Hill, New York, 1971.

Chapter 6 Concrete Placing Equipment

This chapter describes the selection of methods and equipment for concrete dam construction. Aggregate plants, crushing plants, concrete plants, and general purpose equipment are also discussed in Chap. 2. Heavy construction plant requirements applicable to concrete dam construction are covered in this and the following chapter. Chapter 7 describes concrete and aggregate plants, forming methods, placement crews, cleanup crews, curing facilities, and concrete finishing practices to complete the concrete construction system.

Concrete placing equipment is discussed prior to the other components of the concrete construction system because its selection controls the location and selection of the remainder of the system. The type of placing equipment establishes the required capacity of all the other system components, since the main purpose of other units is to furnish sufficient mixed concrete to the placing equipment and to supply it with enough formed areas and placement crews to permit it to operate at maximum capacity.

Equipment for placing mass concrete generally consists of cableways, cranes supported by trestles, and cranes operating from the ground. There are many different types of equipment for placing miscellaneous concrete.

When large concrete projects are constructed, the mass-concrete placing equipment controls the concrete construction system, the construction time, and the concrete cost. It influences and is influenced by the diversion plan, the excavation plan, and the concrete-pouring sequence, with certain exceptions. The methods and equipment used for mass-concrete placement remain quite similar to those used on Grand Coulee and Hoover dams, which were constructed in the 1930s. However, substantial improvement in mass-concrete

126

placing efficiency has since been achieved, with increases in speed, in the area of coverage, and in the lifting capacities of placement equipment.

Placing of lean concrete using rollers or rolled concrete has been successful in several installations and offers considerable promise in the utilization of high-volume continuous placing techniques developed from embankment placing experience.

At some damsites use of preplaced aggregate, with voids filled by the injection of grout, offers an innovative solution for particular site constraints. At other damsites utilization of continuous placement by conveyors often results in substantial acceleration of the placement rate, with consequent substantial reduction in cost. Twenty-four to thirty-six in belts can deliver 400 to 500 yd^3/h, which is equivalent to two to three times the production of major cableways, at a substantial saving in both plant and placement costs. Concrete placement by conveyor has been successfully performed on other mass structures such as bridges, powerhouses, subway stations, building foundations, and spillways on rock- and earth-fill dams.

There have been many new developments in miscellaneous concrete placing equipment. These new developments, however, have had a minor effect on the concrete construction system since miscellaneous concrete placement forms only a small portion of the total cost of a large concrete dam. Therefore, the selection of miscellaneous concrete placing equipment is governed by the criterion that it must function with the mass-concrete placing equipment. However, concrete pumping equipment will undoubtedly be increasingly used in the handling of both mass and miscellaneous dam concrete.

Since this book is written for contractors' engineers who usually do not design equipment but rather rely on manufacturers to design and furnish concrete placing equipment to their specifications, the descriptions of concrete placing equipment are given only to furnish the necessary background for dam construction planning and not in any detail. If design data on this equipment are desired, the reader is referred to equipment manufacturers' literature or other publications.[6.1,6.2]

In these discussions, the terms *mass concrete* and *miscellaneous concrete* are used. Mass concrete is that used in structures of such large cross-sectional dimensions that large volumes of concrete are placed in every pour. It consists of concrete in large concrete dams, in combination dams that contain both concrete and earth- or rock-fill sections, and in earth- or rock-fill dams that contain power intake sections, massive bottom cutoff walls, or other large concrete structures. In this chapter all concrete that is not mass or powerhouse concrete is called miscellaneous concrete. This is concrete in structures of such small cross-sectional dimensions that only a few cubic yards of concrete can be placed in any pour. It is used in small structures, in small dams, or in the facing on rock-fill dams.

CABLEWAYS

Description

A cableway used for placing the mass concrete in a dam consists of a large-diameter track cable which is suspended from two towers and spans the length of the dam. A traveling carriage rides on the track cable, and from this carriage is suspended a concrete bucket that can be raised and lowered as desired. Both of the towers supporting the track cable may be movable, or one tower may be fixed and the other movable. The mobility of a cableway with its traveling towers, the travel of the carriage, and the raising and lowering of the concrete bucket assure that concrete can be placed at any location in a dam.

Cableways are designed to have the capacity to handle various sizes of concrete buckets. Small cableways can handle 4-yd^3 buckets and are known as 12½-ton cableways. Except for one installation, the capacity of the largest cableways is 8 yd^3 or 25 tons. The exception was at Glen Canyon Dam where two cableways were used that had a capacity sufficient to handle 12-yd^3 buckets. Cableways with the capacity to handle 20-yd^3 buckets have been designed but have never been placed in service. Job specifications may limit to 4 yd^3 the amount of concrete that can be dumped at one time. If a dam is constructed under this type of specification, larger-sized buckets can be used, but these buckets must be divided into 4-yd^3 compartments, and the concrete in each compartment must then be dumped separately.

In tower layout and tower travel, with the exception of a few installations, two general types of cableways are used for dam construction. The first type, with one fixed and one movable tower, is called a *radial cableway*. The runway for the movable tower is constructed as a circular arc of constant radius from the fixed tower. This allows travel of the movable tower at a constant track-cable distance from the fixed tower. A typical radial cableway is shown in Figs. 6.1 and 6.2.

At damsites where a radial cableway is planned with the fixed tower located on an abutment that has a steep rock face, the fixed tower can often be replaced with an anchorage excavated into the rock at the proper elevation. Examples of this type of anchorage for the fixed end of a cableway were the cableways used for the construction of Donnells Dam, California, and Yellowtail Dam in Montana.

A cableway used at Salto Osorio Dam on the Iquacu River in Brazil was unusual in that the traveling tail tower consisted of a carriage that traveled upstream and downstream on a concrete wall that followed the existing terrain. A conventional traveling tail tower would have required very extensive excavation. The tail-tower "track" was on the surface of a theoretical sphere, with the top of the head tower at its center. Thus, the span length remained constant, as is required in cableway design.

The other type of cableway is the *parallel cableway*, which has two movable towers operating on parallel straight runways. When these towers travel, they

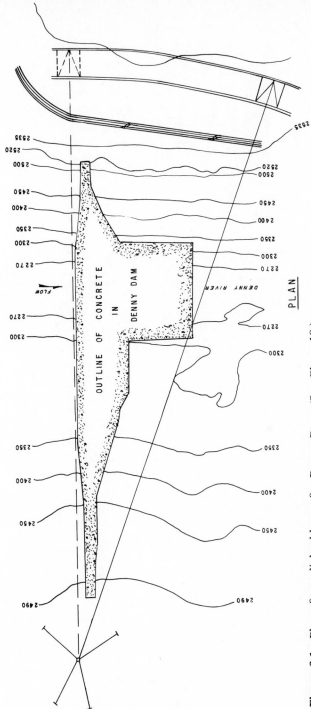

Figure 6.1 Plan of a radial cableway for Denny Dam. (See Chap. 19.)

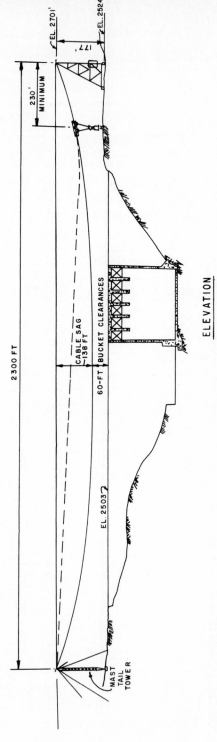

Figure 6.2 Elevation of a radial cableway for Denny Dam. (See Chap. 19.)

130

both must move in the same direction and at the same speed so that the track cable is always at right angles to the runway, with a constant track-cable length. As an illustration, two head towers for parallel cableways are shown in Fig. 6.3.

It is theoretically possible to construct a parallel cableway with the towers operating on parallel runways curved to the same radius or on concentric curved tracks, but as nearly as can be determined, installations of this type have not been made.

As previously stated, there are a few exceptions when dam concrete is placed with cableways that have fixed towers. Sometimes, for economic or other reasons, cableways are installed at an elevation that is too low to top out the dam. The dam may be topped out with a luffing-type cableway with stationary mast towers. Luffing cableways are commonly used for bridge construction. Hook coverage is achieved by movement of the top of the stationary towers as the side guys are shortened and lengthened.

Cableway towers are designated *head* and *tail towers*. The tower at the end of the cableway where the hoist machinery is located, irrespective of whether it is a fixed or movable tower, is called the head tower. The other tower is the tail tower and, depending on the cableway layout, can also be either movable or fixed. All head towers contain sheaves for the hoist and travel lines and facilities for anchoring the track cable. If the head tower is a movable one, then it also contains the drums and motors that control the load and hoist lines, a small air compressor for control of the drum brakes, the motors for moving the tower and supports, and a counterweight to resist the overturning moment caused by the pull of the track cable. The tail tower contains the

Figure 6.3 Two parallel traveling tail towers used for the construction of Detroit Dam, Oregon.

track-cable tightening facilities, sheaves for the travel line, anchorage for the load line, and, if the tail tower is movable, motors for tower travel and a counterweight. If one tower is fixed, it may be a fabricated tower, an A frame, or a mast, or the sheaves may be fastened to structural members anchored into the canyon walls. Fixed towers must be equipped with fairlead sheaves so that the sheave is always centered in the direction of the movable tower. Fixed towers may support cables from any number of movable towers. At Shasta Dam, a 460-ft fixed tower, which supported track cables from seven movable towers, was used. All cableways were of 8-yd^3 capacity with spans varying from 720 to 1,670 lin ft.

If the head tower travels, the hoist machinery is located in the tower. If the head tower is stationary, the hoist machinery may be in the tower or in a separate hoist house. The cableway controls may be located in the head tower or at a control observation point that permits the operator to view both the pickup point and the pour area. At Bullards Bar Dam, California, radio controls were used; the control box was small and portable and, when in use, was set on a pillow to cushion it from vibrations. Also at Bullards Bar, closed-circuit television was used to monitor the carriage and bucket travel.[6.3] On most projects, final spotting of the bucket over the pour area is done by the bellboy, who gives both voice and signal directions to the operator by separate communication systems. Voice signals are used to control the bucket's horizontal movement, and an electronic signal is used to control its vertical position. Figure 6.4 shows the control house for the Guri Dam cableway in Venezuela, located so that the operator had a view of the concrete-bucket dock and also of the pour area.

The movable towers for a cableway are supported on rail trucks that travel on two parallel sets of tracks, constructed at a constant elevation. The trucks under the inclined tower leg must be sturdier than those under the vertical leg, because of the greater force transmitted to the track. Rail sizes used for the runway tracks are approximately 132 lb with 3-in-wide heads. The trucks on each track are propelled by geared motors; their sizes depend upon climatic conditions and tower height, but generally they are 75 or 100 hp each. The track under the inclined tower leg is constructed perpendicular to the leg to permit resistance to the horizontal component of track-cable stress in the tower.

The track cable, often called the *main gut,* is stretched between the head and tail towers and provides a track and support for the carriage as well as for the slackline carriers. The cableway hook is suspended from the carriage and the slackline carriers support the operating cables. Track cables vary in diameter from 1¾ to 3 in for small cableways and from 3 to 4 in for 8-yd^3 and larger cableways. The cable weight varies correspondingly from 22.2 lb/lin ft for a 3-in-diameter track cable to 38.6 lb/lin ft for a 4-in-diameter track cable.[6.1] The cable is made with a track lay, which means that the surface strands are flat and keyed together to resist the wear of the carriage wheels. The size of track cable used for any application depends on the stress in the cable. Stress in the cable is a function of cable sag, cableway capacity, and the cableway span, since the weights of the track, travel, and hoisting cables, carriage, hook, and load enter into the computations. For those interested in computing a track-cable size, formulas are given in cable manufacturers' publications.[6.1]

There are many different methods of rigging the operating cables of a cableway. For those interested in this subject, manufacturers' literature should be obtained and other books reviewed.[6.2] The following is a general description of a cableway rigging that has often been used with satisfactory results.

The operating cable that controls the movement of the carriage is called the *travel line, endless line,* or *conveying line.* (It is not endless, but since both ends are attached to the carriage, it is operated as an endless line.) Both spans of the cable, from the carriage to each tower, are supported by *slackline carriers* which ride on and are supported by the track cable. At the tail tower, the travel line passes through a sheave and then is returned to the head tower; at the head tower both parts of the travel line (the one from the carriage and the one from the tail tower) are passed over sheaves and carried down to the hoist house. There the loop of the cable is wrapped for several turns around the gypsy drum that controls the movement of the cable. Typical cable diameter for an 8-yd^3 capacity cableway is between $1\frac{1}{8}$ to $1\frac{1}{4}$ in.

The operating cable that suspends the hook from the carriage and controls the vertical movement of the hook is called the *hoist line.* One end of the cable is anchored to the tail tower. Its span from tail tower to the carriage is supported by the slackline carriers. At the carriage it is reeved through a sheave down to and through the bucket hook's sheaves, thence back to the carriage and through another sheave. Its span from the carriage to the head tower is again supported by slackline carriers. At the top of the head tower, it is reeved through a sheave and down to the hoist house where the remainder of its length is wrapped on a large-diameter, large-capacity drum. For an 8-yd^3-capacity cableway, this cable diameter will vary from 1 to $1\frac{1}{4}$ in.

At the tail tower, a cable for tightening the track cable is reeved through blocks to get a multipart line. Typical diameter for this line is $1\frac{5}{8}$ in.

If a mast is used as the fixed tower for a radial cableway, it is usually guyed with two backstays of approximately $3\frac{1}{2}$-in-diameter cable and with two forestays of approximately $2\frac{1}{2}$-in-diameter cable. The backstays must have the capacity to resist a stress equal to or larger than that in the track cable. For many applications, the stays are made from used track cable.

To prevent excessive wear on all operating cables, cableways are equipped with large-diameter sheaves and drums. The diameters of these sheaves and drums are determined by the cableway manufacturers. As an indication of their sizes, cable manufacturers recommend that when heavily loaded, fast-moving ropes are used on fixed equipment, sheaves and drum diameters should be 800 times the diameter of the wires in the outer course of the cable.[6.1]

Cable life varies greatly between cableways. The track-cable life shortens as the cableway span increases and the distance from the head tower to load pickup point decreases. Other factors influencing cable life are the diameter of sheaves and drums, method of cable lubrication, load on the cable, frequency of use, type of cable, and the particular cable manufacturer. W. M. Bateman, in the *Handbook for Heavy Construction,* gives the following life for cables: life of a track cable is between 1½ and 2 million tons of payload; travel ropes will place 200,000 yd^3 of concrete; and hoisting ropes need replacement after they have placed from 40,000 to 60,000 yd^3 of concrete.[6.2] These recommendations are for average conditions. In some instances, cables have remained in use much longer, and for long-span cableways, the track cable has required more frequent replacement.

To provide more flexibility in speed control and to give better power-consumption characteristics, direct-current motors with Ward-Leonard controls are preferred over alternating-current motors for driving the cable drums. Motor generators are used for conversion of alternating current to direct current. To permit the carriage to travel horizontally at the same time that the concrete bucket is hoisted or lowered, separate motors supply power to the travel-line drum and the hoist-line drum. Motor sizes vary with speeds of operation. For example, for an 8-yd^3 cableway with an average hoisting speed of 640 ft/min, an 800-hp motor is required to drive the hoisting drum. If this hoisting speed for a loaded bucket is increased to 950 ft/min, two 625-hp motors are used. Similarly, with a carriage travel of 1,600 ft/min, a 500-hp motor is used to control the travel line; if the carriage speed is increased to 2,100 ft/min, a 625-hp motor is used.

The travel line, which controls and moves the carriage, is operated either by a gypsy of approximately 72-in diameter or, preferably, by a multiple-sheave unit. These units are equipped with a weight-set, air-release brake. A gypsy-type drum for driving the travel line requires that the several wraps of wire rope slide on the drum face as the line inhauls or outhauls. This results in a wedging of the incoming line between the flange and the wraps already on the drum, which causes rapid line wear. A *multiple-sheave drive* for the travel line eliminates the sliding and wedging that are so destructive to wire rope. This drive consists of a conventional reversible motor with a pinion bull gear and brake arrangement, which applies power to a drum on which there are typically six separate grooves sized to fit the travel line. A six-groove idler drum is mounted on the same frame, with the axis of this drum tilted slightly from horizontal. The travel line is reeved in six wraps around the two drums, without taking a full turn around either drum. The tilted axis of the idler drum ensures that fleet angle within the drive unit is eliminated. This refinement reduces wire-rope wear.

The hoist line that controls the bucket hook is spooled on a large-diameter, large-capacity drum, which also has a weight-set, air-release brake. Typical dimensions of the drum are 84-in diameter and 84-in length, depending on cableway size and cableway span. The drum is grooved to fit the diameter of the hoist line, and the large cable capacity on the drum allows spooling of all the take-in of the hoisting rope on the first wrap around the drum, thus reducing cable wear.

The travel line and the hoisting line are maintained in elevation across the cableway span by slackline carriers that suspend them from the track cable. Two types of slackline carriers are used. The two carriers at each end of the cableway, which are located outside the area of normal carriage travel, are only moved when the carriage is brought into the tower for repair or servicing. Because of this, these two carriers are of simple design and are often called *old-man* slackline carriers. The remaining carriers are known as *proportional* slackline carriers because of their spacing along the cable and their more complicated design necessitated by their operation in conjunction with the carriage travel. Present-day cableway-carriage speeds would not be feasible without the use of proportional slackline carriers. Older machines used slackline carriers moved by the carriage and spaced by a *button line*. In order to avoid rapid destruction of the slackline carriers, it was necessary to slow the carriage as each carrier was picked up and another dropped off the carriage. In contrast to this, the proportional slackline carriers ride freely on the track cable, propelled by gearing which is actuated by the carriage travel line. When the cableway is pouring concrete, the carriers do not come into contact with the carriage.

Proportional slackline carriers are purchased in sets. A typical set might consist of ten slackline carriers: five for the inhaul side of the span and five for the outhaul side. The carrier nearest the head tower and the carrier nearest the tail tower, in this case, would each be geared to move along the track cable at one-sixth of the carriage speed, the next pair at two-sixths of the carriage speed, and so forth, so that the two carriers nearest the carriage would be geared to move along the track cable at five-sixths of the carriage speed. Thus, the carriers are spaced at constantly varying distances as the carriage travels along the cableway span. None are ever in contact with the carriage except when the carriage is moved all the way to the head tower or all the way to the tail tower. When this occurs, contact between carriers disengages spring-loaded clutches. As the carriage moves away from the towers, the clutches reengage, and the carriers again space themselves along the track cable.

The carriage riding on the track cable has a large number of wheels, from 4 to 16, depending on the size of the cableway. Multiple wheels allow smooth travel of the carriage and reduce track-cable wear by distributing the weight and preventing sharp deflections. The carriage contains provisions for picking up and releasing the slackline carriers and contains two large-diameter sheaves over which the hoisting line passes for control of the concrete bucket.

The carriage used on the cableway for construction of Guri Dam is shown in Fig. 6.4.

There are many makes of concrete buckets available and suitable for cableway use. The main difference in buckets is in the method used to discharge the concrete. This can be done by tripping the release gate by hand, by using an air-activated gate, or by using a gate activated by hydraulic pressure.

The number of buckets in use at any one time depends on specification requirements. Some specifications do not allow the transfer of concrete from one container to another. With this type of specification, concrete must be transported in the same bucket between the mix plant and the cableway and then to the point of pour. The loaded concrete bucket is transported from the mix plant to the point of transfer under the cableway on special self-propelled bucket cars or on trailer-type bucket cars pulled by diesel locomotives. The bucket cars have space for two to four buckets. Along three sides of the bucket spaces are runways at the proper elevation for the operator to hook and unhook the buckets from the cableway. One space on the car is always left empty so that the cableway will have an empty slot for placement of the empty concrete bucket when the cableway carriage returns it to the point of transfer. For bucket pickup the cableway drops an empty bucket into the empty bucket space on the bucket car, picks up a loaded bucket, transports it to the pouring

Figure 6.4 Carriage, bucket dock, transfer track, transfer car, and the control house for the Guri Dam cableway, Venezuela.

Figure 6.5 Changing the cableway bucket using a bucket car. Chief Joseph Dam cableway, Washington. (Washington Iron Works.)

area where the bucket is dumped, and then returns this empty bucket to the bucket car. This procedure is repeated until the last loaded bucket has been hooked onto the cableway. When this has been done, the bucket car returns to the mix plant, while another bucket car takes its place under the cableway. This pouring procedure makes it necessary to hook and unhook every bucket from the cableway, either manually or by an automatic hook. Figure 6.5 shows manual bucket changing for the cableway used to construct Chief Joseph Dam, Washington. This photograph also shows the old-type slackline carriers used with a button line.

Other specifications allow the use of a transfer car. The transfer car is constructed with one or more compartments, each of which is capable of holding the same volume of concrete as the bucket used on the cableway. Each concrete compartment is equipped with fast concrete-dumping chutes or container dumpers for quick loading of buckets at the bucket dock. The transfer car can be either a self-propelled unit or a trailer unit pulled by a diesel locomotive. The transfer car is loaded with concrete at the mix plant and then travels to a position above a bucket dock located alongside the transfer track under the cableway. Here the concrete is dumped from the transfer car into the concrete bucket which has been spotted on the bucket dock by the cableway. The bucket dock is constructed lower than the transfer track, at an elevation that will place the top of the concrete bucket at the proper height to receive the concrete from the transfer car. The dock must be of rigid construction to resist the shock caused by the abrupt stopping of the bucket on the bucket dock. Rubber tires are often used along the dock to cushion this shock. In comparison to the bucket-car method, hauling concrete from the mix plant to the cableway by the transfer-car method saves time because it is not necessary to hook and unhook every concrete bucket to the cableway. Figure 6.4 shows the bucket dock, transfer track, and the transfer car used on the cableway at the original Guri Dam.

Whether a concrete transfer car is used with a bucket dock or whether bucket cars are used, a concrete transfer track is required. Rail-mounted equipment is usually used to transport concrete from the mix plant to the cableway, since it is more efficient and less expensive to operate for this service than is rubber-tired equipment. The transfer track and the bucket dock, if a transfer car can be used, must be of sufficient length that the bucket may be spotted under the cableway when the movable cableway towers are in any position. In order to permit fast switching of the bucket cars, the transfer track is double-tracked with switches at frequent intervals.

Layout

Layout and design of a cableway are, in the majority of cases, the products of a joint effort between the contractor's engineer and the cableway manufacturer. The contractor's engineer must choose between radial- and parallel-type cableways, must locate the runways in plan, must determine the cableway span, and must compute the sag in order to determine the tower height. Next, he must select the size of cableway, i.e., 12, 8, or 4 yd³, and determine how many cableways will be required. This information is then furnished to the cableway manufacturers who will furnish quotes on the cableway or cableways. If the contractor is the low bidder and purchases the cableway, the manufacturer will design and furnish towers, cables, carriage, hook, motors, drums, sheaves, etc., and recommend the weight and position of counterweights.

For estimate preparation, the engineer will have the cableway manufacturer's quotation for the cableway equipment. He must estimate the cost of constructing the tower runways, including any required trestles; the cost of the trackage; the cost of the concrete in the tower counterweight; if a fixed tower is used, the cost of anchors and the tower foundation; and the cost of erecting the cableway towers, cable, cableway machinery, and parts. If the contract is awarded, the portion of cableway design usually handled by the contractor's engineer involves the design of any trestle required in the cableway runway, the design of transfer track, and the design of the bucket dock, if one is used.

To determine the suitability of cableways, to decide between radial or parallel cableways, and to secure other estimating data, the engineer must lay out many possible solutions in plan and elevation. The plan should show the location of the towers and runways; the elevation should show the downstream view of the dam, elevation of the top of the dam, the necessary bucket clearance, cableway span, cable sag, and required tower height. The topography of the dam abutments and the curvature of the dam influence the choice between a radial or a parallel-track cableway. It is necessary to lay out both types with different span distances and with different runway elevations to determine the best solution. The solution selected should be the one which requires the shortest span, the shortest towers, and the least amount of expense in constructing the runways. For straight concrete dams, radial cableways are usually more economical because of the savings in using a fixed tower rather than movable towers. However, if large placing capacities are required, more parallel cableways can be used than radial. Traveling-tower runways may be located in a cut, on a fill, or, if the slope of the ground makes it necessary, on a trestle. An example of runways on a cut and on a trestle are the runways built for the cableways used to construct Detroit Dam, Oregon, as shown in Fig. 6.3.

The movable towers should have sufficient travel to service all the large pours in the dam. It is not necessary to provide cableway coverage for small pours, such as those in the powerhouse, the downstream spillway aprons, or small blocks located at the dam abutments, if tower runway extensions are necessary to service these pours. The expense of runway extensions that require a large amount of fill or excavation or require the construction of a trestle is not justified in such cases, and other means of concrete transport and placement should be used.

In preparing the elevation layout of the cableway, it is accurate enough to allow for a sag in the track cable of 6 percent of the length. (This will probably be changed to a more accurate figure by the cableway manufacturer in his final layout.) From the bottom of the sag to the top of the concrete, approximately 60 ft (18 m) should be allowed for hook and bucket clearance. If the cableway tower heights become critical, the sag can be reduced when concrete is being placed in the top of the dam by using a smaller-capacity bucket during the topping-out process. In determining movable-tower height, the minimum height of a head tower should be approximately 60 ft (18 m) to accommodate the machinery, cable blocks, and counterweight. Head-tower height is also required to furnish concrete-bucket clearance at the concrete-transfer point, to provide hook clearance at yarding areas usually located next to the head tower, and to provide hook clearance when the carriage is brought into the head tower for servicing.

The minimum height required for traveling tail towers is approximately 30 ft (9 m). This is illustrated in Fig. 6.6, which pictures the tail tower used at Yellowtail Dam. Movable towers are preferably kept under 150 ft (45 m) in height, since taller towers create increased operating troubles. It has some-

Figure 6.6 Tail tower for one of the Yellowtail Dam cableways, Montana. (Washington Iron Works.)

times been necessary to use taller movable towers in spite of their disadvantages. In the construction of Guri Dam in Venezuela, it was necessary to use a traveling head-tower height of 210 ft (64 m).

Figure 6.2, illustrating a radial cableway for Denny Dam shows the relationship between cable sag, bucket clearance, tower height, and runway elevations. In this layout, the tops of the tail and head towers are at the same elevation. This is not a requirement, and if site conditions make it necessary, cableways can have towers with their tops at different elevations, so long as bucket clearance is provided for the full length of the dam. The dam shown is of such a size that only one cableway was required, and the most economical type was a radial cableway.

After the cableway tower heights have been determined, the tower widths can be computed. For traveling towers, the tower width determines the distance between the two runway tracks. Towers are designed so that the resultant of the stress in the track cable and the weight of the counterweight passes through the sloping leg of the traveling tower and through the trucks supporting this leg. For preliminary layout purposes, the width of a cableway tower can be taken as 55 or 60 percent of its height. For final estimating purposes, tower widths and tower weights should be secured from the cableway supplier. For preliminary estimating purposes, the weights of towers and trucks are given in Table 6.1.

The weight of a fixed tower common to two cableways is 1.3 times the fixed-tower weight for one cableway.

The weight of a fixed tower common to three cableways is 1.9 times the fixed weight for one cableway.

The weight of A-frame fixed towers is 0.64 times the weight of framed fixed towers.

The weight of mast-type fixed towers is 0.33 times the weight of framed fixed towers.

Running-gear weights for movable towers vary with the stress in the track cable. The running gear will weigh 36 tons for a track cable stress of less than 250,000 lb, 45 tons for stresses between 250,000 and 350,000 lb, and 50 tons for stresses over 350,000 lb.

The counterweight volume and location can be secured from the cableway supplier or can be computed by balancing the moment resulting from the cable stress and its moment arm around the center line of the inside tower track, with the moment computed from the weights of the counterweight and cableway tower and their moment arms around the same point. This computed weight is then increased so that the tower will not overturn when an occasional heavy load occurs on the cableway. The counterweight is usually located directly over the outside cableway tracks. The center of gravity of the tower and the counterweight must fall on the center line of the outside track or fall between the two tracks so that the tower will be stable before the track cable is erected and when the track cable is being replaced. It is important to

TABLE **6.1** **Tower and Running Gear Weights for One 25-ton Cableway**

Height of Tower (ft)	Weight of Traveling Tower (tons)	Weight of Fixed-framed Tower (tons)
30	39	12
40	52	16
50	65	20
60	78	24
70	91	28
80	104	32
90	117	36
100	130	40
110	143	44
120	156	48
130	169	52
140	182	56
150	195	60
160		64
170		68
180		72
190		76
200		80

check this weight resultant when cableways are being altered for reuse at locations other than their original ones. Figure 6.3 shows the shoring required to support the forms for concrete counterweights used for two movable head towers.

The cableway span should be kept as short as possible. The longer the cableway span, the more wear there will be on the cables, the more frequently the cables must be replaced, and the more the work will be delayed because of time lost during cable replacements. Cableways with long spans, such as the 2,912-ft (888-m) span used to pour Guri Dam, Venezuela, and the 2,900-ft (884-m) span used on the cableways for Dworshak Dam in Idaho, have been used, but frequent replacements of the track, traveling, and hoisting cables were necessary.

If more than one cableway is required, the towers may be located on the same or separate runways. If separate runways are used, they should be located so that one cableway will not cross in the path of another. Separate tower runways are warranted when the slope of the ground makes the construction of long runways a major construction problem.

After the type of cableway is selected and its runways are located, the track used to transfer the concrete from the mix plant to the pickup point underneath the cableway can be located. The cableway-layout drawings of Figs. 6.1 and 6.2 show that as the loaded carriage travels along the track cable, it forms a load path that steepens in slope at both ends. Formulas for computing this load path are given in publications of cable manufacturers.[6.1] If the concrete transfer track is located under the steep part of the load path, there will be a sharp deflection in the track cable when a loaded bucket is handled in this vicinity. This will cause severe wear on the track cable and cause additional wear on the travel line. Therefore, the preferred location of the transfer track is where the load path levels out. As a rule of thumb the minimum horizontal distance between the transfer point and the top of the cableway tower should be 10 percent of the span; preferably, the distance should be greater than this. The length of the transfer track should be sufficient to service the cableways when its movable towers are at each end of their travel. The elevation of the transfer track should be as close to the center of the mass of concrete as possible, since this location will result in the least amount of vertical bucket travel when concrete is being placed. Depending on the cableway layout and the type of dam, a transfer-track location at this elevation may be impractical if it interferes with the placement of concrete in a major dam monolith. If this interference occurs, the concrete of this block cannot be poured until the remainder of the dam concrete has been placed. Then, since a portion of the transfer track must be removed to pour this block, its placement may be costly and time-consuming because every bucket of concrete must be transferred from pickup point to the pour by movement of the cableway towers.

The mix plant should be located so that the elevation of its loading point is the same as that of the transfer track and as close to the cableway location as possible. This will allow short and level train travel from the mix plant to the transfer point under the cableway. Because of these different requirements, trial locations for the transfer track and mix plant must be made to determine which location will be most advantageous.

Cableway Capacity

The number of cableways required for a concrete dam is determined by computing the mass-placement rate required to keep the job on schedule and dividing into this the pouring capacity of one cableway. The pouring capacity of each cableway depends on the bucket size, the cycle time required to pour one bucket of concrete, the percentage of time the cableway will be pouring concrete, the hours worked per day, and the days worked per week. The cycle time of a cableway is the time required for vertical and horizontal travel and the fixed time required at the transfer and dumping points. Cycles will vary among dams and among different areas of the same dam because of the variations in travel distance. The use of a computer for determining cycle time saves considerable drudgery. To arrive at the average cycle without using a computer, it is necessary to compute individual cycles for different dam elevations and then find the average by weighing them against the yardage poured for each cycle. To determine the cableway-cycle time, the traveling and hoisting speeds of the proposed cableway must be known.

The following is an example of how cableway-cycle times are computed. This cycle is for a cableway with a horizontal travel speed of 2,100 ft/min and a

Cableway-Bucket Cycle

Load bucket		30 s
Hoist clear of bucket dock		5 s
Horizontal travel:		
Accelerate	5 s	121 ft
Travel	36 s	1,258 ft
Retard	5 s	121 ft
Total	46 s	1,500 ft
Vertical travel:		
Accelerate	5 s	38 ft
Lower	10 s	149 ft
Retard	5 s	38 ft
Total	20 s	225 ft
Therefore, horizontal travel controls:		
Horizontal travel from bucket dock to pour		46 s
Spot and dump bucket		40 s
Return to bucket dock		46 s
Land bucket		10 s
Total		177 s
Change to minutes and seconds		2 min 57 s
Adjust to		3 min

vertical travel speed of 950 ft/min. It is assumed that the center of mass of the dam is located 1,500 ft horizontally and 225 ft below the bucket pickup point and that the accelerating and retarding time for both horizontal and vertical travel is 5 seconds.

The percentage of time that the cableway will be available for concrete pour depends on the amount of lost cableway time and the amount of cableway hook time required for other operations. Cableway lost time occurs when the cableway is being serviced, when cable replacements are necessary, and when the pour is switched from one block to another. As an example of this cableway lost time, the time required to change the track cable of a cableway is approximately 1 week. Much less time is required for replacement of the travel or hoisting cables. Cableway hook time is used for foundation cleanup, for concrete cleanup, for placing forms on blocks and moving them between blocks, for placing reinforcing steel in the blocks, and for setting gates, penstocks, and other materials.

The use of the cableway hook for applications other than concrete pouring becomes critical on jobs where only one cableway is provided. To relieve the cableway of this type of service, provisions may be made for the use of other hooks. Figure 6.7 shows a truck crane operating on a Bailey bridge located on the upstream side of Guri Dam. This truck crane was used to relieve the cableway of some of the hook time required for servicing operations.

When two or more cableways are provided for a project, one may make all the service lifts and pour concrete part of the time, while the others are used solely for concrete pouring. If this is done, the average nonpouring time for the cableway system will still be approximately the same as that for one cableway. For estimating purposes, it can be assumed that cableways will be placing concrete only 75 percent of the time.

After determining the average cycle time, the percentage of time the cableway will be pouring concrete, the number of hours that will be worked per day, the number of days that will be worked per week, and the bucket capacity of the cableway, the cableway's pouring capacity and the required number of cableways can be computed. Using the previously computed 3-min cycle time and assuming that a monthly placement capacity of 100,000 yd^3 (77,000 m^3) is required, the engineer can determine the number of cableways that must be installed in the following manner:

Bucket size	8 yd^3
Average cycle time	3 min
Cycles per hour	20 cycles
Average hourly pouring rate	160 yd^3
Hours worked per day	21 h
Pouring hours per day (75% of time)	15.75 h
Daily average pour (15.75 h)	2,520 yd^3
Weekly average (5-day week)	12,600 yd^3
Monthly average (4.33 weeks)	54,600 yd^3
Number of cableways required	2

Figure 6.7 Truck crane on a Bailey bridge used to assist the cableway, Guri Dam, Venezuela.

The assumed 75 percent availability on concrete pour is an average figure to use for cableway operation. Similarly, 50,000 yd³ (38,000 m³) per month for a 5-day week is a realistic figure to use for average concrete placement for one 8-yd³ cableway. Pouring capacities of 8-yd³ cableways have exceeded this when travel distances were short, mechanical bucket dumping was used, high-speed cableways were installed, and cableway control was good. The following record pours were made at Bullards Bar Dam, California, where two high-speed cableways were employed.[6.3] These cableways had the same traveling and hoisting speeds as those used in the previous computations.

Calendar-month maximum production (7-day week)	Over 180,000 yd³
Weekly maximum production (7-day week)	Over 40,000 yd³
Daily maximum production (3 shifts)	Over 8,000 yd³

If this production is adjusted to give the production of one cableway for a 5-day week, the results are:

Calendar-month maximum production	64,290 yd³
Monthly average hourly pour	138 yd³ (approx.)
Weekly maximum pour	14,385 yd³
Weekly average hourly pour	138 yd³ (approx.)
Daily maximum pour (21½ h)	4,000 yd³
Maximum hourly pour	186 yd³

This monthly maximum production is 1.18 times the average production computed in the sample computation previously shown. This checks quite closely, since the average monthly production should be less than the maximum monthly production.

Production during record months is always greater than the average monthly production, since record months are achieved when the concrete has been placed in each block in the dam's foundation, when there are no major cableway shutdowns, and when the cableway is not needed for handling gates, penstocks, or other structural items. Similarly, the average monthly production rate is achieved only when enough dam monoliths have been started so that adequate pouring area is available, and prior to starting the placement of the narrow pours required to top out the dam. Start-up monthly production should be reduced for several months to allow time to place concrete in all the dam monoliths in the base of the dam. This is known as "getting off the rock." Production at the end of the job will be reduced for several months when only small pours are left in the top of the dam. This slow production at the start and end of the concrete-placement period will cause a concrete-placement curve to take an S shape.

Effect of Bucket Cycle on Mixing-Plant Capacity

To determine the required capacity of the mixing plant, it is necessary to review the cableway-bucket cycle. By examination of the previously computed bucket cycle one readily sees that when pours are made adjacent to the transfer track, 36-second horizontal travel in each direction can be deducted from the cableway cycle, thus giving a cycle time of 1.80 minutes and a pouring rate for one cableway of 267 yd^3/h, or an increase in production of 67 percent. When pours are made at the other end of the dam, and if one assumes that 1,258 lin ft must be added to the carriage's horizontal travel, the bucket cycle would be increased by 36 seconds in each direction, giving a total cycle time of 4.20 minutes or a pouring rate for one cableway of 114 yd^3/h, which is a decrease in production of 29 percent. To maintain the average production rate of 160 yd^3/h for one cableway, the rate of pours made adjacent to the transfer track must exceed 160 yd^3/h to compensate for the reduction in pouring rates when distant blocks are poured. In order to pour at this faster rate, the mixing plant and concrete-transportation method must be of capacities suited to accommodate this faster rate of pour. Also, the mixing plants must have sufficient capacity to permit a mixer to be shut down occasionally when servicing or relining is required.

The capacity of the mixing plant to be installed is a matter of engineering judgment. Except in special applications, the engineer can seldom justify installing a mixing plant of such capacity that the placement equipment can place concrete at its maximum rate when pours are adjacent to the transfer point. Mixing plants are often installed with a maximum capacity that permits the placement equipment to place concrete at 25 percent over the average hourly pouring rate. Then the concrete superintendent may complain that there is not enough mixing-plant capacity to allow him to pour at maximum capacity. However, a 25 percent increase is enough to permit the cableway to maintain the average placement rate, for the percentage decreases in production when making distant pours seldom equal or exceed this. If this 25 percent factor is used to determine mixing-plant capacity in the cableway production example, the mixing plant for the two cableways with an average pouring rate of 320 yd³/h would have a maximum capacity of approximately 400 yd³/h. This checks closely with the mixing plant installed for two cableways at Bullards Bar Dam, where six 4-yd³ mixers were installed in one plant. Since it took 3½ minutes to charge and discharge the six mixers, the total installed capacity was 411 yd³/h. A rule of thumb for determining mixer capacity is that the bucket capacity of the placement equipment should be multiplied by 1½ to determine mixer capacity. If this rule is applied for two 8-yd³ cableways, the mixing plant should have a capacity of 24 yd³, or six 4-yd³ mixers would be required. When both cableways and cranes are used to place concrete simultaneously, combined bucket capacities should be used to determine the required mixing plant capacity.

TRESTLE-CRANE SYSTEM

The trestle-crane method of dam concrete placement consists of pouring concrete from buckets which are swung into position over the concrete pour by cranes supported on the deck of a trestle. The trestle is located parallel to the concrete dam, or the concrete section of any type of dam, with an alignment and deck elevation that allow the cranes to have boom coverage of all the concrete pour areas. Figure 6.8 illustrates this method of mass-concrete placement.

Three types of cranes are used: gantry cranes traveling on rails laid on each side of the trestle, rail-mounted hammerhead cranes, or self-contained cranes such as crawlers or truck cranes supported by the trestle deck. The use of crawler- and truck-mounted cranes is increasing because larger-capacity units are continually being developed. The cranes not only pour concrete, but, like cableways, they perform service lifts such as handling forms, handling lifts required for concrete cleanup, placing reinforcing steel, and handling embedded material. The trestle- and crane-placement method has a flexibility in placement capacity, since capacity can easily be increased by placing more cranes on the trestle. Rail-mounted cranes are preferred, since they travel on a level track and maintain a constant roller-track position so that the swing of the boom is always level. Furthermore, their booms are more rigidly supported than those of crawler units and provide steadier control of the bucket, which allows rail-mounted crane units to pour concrete faster than rubber-tired or crawler units.

Figure 6.8   Trestle and cranes for concrete placement, Libby Dam, Montana. (Washington Iron Works.)

Rail-mounted cranes used for pouring dam concrete from trestles have been gantry cranes, with a few exceptions. Hammerhead, T-shaped cranes were used to pour concrete at Grand Coulee Dam, and the pouring rates for these cranes were excellent. However, because of their high initial cost and the large cost involved in their transportation, erection, and dismantling, new units of this type have not been purchased for use on other dams. Upon completion of Grand Coulee, some of the hammerheads were used in industrial plants. Others were used on Friant Dam, and later these were shipped to India for use on Bakra Dam and are no longer available for dam construction.

Crawler or rubber-tired cranes are used on trestles to pour small dams when such cranes can first be used for other purposes and only later for concrete pouring. For small concrete yardages, the increased placement cost caused by using these dual-purpose cranes is less than the increased capital write-off that would result if gantry-type cranes were purchased. Irrespective of the type of crane used, it is the practice to limit the maximum concrete bucket size to 4 yd^3. Cranes can handle larger buckets at shorter pouring radii, but whether the use of larger buckets can be justified depends on the particular job condition. Concrete buckets can be hooked and unhooked from these cranes manually or automatically. If done automatically, the hook is opened and closed by the pressure of compressed nitrogen, which is activated when the bucket bail contacts the crane hook. The compressed nitrogen is supplied from cylinders carried on the hook, and these cylinders are changed when the nitrogen in them is consumed. Concrete is dumped from the buckets by manually operated gates or by gates that open automatically when pressure is applied. If air-operated gates are used, air is supplied by a compressed-air hose, which is attached to the bucket at the pour area. Pouring procedure requires the crane to drop an empty bucket onto the bucket car, disengage the hook, hook onto a loaded bucket, and move it into pouring position. The concrete is dumped from the bucket, and the empty bucket is returned to the bucket car. This pouring cycle is repeated until the pour is completed.

Layout

If the trestle concrete-placement method is selected for a particular dam, the engineer must select the proper location and elevation of the trestle, the length of the trestle, the number of cranes, the size of cranes required to give adequate boom coverage, the location of the mix plant, and the method of concrete haul. To aid in this selection, a trestle-crane layout drawing is prepared similar to that shown in Fig. 6.9.

Prior to preparing a layout, it is necessary to secure the gantry widths, gantry heights, and boom capacities of different sizes of gantry cranes. Many sizes of gantry cranes suitable for concrete placement are available. Table 6.2 gives examples of the sizes available, and basic data on five gantry-crane models available from American Hoist and Derrick Co. Each of these cranes is equipped with a Model 180A three-drum hoist, which furnishes a concrete-bucket hoisting speed of 220 ft/min. Travel speed of the cranes is 125 ft/min on a level track. Swing speed is 1 rpm. Counterweights consist of rails to form the counterweight tank bottom, and the remainder consist of steel punchings or billets. The radius given is from the center of the crane. The maximum

TABLE 6.2 Typical Gantry-Crane Models

	Crane model				
	101	**152**	**203**	**254**	**305**
Maximum placing radius, ft, using a 4-yd^3 concrete bucket plus 25% impact	50	90	150	180	210
Maximum boom length, ft	120	140	160	180	200
Gantry height, ft	39–66	42–82	42–95	46–117	49–120
Track gauge, ft	16–32	20–36	24–40	28–44	32–48
Counterweight, tons	32–38	50–65	75–100	100–130	120–175
Maximum corner load, tons	46–52	70–82	110–118	142–164	202–228
Weight of gantry rail, lb/yd	100	175	175	175	175

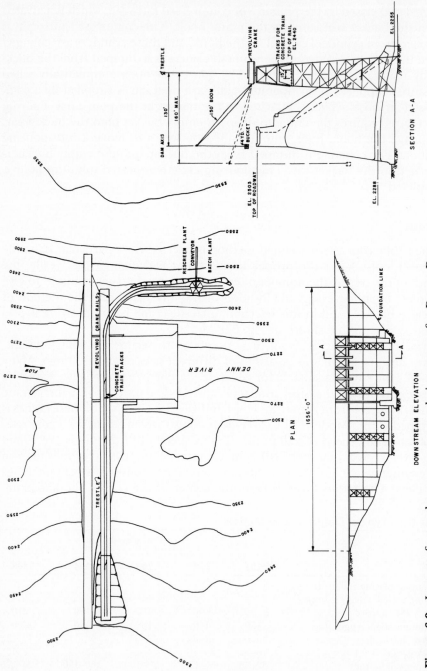

Figure 6.9 Layout of a trestle-crane concrete placing system for Denny Dam. (See Chap. 19.)

corner load varies with track gauge and gantry height. The counterweight varies with the length of the crane tail. Electrically powered cranes are usually used, but in some cases, when electricity has not been available, diesel-driven cranes have been used.

A trestle-crane layout should give consideration to trestle cost compared to crane cost. If trestle height and length can be saved with gantry cranes equipped with high gantries and long booms, it is more economical to use the higher gantries. Their cost will generally be less than the cost of raising the trestle and increasing its length to allow the use of smaller cranes.

The most desirable location and elevation of the trestle deck places the concrete buckets, when they are on the bucket car, as close to the center of mass of the concrete as possible. This location involves the least amount of crane boom swing and the shortest bucket-hoisting distance required to position the concrete buckets over a pour. When narrow concrete dams are constructed, crane coverage can be secured without locating the trestle within the area to be covered with concrete. On such dams, when water is not to be stored behind the dam, a trestle located upstream of the dam with a height level with the center of mass of the concrete will result in the least amount of crane swing and hoisting. If water is to be stored behind the dam before the dam is completed, the use of this area for a trestle is impractical, and the trestle must be located downstream from the dam.

Wide dams often cannot be poured with cranes located on trestles either upstream or downstream of the dam. The trestle must then be located above the dam's sloping downstream face, with its center line as close to the center of the dam's bottom width as possible. The concrete beneath the trestle deck can be poured by drifting the concrete bucket or by pouring through openings in the trestle deck. These openings may be framed into the deck, or the deck may have removable sections.

On an exceptionally wide dam, it may be necessary to provide two trestles in order to furnish crane coverage for the full dam width. This was the case at Grand Coulee Dam, where the trestle decks were located at different elevations. When the trestle legs are located within the dam area, they are encased in the dam concrete as the dam is raised. After the dam is completely poured, the legs can be burned off in recessed pockets in the concrete formed around the legs for this purpose. The remainder of the trestle can then be removed and the pockets filled with concrete to furnish a smooth concrete surface.

Trestle length is controlled by the dam length and the topography of the abutments. Often the trestle will intersect the abutment banks before sufficient trestle length is secured to allow the cranes to service the pours in the end blocks of the dam. Solutions to this problem are to excavate trestle-track extensions in the dam abutments, to raise the trestle, or to pour the end monoliths with other equipment. The topography and the volume of concrete in the abutment blocks will determine the solution to use for any particular dam. Trestle width is controlled by the width of the gantry, or, if gantry cranes are not used, the trestle must be wide enough to permit concrete-haulage equipment to pass by the placement cranes.

To properly estimate the cost of a trestle and crane system, it is necessary to know the volume of excavation required for the trestle foundations, the concrete yardage in these foundations, the weight of steel in the trestle, the board feet of wood decking, the quantity of rails and ties needed for the gantry and transfer tracks, the cost of the electrical distribution system, the cost of "baloney cable" (electric feed cable) needed for the cranes, the required number of cranes and their cost, and the cost of crane erection and crane counterweights.

Similar to a cableway, the most economical method of transporting concrete from the mix plant to the placement cranes is by rail, since rail equipment causes less traffic and less congestion on the trestle than trucks or truck-trailer units. Therefore, concrete is transported from the mixing plant to the placement cranes in rail-mounted bucket cars. The maximum size bucket cars used to date have five bucket spaces.[6.4] A rail haulage system requires that the mix plant be located with its base at trestle height. If such a location is impossible, then truck-trailer units or trucks should be used to haul the concrete.

Capacity

The capacity of a trestle-crane system is determined by the number of cranes provided for the concrete pour. Small dams are often poured with one gantry, but these are exceptional cases, and hook coverage is supplemented with other cranes, such as truck cranes, to take care of other lifting requirements. Dams containing large volumes of concrete may require a large number of placement cranes in order to meet the scheduled production. As an example, for concrete placement in Grand Coulee Dam, enough cranes were installed to set a daily concrete-pouring record of almost 22,000 yd^3 (17,000 m^3). This high rate of pour was made under less favorable conditions than those now existing; since refrigerated concrete was not used, the concrete had to be placed in small 50-ft-square blocks which required frequent moves.

To determine accurately the number of cranes necessary to meet schedule requirements, bucket cycle time must be computed and used to establish the crane pouring rate. Bucket cycle time is composed of the time required to hook the bucket, swing and hoist the bucket, spot the bucket, dump the bucket, and return the bucket to the hauling unit. For preliminary estimating purposes, one can assume that a crane can place approximately 100 yd^3 (77 m^3) of mass concrete per hour and that the crane will be pouring concrete 80 percent of the time. These capacities are slightly higher than the placement rates achieved at Libby Dam, Montana.[6.4] However, since only 24 yd^3 (18 m^3) of mixing capacity was installed, the placement capacity of the six cranes was limited to the maximum production achieved from the mixing plant. These placement yardages are conservative compared to those given by T. G. Tripp in the *Handbook of Heavy Construction* as 144 yd^3 (110 m^3) per hour.[6.2] Placing capacity will be reduced when the crane is placing other than mass concrete. Like a cableway, a gantry crane cannot place concrete continuously, since it must be serviced and make other lifts, but it can pour concrete a greater percentage of the time because there will not be any track-cable replacement delays.

After computing the bucket time cycle for the particular operating conditions to be encountered, the number of cranes required to meet a scheduled rate of pour can be determined in the following manner:

Size of bucket, yd^3	4
Average bucket cycle time, min	2½
Average pouring rate, yd^3 per production hour	96
Percent of time crane will be pouring concrete	80
Average pouring rate, yd^3 per elapsed hour	77
Number of hours worked per day	21
Cubic yards placed per crane per day	1,617
Number of days worked per week	5
Cubic yards placed per crane per week	8,085
Cubic yards placed per crane per month	35,007
Number of cranes required to pour 100,000 yd^3 per month	3

As explained when cableways were discussed, concrete placement equipment cannot maintain its computed average hourly placement rates unless it can place some pours at much faster rates. To accomplish this and to furnish sufficient mixed concrete to maintain production when a mixer is being repaired or serviced, it is recommended that the mixing plant contain enough mixers to supply concrete at an hourly production rate 25 percent greater than the average pouring rate of the placement equipment. For example, if a mixing plant must furnish concrete to three gantry cranes that have an average pouring rate, when placing concrete, of 96 yd^3 (74 m^3) per hour, the mixing plant should be able to produce 360 yd^3 (62 m^3) per hour. Since one 4-yd^3 mixer can produce 80 yd^3 of concrete per hour, five 4-yd^3 mixers will be required. This is the same size mixing plant that would be selected if the previously stated rule of thumb for determining mixing-plant capacity is applied; for example, the cubic yards of installed mixer capacity should be 1½ times the total bucket capacity of the placement equipment.

Bucket-car requirements consist of one car at each crane, one car at the mixing plant, and cars in transit between the mixing plant and the placement cranes. The number in transit depends upon the length of haul.

As previously discussed, the rate of concrete placement will be reduced at the start and completion of a concrete dam, resulting in an S-shaped concrete-placement curve. At the start of concrete placement and until the major part of the dam foundation is covered with concrete, there will not be enough dam blocks on which pours can be made to maintain maximum production. Later, when the dam is topped out, production will again decrease because only small volumes of concrete will be needed for each block. Finally, the high blocks, which will be every other block, will be completed, leaving only one-half the blocks available for concrete placement.

CRANE SYSTEM

When a low concrete dam, or the concrete portion of any other type of low dam, is to be constructed, it is often possible to pour the concrete with gantry or other types of cranes operating from ground level. Gantry cranes can now be obtained with 120-ft-high gantries; this gantry height makes them very adaptable to this type of concrete placement. Concrete-placing procedure with this method is similar to that used under the trestle-crane method, with the exception that the trestle is eliminated. The placement cranes may place concrete from various ground elevations, so that trucks or truck-trailer units may be required to haul the concrete from the mixing plant.

The use of a crane-placement system is illustrated in Fig. 6.10, which pic-

Figure 6.10 Cranes for concrete placement at The Dalles Dam, Columbia River. (Washington Iron Works.)

tures large gantry cranes operating from ground level for placement of concrete in The Dalles Dam and powerhouse on the Columbia River.

Layout

As in the trestle-crane system, before a drawing is prepared for this method of concrete placement, crane dimension and boom-coverage charts should be obtained for the various-sized cranes. The layout should illustrate whether crane coverage may be secured by gantry cranes located on tracks laid at ground level or whether space is available to operate crawler or rubber-tired cranes. Layouts should show how crane coverage will be provided for all concrete structures, since mental pictures of crane coverage may be misleading, and there may be concrete areas that would be extremely difficult or impossible to place in this manner. Gantry cranes can be obtained with different leg lengths to suit specific ground or foundation conditions. After the crane coverage is checked, locations must be selected for the mixing plant and the concrete-haulage tracks or haul roads from the mixing plant to the placement cranes. If possible, the mixing plant should be located near the center of the pouring area to reduce the concrete-haul distance.

Capacity

The number and capacity of the placement cranes and concrete-haulage units and mixing plant size should be computed in the same manner as for the trestle-crane method of concrete pour.

SELECTING THE MASS-CONCRETE PLACING EQUIPMENT

The selection of the mass-concrete placing equipment for any concrete dam is governed by the type of dam and damsite. In this respect, proposed dams and their damsites fall into four groupings.

High Concrete Dams Located in Narrow Canyons.   Damsites in narrow canyons have steep side walls, and most of the canyon floor is occupied by the river. Typical damsites of this type were those for Hoover, Hungry Horse, Shasta, Flaming Gorge, and Bullards Bar dams.

Cableways are advantageous for use at these damsites because their towers and all other major construction facilities can be located outside the dam area and can be erected without interfering with, and before the completion of, the diversion facilities and foundation excavation. This allows concrete placement to start as soon as the dam foundation is excavated. Often the cableway can be operated early enough to assist in the preparation and cleanup of the foundation. Another advantage of a cableway system is that most of the construction facilities will be located above the canyon floor where they will be free from any possible damage from flood flows.

Concrete placement in high dams entails a large amount of vertical bucket travel. Fast concrete pouring requires that the concrete bucket be hoisted or lowered at high speeds and that the placement equipment have sufficient drum capacity to contain the large amount of hoisting cable required. A cableway is particularly suited to this type of operation, since it has fast hoisting speeds and can be equipped with an almost unlimited drum capacity.

When these high, narrow dams contain a large volume of concrete, the concrete-placement equipment must have a large capacity to meet schedule requirements. A cableway system has limited placement capacity, since cableways must span the canyon about as the dam does, which limits the number that can be installed at any damsite. However, this placement limitation can often be overcome, as was done at Shasta Dam where the terrain permitted the operation of seven movable cableway towers which radiated from one central fixed tower. At Glen Canyon Dam the number of cableways that could be used at one time was increased by providing a double set of parallel tracks with two cableway units located on the inside tracks and operating underneath a third cableway with higher towers located on the outside tracks. Pouring capacity was increased also by using cableways capable of handling 12-yd³ buckets (see Fig. 6.11).

When the damsite meets the majority of the conditions suitable for cableway placement, lack of stability of the abutments may make its use impractical. One example of this was at Green Peter Dam, Oregon, which was a typical

Figure 6.11 Cableways at Glen Canyon Dam; three traveling towers on two parallel sets of tracks. (Washington Iron Works.)

cableway site. However, the stability of the right abutment of the dam was indeterminable. The dam contractor elected to use gantry cranes operating from a trestle. Subsequent excavation of the right abutment proved that the choice of this placement method was correct, even though damage was caused to the trestle by a 100-year flood which plugged the diversion tunnel, over-topped the placed concrete, and damaged other construction plant installations.

A perfect damsite for a cableway system occurs when the steep side walls of a canyon flatten above the top of the dam to furnish a level area for the runways for movable cableway towers. One of the best examples of this type of damsite was at Glen Canyon Dam, where the Colorado River cut a narrow channel with almost vertical walls through a flat plain, providing an excellent location for runways for the movable towers of parallel cableways.

In comparison, trestle-crane placement systems have disadvantages for placing concrete in high dams located in narrow canyons. Before a trestle can be erected, the river must be diverted and the foundation completely exca-vated. The erection of the trestle becomes part of the critical path and length-ens the time required for dam construction. The trestle, the cranes, and other construction facilities will be located within the canyon where they will be exposed to flood-flow damage and will congest the limited work areas. Very high dams are often built at damsites in narrow canyons so that a large amount of vertical bucket travel is required for concrete placement. Cranes are less suitable for making these high pours than cableways, since their hoisting speeds are slower and they have limited cable capacity.

Relatively High Concrete Dams Located on Wide, Flat Damsites.   The tres-tle-crane concrete placement system is used for the construction of high con-crete dams, or concrete portions of combination dams, at long, flat damsites that have low abutments. Here the Grand Coulee, Libby, Friant, and Noxon Rapids dams and damsites are typical, as is the second Guri Dam.

Trestle erection will not lengthen the time required for the dam's construc-tion since the damsites are long enough to permit one portion of the dam to be excavated, a section of trestle erected, and concrete placement started while the remainder of the dam is being excavated and the stream is being diverted.

These broad damsites can only be spanned by long dams where concrete placement requires that concrete be hauled long distances between the mixing plant and the pours. A trestle-crane system can be economically adapted to long hauls by increasing the number of concrete-haulage vehicles.

The large volumes of concrete contained in these long dams require that the selected concrete placing system have high production rates. A trestle-crane concrete placement system is preferred for large dams, as any required production can be obtained for a minimum capital expenditure by increasing the number of placement cranes.

Other cost advantages resulting from a trestle-crane system include the use of the trestle to support the water, compressed-air, and electrical distribution lines that service the pour areas, and the use of the trestle deck as a work platform. Compared with a cableway, there will be fewer pouring delays with a trestle-crane system, since there will be no track and operating cable replacement delays. Finally, a trestle-crane system has great flexibility in providing hook coverage to dams of irregular shape.

Conversely, a cableway concrete-placement system is costly to install at wide, flat damsites, and often not enough cableway units can be installed to provide the desired pouring capacities. Cableways must operate across the width of a dam, as compared to a trestle-crane system that operates along its length; therefore, the number of cableways that can be installed to pour long, narrow dams is limited, and the pouring capacity of a cableway system is restricted. If enough cableways can be installed to meet pouring requirements, there will be a high acquisition and installation cost for each as a result of long cable spans and high towers; the latter compensate for low abutment heights and the large cable sags which occur when there are long cableway spans. Acquisition and erection costs for the cableway system will also be high, because the long horizontal distances that the cableway must transport the bucket, between the bucket pickup point and the concrete pour, will increase the cableway cycle time, decrease its production, and require that a large number of cableways be installed to meet the scheduled rates of production. This increase in cycle time and decrease in production for each cableway will cause a proportionate increase in the direct cost of concrete placement.

Low Concrete Dams.   When the topography at the damsite permits and the dam is low enough that concrete can be poured with cranes operating from ground level, this is the most economical concrete placement method. This method minimizes acquisition cost and operating cost. Gantry cranes, crawler cranes, truck cranes, or tower cranes are used for this purpose. Typical of the concrete dams that have been poured in this manner are the low-head spillway dams on the lower Columbia and Snake rivers. This system was used to pour concrete in the powerhouse at Chief Joseph Dam. A large walking dragline was removed from its base and was mounted on a low, specially built gantry which traveled on rails laid on the excavated surface of the intake channel. Because of the dragline's large size, it could utilize a 200-ft boom to make all the powerhouse pours.

Other Dams and Damsites.   Some dams and damsites will not readily fall into any of the three foregoing groups. For these dams and damsites, it is often necessary to prepare comparative estimates that evaluate the acquisition and operating cost of each type of concrete placement system before the preferable system can be selected. If these comparative estimates do not show a definite cost advantage to a particular system, then the system often selected is one that makes use of available equipment owned by the contractor or one that is preferred by the contract manager.

Preplaced-aggregate concrete can have economical advantages in special situations. In these situations clean, coarse aggregate is placed in the forms, and the voids in the aggregate mass are then filled with a fluid sand-cement grout which is pumped into the aggregate from the lowest point of the structure. An interesting solution to raising the Balch Dam in the High Sierra from 93 to 138 ft (28 to 42 m) involved first placing a 7.15- to 10.35-ft-thick layer of Prepakt* concrete on the downstream face during the first short construction season. After thickening, the top 45 ft (14 m) was also placed by the Prepakt* method during the next summer season.[6.5]

In England a successful test has achieved the approval of the National Water Board Authority for alternate bids involving slip forming of upstream and downstream faces for several water-storage dams about 100 ft (30 m) high. In the test, upstream and downstream faces were slip-formed in successive overlapping lifts, with the portion of the dam between faces filled after each lift with rolled lean concrete.[6.6]

The Willow Creek Dam in Oregon was the world's first dam designed and built as a totally rolled compacted concrete structure. This dam contains 400,000 yd^3 (310,000 m^3) of concrete and is 155 ft (47 m) high. The interior mass concrete was designed to utilize only 80 lb of cement and 32 lb of fly ash per cubic yard.[6.7]

Placement of mass concrete by conveyor holds considerable promise. A proposed system was described by Robert F. Oury in the *Proceedings* of the Engineering Foundation Research Conference held at Asilomar.[6.8]

The concept provides for large segregation-free stockpiles with one week's supply of aggregate. These stockpiles are arranged to provide fully live storage with no dead storage reclaim. The concept provides for an integrated computer-controlled conveyor system from crusher to stockpile, to mixer, and thence to final deposit utilizing a mobile conveyor crane boom so that an uninterrupted flow of concrete continues through the placing of the entire monolith.

GURI DAM ENLARGEMENT

The Guri hydroelectric project in Venezuela constructed during the 1960s by Kaiser Engineers International, Inc., was one of the most challenging projects of the decade. This dam is now in the process of being raised from an original height of 180 ft (55 m) to 532 ft (162 m) by Guy F. Atkinson Company, and a comparison of the magnitude of the original project with that of the new

---

* Registered trademark, Intrusion-Prepakt, Inc.

Figure 6.12   Raising Guri Dam, Venezuela. (Guy F. Atkinson Co.)

facility is illustrative of the massive growth in major projects in the last 20 years. Figure 6.12 shows the construction status as of late 1981. Table 6.3 compares the statistics of the original project with those of the more recent one.

Summary of Construction Methods

Concrete Placement in the Dam.   The concrete is placed using whirley cranes equipped with 8-yd³ buckets. The mass concrete is spread with a tractor dozer and vibrated with 6-in vibrators. The concrete buckets are hauled from the batch plant to the cranes with three-bucket self-propelled transfer cars. The cranes service the dam and spillway from a steel trestle which runs the full length of the dam.

Concrete Cleanup.   Surfaces of the existing concrete are cleaned up with a hydro-blasting system.

Formwork.   The panels are metallic, cantilever type. Panel lifting is done using hydraulic cranes.

Setting the Penstock.   Each section, weighing from 70 to 90 tons, is placed using two whirley cranes working in tandem.

Raising of Spillway Gates.   This operation must be done 18 times (twice for each radial gate), using a specially designed lift jack system.

TABLE **6.3 Comparison of the Guri Dam Projects***
Total Dam

	Original Dam	Enlarged Dam
Height	180 ft	532 ft
	(55 m)	(162 m)
Crest length	2,085 ft	37,400 ft
	(634 m)	(11,404 m)
Fill volume	2,000,000 yd³	67,400,000 yd³
	(1,529,000 m³)	(51,500,000 m³)
Concrete volume	1,130,000 yd³	9,000,000 yd³
	(865,000 m³)	(6,880,000 m³)

Concrete Dam Only

Height	348 ft	532 ft
	(106 m)	(162 m)
Crest length	2,772 ft	4,678 ft
	(845 m)	(1,426 m)

Reservoir Capacity

	14,300,000 ac-ft	110,000,000 ac-ft
	(17,639,050,000 m³)	(135,685,000,000 m³)

Powerhouse

	Original Power-house	New Powerhouse
Length	864 ft	1273 ft
	(264 m)	(388 m)
Width	145 ft	259 ft
	(44 m)	(79 m)
Height	210 ft	255 ft
	(64 m)	(78 m)
No. of bays	10	10
Width of each bay	75 ft	92 ft
	(23 m)	(28 m)
No. of installed units	10	10
kW capacity per unit	3 × 218,500	10 × 730,000
(with reservoir at	3 × 270,000	
el. 270 ft)	1 × 340,000	
	3 × 400,000	
Total capacity (kW)	3,005,500	7,300,000
Kind of turbine	Francis	Francis
Volume of concrete in power-	312,000 yd³	585,000 yd³
house and service bay	240,300 m³)	(450,000 m³)

Quantity of Major Equipment Used in the New Concrete Work

Description	Concrete Dam and Spillway	Powerhouse
37 E Clyde whirley crane	8	
32 E Clyde whirley crane	—	3
American 325 whirley crane	—	1
Manitowoc 4600	—	1

TABLE **6.3** **Comparison of the Guri Dam Projects***
Quantity of Major Equipment Used in
the New Concrete Work (*Continued*)

Description	Concrete Dam and Spillway	Powerhouse
Lima 300	1	1
Linden 8000 with Rotec conveyor	1	—
Grove hydraulic cranes	20	—
Concrete pumps	2	1
Washington transfer cars	15	—
Johnson batching plant (240 m³/h)	3	—
Johnson batching plant (120 m³/h)	—	1
Sullair compressors, central system	14 (1400 ft³/min)	
Highway cement hauling units	20	20
150-ton lowbed and tractor	1	1

* Information courtesy of Guy F. Atkinson Company.

THE ITAIPU HYDROELECTRIC PROJECT[6.9]

In 1973 Brazil and Paraguay signed a treaty for the development of the Rio
Paraná, one of the world's largest river systems. The project encompasses 18
generating units which will be operational by 1988, when the installed capacity will be 12,600 MW to be shared equally between the two countries. Figure

Figure 6.13   Main dam, Itaipu hydroelectric project. (International Engineering Co.,
Inc.)

TABLE **6.4**   **Construction Plant and Equipment for the Itaipu Project**

Equipment or Plant	No.	Capacity
Shovels, electric	4	13 m³
Dump trucks	40	60 m³
Concrete batch plants	6	180 m³/h
Clinker grinding mills	2	110 t*/h
Aggregate plants	2	1080 t*/h
Concrete refrigeration plants	2	14.8 × 10⁶kcal/h h
Ice plants	2	24 t*/h
Cableways	7	20 t*
Tower cranes	8	12 t* at 62 m

* Metric tonne

6.13 shows construction work on the main dam, and details of some of the construction plant and equipment are given in Table 6.4

Figure 6.14 shows an overall view of the dam and powerhouse taken from the left of the diversion channel. The project schedule is summarized as follows:

September 1977.   Start concrete in diversion-control structure

September 1978.   Complete excavation in diversion channel, cofferdam closure, and river diversion

Figure 6.14   Itaipu hydroelectric project. (International Engineering Co., Inc.)

August 1979. Start main dam and powerhouse concrete

August 1980. Start spiral case embedment; downstream powerhouse wall completed to el. 108 m

September 1981. Powerhouse structure enclosed

December 1981. Spillway and gates complete

September 1982. Storage in reservoir begins

Late 1982. Startup tests on first generator

Mid-1983. Start operation of first generator

October 1988. All 18 generating units operational

MISCELLANEOUS-CONCRETE PLACEMENT EQUIPMENT

The term *miscellaneous concrete* is used to describe concrete located separate from mass concrete and of such a small quantity that the purchase of large-capacity equipment for its placement is not justifiable. Specifically, miscellaneous concrete is concrete that cannot be placed with mass-concrete placement equipment. Included in this category are the small end blocks of concrete dams, small concrete dams, slab facing of some rock-fill dams, diversion-tunnel linings, the backfill of diversion conduits in dams, training walls, spillway floors, fishladders, intake structures, and any small isolated concrete structure connected with dam construction.

Only minor amounts of miscellaneous concrete are included in dam contracts. Individual pours are small, and there is usually a large amount of standby and delay time connected with each pour, resulting in low hourly pouring rates. This restricts the expenditures that can be justified for the purchase of concrete placing equipment. Because of the great variety in the types of structures containing miscellaneous concrete, there are many ways that it can be placed, and new placement methods are constantly being developed. Methods commonly used for placing miscellaneous concrete are:

1. Direct placement from agitator or dumpcrete trucks
2. Placement with truck or crawler cranes
3. Conveyor-belt placement
4. Placement with air guns
5. Placement with pumps
6. Slip-form placement
7. Tower-crane placement
8. Placement with guyed derricks
9. Placement with stiffleg cranes

Refer to Chap. 15 for methods and information applicable to this type of work.

REFERENCES

6.1 *Wire Rope Handbook for Western Rope Users,* U.S. Steel Corp. (Pittsburgh, Pennsylvania), 1959.

6.2 John A. Havers and Frank W. Stubbs, Jr., eds., *Handbook of Heavy Construction,* McGraw-Hill, New York, 1971.

6.3 "California's Bullards Bar Dam," *Western Construction,* July 1968, p. 29.

6.4 "Libby Dam Sheds Its Electric Blanket," *Contractors and Engineers Magazine,* April 1970, p. 56.

6.5 *General Applications,* Intrusion-Prepakt, Inc. (Cleveland, Ohio), 1978.

6.6 "Slipforming Successfully Used for Unusual Applications," *Rocky Mountain Construction,* May 15, 1981, p. 18.

6.7 "Universal Use of Rolled Concrete in Dams," *Civil Engineering,* September 1981, p. 10.

6.8 Robert Foury, "Ribbon Batch and Conveyor Placing System for Concrete Dams," *Rapid Construction of Concrete Dams,* American Society of Civil Engineers, New York, 1970, pp. 281–295.

6.9 "The Bi-National Itaipu Hydropower Project," *Water Power and Dam Construction,* October 1977.

Chapter 7 The Concrete Construction System

Concrete placement equipment, its production capacities, and its selection are discussed in Chap. 6. This knowledge is a prerequisite for understanding the present chapter because concrete placement equipment controls the selection of the other units that complete the concrete construction system.

This chapter describes the remaining components of the concrete construction system. The primary purpose of these remaining components is to supply sufficient mixed concrete, formed areas, placement crews, cleanup crews, curing facilities, and concrete finishers to permit the placement equipment to place concrete continually, at scheduled production rates.

CONSTRUCTION SYSTEM SELECTION

That a concrete construction system is complex is shown by Fig. 7.1, a flow diagram for concrete dam construction. Selection of the components of a concrete construction system is accomplished by establishing work quantities, preparing a construction schedule, and then choosing system components that will operate efficiently within the system and that will have the capacity to meet the scheduled rate of production.

If the construction of the project requires the placement of large quantities of concrete, the system concept is to reduce labor cost by the use of equipment. If only small quantities of concrete must be placed, the system concept is to reduce equipment cost by performing more work with hand labor. When equipment and labor cost are balanced properly, a low cost estimate will result.

After all units of the concrete construction system have been selected, a final review should be made to determine whether the essential facilities have

166

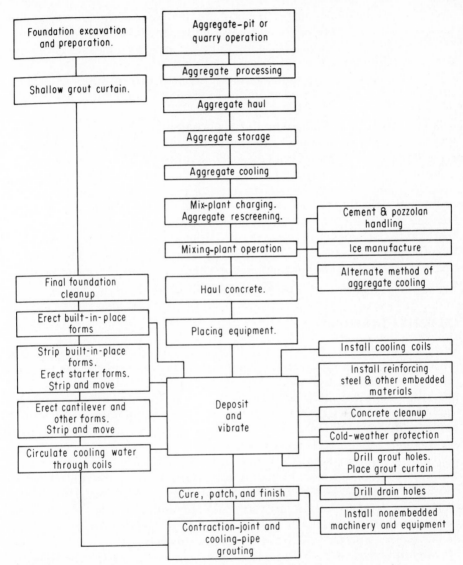

Figure 7.1 Work-flow diagram of a concrete system for dam construction.

been provided. Then the capacities of all plant and equipment units should be checked against the capacities required to keep the work on schedule.

To provide background for the selection of the components of the concrete construction systems, descriptions are given of the preparation of quantity takeoffs, the scheduling of concrete construction, the performance of each phase of concrete construction, and the equipment and plant units used to perform this construction. The order followed is that shown in Fig. 7.1:

aggregate production, haul, rescreening and mix-plant charging; cement and pozzolan handling; aggregate and concrete cooling; concrete mixing (concrete-haul and concrete-placement methods are described in Chap. 6); foundation and concrete cleanup; placement of embedded items; concrete forming; concrete depositing and vibration; cold-weather protection; curing; patching and finishing; installation of permanent facilities; and drilling and grouting.

After construction equipment and plant types have been selected, the required number of operating units is established by dividing the production capacity of each into the maximum scheduled rate of concrete placement.

Equipment availability is not used to increase the number of operating units to arrive at the total equipment requirements, in figuring for large plants and equipment units, because it costs less to shut down operations when a large equipment or plant unit, such as a cableway, is out of service than to provide spare units. The converse is true for smaller equipment units. As previously discussed, equipment availability on heavy construction projects is between 70 and 80 percent, with the remainder of the equipment under repair or being serviced, and therefore unavailable for productive use.

QUANTITY TAKEOFFS

Quantity takeoffs are made to check the work quantities listed in the bid schedule and to establish work quantities which are not listed in the bid schedule but which form a part of the cost of the listed work items. Such work quantities commonly include: the area of rock foundation that must be cleaned prior to being covered with concrete; the area of concrete that must be cleaned prior to the placement of the next pour; the area of each type of surface finish; the area of construction joint treatment; the surface contact of each type of concrete form; the area of any required form lining; the area of each type of form that must be built or purchased; the amount of blocking or shoring that must be used to support soffit or other types of forms; the trusses that may be required to support forms; the quantities of walkways, ladders, and stairs that are required to provide access to and from the pour and forming areas; and the volume of overbreak concrete.

Overbreak concrete is designated as nonpay concrete in the specifications. It occurs when the plans show concrete pay lines and the specifications state that the concrete required to fill any excavated area that extends past these pay lines will not be measured as pay quantities. Specification writers insert this clause in the specifications to force the contractor to use careful blasting techniques and to avoid excavating more rock than is necessary, but it means that the quantity takeoff must list the estimated volume of anticipated overexcavation, which is the same as the anticipated overbreak concrete. The estimator places the cost of this overbreak concrete against the pay concrete bid items, as this is the only way the contractor will be reimbursed for this work.

Form quantity and concrete takeoffs should be made simultaneously, using the same personnel, in order to eliminate errors in establishing form ratios. (*Form ratio* is the square feet of form contact for 1 yd^3 of concrete.) If concrete and form quantities are computed from the same concrete dimensions and an error is made, this error will not change the form ratio, since it will be a compensating error for both quantities. Estimators use form ratio to establish form cost for each concrete bid item.

Form cost is such a large portion of the total cost of concrete that the determination of the form ratios is necessary for proper concrete-cost estimating. The form ratio is a measure of the form cost and placing cost of each concrete bid item. The thinner and more complicated the structure, the greater will be the number of square feet of forms per cubic yard of concrete, the greater will be the form ratio, the greater will be the form cost for each cubic yard of concrete, and the longer it will take, and therefore the more costly it will be, to place a cubic yard of concrete. Also the ratio of the form cost to total direct concrete cost increases with an increase in form ratio, since form cost is a major cost for most concrete structures. When mass concrete is poured with a form ratio of between 1 and 2, the form cost will be between one-eighth and one-fourth of the direct cost of concrete. This proportion increases as the form ratio increases and may constitute four-fifths of the direct cost of concrete for concrete structures with a form ratio of 80.

The form takeoff should show the square feet of forms that should be purchased, the square feet of forms that must be job-made, and the number of times each type of form will be reused. The form takeoff should list the different types of forms: built-in-place forms, lift-starter forms, cantilever upstream face forms, cantilever downstream face forms, transverse and longitudinal cantilever bulkhead forms, collapsible gallery forms, curved-panel forms, straight-panel forms, soffit forms, stairway forms, special-shaped forms, and block-out concrete forms. Listings should be made of the square feet of form contact that will be handled by each type of form and the number of times each form will be reused.

Before the form takeoff is started, the minimum number of form reuses that will justify the use of purchased forms should be established. If steel forms are economical when a form can be used four times, this should be reflected in the form takeoff. The form takeoff should make maximum use of purchased steel forms. This includes using special steel forms for special shapes, using high panel forms for wall pours, and using collapsible forms for galleries and other interior forms. Slip-form usage is economical for spillway slopes, for towers, for shafts (both vertical and inclined), and for concrete facing on rock-fill dams. Sonotube forms are economical for circular shapes. Plastic forms are being developed for many applications.

Many points require special attention when preparing the form takeoff for a concrete dam. Among these is the amount of built-in-place forms that will be required to make the pours that are placed on the rock surface of the dam foundation. This number should be determined as accurately as possible, since the construction of these forms is costly. Another item is the block-out concrete, the forms used to block off these areas, and the forms later required to contain the block-out concrete when it is poured. *Block-out concrete* is concrete that encases steel gate guides or other steel parts that must be accurately placed and aligned. This concrete is not placed at the same time as the major pours, since at that time the space it occupies is formed off. After the major portion of the concrete structure is poured, the gate guides or other steel parts are set and aligned, and the block-out concrete is then formed and poured. Because of the small quantities involved, the difficulty in placing, and the high form ratio, the cost of placing a cubic yard of block-out concrete is very high. However, since each project has only a few cubic yards of block-out concrete, its total cost is not a major item.

The concrete-pouring schedule and sequence should be established before the form takeoff is made, so that maximum form reusage can be planned. This is of prime importance when it is planned to complete one section of a dam before the remainder is started, since forms from the first section can then be reused on the remainder of the dam. A concrete-pouring diagram will help in form scheduling. The preparation of this diagram will be discussed under scheduling of concrete construction, which is the next subject in this chapter. By using this placement diagram and the dimensions of each dam monolith, the required number of dam forms can be determined. Since the time available for estimate preparation is limited, the form takeoff often must be started before the concrete schedule has been prepared. In this case the form takeoff is made using an assumed schedule and corrected later to fit the final schedule.

To properly prepare a form takeoff, it is necessary to understand form types and how these forms are used in dam construction. Dam monoliths are poured by starting the pour in alternate blocks across the bottom of the dam. The first lift of concrete placed on the rock surface is contained by *built-in-place forms* that are braced against the rock and anchored to steel hooks placed in holes drilled into the rocks. These forms must be built in place since they must be hand-tailored to the rock surface.

The next few pours are formed with *lift-starter forms*. These are noncantilevered forms anchored at the bottom to she-bolts placed in the previously placed concrete lift. The tops are anchored to inclined tie rods fastened to anchors placed in the surface of the concrete during the previous pours. At the top of the forms, she-bolts are installed so that they will be imbedded in the pour and will be available to anchor the bottom of the forms when they are raised to contain the next pour. Lift-starter forms are used one or two times for each dam monolith until the concrete surface is raised sufficiently in elevation that anchorage space is available for the cantilever legs of the regular forms. Only enough lift-starter forms are required to form one or two monoliths, because after they have been used on one monolith, they can be moved to the next one. These forms can be shop-built wood forms or purchased steel forms.

Cantilever forms are then moved onto the block and are anchored at both the top and bottom of the cantilever by she-bolts placed in the previous two pours. She-bolts are located at the proper spot near the top of the form and encased in the concrete pour to provide anchorage for the cantilevers when the forms are raised. These forms are used for the remainder of the concrete in the monolith. Each dam monolith is formed with three types of cantilever forms: upstream face forms, transverse and longitudinal bulkhead forms, and downstream face forms. If the dam has variable curvature of its downstream face, the cantilever downstream face forms must have adjustable surfaces to fit this change in curvature. When 7.5-ft (2-m) -or-higher lifts are used and the dam has an inclined downstream face, the downstream forms are hinged at the middle. This allows the top section of the forms to be swung back during the placement of the first half of the concrete pour, which permits the concrete bucket to be spotted nearer the downstream face of the pour.

High-gravity dams may be so wide at the base that the lower section of each monolith must be divided into an upstream and a downstream pour. These pours must then be separated by longitudinal bulkhead forms. To provide space for cantilever forms, a 10- to 15-ft (3- to 4.5-m) differential in elevation between the concrete surfaces of the two pours must be maintained.

After the alternate dam monoliths have been poured to a height that gives clearance for the cantilevered bulkhead forms, pours can be started in the remaining monoliths. They are formed like the first group of monoliths, except that transverse bulkhead forms are not required. These remaining monoliths must be kept from 10 to 15 ft (3 to 4.5 m) below the lowest section of the first group of monoliths so that cantilever-form clearance can be maintained. If the alternate block system of pouring is not practical, monoliths can be poured in a stepped fashion to provide cantilever anchorage clearance. The difference in block elevations and the use of cantilevered steel-panel forms are very well illustrated by Fig. 7.2 of Glen Canyon Dam. Concrete forms, form hardware, and dam-forming techniques are amply illustrated in form hardware catalogs.[7.1] Figure 7.3 shows the formwork on the main dam on the Itaipu hydroelectric project on the Paraná River on the Brazil-Paraguay border. In this photograph, forming for the power intake structures is beginning.

When a pour is completed, small hydraulic cranes are used to support and raise the forms while they are being disconnected from their position and raised and connected to the next set of she-bolts. These small cranes are moved from pour to pour by the concrete placing equipment. Forms must be designed so that form stripping, raising, cleaning, and alignment can be done with a minimum crew. The forms must be strong enough to stand the stress of many reuses, to withstand the shock of occasional contact with the loaded concrete bucket, and to support walkways. The labor required to anchor and

Figure 7.2   Concrete forming at Glen Canyon Dam, Colorado. (Washington Iron Works.)

align cantilevered steel-panel forms has been reduced by using forms with quick and efficient anchoring devices, such as the coffin-handle anchorages used on the forms at Libby Dam in Montana,[7.2] or by using forms that contain self-raising jacks, such as those used at Dworshak Dam in Idaho.[7.3] The extra cost of purchasing self-raising forms is justifiable when they are used on a high dam which permits many form reuses.

The number of steel dam forms required to keep concrete forming ahead of concrete placement can be readily computed if concrete placement is started in the lowest dam block and pours are made across the total width of the dam as the dam is raised in elevation. Face forms are the controlling factor in form purchase since a greater area of face forms than of bulkhead forms is usually required. The length of face forms necessary to maintain a constant pouring schedule and to allow placement in each block increases as the dam is raised in elevation. Often the pouring schedule can be arranged so that end abutment blocks are poured after the center blocks are completed. Then the upstream and downstream face forms from the center blocks can be used for forming these blocks. This scheduling problem may require a study comparing the savings in form purchase to the additional cost of prolonging the schedule before the best solution can be determined. After the quantity of face forms has been determined, sufficient bulkhead forms should be purchased so that bulkhead forming can match dam-face forming.

A greater number of bulkhead forms are required per dam monolith near the dam's foundation than are required per monolith when the dam approaches its final height. As the dam approaches its final height, a larger number of monoliths can be poured at one time, a fact which must be taken into account when bulkhead form requirements are reviewed. The number of monoliths that can be poured at one time depends on the lineal feet of face forms purchased. To find the maximum length of bulkhead forms required, the amount needed at each dam elevation to balance the length of face forms purchased can be determined, and a curve can be drawn, with dam elevation as one ordinate and lineal feet of bulkhead forms as the other ordinate. The maximum length of bulkhead forms required can be taken from this chart.

If sectional concrete placement is used for pouring the dam, it is necessary to examine a pouring diagram to determine the form requirements. Pour diagrams are discussed in the following section.

When the dam concrete has been placed to the elevation of the dam galleries, gallery forms must be erected on the pour. Gallery forms should be collapsible so that they can be readily stripped and reused. When forms must be hand-tailored to fit surfaces, such as vertical form surfaces with projecting embedded materials spaced at irregular intervals, such forms must be built in place.

Figure 7.3 Formwork on the bi-national Itaipu hydroelectric project, Brazil and Paraguay. (International Engineering Co., Inc.)

The extent to which the other types of forms are used varies with each project. If concrete pouring is scheduled in a stepped fashion, special steel-shaped forms used on one monolith can be stripped and used on the next monolith and reused as required along the length of the dam. Models that can be disassembled by pours have been used to solve complicated forming problems encountered in miscellaneous structures.

When the forms and concrete quantities are to be computed for large, regular-shaped structures, a considerable amount of takeoff time can be saved by a computer takeoff. A computer takeoff will save in time and manpower when the forms and concrete yardages must be computed for a thin, double-curvature arch dam. The computer can be programmed to tabulate the number of forms and amount of concrete for every pour in the shape of a pour diagram, with each pour in its proper location and elevation on the dam cross section. The programming of a computer to take off forms for interior openings such as galleries, etc., is so complex that this work is done by hand. Using a computer to take off form quantities for irregular or special structures is seldom justifiable, since more time may be spent programming than is saved on the takeoff.

After the form takeoff has been completed, it should be reviewed to determine if forms have been used to maximum advantage. Form manufacturing companies will review proposed projects, give a quotation on the forms, and make recommendations on the number and kinds of forms required. This information can be used to check the contractor's takeoff.

WORK SCHEDULING

Before the components of a concrete construction system can be selected, it is necessary to prepare a concreting schedule so that the scheduled hourly concrete-placement rates will be available. Then the maximum placement capacity of concrete in cubic yards per hour can be established, which will control the capacity of many of the system components and can be used to establish the tons of aggregate required per hour, which in turn controls the capacity of the aggregate plant and the aggregate-handling facilities.

The construction schedule for a concrete dam must allow time for work access, for the mobilization of plant and equipment, for the construction of diversion facilities, and for the excavation of the dam foundation. Initiating concrete placement differs with different diversion methods and sequences of performing the dam excavation. These subjects are discussed in earlier chapters.

Scheduling the dam foundation excavation, the construction of the diversion facilities, and the time required to mobilize and erect the concrete construction equipment will establish the date that concrete pouring can start. The date that concrete pouring must be completed is established by allowing time at the end of the schedule for the completion of any work that must be done after all concrete is placed. After these control dates have been determined, the best method of scheduling concrete placement is to use a pour diagram. A typical concrete-pouring diagram consists of an elevation of the dam showing each pour, the concrete yardages in each pour, and the date that the pour can be placed. Instead of showing dates of pour placement, the pours made in every month can be colored. Figure 19.7 shows a typical concrete-pouring diagram.

Pour scheduling must allow for a slow start of concrete placement, since, at the start of each monolith, time must be spent on rock cleanup and considerable time is needed to construct built-in-place forms. Alternate blocks are poured until they are approximately 15 ft (4.5 m) above the intermediate blocks. This is discussed under form takeoff in the previous section.

Pouring will be slow and limited in volume until enough monoliths have been started to provide sufficient pouring area. With a five-day work week, seldom can more than one lift per week be placed on the same monolith. Time between lifts is required for concrete cleanup, for raising and setting forms, for installing embedded material, and for allowing the concrete to develop enough strength to support the raised form.

After sufficient monoliths are started, the number of pours that can be made in one day is controlled by the maximum pouring rate that can be achieved and not by the one-lift-a-week criterion. The rate that concrete can be raised in each monolith is delayed when gates, penstocks, gallery forms, etc., must be installed.

As the concrete is raised in the dam, the amount of concrete in each pour will decrease until the criterion of one lift a week again controls the rate of concrete placement. This will continue until the top pours are reached.

Topping out of the dam is slow, since special forms must be installed for the dam's parapet and sidewalks and the quantities of concrete finish increase. After each high alternate block has been topped out, the low alternate blocks must be raised and topped out.

Completing the pouring diagram establishes the desired maximum pouring rate to use when computing the required number of concrete-placing units. The production capacity of the different types of placement equipment units is covered in Chap. 6, where it is explained that the mixing plant's capacity can be computed by multiplying the bucket capacity of the placing equipment by 1.5. The number of tons of aggregate that must be produced in one day is computed by multiplying the maximum daily production of mixed concrete by 1.85. These production rates are then used in selecting the equipment and plant units required to complete the concrete construction system.

AGGREGATE PRODUCTION

Aggregate-production facilities should be selected and arranged to produce aggregates to specification requirements at a production rate that will equal the mixing plant's demand. Depending on the specifications and the suitability of the aggregate source, concrete aggregates may be *natural aggregates* produced from gravel deposits or *manufactured aggregates* produced by crushing and processing quarried or excavated rock. Specification requirements for aggregates differ because contracting agencies and owner's engineers differ among themselves in their concrete technology.

One of the major differences in concrete technology of importance to a contractor is whether the specifications permit the use of natural aggregates that contain reactive elements. If concrete is made from reactive aggregates, the alkalies released by hydration of the cement will have a chemical interaction with siliceous-reactive aggregates to form alkali silica gel. Over a time period, the concrete may be broken by the osmotic swelling of this gel. When concrete is made from reactive aggregates, formation of this gel can be prevented by using low-alkali cement or by including in the mix certain pozzolanic materials that neutralize the alkali produced during the cement's hydration. This type of pozzolan can be produced from certain volcanic materials, from pumice, and from Monterey shale. Fly ash is also a suitable type of pozzolanic material to use for this purpose. Some concrete technicians will permit the use of reactive aggregates when this type of pozzolanic material is added to the mix. Under this type of specification, when suitable local gravel deposits are located adjacent to the projects, they can be used as an aggregate source. Such gravel deposits are the most economical source of aggregates, because it costs less to excavate gravel and install and operate a natural-aggregate processing plant than it costs to operate a quarry and install and operate a plant for manufacturing aggregates from quarried rock. Aggregates produced from stockpiled rock from the dam's excavation are less costly than aggregates produced from quarried rock, since the cost of stockpiled and reclaimed excavated rock is less than the cost of quarrying rock. But rock from the dam foundation is seldom suitable for aggregates, so this must be classed as a special application.

Other specifications prohibit the use of reactive aggregates in the concrete mix. Under this restriction, when any local gravel deposits contain reactive materials, they cannot be used as an aggregate source. Instead, aggregates must be manufactured from approved rock quarries containing nonreactive rock. At many damsites, these approved quarries have been located a long distance from the dam or on top of steep, inaccessible hills. The use of manufactured aggregates has often greatly increased the cost of concrete construction.

Plant Layout

A comprehensive knowledge of aggregate-processing equipment, its application, and its production rates is a prerequisite for the layout and design of an aggregate plant. Because aggregate production is a specialized field, the prime contractor often subcontracts this production to aggregate contractors, thus allowing him to concentrate his efforts on the major work items. If an aggregate plant is required and the estimator lacks experience in aggregate-plant design, either a consultant should be retained or an aggregate equipment distributor should be contacted to design the plant, advise on its capital and operating cost, and establish operating procedures. Engineers who are interested in aggregate-plant design should review Chap. 2 and other references.[7.4,7.5,7.6,7.7,7.8]

Equipment manufacturers' literature contains tabulations on every phase of aggregate production. These include the horsepower required, the tons of aggregate crushed per hour, and the crushed-products screen analysis for different sizes and types of crushers; the size, horsepower, and capacity of conveyor belts, vibrating screens, log washers, scrubbers, heavy-media equipment, rod mills, classifiers, dehydrators, conical separators, etc.; the tons of each size of aggregates required for a typical yard of mass concrete; and the quantities of aggregates that can be stored in various-shaped stockpiles. Manufacturers will lay out and quote on stationary aggregate plants, on portable aggregate plants that have all major equipment mounted on rubber-tired dollies, or on any equipment component of the plant. The use of portable plants for aggregate production is rapidly expanding because of the excellent resale values based on their mobility.

The specification restrictions on the concrete aggregates should be reviewed prior to preparing an aggregate-plant layout. Particular emphasis should be placed on determining whether the specifications call for the performance of unusual items of work, so that these requirements can be incorporated into the plant's design. For example, some specifications have required that minus 200-mesh rock dust be produced for use in the concrete mix. Requirements such as this deserve special attention, since this rock dust must be ground as fine as cement, and large quantities of this material cannot be produced in a regular aggregate manufacturing plant but only in plants that have grinding mills similar to those used for cement manufacturing.

The first step in aggregate-plant design is to determine the number of tons per hour of sand and of each size of aggregate that must be produced to meet schedule requirements. The total amount of sand and aggregates used in one cubic yard of concrete varies between 1.65 and 1.95 tons, depending on the specific gravity of the aggregate and the maximum size of aggregate used in the mix. When the mass concrete contains 6-in aggregate, the average consumption of sand and aggregates per cubic yard of concrete will be approximately 1.85 tons. If the aggregate plant is operated the same number of daily shifts as the mixing plant, its required production in tons per hour is 1.85 times the cubic yardage of concrete scheduled to be produced per hour by the mixing plant. If the aggregate plant is to be operated a fewer number of shifts than the mixing plant, its production should be increased proportionately. This total aggregate production in tons per hour should be divided into the tonnages required for sand and each size of aggregate by applying mix proportions to tonnage requirements.

After the aggregate tonnage requirements are determined, an aggregate-plant flow chart can be prepared showing the tons of plant feed, the relative location of each piece of equipment, the tonnages that will be processed by each piece of equipment and transported by each belt, and the tonnages of each size of aggregate produced. If the flow chart indicates that the production in the individual aggregate sizes differs from the mix-plant requirements, equipment must be added to crush and process the extra production in one size to make up the shortage in production of a smaller size. A flow chart for a manufactured aggregate plant is shown in Fig. 7.4.

The aggregate plant should be located strategically between the aggregate pit or quarry and the mix plant, so that hauling cost will be held to a minimum. Unique conditions in the location of aggregate source and plant have resulted in equally unique layouts. At Detroit Dam, Oregon, the source of aggregates was a quarry located high above the left abutment of the dam. Since there were no level areas between the quarry and the mix plant, shelves were excavated in the hillside, and the aggregate plant and the timber-crib bins used for aggregate storage were located on these shelves. The only alternate location would have been on level ground across the river. This alternate location would have resulted in a long haul from the quarry to the aggregate plant, with a return haul from the aggregate plant to the mix plant almost as long.

At Dworshak Dam, Idaho, the quarry was situated at a location that allowed the primary and the secondary crushers to be placed in an underground chamber that was excavated directly below the quarry.[7.9] Quarried rock was shoved from the quarry floor into a 20-ft-diameter vertical shaft. An apron feeder located at the shaft's bottom was used to feed the primary crusher. The discharge from the primary crusher was then crushed to 6 in minus in two secondary crushers. This crushed product was small enough to be transported by conveyor belts through a tunnel to the left abutment of the dam where the remainder of the aggregate plant was located. The glory-hole shaft between the quarry and the primary crusher eliminated the need for a truck load-and-haul operation from quarry to crusher, with a resulting savings in hauling cost.

At locations that have unrestricted areas, the aggregate-plant layout may encompass several acres. Separate stockpiles are often used for sand and each size of aggregate. When large stockpiles are desired, they can be readily obtained by using a truss-supported stockpiling conveyor. The truss for this conveyor pivots about the charging end, while the other end of the truss is a frame supported on wheels that travel on curved rails. The conveyor travels and pivots on these wheels, forming a curved stockpile.

To give some idea about the cost of providing and operating these plants, examples of both natural-aggregate and manufactured-aggregate plants are described below. However, these descriptions are not accurate enough to be used as a basis for aggregate-plant design.

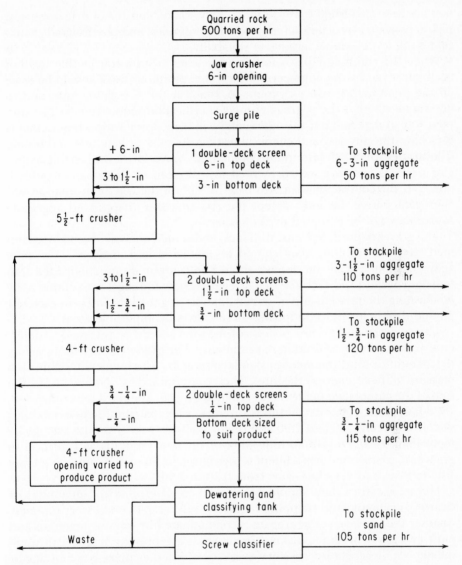

Figure 7.4 Flow diagram of a manufactured-aggregate processing plant.

Natural-Aggregate Plants

A natural-aggregate plant must be capable of producing aggregates that will meet specification requirements either from an approved gravel source or from a source that has had enough development work done to justify its use as the basis of a bid. If the specifications designated *approved aggregate deposits,* the deposits have been explored by the owner's engineers and this information has been furnished to the bidders. Exploration consists of drill holes, test pits, exploratory trenches, screen analysis of samples, and physical and chemical testing to determine aggregate suitability.

When the contractor plans to use aggregates from a source that has not been approved by the engineers, sufficient exploratory work should be done on the deposit to prove its extent, to establish the size gradations, and to determine whether the aggregates meet specification requirements. The simplest way to determine if the aggregates will meet specification requirements is to have the physical and chemical testing done by private material-testing laboratories. Detailed descriptions of methods of exploring and testing aggregate deposits are very well presented in a Bureau of Reclamation publication.[7.7] If the contractor wishes to base his bid on the use of an unapproved aggregate source, he must accept the risk that the unapproved aggregate source may not be accepted by the owner.

The gravel deposit, test pits, drill logs, and screen analysis should be examined, and the resulting data should be tabulated and analyzed. The total aggregate tonnage in the deposit should be computed and subdivided into tonnages of sand and each size of aggregate. These tonnages are computed by establishing the relationship of screen analysis to gravel thickness in each test pit and drill hole, and multiplying these results by applicable areas.

The *gradation* and *fineness modulus* of the sand should be reviewed to determine if they meet specification requirements. The gradation is determined by the amount of sand retained on standard screens. The fineness modulus is a measure of the fineness of the sand. This is determined by dividing by 100 the sum of the cumulative percentages retained on U.S. Standard sieves nos. 4, 8, 16, 30, 50, and 100. Specifications often require that the fineness modulus shall be between 2.30 and 2.90 and that it shall not vary from the average by more than 0.15. To give an indication of fineness modulus with respect to grain size, coarse sand has a fineness modulus of 2.50 to 3.50, fine sand from 1.00 to 2.50, and very fine sand from 0.50 to 1.50.

The exploratory data should be studied to determine the quantities of clearing, grubbing, and stripping required to develop the necessary tonnage; whether the top layer of aggregate is contaminated by organic materials and must be wasted; whether unsuitable materials are present in lenses through the deposit, necessitating selective excavation; whether there is a shortage or surplus of sand or any size of aggregate; and whether there is oversized material in the deposit that must either be scalped out at the gravel pit or crushed at the processing plant.

It is necessary to provide loading and hauling equipment to excavate material in the aggregate deposit and transport it to the plant. The equipment used to excavate the dam foundation may be available for aggregate-pit excavation, provided the foundation excavation is completed before aggregate production is to start. A dragline will be required if the gravel lies below the water surface.

If the gravel deposit contains aggregates suitably sized to meet the concrete mix demands and sand of the proper gradation and with the right fineness modulus, then the aggregate plant only needs to be a screening plant. A layout of this type of plant is illustrated by Fig. 7.5. This plant was laid out in a straight line with single-deck screens, which were located on towers over stockpiles and provided with conveyors to transport the material that passes through each screen to the next screen. The flow of materials can be readily followed in this type of plant. The other type of plant contains a central screening tower supporting multideck screens. Conveyor belts fan out from this tower and transport the sized material to stockpiles arranged around the tower. The free fall of the large aggregate into stockpiles is restricted by the use of rock ladders, which limit the vertical drop of the aggregates to between 2 and 3 ft (0.5 and 1 m), thereby eliminating size segregation and aggregate breakage.

The capacity of the plant must be sufficient to meet the mixing plant's demand. Also, the surge pile should have sufficient live-storage capacity to permit the aggregate plant to operate continuously and to be independent of the receipt of plant feed from the haul trucks. Wide conveyor belts are required to handle 6-in aggregate; therefore, the capacity of these wide belts often establishes plant and screening capacity. The plant may be sized so that it must operate as many shifts as the mix plant or so that one or two shifts a day furnish enough aggregate for three shifts a day of mix-plant production.

If a large tonnage of aggregate is to be handled, reclaiming from stockpiles is done by conveyor belts located in reclaim tunnels underneath the stockpiles. Batch feeding of the sized aggregates onto the reclaim belts is controlled by feeders located under gates placed in the roof of the reclaim tunnel. These belts can discharge either directly into aggregate-haulage trucks or into bins equipped for automatic loading of trucks. When the aggregate tonnage is small, the cost of installing a belt reclaiming system seldom can be justified; consequently, front-end loaders are used for aggregate reclaiming.

If the gravel deposit does not contain well-graded suitable materials, if there is a deficiency of some aggregate sizes, or if material is lacking in other qualities, additional equipment must be added to remedy these situations. If roots, sticks, and other similar materials are present in the gravel, a log washer may be required. When the aggregates are coated with objectionable material, it may be necessary to add a scrubber to the plant to remove this coating.

When the gravel deposit does not contain sufficient large-sized aggregates, one method of securing more quantities of this size is to crush the oversized material from the pit. If there is not a sufficient quantity of oversized material to make up this deficiency, it may be possible to make up this size shortage from other aggregate sources. If this is impractical, excess gravel must be processed through the plant until sufficient larger sizes are produced. This increased plant feed will result in an overproduction of the other aggregate sizes, which must be wasted.

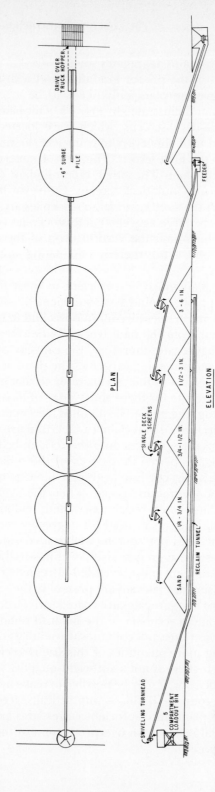

Figure 7.5 Straight-line arrangement of an aggregate-screening plant.

When there is a shortage of smaller-sized aggregates, extra production of the smaller sizes can be achieved by crushing larger-sized aggregates. This requires the addition of a cone crusher and of recirculating conveyors which feed the crusher. When specifications do not permit crushing of aggregates and if shortages occur in any size of aggregate or sand, enough additional gravel must be processed to make up this shortage, and the corresponding excess of production in other sizes of aggregates must be wasted.

The deposit may contain small-sized particles of unsuitable, lightweight materials with specific gravities of less than 2.50. In this case the aggregate plant must be equipped with *sink-float* processing equipment to remove this unsuitable material. Sink-float equipment is drum-shaped and contains a heavy-media mixture of water, magnetite, and ferrosilicon. The light, unsuitable aggregates float on the media and are discharged over the lip of the drum and wasted. The heavier, suitable aggregates sink and can be removed and used. Other methods can be used for removing lightweight aggregates, such as other types of heavy-media separation, hydraulic jigging, and elastic fractionation. (Separation of the lightweight aggregates by *elastic fractionation* is accomplished by a process that makes use of the principle that denser aggregates will rebound a greater distance when they are dropped on a steel plate.)

If the gradation and the fineness modulus of the sand do not meet specification requirements, this can often be corrected by splitting the sand into several sizes and reblending the sizes to the desired gradation. This necessitates the addition to the aggregate plant of sand classifiers, sand pumps, settling ponds, additional belts, additional stockpiles, and reblending facilities. Depending on the complexity of the required splitting, sand can be separated into different sizes by the use of screw classifiers, rake classifiers, hydraulic classifiers, or hydraulic sizers. Water settling ponds are required, since specifications seldom permit the discharge of the processing water into drainage systems.

If gradation and fineness modulus requirements cannot be met by separation and reblending of the sand, crushing or grinding of the sand may be required. This can be done by adding a rod mill or a sand crusher.

When there is a shortage of sand, when there is a lack of coarse sand, or when the sand that is present in the deposit is unsuitable and must be wasted, sand must be manufactured from any excess in production of any size of aggregate. This necessitates the addition of sand-manufacturing equipment to the processing plant. This sand-manufacturing equipment consists of cone or Gyrasphere crushers and recirculating belts to furnish the plant feed, rod mills or sand crushers to produce the sand, sand classifiers, sand pumps, and settling ponds. Since these facilities are similar to those required for the manufacture of sand from quarried rock, more detailed descriptions of their use will be given under manufactured-aggregate production.

Manufactured-Aggregate Plants

Concrete aggregates are manufactured either from quarried rock or, in special cases, from rock excavated from the dam foundation or other required locations. Quarried rock is used when the specifications make its use mandatory or when the haul from the closest suitable gravel deposit to the jobsite is so long or difficult that it makes the use of natural aggregates uneconomical.

If aggregates are to be manufactured from required rock excavation, a large stockpile is a necessity, since aggregate usage occurs after the major portion of the excavation has been performed. If the haul to the rock stockpile is greater than to a disposal area, the extra haul cost should be charged to aggregate production. Selective loading of the excavated rock may be required, since the stockpiled rock should be of sizes that can be handled by the primary crusher; otherwise secondary drilling and blasting will be required.

If aggregates are to be manufactured from quarried rock, the quarry site should be inspected so that the quarry development can be planned and suitable equipment can be selected. The development of a rock quarry and the production of broken rock is described in Chap. 9.

The shovel used in the quarry should be sized to fit the opening of the primary crusher, so that the shovel cannot pick up larger rocks than the crusher can handle. Equipment manufacturers list the size of shovel that should be used with each size of crusher. To obtain efficient loading and haulage operation, truck size should also be matched to shovel size, using the criterion that only three or four shovel passes should be required to load a truck to capacity. Since the development of the large, front-end loaders, economics often dictate the use of loaders instead of shovels in many rock quarries.

An aggregate-manufacturing plant is composed of two plants: a rock plant and a sand plant. Quarried rock is the feed for the rock plant; the sand plant feed is minus ¾-in material that is produced by the rock plant. If an insufficient amount of minus ¾-in rock is produced in the primary and secondary crushing of the quarried rock, additional sand-plant feed can be secured by using cone or Gyrasphere crushers to crush an excess production of any size of aggregate. The design of both plants is established by preparing flow charts. Figure 7.4 shows the flow chart for the aggregate-manufacturing plant used in the example estimate in Chap. 19.

A typical rock plant starts with a drive-over truck dumping bin equipped with grizzly bars to reject oversized material. This oversized material must be drilled and shot or broken with a headache ball before it can be placed back into the system. From this bin, fines fall through grizzly bars directly onto a conveyor belt, and the coarse material is fed into a primary crusher, or in some instances, both the fine and coarse material are fed to the primary crusher. The primary crusher can be either a jaw or a gyratory type. The feed to the primary crusher is controlled by a feeder equipped with a magnet to remove shovel-bucket teeth or other pieces of iron. The larger the crusher, the larger the rock that the crusher can handle and the less the amount of secondary drilling and shooting required in the quarry. At Detroit Dam, Oregon, where aggregate was manufactured from quarried rock, the primary crusher was a 62- by 84-in Blake-type jaw crusher. This crusher had a large enough opening to permit the use of coyote blasting in the diorite-rock quarry and to minimize the amount of secondary drilling and shooting, thus reducing the quarrying cost and, correspondingly, the aggregate cost.

If the aggregate plant is designed for low production, the discharge opening of the primary crusher is set to produce minus 6-in material, and secondary crushing is not installed. When high-production aggregate plants are installed, the primary crusher is set with a larger discharge opening, i.e., from 12 to 15 in, and one or more secondary crushers are installed to reduce the rock to minus 6 in. These secondary crushers are usually of the gyratory type.

The crushed rock is conveyed from the crushing unit to a raw-material surge pile of sufficient size to permit the remainder of the aggregate plant to operate independently of the rate of rock production in the quarry. From the surge pile, the rock is conveyed to vibratory screens, where the aggregate is separated into sizes. The screened products are then placed in separate stockpiles.

The screening section of the plant can be laid out in a straight-line pattern, or with a central screening tower. The difference between these two is explained in the former section on natural-aggregate processing plants. Similar to a natural-aggregate plant, rock ladders are used beneath the discharge end of each large-aggregate stockpiling conveyor to limit the free fall of the rock into the stockpile, so that size segregation and aggregate breakage will not occur.

The aggregate stockpiles are often equipped with recirculating belts which reclaim any excess production of any size of aggregate and feed it to tertiary crushers to produce a finer-sized aggregate or feed for the sand plant. If excess production is not reclaimed from the stockpiles, it may be split at the screen discharge or distributed between the stockpile and the tertiary crushers from a storage bin located beneath the screens. A typical installation of a tertiary Gyrasphere crusher is shown in Fig. 7.6. Figure 7.7 shows a portable aggregate plant equipped with screening and crushing equipment.

Two types of sand-manufacturing plants are in common use, a rod-mill plant and a sand-crushing plant. If a rod mill is used, its feed is minus ¼-in material produced by cone or Gyrasphere crushers. A rod mill can produce sand to meet specification requirements by varying the maximum size of the feed to the rod mill, the speed of rotation of the rod mill, the length of time that the material is in the rod mill, and the rod charge in the mill. Not only does the rod mill control the sand gradation, but it controls the rate of sand production. The history of many rod-mill-type sand-manufacturing plants is that insufficient rod-mill capacity was initially installed; consequently, in order to make sufficient sand, additional rod mills were required.

A recent development in sand manufacture is to use a sand crusher such as

Figure 7.6 View of an aggregate plant showing screening towers and a Gyrasphere crusher. (Smith Engineering Works.)

the Gyradisc crusher (Nordberg) in a closed crushing and screening circuit. These crushers have fine adjustable discharge openings; however, fine sand particles are produced by the layers of particles acting upon each other to accomplish reduction. Feed for the plant is the excess in production of 3/4- by 1/4-in material and a portion of the sand that passes a no. 4 screen and is retained on a no. 12 or 14 screen (see Fig. 7.5). By varying the discharge

Figure 7.7 Portable aggregate plant equipped with a jaw crusher. (Pioneer Division, Porter, Inc.)

opening of the crusher, the size of the top sand screen—no. 4 or 6, and the size of the bottom screen—no. 12 or 14, the required gradation can be produced from many types of rock. The feed should have less than 4 percent total moisture and less than 7 percent surface moisture. This sand-manufacturing process has produced concrete sand successfully at many concrete dams, including Green Peter, Oregon, and Dworshak, Idaho. When its use is applicable, the cost of sand production will be less than that of utilizing a rod mill.

After sand is produced, it must be dewatered and sized. Often dewatering and sizing can be done in single-stage screw classifiers. In some instances, it is necessary to separate the sand into different sizes in a series of screw classifiers, or in hydraulic sizers, or in hydraulic classifiers. These sizes of sand are separately stockpiled; when reclaiming is done, they are blended on the reclaim belt to produce sand of the required gradation.

The wastewater from sand plants must be clarified by using thickeners or by retaining the water in large settling ponds. Multiple sand stockpiles may be necessary when specifications require that the sand remain in free-draining stockpiles for a specified time period before usage.

The method used for reclaiming the aggregate and sand from the stockpiles and for truck-loading this material depends on the tonnages to be handled. Conveyor-belt reclaiming is done when large tonnages are involved, and front-end loader reclaiming when only small tonnages are handled.

The foregoing descriptions of aggregate and sand manufacturing plants are very general. If the plant feed has special characteristics, other types of equipment such as roll crushers, reversible impactors, hammer mills, etc., may be required.

AGGREGATE HAUL

The processed aggregates must be transported from the aggregate plant to the mixing plant and charged into the mixing plant's batching bins. When the aggregate plant is located adjacent to the mixing plant, the aggregates are transferred between the two plants on a conveyor belt. When there is a considerable distance between the two plants, the most economical method of making this transfer is with bottom-dump trucks, since their ratio of payload to truck weight is excellent. At Bullards Bar Dam on the Yuba River, the concrete aggregates for a plant with six 4-yd^3 mixers were hauled with one primary mover pulling two 60-ton bottom-dump trailers.[7.10]

When there is a long distance between the aggregate plant and the mix plant, it is impractical to attempt to schedule the proper arrival of aggregates sized to comply with the mixing plant's demands. Instead, aggregate storage and reclaiming facilities are installed adjacent to the mixing plant to provide a surge storage for recharging the mixing plant's batching bins. These aggregate storage and reclaiming facilities usually consist of a drive-over truck receiving bin and conveyor belts to transport aggregates to the stockpiles. The elevated section of the conveyor belt that is located over the stockpiles must be equipped with a traveling tripper or a plow so that each size of aggregate can be discharged into the proper stockpile, with rock ladders limiting the free fall of the large aggregates into the stockpiles. As in aggregate plants, reclaiming is done by conveyor belts if large tonnages must be handled and by front-end loaders if the tonnages handled are minor. The reclaimed aggregates pass through a rescreening plant if one is located near the mixing plant; otherwise, they feed the mixing plant's charging conveyor.

RESCREENING AND MIX-PLANT CHARGING

Some specifications require that the coarse aggregates be rescreened to eliminate any size degradation before they are charged into the mixing-plant bins; if this is specified, a complete screening plant is required, equipped with spray nozzles to reduce dust, with dewatering equipment, and with settling ponds to clarify the water used in the screening process. This rescreening plant can be located on top of the mixing plant or on the ground adjacent to the mixing-plant charging conveyor. The latter location usually results in the least capital expenditure.

Charging of the mixing plant's aggregate storage bins is accomplished by repetitive batch loading of the sand and each size of aggregate on an inclined conveyor belt. The belt extends from ground elevation to the top of the batch plant or rescreening plant. The required capacity and method of operation of this belt are described when mixing plants are discussed.

CEMENT AND POZZOLAN HANDLING

Since cement is a required ingredient in all concrete, it is necessary to provide facilities to handle its receipt, storage, and transfer from the storage silos to the mixing plant's batching silos. Similar facilities are a necessity for pozzolanic materials when they are used in the mix. Pozzolanic materials are often a required ingredient in dam concrete, since they can replace part of the cement, act as a cement dispersing agent, and improve the workability of the concrete. When the concrete workability is improved, the water-cement ratio can be decreased, which increases the strength of the concrete. Also, some types of pozzolanic materials will combine with the free alkali in the cement to neutralize its reaction with reactive aggregates.

If interground cement and pozzolan can be purchased and this mixture will satisfy the specifications, its use is economical, since only one set of handling facilities will be required.

If the cement and pozzolan are delivered by truck to the mixing-plant area, the installations necessary to store and handle them will be of a minimum. Modern cement and pozzolan delivery trucks contain pressure equipment for truck unloading. The pressure system forces the cement out of the truck, through pipes, and into the storage silos. With this type of delivery, the facilities required at the site are storage silos; screw conveyors beneath the silos, which transfer the material from the silos into bucket elevators; and bucket elevators, which discharge the cement into the top of the mixing plant batching silos. A pressure system may also be used for transfer from storage to the batching silos.

Separate silos and handling facilities are required for cement and pozzolan unless they are interground and delivered in a mixed condition. Sufficient storage silo capacity should be installed so that the mixing plant will operate independently of the delivery of cement and pozzolan. The minimum storage to be installed varies with the distance from the project to the location of the cement source, the job location, the ease of delivery, and weather conditions which may affect delivery.

If the dam is at a location where the cement and pozzolan delivery trucks discharge the load by gravity through a bottom gate and are not equipped with a pressure discharge system, then a receiving hopper, screw conveyor, and bucket elevators are required to receive and place the cement and the pozzolan into storage silos.

If the cement and pozzolanic materials are delivered by rail, the contractor must supply and operate pumps to transfer these materials from railroad cars into trucks for haulage to the mixing-plant site.

The mixing plant may be in such an inaccessible location that cement-delivery equipment will not have access to the mixing-plant site. In this event, the cement is often pumped from the point of receipt, through a pipeline, to silos at the mixing plant. At Detroit Dam, Oregon, cement was received on railroad cars at a siding located on the right abutment, directly across the canyon from the mixing plant, which was located on the left abutment. The railroad cars were unloaded with pumps that pumped the cement through a pipeline into storage silos located on the left abutment, adjacent to the mixing plant. The pipeline was suspended from a cable stretched across the river canyon.

AGGREGATE AND CONCRETE COOLING

If cold concrete is placed in a dam and if, when necessary, the temperature of the concrete is controlled after placement, the heat of hydration of the cement will not raise the temperature of the concrete above its final temperature, which is approximately the same as the average temperature of the locality. If the temperature of the concrete never rises above its final temperature, it will not drop in temperature and will not be subject to shrinkage; if it does not shrink, grouting becomes a minor consideration. Also, the concrete will reach its final temperature in a relatively short time, as compared to the years required when its temperature is not controlled. If the temperature of mass concrete is controlled, concrete can be placed in large blocks and the contraction joints need not be grouted, or they can be grouted shortly after concrete placement of each grouting lift has been completed. Whether contraction-joint grouting is necessary depends on the type of dam and the requirements specified by the contracting agency.

The placement temperature of the concrete and the extent of temperature control that must be maintained after the concrete has been placed vary with the amount of heat generated by the hydration of the cement and the rate at which heat is dissipated from the pour. The heat generated in the pour varies with the quantity and type of cement used in the mix. Heat dissipation from the pour varies with the pour height, pour dimensions, time interval between pours, type of aggregate, type of forms, length of time the forms are in contact with the pour, concrete curing method, and temperature of the locality. Concrete should not be placed at such cold temperatures that freezing will occur. When concrete pours are massive (preventing rapid heat dissipation) or when placed in extremely hot weather, the temperature rise of the poured concrete cannot be prevented solely by the use of cold concrete. Under these conditions, additional concrete cooling must be done after the concrete is placed in the form. This additional cooling is achieved by pumping chilled water through cooling coils located on the top of the rock foundation and on top of each pour. The average length of time that this chilled water must be circulated through each cooling coil to control the temperature rise varies, but often is approximately 12 days. Near the dam foundation where heat generation is greater than heat dissipation, pour heights are reduced so that the cooling coils can be installed at closer intervals. When the dam concrete reaches an elevation that contains smaller pours, more heat will be dissipated, so that cooling coils may no longer be required and the use of cold concrete may control the temperature rise in the concrete.

At some locations the river water may be cold enough when circulated through the cooling coils to cool the concrete; otherwise refrigerated water is required, as is the refrigeration equipment needed to produce it. Control of the concrete temperature with cooling coils is based on the use of cooling-water velocities of 2 ft^3/s (5 gal/min) through 1-in-diameter cooling coils. The tonnage of refrigeration required to chill this water is determined by establishing the maximum amount of water that must be chilled. This can be computed using the construction schedule to establish the maximum number of pours that must be chilled during one time period.

Specification restrictions on concrete placement during hot weather differ depending on the contracting agency, the size of the concrete structure, and the temperature at the locality. Specifications may require that aggregate piles be shaded from the sun, that they be sprinkled to permit evaporative cooling, that aggregate conveyor belts be shaded, that concrete placement not be done during the summer months, that concrete placement be restricted to night-time pouring during the summer months, that concrete be placed at a specified temperature, and that riverwater or chilled water be circulated for a defined time period through cooling coils placed in the concrete. The specifications may limit the maximum and minimum time for lift raising and may designate the time for form removal. Specifications should always be reviewed to determine the restrictions on concrete placement because of temperature conditions and how these restrictions will affect the construction schedule.

The temperature of mixed concrete can be controlled using many different methods. The simplest methods are those that cool the large aggregates and utilize ice in the concrete mix. The cooling of sand, cement, and pozzolan is undertaken only when it is a necessity, since their cooling is difficult and expensive. Sand can be cooled by using the vacuum process to provide effective evaporative cooling. Sand, cement, and pozzolan can also be cooled in commercial-type coolers.

If the aggregate storage piles are shaded, temperature rise will be minimized. Evaporative cooling of aggregates to wet-bulb temperatures can be accomplished by spraying water on the aggregate storage piles. Large aggregates can be cooled by spraying chilled water on them while they are transported by a covered conveyor belt. Figure 7.8 shows an installation of this type.

The simplest and easiest method of reducing the temperature of concrete is to replace part of the free water in the mix with ice. This method provides

Figure 7.8 Cooling concrete aggregates; two enclosed 60-in belts with chilled water sprays. Dworshak Dam, Idaho. (Lewis Refrigeration Co.)

enough cooling action when the concrete temperature must be lowered only a few degrees. Since the ice that can be used in the mix is limited to approximately 80 percent of the free mixing water, its cooling action must be supplemented by other methods of cooling when large temperature reductions are required. The required temperatures can often be obtained by cooling aggregates and by using ice as a varying mix ingredient to control the final temperature. A simpler rule is that the hotter the day, the greater the amount of ice used in the mix. If ice is used in the mix, it is necessary to provide ice-making machinery, storage bins, and a method of conveying ice to the mixing plant, ice bins in the mix plant, and an ice batcher.

Large aggregates can be cooled by immersing them in chilled water. This is a batch-type process. The aggregates are placed in watertight bins, and the bins are sealed and flooded with chilled water. The aggregates are retained in these bins until their temperature is lowered to that of the chilled water. The water is then withdrawn, the bottom of the bin is opened, and the aggregates discharged onto a conveyor belt for transportation to the mixing-plant charging bins. With proper scheduling of cooling operations and mixing-bin charging, aggregate cooling can be accomplished with one bin for each size aggregate. This method of aggregate cooling was used at Detroit Dam, Oregon.

Large aggregates can be cooled in the mixing-plant batch bins by the upward circulation of cold air through the aggregates. When this cooling method is used, the mixing-plant batching bins should have capacity to allow the aggregates to be cooled for approximately 3 hours. Since each bin is charged in succession, to achieve this cooling capacity the bins must have a capacity sufficient to run the mixing plant from 4 to 5 hours. The ammonia compressors required for this cooling process are installed adjacent to the mixing plant. Ammonia piping, pressure controls, and ammonia expansion chambers connect the compressors to heat exchangers that are located on the plenum chamber erected around the mixing plant's batching bins. These heat exchangers cool the air in the plenum chamber. Fans are located in this chamber to force the cold air upward through the aggregates in the batching bins and back to the plenum chamber where it is again cooled. This type of cooling has been used on many projects, including Mossyrock Dam in Washington and Glen Canyon Dam on the Colorado River. Figure 7.9 shows the installation of heat exchangers on a mix plant at Dworshak Dam, Idaho.

Another method of cooling aggregates is to use a partial vacuum to expedite evaporative cooling. Aggregates are placed in a large, airtight bin connected to a chamber containing a steam jet. The action of this jet draws air and water vapor from the bins, producing a partial vacuum and causing the moisture on the aggregate to vaporize, thus cooling it. Approximately 30 to 45 minutes are required to cool the aggregates in each bin. The bins are baffled inside, and these baffles serve as rock ladders, protect the walls of the

Figure 7.9 Heat exchangers mounted on the plenum chamber of the exterior of a mixing plant, Dworshak Dam, Idaho. (Lewis Refrigeration Co.)

bin, and distribute the aggregate throughout the height of the bin. Bins may be located adjacent to the mix plant or may be part of the mixing-plant structure. Before being charged into the bins, the aggregates and sand should be dampened until the minimum moisture content of the aggregates is 1.5 percent and of the sand is 3.0 percent.

When sand is cooled by the vacuum process, the sand bins are equipped with special baffles that distribute the sand so that air spaces are provided in the centers of the bins and next to the bin walls. Another method of constructing a sand-cooling bin for this cooling process is to separate the bin into lower and upper compartments. When the bin is sealed and placed under vacuum, a sand meter opens and lets the sand flow at a measured rate from the upper half of the bin into the lower half. This sand flow permits the moisture to boil off and cools the sand by evaporation.

The number of aggregate bins required for vacuum cooling can be reduced by mixing the various sizes of aggregates in the same proportions as they are used in the concrete mix before they are charged into the cooling tanks. After the mixed aggregates are cooled, they are resized by an aggregate rescreening plant located on top of the mixing plant. This premixing of the aggregates makes it possible to cool all the aggregates in three cooling bins. While one bin of premixed aggregates is being cooled, another aggregate bin can be charged, and the cooled aggregates can be drawn from the remaining bin. Two additional bins are required when the sand must be cooled. In addition to the bins, steam generators, steam jets, and conveyor belts are required. The vacuum-cooling method has been used on many projects, including Dervini

Kan Dam in Iraq, Hartwell Dam in Georgia, Savannah River Dam in Georgia, and Green Peter Dam in Oregon, as shown in Fig. 7.10.[7.11]

The selection, design, and layout of a suitable refrigeration and cooling plant is such a complex problem that it is recommended that the planning engineer contact either refrigeration-plant equipment suppliers or competent consultants to assist in laying out the cooling plant.

When a consultant or a refrigeration-plant manufacturer recommends the method and extent of cooling, the planning engineer should always check the results. The computations that are made to check refrigeration rquirements are illustrated in Chap. 19, in the discussion of the selection of the plant and equipment.

CONCRETE MIXING

The concrete-mixing plants used for dam construction differ in design because of differences in specification requirements, and they differ in capacity because of differences in project requirements. Concrete mixers must have a capacity of 4 yd^3 or greater when 6-in aggregate is used in the concrete mix. The hourly production rates of the mixing plants will vary from 60 yd^3/h, when they are equipped with one 2-yd^3 mixer, to 410 yd^3/h, when they are equipped with six 4-yd^3 mixers. The control boards become complex when many different concrete mixes are required, and supports often must be

Figure 7.10 Mixing-plant and vacuum-cooling aggregate tanks, Green Peter Dam, Oregon. (C. S. Johnson, Division of Koehring Co.)

enlarged to take care of the weights of rescreening plants placed on top of their bins or of aggregate cooling equipment hung on the side of the plant. See Chap. 2 for additional information on concrete mixing and batching plants.

It was previously explained that the total cubic-yard capacity of the concrete mixers should be 1.5 times the bucket capacity of the placement equipment in cubic yards. This means that if the total bucket capacity of the placement equipment is 16 yd^3, then six 4-yd^3 mixers should be installed. The hourly capacity of the mixing plant can be determined by computing the number of cycles per hour made by each mixer and multiplying that by the mixer capacity and number of mixers. The number of cycles per hour depends on charging time, mixing time (which is a specification requirement), and mixer discharge time.

Typical hourly production rates for mixers of different size are given in Table 7.1.

TABLE 7.1   Hourly Production Rates for Mixers

	2-yd^3 Mixer	3-yd^3 Mixer	4-yd^3 Mixer
Charging time	¼ min	¼ min	¼ min
Mixing time	1½ min	2 min	2½ min
Dumping time	¼ min	¼ min	¼ min
Mixing cycle	2 min	2½ min	3 min
Cycles/h	30	24	20
Yd^3 production/h	60	72	80

The mixing plants cannot be scheduled for constant mixing of concrete at maximum-capacity production rates, since there are times when one or more mixers have to be removed from service for repair or relining.

When six mixers are arranged in a circle in one plant, the time required to go around the circle and charge each mixer may be greater than each mixer's cycle, and this factor may control production. As an example of how hourly mix-plant production is determined, a 4-yd^3 mixer will produce 20 batches per hour when it is operated in conjunction with five mixers or less; under these conditions it has an hourly production of 80 yd^3/h. When it is operated with six mixers, the charging cycle controls the number of mixing cycles, reducing them to 17 and the hourly production to 68 yd^3/h per mixer.

Specifications for different projects may differ in mixing-plant requirements, such as whether accumulative or individual weighing and measuring equipment must be used; what kind of sampling equipment, and what amount of control equipment should be installed; whether rock ladders are necessary; and what type of bins is preferred. The number of different types of concrete mixes that are required influences the size and number of mixers.

For rapid plant operation, computerized control for a rapid change of mix ingredients is desirable.

Aggregate cooling requirements may necessitate the use of large storage bins and the installation of plenum chambers, heat exchangers, fans, etc.

Specifications often require that aggregate rescreening facilities be placed adjacent to or on the top of mixing plants. When located on top of the mixing plants, they of course increase the height of the plant, and this necessitates the use of a longer charging conveyor and increases the load on the bin supports.

When there is a small quantity of concrete to be placed, the specifications may permit the use of a *minimum mixing plant.* A typical minimum plant consists of low, open-type aggregate bins; a cement silo equipped for truck delivery of cement; a weigh batcher that successively accumulates the charge for the mixer as it travels under the aggregate and sand bins and the cement silo; a conveyor belt that receives the charge from the weigh batcher and charges the mixers; and a small mixer of either the tilting type or the rotary type. Since the aggregate bins are of a low, open type, they can be charged with aggregates by using a rubber-tired, front-end loader. When only one type of mix is required, when belt conveying of the batched material to the mixers is permissible, and when concrete must be produced in large quantities, this type of mixing plant can be secured with a tilting mixer that has a capacity up to 12 yd.^{3} However, because of restrictions concerning the method of charging mixers, because different types of concrete mixes may have to be produced almost simultaneously, and because individual weigh batching may be required, this type of plant is seldom used in concrete dam construction. When many mixes are required, when individual weigh batching is needed, and when mixer charging must be done without producing dust, a stationary plant with more complex weighing, measuring, and batching equipment is required.

Dams containing large volumes of concrete of many different mixes may require mixing plants that contain multiple mixers. The mixing plant used for the construction of Glen Canyon Dam contained six 4-yd^3 mixers. It also included a rescreening plant, ice batching facilities, and facilities for air cooling of large aggregates. Six 4-yd^3 mixers is the greatest number of mixers that can economically be located in one plant. Two mixing plants have been installed on dam projects that required high rates of concrete production. At Grand Coulee Dam, two mixing plants were installed, each containing four 4-yd^3 mixers. At Dworshak Dam, two mixing plants were used, one containing six 4-yd^3 mixers and the other four 4-yd^3 mixers.[7.9]

A schematic layout of a mixing plant that includes an aggregate charging conveyor, bucket elevators and chutes for charging the cement and pozzolan silos, aggregate rescreening facilities, batch bins, aggregate-cooling facilities, ice batching, six 4-yd^3 tilting mixers, and two wet-batch hoppers is shown in Fig. 7.11. This type of plant will meet most specification requirements, since it is equipped with self-cleaning batch bins, individual automatic measuring and

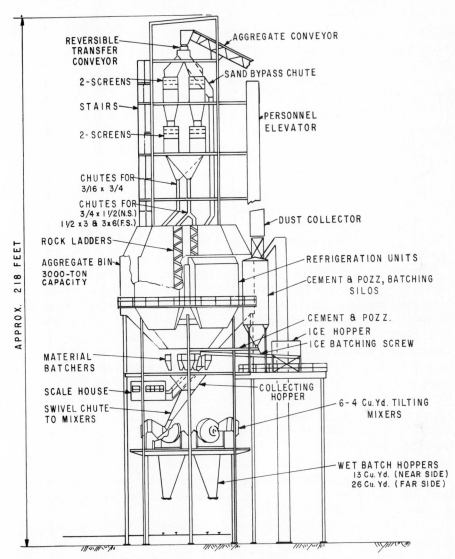

REVERSIBLE TRANSFER CONVEYOR

AGGREGATE CONVEYOR

2-SCREENS

SAND BYPASS CHUTE

STAIRS

PERSONNEL ELEVATOR

2-SCREENS

CHUTES FOR 3/16 x 3/4

CHUTES FOR 3/4 x 1 1/2(N.S.) 1 1/2 x 3 & 3 x 6(F.S.)

ROCK LADDERS

DUST COLLECTOR

AGGREGATE BIN 3000-TON CAPACITY

REFRIGERATION UNITS

CEMENT & POZZ, BATCHING SILOS

CEMENT & POZZ. ICE HOPPER

ICE BATCHING SCREW

MATERIAL BATCHERS

SCALE HOUSE

COLLECTING HOPPER

SWIVEL CHUTE TO MIXERS

6-4 Cu.Yd. TILTING MIXERS

WET BATCH HOPPERS 13 Cu.Yd. (NEAR SIDE) 26 Cu.Yd. (FAR SIDE)

APPROX. 218 FEET

Figure 7.11 Schematic diagram of a concrete-mixing plant.

weighing equipment, equipment for automatic measuring of concrete consistency, push-button control of the scales so that any one of the multiple of mixes can be weighed simultaneously for mixer charging, batch-type tilting mixers, and automatic recording of each operation. This schematic layout is of an open-type plant, but contractors place siding over the lower part of the plant to protect the operators and control equipment and to provide protected areas for concrete testing and office space.

The flow of concrete materials through mixing plants of this type is started by loading the charging conveyor with successive charges of sand and each size of aggregates, leaving empty spaces between each size so that there cannot be size intermixing. Because the charging conveyor must operate as a batching conveyor, the maximum tonnage it can handle will be approximately 70 percent of its theoretical capacity. To illustrate, six 4-yd^3 mixers will mix 410 yd^3 of concrete per hour requiring 760 tons of aggregate per hour. At 70 percent efficiency, this will increase to 1,090 tons per hour. This establishes a charging conveyor size of 42 in or greater, and it must operate at a speed of 250 ft/min.

The charging conveyor transports the sand and aggregate to the top of the mix plant where the coarse aggregate passes through a rescreening plant. The purpose of this plant is to eliminate any size degradation in the aggregate. The rescreening plant must be sized to take the production of the charging conveyor; for a plant containing six 4-yd^3 mixers, it should have a screening capacity of 1,090 tons per hour.

After passing through the screens, the aggregates are chuted into the batching bins. These bins are equipped with rock ladders to restrict breakage and prevent segregation. Five sand and aggregate bins are arranged around center silos for cement and pozzolan. Bins are sized in accordance with the percentage of sand and each size of aggregate used in the concrete mix. Cement and pozzolan are charged into their respective batching silos by bucket elevators and chutes. The silos and bins should contain enough cement, pozzolan, sand, and aggregate to supply the mixers for 2 or 3 hours. This permits the mixing plant charging conveyor to transport one size of aggregates for 15 to 30 minutes.

For the mixing plant containing six 4-yd^3 mixers, 3 hours of bin storage would require sand- and aggregate-storage bins to have 2,200 tons of storage capacity. If cold air is circulated through the mixing plant's storage bins to cool the aggregates, it often takes 3 hours to cool them to the desired temperature. This necessitates increasing the bin sizes to 4- or 5-hour capacity, which is a capacity of from 3,000 to 3,600 tons.

The required weights of aggregates, sand, cement, and pozzolan for charging the mixers are released from the bins, individually weighed, and dumped into a collecting hopper; then the required amounts of ice, water-reduction agent, and water are added. This collecting hopper has a swivel chute that allows it to progressively charge the tilting mixers arranged in a circle beneath it. At the end of the mixing cycle, the mixers discharge the mixed concrete into one of two wet-batch hoppers, and these hoppers discharge the concrete into concrete buckets or into transfer cars, depending on the type of haulage equipment used.

Mixing-plant manufacturers will quote and supply complete mixing plants that will comply with various specifications. When requested, they will quote on a complete mixing plant, including charging conveyors, cement- and pozzolan-storage silos and bucket elevators, and aggregate rescreening and cooling plants. When a request is made for a mixing-plant quotation, it should specify the number of mixers, include the specification requirements, designate whether pozzolan is to be used in the mix, list the number of additives that will be used in the mix, and specify if ice batching is required or whether cooling is to be incorporated into the plant. The estimator can then use the manufacturer's quoted price for the complete mixing plant and storage silos in his cost estimate. To this he must add the cost required for site preparation, foundation construction, distribution of utilities to the plant, construction of a bucket cleaning yard, plant erection, furnishing and installation of the housing around the mixing plant, and any required insulation.

FOUNDATION CLEANUP

All rock foundation surfaces must be cleaned prior to being covered with concrete. Cleaning consists of the removal of unsound rock, debris, mud, and running or standing water. The equipment required for this work consists of air and water jets, hand tools, and skips to contain the excavated material. Cranes may be required to handle and dump these skips. Air and water must be supplied to the cleaning equipment. See Chap. 5 for details of foundation preparation, which is quite similar to core cleanup requirements on earth-fill dams.

CONCRETE CLEANUP

The surface of all concrete pours must be cleaned, and standing water must be removed before the next concrete lift can be placed. All laitance and loose or defective concrete must be removed, and the larger aggregates partly exposed to provide good concrete bonding. This can be done from 4 to 14 hours after the concrete has been placed, using high-pressure air and water cutting jets. Air and water cutting devices with pressures over 6,000 lb/in^2 have been developed. This high-pressure cutting is so effective that it can replace the more expensive sandblasting procedure; sandblasting equipment may, however, be required by the specifications.

Cleanup crews will need hand tools, air and water jetting equipment, and sandblasting equipment. Utility distribution to each pour area is also required.

PLACEMENT OF EMBEDDED ITEMS

The reinforcing steel and all other embedded materials must be placed in a pour before concrete placement can start. The other embedded materials may be steel penstocks, air-vent piping, pipe drains, concrete cooling coils, grout pipe, metal grout seals, water pipe, trashrack anchors, anchor bolts, electrical conduit, electrical outlet boxes and fixtures, pipes through which drilling is performed, manhole and hatch-cover frames, ladder rungs, grating guides, etc.

So that pours will not be unnecessarily delayed, it is essential that all the embedded materials be installed and checked before form erection and concrete cleanup are completed. Pour drawings showing the dimensions and survey control points for each pour are prepared prior to concrete placement. These drawings also show the location and quantities of embedded materials. All embedded materials must be transported to and placed inside the formed area. A cableway or crane hook is needed for lifting and placing penstock sections. Pours may be delayed for a week or more when penstock or other large embedded materials are installed and fabricated. Other pieces of equipment required are rod benders, welding machines, and hand tools.

Often the erection of reinforcing steel, embedded conduit, penstocks, etc., is done by subcontractors, but the contractor usually must supply the hook required to move the material to the pour.

CONCRETE FORMING

Concrete forms are required to maintain freshly placed concrete in the desired shape until it develops enough strength to be self-sustaining. The cost expended on the erection of these forms is one of the largest components of the total direct cost of the concrete. Form materials, form oil, form hardware, and form labor are the major items of form cost. *Form hardware* is a general term designating form anchors, form ties, tie clamps, and form fasteners. Form oil and form hardware are expenses per square foot of form contact. The cost of form materials and forming labor has increased so sharply that economical form cost can only be achieved when the expenditures for these items are minimized. The cost of these two items can be reduced by using purchased metal forms instead of shop-built wood forms and by scheduling concrete forming in such a way that the greatest reuse of forms is obtained, while consistently providing sufficient formed areas to permit the scheduled rate of concrete placement.

The facilities required to construct and erect forms consist of a carpenter's shop complete with saws, a staging area for form construction, form storage areas, a truck to move forms, small hydraulic cranes for lifting dam-monolith forms, and powered hand tools.

CONCRETE DEPOSITION AND VIBRATION

After the concrete or foundation cleanup has been completed, the forms have been set, the embedded materials and reinforcing steel have been placed, and the pour has been approved, then concrete can be deposited in the formed area.

Mass-concrete placement is started at one end of the pour by covering the foundation or the previously placed concrete with a 3/8-in layer of mortar that is worked into any irregularities. Next, a horizontal layer of concrete is placed across the downstream end of the pour. Then successive horizontal layers of concrete are placed, bringing a short length of pour to its full height across the width of the block. The upstream face of the pour is stepped, with the bottom layer extending farther into the pour and each succeeding layer extending a shorter distance into the pour. Concrete placement is continued in an upstream direction in this stepped manner until the pour is complete, keeping the unconfined upstream edge of the successive layers of fresh concrete as steep as practical. As a bucket of concrete is deposited, vibrators are used to level and compact the concrete. This is performed using large vibrators with vibrating speeds of 6,000 rpm and heads 4 in or more in diameter. Compressed air, water, and power must also be distributed through the pouring area.

Placing structural concrete is slow, since pours are small and accurate bucket spotting is required to discharge the concrete into the narrow form openings. More time is required for concrete vibration in this type of concrete placement, which necessitates that more time elapse between the dumping of buckets. Smaller buckets may be used in lieu of larger buckets suitable for mass-concrete pours.

COLD-WEATHER PROTECTION

Concrete placed during cold weather must be protected from freezing. The effort and cost of providing cold-weather protection increases as the temperature decreases. Mass pours do not require as much cold-weather protection as do thin reinforced pours, because the heat of hydration of the cement is not readily dissipated from mass pours and helps control the temperature. When the concrete-placement schedule is prepared, mass pours can be scheduled in cold weather, but structural concrete should only be scheduled in warm weather. If it is necessary to place structural concrete when temperatures are below freezing, the cost of pouring and protecting this concrete may almost double warm-weather placement cost.

If concrete is placed when the mean daily temperature is near 40°F (4°C) and above, the specifications often require that it be protected against freezing temperatures for at least 48 hours after placement. Specifications also generally require that when the mean daily temperature is below 40°F (4°C), concrete be placed at a temperature not less than 50°F (10°C), be maintained at this temperature for 72 hours, and be protected from freezing temperatures for an additional 3 days.

During cold weather, steam jets are used to remove all ice, snow, and frost from the interior of the forms, from the reinforcing steel, and from other embedded materials before concrete placement is started. If concrete pours are on the rock forming the dam foundation, the rock must be in an unfrozen state. Foundations can be covered with straw or other materials to help prevent their freezing. Steam may be required, however, to thaw the foundation if it becomes frozen. If cooling coils have been installed between concrete lifts, steam can be passed through these coils to remove snow and ice before pouring and to maintain the proper temperature in the concrete after it is placed. Steam is also used to heat the aggregates prior to their use in the mix. With preheated aggregates and warmed mixing water, mixed concrete can be produced with the desired temperature. When mass pours are made, pour protection can be provided on the sides of the pour with form insulation, and the forms may be left on the pours until temperature protection is no longer required. The top of the pour may be protected from a temperature drop by using heat lamps, or if the area is covered, the temperature of the covered area can be controlled by the use of piped steam or salamanders (oil heaters). During cold-weather pouring, the water-cement ratio of the concrete may often be lowered, and if the concrete is not subject to sulfate attack, its freezing temperature can be lowered by using 1 percent calcium chloride in the mix. When concrete is placed during extremely cold temperatures, steam curing is often substituted for water curing. When thin structural pours are made, either a greater amount of form insulation is required or steam must be applied to the form surfaces. The amount of insulation required for different thicknesses and types of pours is tabulated in various publications.[7.7]

Concrete placement during the winter months may be suspended if extremely low temperatures are anticipated. This occurred at Libby Dam where concrete was not placed from mid-November to the first of March, since during this period temperatures reached as low as $-20°F$ ($-30°C$). To protect the concrete, the contractors installed 1½ to 2 in of urethane insulation over the tops of the monoliths, on the upstream and downstream sides of the top two pours, on exposed bulkhead joints, and over openings in the dam concrete. On the tops of the 19 highest blocks heating cables were installed between the concrete and the insulation that maintained temperatures of 50° to 56°F (10° to 13°C) throughout the winter. A total of 12½ acres of insulation, 100 electric heating units, and 50 miles (80 km) of heating cable, wire, and conduit were installed. The power consumption for heating was 1,900 kW.[7.12]

CURING

The amount of water contained in fresh concrete is more than enough for hydration of the cement. This contained water must be maintained or replenished in the concrete during the early rapid stages of cement hydration. This control of the contained water is called *concrete curing,* and is accomplished by keeping the concrete moist, by sealing exposed surfaces with a sealing compound, or by using steam when concrete is placed during periods of cold weather.

The curing required for each type of concrete and the length of the curing time is delineated in the specifications. Requirements for moist curing will differ in regard to both the required application and the length of the curing time, depending on the type of concrete surface, the type of forms, and the length of time that the forms are left on the pour. Flat concrete surfaces are often cured by ponding or by being covered with wet sand, wet earth, or wet burlap. Vertically formed surfaces are often cured by using a perforated pipe or soaker-hose system that trickles water down the face of the forms or concrete. Another moist curing method for both flat and vertical surfaces is the use of sprinklers to continually wet the concrete surfaces.

PATCHING AND FINISHING

The amount of patching and concrete repair that must be done to the cured concrete is controlled by the adequacy of the concrete placement. All imperfections in the exposed concrete surface must be corrected. Any concrete that is damaged, honeycombed, or fractured, that has excessive surface depressions or is otherwise defective must be removed and built back up to the desired surface with dry-pack mortar or concrete. Ridges and abrupt protruding irregularities must be removed by bush hammering or grinding. The permissible amounts by which the concrete surface can vary from a true plane are set forth in the specifications. The owner's engineer uses a template to check these variations. Unsightly stains on exposed surfaces may have to be removed by grinding, sandblasting, washing with a solvent, or using additional sealing compounds. Holes greater than $\frac{1}{4}$-in (0.6 cm) in diameter left by the removal of form hardware must be filled with dry-pack mortar.

The kind of finishing required for different concrete surfaces is defined in the specifications. The amount and kind of finish should be tabulated when the form and concrete takeoffs are made. Finishing of unformed surfaces will vary between screeded surfaces, floated finish, steel trowel finish, or special finishes such as hardened floor topping or terrazo. Finish of formed surfaces is seldom required. The work involved after form stripping is concrete repair and patching, consisting of the repair of defective concrete, the removing of excessive irregularities, and the filling of the holes left by form hardware. If the formed concrete surfaces are not up to specification requirements, sack rubbing of the surface may be necessary.

When the formed concrete surface is subject to special wear, such as hydraulic erosion on a spillway slope or the concentrated wear on stair risers, a special stoned finish may be required. Special finishes on formed surfaces may also be required for architectural reasons.

INSTALLATION OF PERMANENT FACILITIES

Dam construction includes the installation of equipment and facilities required to control the water at the damsite or to service these water-control facilities. Typical items of this nature are gate and bulkhead guides, gates and bulkheads, penstock sections, navigation lock gates and operating mechanisms, elevator guides, elevators, cranes, etc. The construction schedule must allow sufficient time to install, fabricate, and align these items. Installation time is extensive when large spillway gates or navigation lock gates must be installed. The installation of permanent machinery and equipment cannot commence until the surrounding concrete has been placed. Therefore, their installation must be correlated with the concreting schedule. A level plant area is required for site assembly, fabrication, and storage of these items. Construction equipment required consists of welding machines, cranes, and heavy-duty transports. Adequate crane capacity is needed for erection and assembly of these items in their final position. The fabrication and installation of these items require that special crafts be employed; prime contractors often purchase these items on an installed basis, or if the owner supplies these items, the prime contractor may subcontract their installation.

The concrete immediately surrounding the guides for gates, bulkheads, and similar equipment is not placed when the dam blocks are raised, but is blocked out for future placement. When all the surrounding concrete has been placed, the guides are erected and aligned in these blocked-out areas. After alignment is completed, the concrete encasing these guides is placed. Using this procedure, gate-guide alignment can be accomplished to the accuracy specified. A large amount of fitting and welding is required during erection of all these items, and the extent of this work can only be determined by making quantity takeoffs.

If the dam contains a navigation lock, the fabrication and erection of the lock gates and closure mechanism is a major problem, and large heat-treating ovens may be required. Another specialty work item is the fabrication and erection of the permanent cranes required to handle trashrack sections or stop-log bulkheads.

DRILLING AND GROUTING

Drilling and grouting for concrete dam construction must be started as soon as the dam foundation is excavated and cannot be completed until all major concrete pours are made. The work that must be done after the foundation has been excavated is the placement of the shallow grout curtain as described in Chap. 5.

The main grout curtain is placed after the concrete has been poured above the dam foundation galleries. When the foundation galleries are constructed, 1½- and 3-in sections of pipe are embedded in the gallery floor, through which the grout holes and drain holes are later drilled. Drilling and grouting are started by drilling 1½-in grout holes to the prescribed depth. A piston-type grout pump capable of operating at high pressures is used to inject the grout. Figure 5.5 depicts a grouting setup. After the grout curtain has been placed, 3-in drain holes are drilled on the downstream side of the grout curtain; these holes remove any water that passes through the grout curtain and discharge the water into drainage ditches located along one side of the foundation gallery. Grouting is also required to fill the contraction joints and the embedded cooling pipes upon completion of chilled-water circulation. Grout piping and grouting connections to the contraction joints are embedded in the concrete as the dam is constructed. Metal grout seals, which retain the grout, are placed near the downstream and upstream edges of the contraction joints, and at approximately 50-ft (15-m) intervals, horizontal grout seals are installed to form separated grout lifts. After concrete placement has been completed in each grouting lift and the concrete has been cooled, the contraction joint and the embedded cooling pipes can be filled with grout.

Drilling and grouting are specialty items usually subcontracted to drilling and grouting contractors. When the prime contractor does this work with his own forces, he must purchase or rent diamond drills and grout pumps and distribute grouting sand, cement, water, compressed air, and electricity to this equipment.

UTILITY DISTRIBUTION AND SERVICE FACILITIES

Water, power, and compressed air must be supplied and distributed to the various plant and construction areas. If any work is performed on a two- or three-shift basis, it is necessary to illuminate the work areas. Dam areas may be illuminated by stringing floodlights on a cable spanning the damsite or by placing banks of floodlights on elevated towers.

The service facilities required are compressor plants, water-pumping plants, a machine shop, a rigging loft, craft buildings, toolboxes, service cranes, service trucks, and pickups. Some of these facilities may be available from other construction operations, such as the service facilities provided for the dam excavation.

REFERENCES

7.1 *Forms and Rock Bolt Engineering Catalog No. 1968–69,* Williams Form Engineering Corp. (Grand Rapids, Michigan), 1980.

7.2 "First Concrete at Libby Dam," *Western Construction,* October 1968, p. 33.

7.3 "Dworshak Dam Concreters Crank Up for 200,000 cu yd per month," *Engineering News-Record,* December 5, 1968, p. 32.

7.4 Donald D. Barnes, ed., *Aggregate Producers Handbook,* 2d ed., Smith Engineering Works, a division of Barber-Greene Co. (Milwaukee, Wisconsin), 1960.

7.5 *Crusher Slide Rule,* Smith Engineering Works, a division of Barber-Greene Co. (Milwaukee, Wisconsin).

7.6 *Heavy Media Separation,* Wemco, a division of Western Machinery Company (San Francisco, California).

7.7 *Concrete Manual,* 7th ed., U. S. Department of the Interior, Bureau of Reclamation, Denver, Colorado, 1963.

7.8 John A. Havers and Frank W. Stubbs, Jr., eds., *Handbook of Heavy Construction,* McGraw-Hill, New York, 1971.

7.9 "Concreting Starts at Dworshak Dam," *Western Construction,* August 1968, p. 34.

7.10 "Truck Takes Over as Concrete Plant's Lifeline," *Construction Equipment and Materials,* February 1969, p. 34.

7.11 "New Way to Cool Aggregates—by Steam, Savannah River Dam," *Engineering News-Record,* April 9, 1959, p. 38.

7.12 "Libby Dam Sheds Its Electric Blanket," *Contractors and Engineers Magazine,* April 1970, p. 56.

Chapter 8 Earth-Fill Embankment Construction

This chapter describes the construction of earth-fill embankments. Systems used in the construction of this type of embankment differ since individual embankments have different slopes, a different number of fill zones, and different types of fill. For economic reasons, earth-fill embankments are designed to suit each foundation condition and to utilize fill from adjacent borrow pits. This chapter describes the construction of a typical earth-fill embankment containing a zone of impervious fill to prevent water passage through the embankment; a downstream pervious drainage zone to remove any water seepage; a downstream zone of random fill for dam stability; an upstream random-fill zone to provide dam stability when the level in the reservoir is lowered; and an upstream surfacing of riprap to protect the fill from wave action.

Some of the work required to construct an earth-fill embankment has been described in previous chapters. Briefly, the planning and construction of any required diversion facilities are described in Chap. 4; the scheduling and performance of foundation excavation and preparation are discussed in Chap. 5, as is the proper synchronization of fill placement in the embankment with foundation excavation; and the planning and construction of any associated concrete structures are described in Chaps. 6 and 15.

The facets of earth-fill embankment construction presented in this chapter include the planning required for the excavation of fill materials from borrow pits and the processing of these materials, transporting the fill to the embankment and spreading it on the embankment, control of the moisture content of the fill and compaction of the fill, placement of any required riprap surfacing, and installation of settlement- and pressure-recording instruments.

207

Other aspects of earth-fill embankment construction are dealt with elsewhere. The development and operation of quarries that may be required to produce pervious material and riprap are described in Chap. 9. When earth-fill dams contain random zones of coarse gravel, transition zones are located between the impervious zone and the gravel fill. (This transition zone material is also required on rock-fill dams with impervious cores.) The method of producing transition-zone material is discussed in Chap. 2.

CONSTRUCTION SYSTEM SELECTION

Economical construction of earth-fill embankments is accomplished by using construction equipment to perform the major work and limiting hand labor to the performance of a few minor tasks. For this reason, the cost of constructing earth-fill embankments mainly comprises the operating, maintenance, and depreciation costs of the construction equipment. To minimize the equipment operating and maintenance costs, a construction system should be selected that utilizes large equipment units. The larger the equipment units, the greater the productivity and the less the amount of operating and maintenance labor required for each work unit. The depreciation resulting from the large capital expenditures required to purchase these large equipment units will be small per work unit, since it can be amortized against the large fill quantities required for most embankments.

Many variables must be taken into consideration prior to the selection of the construction system. Earth-fill embankments are constructed using many different types of equipment since they are composed of different types of fill; borrow pits vary in depth and in characteristics; haul roads are of different lengths and grades; embankment zones vary in number and width; and there are differences in the lift thicknesses of embankments.

When the hauls are short, and unless excessive haul grades are encountered, transportation cost is low, and loading cost controls equipment selection. As haul distances increase, the transportation cost also increases until it becomes the major consideration. Since long haul distances are required for the construction of most embankments, the method of haulage becomes the critical item in selecting the construction system.

Construction system selection cannot be made until quantity takeoffs of embankment quantities are completed and tabulated in the order that they will be placed in the embankment. The borrow pits must be investigated, and a determination made whether processing of the borrow-pit materials is required. A flow diagram and a layout of any required processing plant must be prepared, and the placement of the fill into the embankment must be scheduled. Finally, the haul roads must be laid out and the haul distances and grades determined. When all these data have been assembled, haulage cycles can be computed and hourly production figures established. These figures will control equipment selection.

QUANTITY TAKEOFFS

Before quantity takeoffs are started, the relationship between bank yards, loose yards, and embankment yards should be established. Payment is received for fill measured in the embankment, termed *embankment cubic yards.* The yardage hauled from borrow pits to the embankment is *loose cubic yards.* The yardage loaded by the loading equipment is *bank cubic yards.* There are such wide differences in the relationship of these units of measurement for most fill materials that accurate determinations of their relationship can only be established by performing field tests. A very rough approximation of the relationship among these three units of measurement is given in Table 8.1.

Using this table it can be seen that 1 yd^3 of common earth in the borrow pit will produce 1.25 yd^3 of loose earth in the haul trucks and 0.80 yd^3 of compacted earth in the embankment.

Quantity takeoffs of the materials in an embankment are prepared to check the bid quantities and to determine the required quantities of each type of material in each foot of elevation in the dam embankment. Graphs should be prepared with embankment elevations as one ordinate and tons or cubic yards of fill as the other. These curves are helpful in determining the amount of each type of fill that must be placed between any two embankment elevations. When different longitudinal sections of the dam are scheduled to be placed at different times, the quantity takeoffs should be computed so that separate tabulations can be prepared for each embankment section. Embankment fill curves should be prepared in a similar manner.

BORROW-PIT INVESTIGATION

Borrow pits should be investigated to determine whether they contain sufficient quantities of fill for each embankment zone. The drill logs and inspection pits should be examined to determine whether fill materials lie in stratified layers, which may necessitate that they be blended during excavation. When blending is required, the loading equipment selected must be of the type that can operate in this manner. It should be determined whether the borrow-pit materials require processing to comply with specifications for fill materials. Processing may consist of scalping, screening, washing, blending, or any com-

TABLE 8.1 Comparison of Bank Cubic Yard, Loose Cubic Yard, and Embankment Cubic Yard

	Bank Yd^3	Loose Yd^3	Embankment Yd^3
Clean, dry sand or gravel	1.00	1.14	0.88
Clean, wet sand or gravel	1.00	1.16	0.86
Loam and loamy soil	1.00	1.20	0.83
Common earth	1.00	1.25	0.80
Dense clay	1.00	1.33	0.75

bination of the four. At some locations it may be necessary to open quarries to produce riprap; at other locations field stones must be gathered for this purpose. When gravel deposits are not present, it may be necessary to open quarries to produce this material.

When specifications state that the inspection personnel will judge the suitability of each load of material before it is placed in the embankment, production will drop, cost will increase, and claims may be presented by the contractor.

The moisture content of the impervious borrow pits should be reviewed. Some impervious borrow pits are so wet that the material must be dried by harrowing, aerated by passing it through pug mills, or dried in gas-heated dryers. When the impervious borrow pits contain dry material, it may be necessary to raise the moisture content of this material by extensive sprinkling. If there is any indication that there will be difficulty in maintaining the moisture or in handling the impervious fill material, a qualified consultant should be retained to study this problem. Sometimes these problems become so acute that specifications must be changed before embankment placing can be performed.

Impervious materials that may be difficult to handle are those containing appreciable amounts of the mineral halloysite or cohesionless silts and fine, uniform silty sands. Even good impervious material such as a well-graded mixture of sand, gravel, and claylike or silty fines may be subject to coarse-particle separation when it contains too many gravel-sized particles.

If gravel is to be excavated below water level, it should be examined or tested to determine whether it contains so many fines that it will not readily drain. If it is of this character, it may be necessary to windrow the material and permit it to drain before it is placed in the embankment. At Navajo Dam, New Mexico, the gravel that was excavated below the water surface and hauled directly from the borrow pit to the embankment had such fluid characteristics that equipment flotation on the fill could not be maintained.

PROCESSING BORROW-PIT MATERIALS

If processing of fill materials is needed, a flow diagram and a proposed layout of the processing plant should be prepared. Processing varies from the scalping of large rocks from impervious materials to screening, washing, and reblending of fill materials. The most suitable location for a processing plant is determined by evaluating all site conditions. If wet processing is required, the separation or settlement of fines from the wash water will influence the location of the processing plant. Specialists should be retained to assist in the design of extensive processing plants when such plants are required. Detailed discussion of processing plants is not presented here, but since the layout of a processing plant is similar to that of an aggregate plant, Chap. 7 should be referred to.

One of the largest earth-fill embankment processing plants was installed at Portage Mountain Dam, Canada. This plant processed fill materials for five of the zones in the dam embankment. Portage Mountain Dam contained 57,500,000 yd^3 (44 million m^3) of fill, of which 27,500,000 yd^3 (21 million m^3) required processing for the dam's impervious core. The prime source of fill material was pit-blended sand and gravel from a glacial moraine. This material did not require processing for use in the main random shell zone but had to be processed and blended for use in the other zones. The local silt material was too wet and had the wrong characteristics to be used for impervious fill. Impervious fill was produced by blending 89 percent minus ⅜-in moraine material with 11 percent of the local silt. Other processed material was 3- by ⅜-in washed gravel and two sizes of washed sand. Washed gravels and sands were blended in different proportions to produce the specified fill materials for the transition, filter, drain, and pervious shell zones.[8.1]

The processing plant consisted of a dry-screening plant, a wet-screening plant, and a washing plant. Part of the sand and gravel from the glacial moraine was feed for the dry-screening plant. This plant contained twelve 8- by 20-ft double-deck screens which had a screening capacity of 7,200 tons/h and which produced minus ⅜-in material. Most of the minus ⅜-in material was blended with the local silt to produce impervious core material, and the remainder was used as part of the feed for the washing plant. The plus ⅜-in material was used as feed for the wet-screening plant.

The basic screens for the wet-screening plant were two 8- by 20-ft double-decked screens. The top screen had 3-in openings and the bottom screen had ⅜-in openings. This plant produced washed 3- by ⅜-in gravel. Oversized material and excess production of washed 3- by ⅜-in gravel was used in the random shell zones. Minus ⅜-in material from this plant and minus ⅜-in dry-screened material from the dry-screening plant were the feed for the washing plant. The washing plant produced two sizes of washed sand that were used for blending purposes.

Another large material processing plant was installed at Tarbella Dam, Pakistan. This plant used a computer to control the flow of material.[8.2]

Sugar Pine Dam near Auburn, California, was one of the first dams in the United States to incorporate 5 percent bentonite by weight into parts of the impervious core. Powdered bentonite was mechanically mixed for a 10-ft-thick blanket along the bottom and upstream face of a cutoff trench and in the top 33 ft of the core. The higher plasticity of this mixture is expected to provide better resistance to differential or tensile cracking and to cracking from seismic movement. The more plastic material is also expected to give additional protection against piping.[8.3]

HAUL-ROAD LAYOUT

Haul roads should be laid out from the borrow pits to each height of embankment so that haul distances, road grades, and road-construction quantities can be determined. Errors made in computing truck turnaround cycles seldom result from poor judgment in regard to equipment performance but are made because haul roads had to be constructed differently from those initially planned.

The haul roads are often planned for travel of loaded trucks up one of the dam abutments, across the dam as the trucks dump their loads, and down the other abutment. This permits a circular flow of the haul trucks back to the borrow pits. If more than one type of haulage unit is used, the haul roads should be wide enough to permit haul trucks to overtake and pass each other; otherwise, all haulage units will be restricted to the maximum speed of the slowest unit.

Haul roads should have a well-maintained, dust-free surface so that maximum truck speed can be obtained and tire wear can be minimized. Grades should be held to the practical minimum. Continual grading and sprinkling of the haul roads are of prime importance. At San Pedro Dam, California, where 120-ton bottom-dump trucks were used to transport the dam fill, the contractor laid out the haul roads so that grades were under 3 percent. Rainbird sprinklers were used in place of water trucks to keep road surfaces free of dust.

EMBANKMENT SCHEDULING

The scheduling of the placement of fill into the embankment is of maximum importance to embankment construction-system selection, because maximum equipment requirements are controlled by the fill quantities scheduled for daily placement.

The first consideration in embankment scheduling is determining the date that fill placement will start and the date that it must be completed if all project work is to be finished prior to the specified completion date. The embankment construction schedule must allow time for providing access to the project; for plant, equipment, and personnel mobilization; for the construction of any required diversion facilities; and for foundation excavation. (Scheduling of foundation excavation and grouting is discussed in Chap. 5.)

Other factors that control the initiation of embankment placement are, first, whether the specifications allow sectional construction of the embankment and, second, whether the embankment is long enough to warrant sectional construction. If there is a long section of embankment on one or both abutments and if specifications permit, considerable yardage can be placed in these sections while the diversion facilities are being constructed and the bottom of the dam is being excavated.

The dates when embankment placement can start and when it must be completed will establish the number of calendar days available for embankment placement. These calendar days must then be reduced by the number of calendar days when weather will be unsuitable for embankment placement. If the embankment is wide, if the random-fill zone is pervious, and if specifications permit fill placement at different elevations across the embankment width, fill placement in the embankment can be scheduled so that it can be performed during part of the rainy seasons. This is accomplished by expediting the placement of the impervious zone during dry weather so that it is raised in elevation above the surrounding zones, thus providing space for placement of the pervious zones during the wet weather. Pervious-zone placement can continue during the wet weather until the surface of the pervious zones approaches the elevation of the previously placed impervious zone.

The method selected for placing fill into the embankment establishes what control weather will exert on embankment placement. When the method is selected, the weather records should be reviewed to determine how many calendar days each year are suitable for embankment placement. If other earth-fill projects have been constructed in the surrounding area, research of their construction will provide data on this subject. In some heavy rainfall areas, only 100 days out of the year have suitable weather for impervious placement. In other heavy rainfall areas, the contractor has had to erect tents over the embankment so that he could finish the embankment within the contract period.

Another weather control over embankment placement is the number of days that heavy freezing may occur. Frozen material should not be placed in an embankment, nor should fill material be placed on the frozen surface of an embankment. When the surface of the embankment is frozen to a shallow depth, embankment placement can continue if this frozen material is removed by scrapers or other equipment. Similarly, frozen material must be removed from the surface of the borrow pits so that unfrozen fill material can be delivered to the embankment.

After a determination has been made of the total number of calendar days available for embankment placement, the maximum number of working days when maximum fill placement can be achieved should be determined. Calendar days are reduced to working days by establishing whether the project will be worked 5 or 6 days a week. If a 5-day workweek is used, overtime payments will be held to a minimum, but construction equipment requirements will be at a maximum. If a 6-day workweek is used, overtime payments will be at a maximum, but construction equipment requirements will be at a minimum. To properly select the number of working days per week, comparative cost estimates may be required. At some locations the supply of skilled labor may be short, and it may be necessary to work 6 days a week so that labor will be attracted by the overtime premiums.

When the total number of working days has been established, the number of maximum productive working days is determined by reducing the total workdays by the number of low productive working days that are necessary at the start and at the completion of the embankment. When work is started, fill placement will be slow since the rock foundation under the impervious zone must be covered with hand-compacted fill until a sufficient area has been developed to provide working space for mechanical compactors. Embankment placement will continue to be slow until enough length of fill is constructed to allow rapid and continuous dumping by the haulage vehicles. Fill placement will be slow while the embankment is being topped out, since the embankment will be so narrow that it will restrict the rapid and continuous dumping of the haulage units. Also, the impervious zone will occupy most of the embankment width and the more easily placed random zone will have become very narrow. Therefore, maximum embankment-placement rates can only be scheduled when fill is placed in the major part of the embankment, since the random fill zones will then be wide and there will be unlimited area to dump haulage vehicles and for rapid fill spreading and compaction.

The maximum daily rate of fill placement is computed by subtracting the number of low productive starting and finishing working days from the total available working days, by subtracting the fill placed during these low productive days from the total fill, and then by dividing the remaining fill by the remaining working days.

After the maximum daily production is determined, the number of shifts that will be worked per day and the number of hours that will be worked per shift must be determined before hourly production rates can be established. Two different methods of scheduling daily working hours are used for embankment construction. One method is to work two 10-hour shifts per day and to service the construction equipment between shifts. If overtime pay is time and one-half and if multiple-shift work is 7 hours' work for 8 hours' pay, then 18 hours will be worked a day for an equivalent of 22 hours of straight-time pay, or a pay/work ratio of 1.22.

The other method of scheduling daily working hours is to work three 8-hour shifts and to service the construction equipment while it is being used. This results in 21 hours' work for an equivalent of 24 hours of straight-time pay, or a pay/work ratio of 1.14. This second method of scheduling will reduce overtime payments, but hourly production will also be reduced because of lost time while equipment is being serviced. On the other hand, roadside servicing of haulage equipment has become so fast and efficient that very little equipment time is lost and many large earth-fill projects are constructed in this manner.

After the hours that will be worked have been established, the maximum hourly fill-placement rates are computed by dividing the working hours a day into the maximum daily production requirements. This hourly placement rate is then used to establish the required number of operating construction-equipment units.

The detailed scheduling of embankment construction should be planned so that all excavated material to be used in the dam embankment is hauled directly from the point of excavation to the embankment without being placed in storage stockpiles. The use of storage stockpiles is expensive, since reclamation of material from these stockpiles costs almost as much as it does to load the fill material into haulage units in the borrow pits. Direct placement of fill cannot always be done when conveyor belts or trains are used to transport the fill, since surge piles are often required by these transporation systems. However, surge-pile storage is not as expensive as stockpile storage, since surge-pile reclaiming is usually done with belts and therefore is not expensive.

An embankment-placing schedule should permit the piezometer tubes and settlement instruments to be installed without delaying embankment placement. Piezometers are often required in the dam foundation and in the embankment, with piezometer tubing running to a central station. Vertical- and horizontal-movement measuring installations may be required in the fill. It may be necessary to install foundation settlement base plates on the embankment foundation. Specifications may call for surface settlement points to be installed on the embankment and for temporary control points outside of the embankment area. This work is not difficult and the total cost of installation is relatively minor. These installations necessitate the use of a large amount of hand labor.

HAULAGE EQUIPMENT

Prior to discussing the selection of embankment-construction equipment, descriptions will be given of the available equipment and the advantages that apply to the use of each type of equipment. Because the equipment used for transporting the fill material often controls the selection of other embankment equipment, it will be discussed first.

Bottom-Dump Off-Highway Trucks

This type of truck provides the most versatile method of hauling fill material. It has the best payload-to-truck weight ratio of any off-highway type of haulage vehicle. Most earth-fill embankments have been constructed using bottom-dump haulage of fill material. These trucks can be loaded by any top-loading method, and they are economical for haul distances exceeding 7,000 ft (2,100 m), provided that excessive grades are not encountered. They are suitable for use on all but the smallest of embankments, and they are particularly advantageous for transporting fill material from scattered and widely separated borrow areas.

Bottom-dump off-highway trucks can be secured in sizes up to 120-ton capacity, and manufacturers are constantly producing larger units. Figure 8.1

Figure 8.1   A 15-yd³ front-end loader and a 120-ton bottom dump truck. (K. W. Dart Truck Co.)

shows a 15-yd³ front-end loader loading a 120-ton bottom-dump truck. Maximum truck size is controlled by tire capacity and not by engine or truck size. On long hauls, tires heat and therefore the tire wear increases. This tire heating can be reduced by using smaller dual tires. Large single tires are advantageous to use when truck flotation on the embankment is a problem.

Bottom-dump trucks were used on the construction of San Luis Dam, California. This was an earth-fill embankment-type dam. The 65 million yd³ (50 million m³) of fill in the embankment was hauled from adjacent borrow pits by 100-ton bottom-dump trucks. These trucks were loaded by towed belt loaders during the process of preparing borrow pits for a wheel excavator. Fill from the prepared borrow pits was loaded into the bottom-dump trucks by a large Bucyrus-Erie wheel excavator which could load a truck in 45 seconds.

Scraper Haulage

Scraper haulage of fill material is more expensive than bottom-dump transportation because scrapers have a lower payload-to-equipment weight ratio. However, scrapers are economical to use on hauls of under 7,000 lin ft (2,100 m) because the low cost of scraper loading outweighs the more expensive haulage cost.

Scrapers are powered by wheeled tractors that come in many size ranges with different power units. Scrapers can be powered by either single or dual engines and can be secured with a bowl capacity varying from 11 to 44 yd³.

For single-engine scrapers, the loaded weight to horsepower ratio increases as the scraper size increases. With good haul roads at minimum grades, the lowest haulage cost is achieved by the use of large-capacity single-engine scrapers. When grades increase, it is often necessary to match smaller-size scrapers with more powerful tractor units. When extreme grades are encountered, the lowest cost is achieved by using twin-engine scraper units, with one engine driving each axle. This type of scraper is more costly to purchase and operate, but there will be such a time saving on steep hauls that lower unit cost will result.

Elevating, self-loading scrapers are suitable for embankment construction when only a few scrapers are required or when scrapers and pushers cannot be properly balanced. Since elevating scrapers contain equipment that allows them to self-load, they are less efficient transportation units because of the additional weight they must transport. The weight of the elevator will reduce the carrying capacity of the scaper from 2 to 3 yd^3. Figure 8.2 shows 30-yd^3 tandem twin-engine scrapers.

The most economical type of scraper to use for any haul condition usually cannot be selected without making comparative cost studies between different types. Computers can be used to eliminate much of the detailed computation required for this type of study.

Off-Highway Rear-Dump Trucks

The off-highway rear-dump truck is a sturdy truck designed to take the shock of having a shovel dump large rocks into the bed or having a shovel hit the truck with the bucket. The truck's low payload-to-truck weight ratio results in high direct cost when it is used for transporting common fill material. Savings in equipment write-off may outweigh this high direct cost when rear-dump trucks have been used on rock excavation and are available for use on embankment construction. If these trucks have been obtained for other uses and are available, they are often used to haul the small volumes of fill required for narrow embankment zones. Typical zones of this type are those consisting of sand or other granular materials located between the impervious core and coarse gravel zones. Rear-dump trucks have more horsepower per ton of

Figure 8.2   Tandem 30-yd^3 twin-engine scrapers. (Caterpillar Tractor Co.)

loaded capacity than other types of trucks; thus they may be economical for transporting common material over steep haul roads.

On-Highway Bottom-Dump Trucks

The standard on-highway bottom-dump truck has narrow tires, a low horse-power to weight ratio, and a high payload-to-truck weight ratio. They are designed for highway haulage, cannot maintain speed on grades, and have difficulty maintaining flotation on an embankment. Their use is economical when they haul fill over existing highways or over flat, well-maintained construction roads. The longer the haul, the more economical is their use. As an example, four 70-ton-capacity on-highway bottom-dump trailers have been used as trailer trains hauling fill over a level roadbed. Each train was powered by a 735-hp tractor unit supplemented with a 318-hp convertible dolly inserted between the first and second trailers. The train was loaded by a portable belt loader. Turnaround time for the train on a 9.1-mile haul was 57 minutes.[8.4]

Bottom-dump on-highway trailer units can be obtained with wide single tires that give good flotation on fill. Often when scrapers are used to transport fill material on long, level hauls, one of these lightweight units is hitched behind the scraper, since the scraper will have sufficient power to pull this extra load. To do this, it is necessary to have a system for top-loading the trailer units in the borrow areas. These trailer units also have been used behind off-highway bottom-dump units to increase their capacity on long hauls. Since the tractor for the bottom-dump unit usually does not have enough power to pull both trailer units at high speeds, a remotely controlled power unit is inserted between the two trailers.

Rail Haulage

Rail transportation of fill has very limited application. It is suitable for use when the fill material is located in deep, continuous deposits; when hauls exceed 5 miles (8 km); when haul grades are suitable for rail haulage; when a large hourly placement capacity is required; and when there are sufficient quantities to economically amortize the rail system. When rail transportation is utilized, separate excavation and transportation systems must be used in the borrow pits to supply fill to train loading stations, and a separate system is required at the damsite to transport the material from the train discharge point onto the dam embankment.

Rail transportation was used to construct an earth-fill embankment for Oroville Dam on the Feather River in northern California. The main fill in the dam was 60 million yd³ (46 million m³) of dredger tailings located approximately 12 miles (19 km) downstream of the dam. A single-track railroad line which passed the damsite and was located adjacent to the borrow pits was made available to the contractor because the Western Pacific had replaced it with a relocated main rail line.

A wheel loader was used in the borrow pit to load dredger tailings onto a 54-in portable collecting conveyor belt. This belt transported the tailings to a train-loading station, where it deposited the tailings on a large shuttle conveyor. The shuttle conveyor stockpiled the tailings over a reclaim tunnel of a diameter adequate to permit a train to pass through it. The tunnel was long enough to contain 10 cars, and loading gates were spaced so that all 10 cars could be loaded at one time. The existing rail line was extended to the train-loadout station and was double-tracked between the station and the dam site. Hauling from the borrow pit to the dam was done by four trains that operated at 20-minute intervals. Each train consisted of diesel locomotives pulling 42 gondola cars of 70 yd³ capacity.

At the damsite, each set of locomotives dropped its string of cars in front of a double-rotary car dumper and then hooked onto an empty string of cars for the return trip to the loadout point. A winch and cable were used to pull the parked loaded string of cars into the car dumper. The car dumper dumped two cars at a time on 60-second intervals. The dumped tailings were discharged into a concrete hopper that fed a conveyor which transported the material to the main stockpile.

At the stockpile, a traveling stacker stockpiled the material over a 1,600-lin-ft reclaim tunnel. Reclaiming from the stockpile was done using a belt conveyor located in the reclaim tunnel. This conveyor belt transported the material up the dam's right abutment. As the dam was raised in elevation, the conveyor was successively extended up the abutment. The conveyor belt discharged the fill onto a portable conveyor located on the top of the embankment. Portability of this belt was achieved by supporting it with cables suspended from an A frame constructed on the undercarriage of a shovel. The portable conveyor belt discharged the fill into truck-loading bins that could load two 100-ton bottom-dump trucks simultaneously. These trucks distributed the fill across the embankment. As the embankment was raised in elevation, a higher foundation was prepared for the truck-loading bins. They were then skidded from their old location onto this higher foundation, and the portable belt was relocated so that it discharged into the relocated bins. Capacity of this system was 4,500 yd³ (3,500 m³) maximum per hour, with an average production of 3,500 yd³ (2,700 m³) per hour.[8.5,8.6]

Conveyor Belts

Conveyor belts provide another specialized method of fill transportation which is used when the fill occurs in large deposits, when high hourly production is required, when the yardage in the embankment is sufficient to economically amortize the cost of the system, and when haul distances are not so great as to result in too large a capital wire-off for the cubic yards of fill in the embankment. Conveyor belts are economical to use when material must be transported over steep terrain. Three earth-fill dams where conveyor belts were used for transportation of fill material are Trinity Dam, California; Portage Mountain Dam, Canada; and Warm Springs Dam, California.

Trinity Dam contained 10 million yd^3 (7.7 million m^3) of impervious material. The source of this material was on the top of hills surrounding the dam. Since truck transportation of this material down the steep haul grades presented a definite problem in truck braking capacity, the contractor elected to use conveyor belts for the haul. The material was excavated in the borrow areas by scrapers which transported the material to drive-over belt loading stations where they dumped their loads. Grizzly bars at these loading stations were set at 24 in to reject oversized material. Material over 8 inches in size was crushed to minus 5 inches in roll crushers. A flight of conveyor belts transported this crushed material and the minus 8-in material down the hill. The belts were 48 in and 42 in wide; the total flight length was 10,145 ft (3,090 m); and the drop in elevation was 1,000 ft (300 m). The capacity of the system was 1,850 yd^3 (52 m^3) per hour. The longest flight was 1,895 lin ft (580 m). The belts discharged into three truck-loading bins where water could be added to increase the moisture content of the impervious material. Trucks were used to transport the fill from the bins to the embankment.

A large conveyor-belt system was used for the construction of Portage Mountain Dam, British Columbia, Canada.[8.1] Portage Mountain Dam contained 57 million yd^3 (44 million m^3) of fill. The fill materials available for embankment construction were primarily sand and gravel from a glacial moraine located approximately 4 miles (6 km) from the dam. The terrain between the moraine and the dam had considerable grade differences. Weather conditions permitted embankment placement during only 6 months of the year. To meet the construction schedule, fill had to be placed in the embankment at a rate of 2.5 million yd^3 (1.9 million m^3) a month. In order to move this amount of fill material from the moraine to the damsite over the existing grades, the contractor installed a conveyor-belt system that had a capacity of 12,000 tons per hour.

Material was blended in the pit by using large bulldozers to shove the material down steep slopes to belt loaders. The belt loaders discharged onto 48-in shuttle conveyors; these shuttle conveyors fed 60-in gathering conveyors; and the gathering conveyors fed a 72-in main conveyor 15,000 lin ft (4,500 m) long. Two belt loaders were used to feed each gathering conveyor, and two gathering conveyors fed the main belt.

Near the dam, the main belt discharged into two surge piles. One surge pile was used for storage of unscreened material, which was used in the random-fill zone. The other surge pile furnished the feed for the dry-screening plant. Part of the product of the dry-screening plant was wet-screened. Two 60-in conveyor belts were used to transport the fill material from the unscreened surge pile and the screening plants to two truck-loading bins located on the dam embankment. From the bins, the material was transported and distributed on the embankment in bottom-dump off-highway trucks.

Warm Springs Dam is a 317-ft (97-m) -high, 3,000-ft (915-m) -long rolled earth dam. The main borrow area is located on a hill 800 to 1,200 ft (240 to 370 m) above the dam foundation. The main downhill conveyor is 72 in wide by 3400 ft long and terminates in a 1,150-ton surge bin. The fill is transferred from the bin to either of two smaller belts that transport it to double-hopper mobile bins via 54-in-wide, 250-ft-long radial stackers. The fill is then taken to the damsite in 110-ton haulers. When running at capacity, the conveyor system will deliver 8,000 tons of fill per hour. The mobile bins at the damsite will deliver a load every 40 to 50 seconds.

A total of 13,000 hp is required for the material handling system. Once the system is moving, it only requires 1,500 hp at idle speed. When the conveyor system is loaded and reaches operational speed of 800 ft/min, the drive motors become generators and retard the motion of the belt, generating 4,400 kW which is plugged into the project power grid.[8.7] Figure 8.3 shows an overall view of the conveyor system.

Figure 8.3 Conveyor system, Warm Springs Dam, California. (Auburn Constructors, A Joint Venture.)

LOADING EQUIPMENT

Scraper Loading

Scraper loading is less costly than using loading equipment to top-load trucks because the scraper participates in the loading. The extent of this participation is dependent on the type of scraper. However, as haul distances increase, the additional cost of scraper haul makes their use uneconomical.

Crawler tractor-drawn scrapers are usually self-loading but are not used in embankment construction because they are slow and have limited capacities. Self-loading elevating scrapers are self-contained units that do not require pushers and operate individually. The elevator reduces the payload-to-scraper weight ratio and thus increases transportation cost. As haul distances increase, this increase in haul cost soon becomes more than the savings in loading cost. Self-loading scrapers have application for small-yardage embankment construction when haul distances from borrow pits are short, when a small number of scrapers are used, or when more scrapers are used than the pushers can load.

Scraper trains consist of a number of scrapers connected together and controlled by one operator. They are self-loading because the operator can use all the power in the train to load each scraper in succession. To date, scraper trains have had slow operating speeds, and if one unit breaks down, the whole train will stop. These units have not been used for embankment construction, but have been used on short-haul applications such as canal construction.

One helpful new development is a quick-connection attachment that can be installed on scrapers. This permits scrapers to be self-loading, for they can be automatically connected in pairs during the loading operation and can use the power of both towing units to load each scraper. After loading, they are automatically disconnected and operated as separate units. During the loading operation, the front scraper is loaded first and the rear scraper last. The rear scraper can be loaded with a greater load than the front scraper, since when it is being loaded, the power unit for the loaded front scraper can exert more tractive effort because of the load on its drive wheels. On one construction project, two quick-connected scrapers have been used to load a third scraper.[8.8]

The majority of scrapers are push-loaded by crawler-type tractors. The larger the scraper, the larger the pusher must be. For small scrapers in good material, a 180-hp crawler tractor may supply sufficient pushing power. For 40 yd^3 struck-capacity scrapers in tough loading, the equivalent of three 385-hp pushers may be required. Crawler-type pushers develop better tractive effort than wheel-type pushers. Dual crawler-type pushers, controlled by one operator, can engage scrapers quicker and will load approximately 25 percent more scrapers than two pushers operated as independent units. It is advantageous to use wheel-type pushers in abrasive sand (since track wear is eliminated) and for pushing applications where their greater speed can be utilized.

Front-End Loaders

Common materials can be rapidly loaded into trucks by wheel-mounted front-end loaders. The capacity of this type of front-end loader is constantly increasing, and they are now available with up to 24-yd^3 buckets. As an example of their loading capacity, a 15-yd^3 front-end loader has loaded 120-ton bottom-dump trucks with dredger tailings (large gravel) in five passes, taking approximately 3 minutes. Both the capital cost and the operation cost of this type of loader are minimal compared to those for shovel operations.

Towed-Belt Loaders

Older plow-type belt loaders are equipped with 54-in belts. Their loading capacity is between 1,000 and 1,200 yd^3/h when two large tractors are used to pull the loaders in good material on long, flat runs. Figure 8.4 shows this type of loader. Plow-type belt loaders manufactured by Euclid, Inc., are equipped with 72-in-wide belts and have a maximum loading capacity of 3,000 bank yd^3/h. This is an economical method of loading bottom-dump trucks from suitable borrow areas. Production drops sharply if borrow areas are short, since frequent turning of the loader will be necessary. This type of loading equipment should not be used when selective loading or blending is required in the borrow pit.

Figure 8.4   Towed-belt loader. (Earth-moving Equipment Division, General Motors Corp.)

Elevating-Belt Loader Attachments

Elevating-belt loader attachments can be secured to a grader or a large-type tractor. Production is less than one-half that of the older type of towed-belt loaders, but equipment cost is reduced.

Belt Loaders

Belt loaders are used for loading common material from deep deposits located on steep slopes. Figure 5.2 shows this type of loader in use. The capacity of belt loaders is dependent on the size of the belt, the number of bulldozers feeding the belt, and the distance the bulldozers must push the material. As discussed in Chap. 5, belt loading has been done with production rates of 1,800 yd³/h (1,400 m³/h). As the material is loaded out, the distance that the bulldozers must push their load is increased. To keep this distance to reasonable limits, the belt loader must be relocated as often as once a day. This takes approximately 1 hour for three bulldozers. The loaders are usually moved on nonshift hours at overtime rates so that they will always be available for truck loading during the regular shift.

Wheel Excavators

Wheel excavators have been used to load out borrow-pit material at Oroville Dam and at San Luis Dam. At Oroville Dam, a wheel excavator loaded dredger tailings onto a gathering conveyor belt. It had a maximum capacity of 4,500 yd³/h (3,500 m³/h) and an average capacity of 3,500 yd³/h (2,700 m³/h).

Figure 8.5   Wheel excavator used at San Luis Dam, California. (Bucyrus-Erie Co.)

Figure 8.6   Wheel excavator. (Barber-Greene Co.)

At San Luis Dam, a wheel excavator loaded random material from a long, previously prepared bank into 100-ton bottom-dump trucks, loading each truck in 45 seconds. This wheel loader is shown in Fig. 8.5.

The continuous banks required for the operation of this loader were prepared by using towed-belt loaders to load the shallower material into the trucks. The first cost of large-wheel excavators is so large that economical equipment write-off can only be secured where large volumes are to be handled. This was possible at these two dams, since Oroville Dam contained approximately 60 million yds^3 (46 million m^3) of dredger tailings and San Luis Dam contained approximately the same amount of random fill.

Small wheel excavators are being developed.[8.9] One manufacturer is producing a wheel excavator of a smaller size using a ditcher principle for the excavating wheels. This type of wheel excavator, utilizing two wheels as pictured in Fig. 8.6, has achieved a loading capacity of 3,000 loose yd^3/h.[8.10]

Power Shovels

Power-shovel loading is done when the shovel has been used for rock excavation and is available for use on common excavation. In some cases power-shovel loading may be used to break out and load hard-packed material in confined areas.

Draglines

Large draglines are used to excavate large quantities of material from below the water surface. Truck loading can be expedited by having the dragline discharge its load into a portable truck-loading bin. When the dragline must raise the bucket high enough to dump into the bin, its cycle time increases and production decreases. This decrease in production can be eliminated by having the dragline discharge its bucket into a low hopper, with a conveyor belt transporting the material from the hopper to the top of the truck-loading bin.

Sometimes materials that have been excavated in the wet contain enough fines to restrict their rapid drainage; if these materials were to be placed directly in the dam embankment, it would be difficult to maintain equipment flotation. Thus, when this situation occurs, the dragline must windrow the material, and it must be given time to drain before it is hauled to the embankment. After it has drained sufficiently, it can be loaded out by the dragline, by front-end loaders, or by other loading equipment. Figure 8.7 shows a dragline working in a wet location.

Truck-Loading Bins

Truck-loading bins, charged by either conveyors or draglines, are often used to top-load haulage units rapidly. There is no reason why portable truck-loading bins cannot be used with many other types of excavating equipment. As an example, front-end loaders could dump on portable belts that discharge into loading bins. Less truck spotting and loading time would be required, and fewer trucks would be used.

Figure 8.7   Dragline at work. (Bucyrus-Erie Co.)

Scraper Top-Loading Trucks

Scraper top-loading of haul trucks has been done when the borrow pit is located a long distance from the embankment. The scrapers are used to excavate the material in the borrow pit and to transport it to a loading bridge where they can discharge the load into trucks. The main disadvantage of this system is that it is difficult to key scraper capacity to truck capacity.

A good application of scraper loading and bottom-dump hauling occurred at a wet, impervious borrow pit located a long distance from Hell Hole Dam on the Rubicon River, California. The material was dried in the borrow pit by digging drainage ditches, exposing extensive surface areas, and then aerating the material by harrowing. The top layer of dried material was then harvested with scrapers that dumped their loads adjacent to a portable belt loader. Two bulldozers then fed this material to a belt loader which top-loaded bottom-dump haulage units.

EMBANKMENT EQUIPMENT

After the fill material has been transported to the embankment, it must be spread to the desired lift thickness, brought to nearly the optimum moisture content, and compacted to the required dry density.

Spreading Equipment

The depth to which each layer of fill material can be placed in the embankment varies between 4 and 24 in (10 and 60 cm). This depth is dependent on the type of fill material, on the type of compactor, and on the number of compactor passes. Fill spreading on the embankment may be done with either crawler-type or wheel-type bulldozers, with blades mounted on the front of compactors, and with motor graders. The amount of equipment required for spreading the fill varies with the type of haulage equipment. Since scrapers can discharge their load to any lift thickness, fill spreading will be at a minimum when they are used. Bottom-dump trucks discharge their loads in relatively deep, narrow windrows, and considerable fill spreading must be done when they are the haulage vehicles. Rear-dump trucks deposit their loads in relatively deep piles, which entails the maximum use of spreading equipment. Figure 8.8 shows a 180-hp motor grader.

Equipment for Controlling Moisture

The moisture content of the fill material is controlled by the use of large sprinkler-tank trucks of from 8,000- to 12,000-gal capacity. Figure 8.9 shows this type of truck.

Figure 8.8   Motor grader. (Caterpillar Tractor Co.)

These trucks are pulled by the same type motive unit as is used on scrapers and bottom-dump trucks. Some contractors have converted used bottom-dump trucks to this service. A separately driven gas or diesel 6- to 8-in pump is used to furnish pressure for the sprinkling system. In order to maintain the tank truck in continuous operation, its water tank must be refilled in as short a time as possible. This can be accomplished by having elevated storage water tanks of larger or of similar capacity outfitted with a large-diameter, drop-valve truck filling spout. There is always sufficient time between tank-truck fillings to replenish the water in the storage tanks with relatively small-capacity water pumps or water lines. At some dams where large amounts of water have been required in the embankment, water has been distributed on the embankment by fast-connecting pipelines equipped with sprinklers. This requires the continuous relocation of the distribution piping as the embankment is raised in elevation.

Figure 8.9   A 12,000-gal sprinkler truck. (Southwest Welding and Manufacturing Co.)

Compaction Equipment

Compaction of the embankment is handled in different ways by specification writers. One method is to specify the type of compactor and the number of compactor passes. Another method is to specify the required dry density and place the responsibility on the contractor to determine the type of compactor and the number of compactor passes that will produce this density.

The type of compactor used and the number of compaction passes required varies with the type of fill material and the thickness of lifts. The more pervious and granular the fill material, the thicker the lifts and the fewer compaction passes required. Since the impervious zone must be fine-grained plastic material to provide an efficient water barrier, this zone must be placed in the thinnest layers and receive the most compaction. If granular, gravelly material is used in the outer shell zones of the embankment, it often can be placed in 2-ft (0.6-m) layers and compacted with crawler-tractor treads. Compaction is quite a complex subject on which a variety of information has been published.[8.11,8.12]

There is a wide variety of compaction equipment available for embankment construction and a wide choice in motive power. Compactors may be of the towed-drum type; compaction-drum-type wheels can replace standard wheels on wheeled tractors; single drums can replace the rear wheels of wheeled tractors; and motorized units can be obtained composed of two, three, or four compaction drums which are capable of movement in either direction. The wheeled-tractor units and some of the motorized-drum units can be outfitted with blades for spreading the fill material. The operating speed of compaction equipment varies between 3 and 10 mph (5 and 16 km/h), depending on the type of unit.

Following are listed the different types of compactors, the types of fill material best packed by each type of compactor, and the thickness of fill layers suitable for each type of compactor:

Sheepsfoot rollers are used for compacting clays, clay and silt mixtures, and gravel with clay binders. Lift thickness varies between 6 and 12 in (15 and 30 cm).

Tamping-foot compactors are used on sandy clays and silts with clay binders. Lift thickness varies between 7 and 12 in (18 and 30 cm).

Pneumatic, small-tired compactors are used on sandy silts, sandy clays, and gravelly sands. Lift thickness varies from 4 to 8 in (10 to 20 cm).

Pneumatic, large-tired compactors are used on sandy silts, sandy clays, and gravelly sands. Lift thickness can be as much as 24 in (60 cm). Figure 8.10 shows a compactor of this type.

Vibratory compactors can be obtained with smooth steel drums, sheepsfoot drums, and tamping drums. Lift thickness varies between 3 and 12 in (8 and 30 cm).

Figure 8.10 A 100-ton rubber-tired compactor, Portage Mountain Dam, Canada. (Southwest Welding and Manufacturing Co.)

Grid rollers are used on granular soils. Lift thickness varies between 5 and 10 in (13 and 25 cm).

Steel-wheel rollers are used on sandy silts and other granular materials. Lift thickness varies between 4 and 8 in (10 and 20 cm).

Segmented wheel-rollers are used on wheel-type tractors equipped with dozer blades. This type of compactor is useful in pulverizing material.

Tractor treads on large crawler tractors are used in compacting pervious gravels. Lifts may be placed as thick as 2 ft (0.6 m).

Hand-held compactors are required to compact fill material placed in thin layers (3 or 4 in) on and adjacent to the rock foundation and adjacent to any structures. Hand compaction must be done until there is sufficient compacted working area so that machine compaction can be used. Additionally, because mechanical compactors cannot operate adjacent to the abutment, hand compaction must be continued up the rock surface of each abutment so that the impervious fill is compacted into all the irregular rock pockets. The more frequent the occurrence of sharp irregularities in the dam foundation, the greater the amount of hand compaction that will be required. In order to estimate this work quantity, it is necessary to visualize the appearance of the final prepared rock surface.

Equipment Used for Placing Riprap and Rock-Slope Surfacing

The riprap and rock surfacing on the embankment slopes can be placed as the embankment is raised, or it can be placed after the embankment is virtually completed. If it is placed as the embankment is being raised, the easiest method calls for use of a backhoe or a crane to place this material on the slope. The crane can work off the top of the embankment and can obtain the rock from small, truck-dumped stockpiles located on the embankment and adjacent to the slope. The placement crane or backhoe must have lifting capacity sufficient to handle the largest size of rock.

If riprap and rock surfacing are not placed until the embankment is completed, one of two methods is generally preferred for placing this material. One method is to use rear-dump trucks to dump the surfacing material over the edge of the embankment. Spreading and leveling of the surface material is done with bulldozers operating up and down the slopes. The other placement method is to construct haul roads in the embankment slopes so that trucks can haul surfacing material to cranes located on these haul roads. The roadways are progressively leveled out by the cranes as they place the surfacing material on the embankment slopes.

During the past decade, soil cement has partially replaced riprap as an effective means of providing slope protection, particularly where sound, durable rock is not economically available. The use of soil cement was pioneered by the Bureau of Reclamation with a test section in the Bonney Dam reservoir, Colorado. Since that time, this form of slope protection has been used effectively on dams of up to 200 ft (60 m) or more in height.

EQUIPMENT SELECTION

Before the construction equipment is selected, it should be determined if equipment requirements can be reduced by using equipment made available from other dam construction operations, such as equipment used for excavating the dam foundation. Priority in equipment selection should then be given to equipment that will be suitable for use on other project work, either before or after it is required for embankment construction.

Remaining equipment selections should be of types and units that will maintain the required rate of fill placement in the most economical manner. The advantages of each type of loading, transporting, and embankment-placing equipment have already been discussed in this chapter. Final selection of equipment types must be based on judgment, experience, and, if required, on comparative cost estimates.

Comparative Cost Estimates

If comparative cost estimates are used to determine the most economical type of equipment, the comparative estimate should show the hourly write-off cost and the hourly operation cost of the equipment. Table 8.2 shows how comparative cost estimates are prepared; this example compares the cost of using two different loading methods to top-load 100-ton off-highway bottom-dump trucks. One method of loading is to use a portable belt loader fed by four bulldozers. The other loading method is to use two $8\frac{1}{2}$-yd^3 front-end loaders and to provide a bulldozer to clean up and push material to the front-end

TABLE 8.2 Comparative Cost Estimate

	Front-end Loading System	Belt Loading System
Estimated capital cost:		
Belt loader	$	$ 250,000
Bulldozers	320,000	1,280,000
Front-end loaders	670,000	
Total	$990,000	$1,530,000
Useful life, hours	10,000	10,000
Write-off/h	$ 99.00	$ 153.00
Estimated direct cost/h:		
Operators	$ 90.00	$ 150.00
Truck spotter	20.00	20.00
Maintenance labor	30.00	40.00
Repair parts and supplies	25.00	30.00
Total direct cost/h	$ 165.00	$ 240.00
Direct cost plus write-off/h	$ 264.00	$ 390.00
Cost/bank yd^3 at 1,000 yd^3/h	0.264	0.390
Cost/embankment yd^3 at 800 yd^3/h	0.330	0.488

loaders. This cost comparison indicates that front-end loading will be the preferable loading method.

Selecting Borrow-Pit Processing Equipment

If the borrow pits contain oversized material that is not permitted in the fill, this oversized material should be removed by screening and wasted in the borrow pits. When borrow pits contain minor amounts of oversized material, removal can be accomplished by using portable screens. If belt loaders are used for borrow-pit loading, oversized material can be removed by screens installed on the discharge end of the loading belt. If other types of loaders are used, oversized material can be removed on grizzly-type screens mounted on portable frames, constructed to permit gravity truck loading of the screened material.

When the borrow pits contain large quantities of oversized material, a scalping screening station is installed at the borrow pits. These stations are arranged so that the pit trucks can dump their loads directly onto a grizzly, and from the grizzly the screened material is transported by conveyor belt to truck-loading bins or directly into haul trucks. The oversized, rejected material can be loaded into trucks in a similar manner.

If processing of fill material is necessary, a processing plant must be designed. A flow diagram furnishes the required data for processing-plant design and for selection of screen size. The design of a processing plant is quite similar to the design of a natural aggregate plant.

Determining the Number of Loading Units

After the type of loader has been selected, the number of loading units required to maintain the scheduled production must be determined. This is accomplished by dividing the loader's hourly production capacity into the scheduled hourly rate of fill placement. The best method of determining loading capacity is from past job records or from work experience. If these are lacking, the production capacities of different loading units can be found in tables published by equipment manufacturers.

The tables are based on the equipment working 60 minutes of every hour, on the theoretical capacity of the loading bucket, and on the maximum efficiency obtainable from the operator and from the equipment. This necessitates that production rates be reduced because such conditions will never occur. The 60-minute hour should be reduced to a 50-minute hour to take care of unavoidable delays. The bucket capacity must be reduced to that obtainable for each type of material. An analysis must be made of the loading cycle, since it varies with the shovel swing or the distance a front-end loader must travel. Production must be reduced because the machine and operator will not operate at 100 percent efficiency. All these adjustments are explained by equipment manufacturers in the fine print of the text accompanying their tables.

Approximations of the production that can be achieved by different loading equipment are included in Table 8.3 as an aid to individuals who do not have access to other information. When the angle of swing increases to 180°, shovel production will be 70 percent of that shown in the table. Production decreases with either a decrease or an increase in the optimum bank height in accordance with Table 8.4. Optimum bank height occurs when the bank is level with the shovel's dipper-stick pivot shaft.

Rubber-tired, articulated, front-end loader production will be the equivalent of the production of a power shovel that has a bucket two-thirds the size

TABLE 8.3 Shovel Production in Bank Yd³ per Hour; 70% Efficiency, 90° Swing

Bucket Size (yd³)	Optimum Bank Height			
	Sand and Gravel	Common Earth	Hard Clay	Wet Clay
½	60	55	45	30
1	140	110	85	65
2½	260	210	180	140
3½	360	320	280	200
4½	440	390	310	240
6	560	500	400	300
8	750	650	550	420
10	900	800	650	500
14	1,200	1,000	860	660

TABLE 8.4 Adjustments
to Shovel
Production for
Bank Height

	Factor
Optimum height	1.00
80% of optimum	0.98
60% of optimum	0.91
40% of optimum	0.80
120% of optimum	0.97
140% of optimum	0.91
160% of optimum	0.85

SOURCE: *Caterpillar Earthmover Performance Handbook,* Caterpillar Tractor Company.

of the front-end loader. This is a rough approximation, since loading capacity varies with the distance the loader must travel between the bank and the haulage unit.

Dragline production varies with the type of dragline. Compared to shovel production, cycle time increases more with an increase in equipment size. The production of draglines with a bucket size of 1 yd³ or smaller will be approximately 80 percent of the production of a shovel with an equivalent size bucket. This relationship reduces to 75 percent for 2½-yd³ draglines, to 65 percent for 3½-yd³ draglines, and to 58 percent for draglines of 4½-yd³ size and larger. When draglines are excavating material from below a water surface, production rates should be reduced another 30 to 50 percent.

Towed-belt loaders operating in a suitable long borrow pit can load 1,000 bank yd³/h if they are the older type equipped with 54-in belts; if they are the newer type, equipped with 72-in belts, they can load 3,000 bank yd³/h. This capacity is reduced when the borrow pit is shorter and more equipment turning is required.

A 60-in belt loader loading large bottom-dump trucks with free-running material from a sidehill deposit, with four D-9s pushing the material into the belt loader, has a maximum loading rate of 1,800 bank yd³/h.[8.13] Any change from optimum loading conditions will rapidly decrease this loading rate.

Determining the Number of Haul Units

The number of operating trucks or scrapers required on the haul from the borrow areas to the embankment is established in the following manner. After the average turnaround cycle for the haulage unit is determined, the cycle time divided into a 50-minute hour gives the number of cycles per unit per hour. The number of cycles per hour times the haulage-unit capacity gives the haulage unit's hourly production capacity. This number divided into the scheduled hourly production gives the required number of operating units.

The computation of the average cycle time becomes quite complex since hauls lengthen, grades increase, and cycle times increase as the embankment is raised in elevation. Computers are helpful in establishing cycle time. The number of cycles per hour should be computed on a 50-minute-hour to reflect job efficiency. Detailed explanations of the computations required to determine cycle time are given in Chap. 5.

If scrapers are used on the excavation, the balancing of the scraper fleet to the pushers can be done by dividing the pusher cycle into the scraper cycle to find the desired number of scrapers for each pusher. Fractions should be disregarded and not rounded off.

A typical pushing-cycle computation is as follows:

Scraper loading time	0.90 min
Pusher return	0.45 min
Allowance for stopping and turning	<u>0.25</u> min
Total pusher cycle	1.60 min

If the scraper cycle is 10.5 minutes, then this, divided by 1.60, is 6.6. Therefore, one pusher will be required for every six scrapers.

If conveyor belts are used for transporting material, their capacities can be obtained from published handbooks.

If trains are used for fill haulage, their capacity is dependent on haul lengths and grades, engine horsepower, number of cars per train, capacity of each car, method of loading, and car dumping method.

Determining Embankment Equipment Requirements

The determination of the number of equipment units required for spreading the fill, for obtaining the proper moisture content, and for compacting the fill is computed from the rate of fill placement, the type of fill material, the lift heights, the number of compaction passes required, the type of compactor, and the speed and width of the compactor.

These are all straight-line computations and should present few problems. Most embankment spreads are organized with one spread of equipment to handle the impervious core and another spread to handle the remaining zones.

Total Equipment Requirements

The foregoing calculations can be used to establish the number of equipment units that must be in operation to maintain the scheduled rate of production. Total production equipment requirements will exceed this number by one-fourth or one-third since equipment units will only be able to work from 70 to 80 percent of the time. This is termed *equipment availability*. Equipment units will be unavailable between 20 and 30 percent of the time because they must be pulled out of production for servicing and for repair.

After the fleet of production equipment has been determined, equipment must be added to this production fleet for road building, for keeping the space clean around excavating units, for road maintenance (i.e., water wagons and graders), for equipment servicing units (i.e., grease and fuel trucks), for tire changing, and for transporting labor and supervisory personnel (i.e., pickups and man-haul trucks). Provisions for water distribution, job lighting, repair shops, tire shops, a rigging loft, and craft-tool storage facilities must also be made. If some of the servicing equipment or plant facilities have been furnished with other construction systems, duplication is often unnecessary.

INSTRUMENTATION REQUIREMENTS[8.14]

The purpose of instrumentation is to observe the behavior of the embankment throughout its life. Today all major dams include instruments to monitor behavior, and a large number of dams initially constructed without instrumentation have had monitoring devices added during recent years. Duplication of measurement by employing several types of instruments to confirm results is generally followed today. Settlement can be measured with cross arms and hydraulic leveling devices; pore pressure can be measured by open-type and pneumatic piezometers; and horizontal displacements can be measured by inclinometers and linear extensometers. Major types of instruments normally utilized for dam embankments are described below. The estimator must assess not only the effect of the individual installation effort but must also assess the effect that installation of the various devices will have upon the overall production cycles.

Cross Arm.   The cross-arm settlement device is utilized in both compacted soil and in rock-fill masses. It features an open tube 3 to 4 in (8 to 10 cm) in diameter with a torpedo weight attached to a tape to measure depths. Precision of the order of ±1 mm can be achieved in depth measurement to the surface.

Inclinometer (Slope Indicator).   This device has made it possible to measure both settlements and horizontal displacements over intervals of the order of 2 m. Precision is of the order of magnitude of the cross arm. The device features an open tube and a torpedo and can be used at inclinations of 30° or more from the vertical; accuracy is diminished at casing inclinations over about 7°.

Hydraulic Leveling Device.   This type of instrument is usually applied to measure settlements inside the embankment at specific locations of interest or to verify the performance of inclinometers. At Tarbella Dam, Pakistan, double-tube fluid settlement devices with leads several thousand feet in length were successfully used.

Linear Extensometer. These instruments measure displacements between two reference plates by means of electrical resistance. Readings can be of the order of accuracy of ±0.1 percent. Installing these devices in a direction other than the horizontal is difficult and time-consuming.

Piezometer. Three types are generally used for dams: the open tube, the pneumatic, and the electrical resonant wire. All piezometers measure the water level either directly or through measurement of the hydraulic head. They present an excellent method of measuring the hydraulic gradient through the dam. Vibrating wire piezometers were used very successfully at Tarbella Dam, where leads have had to be extended up to 10,000 ft (3,000 m) in length. Both pneumatic and hydraulic types were installed at Sugar Pine Dam to monitor the accuracy of the pneumatic piezometers.

Pressure Cell. This type of device, representing one of the simplest methods of measuring total normal stresses, was developed in Mexico in 1969. Its dependability has been questioned, and further research is necessary to determine the limitation of the device and the proper procedure for installing it in the embankment.[8.14]

Sugar Pine Dam in California, designed by the Bureau of Reclamation, contains eight different types of instruments. A total of 120 are in place, representing a 300 percent increase over any pre-Teton dam experience. The instruments will permit extensive monitoring of embankment stresses and of horizontal and vertical displacements and pore water pressures in the embankments, abutments, and foundations. Instruments will also record information on embankment seepage and on seismic motion.[8.3]

REINFORCED EARTH DAMS

A patented earth reinforcement system was developed by French architect and engineer Henri Vidal. It was first used in the United States in 1979 for a flood-control dam in Austin, Texas. The 344-ft (105-m) -long, 33-ft (10-m) -high Woodhollow Dam has a clay core keyed into bedrock. Its tiered downstream face was formed by layering galvanized steel strips and a special granular fill behind a facing of interlocking concrete panels. Random fill was used for the first tier followed by an 11-ft (3-m) -high reinforced earth wall benched back to a 13-ft (4-m) -high wall. The landscaped tiers allow a shorter toe-to-toe width than a conventional earth fill, an important consideration at a tight site.[8.15]

REFERENCES

8.1 W. Irvine Low, "Portage Mountain Dam Conveyor System," *Journal of the Construction Division,* ASCE, September 1967.

8.2 "At Tarbella Dam—Computer Controls 15 Processing Plants," *Contractors & Engineers Magazine,* November 1969, p. 47.

8.3 "Dam Tests New Seismic Design," *Engineering News-Record,* January 22, 1981, p. 74.

8.4 "Earthmover Train Hauls 308 Tons per Trip to Highway Grade," *Construction Methods and Equipment,* September 1969.

8.5 "Big Wheel Replaces Dragline at Oroville," *Engineering News-Record,* May 14, 1964, p. 26.

8.6 "It's Push-button Earth Moving at Oroville," *Engineering News-Record,* October 8, 1964, p. 74.

8.7 "Warm Springs Dam," *California Builder & Engineer,* December 31, 1979, p. 12.

8.8 "Three Scrapers in Tandem Make Fast Work of Moving Earth at Building Site," *Construction Methods and Equipment,* September 1969, p. 66.

8.9 "New Developments in Earth Moving, Part 2: Wheel Excavators," *Construction Methods and Equipment,* February 1965.

8.10 David C. Etheride, "Excavator Feeding Two Lines of Bottom Dumps Never Stops in Work on 4-Million-Yd Job," *Construction Methods and Equipment,* April 1970, p. 61.

8.11 A. W. Johnson and J. R. Sallberg, *Factors That Influence Field Compaction of Soils, Compaction Characteristics of Field Equipment,* Highway Research Board Bulletin 272, publ. 810, National Academy of Sciences, National Research Council, Washington, D. C., 1960.

8.12 *Handbook of Compactionology,* American Hoist and Derrick Co. (St. Paul, Minnesota), 1968. *Caterpillar Earthmover Performance Handbook,* 10th ed. Caterpillar Tractor Co. (Peoria, Illinois), 1979.

8.13 "Modified Equipment Moves 45,000 Yards per Shift," *Roads & Streets,* April 1969, p. 48.

8.14 Stanley D. Wilson and Raul J. Marsal, *Current Trends in Design and Construction of Embankment Dams,* ASCE, New York, 1979, pp. 73–76.

8.15 "Unique Dam Survives Test," *Engineering News-Record,* June 18, 1981, p. 51.

Chapter 9 Rock-Fill Embankment Construction

This chapter describes how the construction of rock-fill embankments is planned, scheduled, and performed. The many differences in the construction of these embankments result from the use of two basic types of water barriers.

When impervious fill is located near the damsite, the embankment is usually designed to have an impervious core for a water barrier. Embankments designed in this manner contain a main rock-fill zone for dam stability; a sloping, impervious core placed upstream of the main rock fill, which is inclined to the natural slope of the rock and permits dam stability to be obtained by a minimum of rock fill; a narrow rock-fill zone on the upstream side of the impervious zone, which serves as protection and support for the impervious zone when the water in the reservoir is lowered; and narrow zones of graded granular material, which function as fill transitions between the impervious core and the rock fill.

When rock-fill embankments are constructed in high, mountainous areas, impervious fill is often impossible to obtain economically. In this case, rock-fill embankments are designed with a thin, concrete slab functioning as the water barrier. This type of embankment contains one main rock-fill zone with a selectively placed, narrow rock facing on its upstream slope. This rock facing supports the concrete slab that forms the water barrier.

Some of the work required for the construction of a rock-fill embankment has been described in other chapters in this book. The planning and construction of any required diversion facilities are included in Chap. 4. The planning and scheduling of the excavation and preparation of the dam foundation are covered in Chap. 5. The scheduling of embankment placement should be correlated with the sequence of performing the foundation excavation, as also described in Chap. 5. The construction of any associated concrete structure is covered in Chap. 15. The construction of the impervious zone of a rock-fill dam is similar to the construction of the impervious zone of an earth-fill dam, as is described in Chap. 8.

The descriptive information still lacking for rock-fill embankment construction is presented in this chapter, which covers the development and operation of rock quarries, the production of granular materials for the transition fill zones, the construction of haul roads, the loading and transportation of fill materials from the quarries and borrow pits to the embankment, and the placement of the fill in the embankment.

CONSTRUCTION SYSTEM SELECTION

The selection of the most economical system for constructing rock-fill embankments is accomplished by preparing takeoffs of embankment fill quantities, planning the development and operation of the rock quarry, planning the production of filter materials, selecting the method of transporting the fill materials to the embankment, laying out the haul roads, planning the placement of fill in the embankment, scheduling the construction of the embankment, and determining the type and number of construction equipment and plant units required to maintain the scheduled production.

Since the rock quarry is the source of the rock fill that forms the major portion of a rock embankment, the quarry development and operation is the controlling component of the embankment-construction system. (For information on presplitting and other controlled blasting techniques that have applications in quarrying, refer to Chap. 5.)

QUANTITY TAKEOFFS

Quantity takeoffs are made to check the bid quantities, to establish work quantities that are not listed in the bid schedule, and to furnish a breakdown of quantities to use for construction scheduling, for establishing hourly production rates, and for selecting the number and size of the equipment units. Quantities should be tabulated so that the quantities of fill between any two embankment elevations are available for the entire embankment and for any planned sectional construction of the embankment. An excellent method of tabulating this information is to prepare embankment fill curves with the embankment elevations as one ordinate and fill yardages as the other.

Before the quantity takeoffs are started, the relationship between embankment yardage, loose yardage, and bank yardage should be established. These relationships are used in determining the yardages that must be excavated in quarries and borrow pits and in computing the required number and sizes of the drilling, loading, and hauling equipment units. The only true method of establishing these relationships is by field testing. Approximations of these relationships for impervious and granular material are given in Chap. 8. To assist in initial planning, Table 9.1 lists approximations of these relationships for various types of rock.

ROCK-FILL SOURCE

The source of rock for the embankment fill may be a rock quarry or a large rock cut required for the dam spillway or power-intake channel. When the rock is obtained from a quarry, the quarry may be a coyote quarry or one of the three types of blast-hole quarries. When the rock fill is secured from a large cut, the development of the cut is quite similar to the development of a blast-hole quarry, except that the cost of excavating the rock exceeds what it would cost to obtain it from a quarry. This increased cost is caused by more restrictive face development and haul-road construction and by the necessity of presplitting to develop the nearly vertical faces in the rock cut. This extra cost is justifiable, since the cut will serve as a permanent facility of the dam as well as a rock quarry.

QUARRY PLANNING

Prior to planning the development of a quarry, all available data on the proposed quarry should be assembled and analyzed. These data should include specification requirements, amount of stripping required, type of rock, rock boundaries, location and quantities of any unsuitable rock, location of any major shear planes or fault zone, whether rock slippage along shear or fault planes will be a problem, amount of jointing or stratification in the rock, weight of the rock, rock swell factor, embankment-shrinkage factor, estimated drilling speeds, *powder factor* (pounds of explosives required to break one bank cubic yard of rock), and how the rock will break (size of pieces).

TABLE **9.1** **Estimated Relationship between Bank Yards, Broken Yards, and Embankment Yards**

	Bank		Broken		Embankment	
	yd³	lb/yd³	yd³	lb/yd³	yd³	lb/yd³
Granite	1	4,500	1.5–1.8	3,000–2,500	1.35–1.65	3,250–2.750
Limestone	1	4,200	1.65–1.75	2,550–2,400	1.5–1.6	2,800–2,630
Sandstone	1	4,140	1.4–1.6	2,950–2,600	1.2–1.4	3,450–2,950
Shale	1	3,000	1.33	2,250	1.2	2,500
Slate	1	4,590–4,860	1.3	3,500–3,750	1.2	3,800–4,050
Traprock	1	5,075	1.7	3,000	1.5	3,400

Planning a quarry is started by determining whether the rock deposit should be broken by explosives placed in *coyote drifts* (4- by 6 ft-drifts excavated in the quarry face) or by explosives placed in drilled blast holes. Coyote quarry planning consists of developing a suitable quarry face, quarry floor, and a coyote-drift system. Blast-hole quarry planning entails developing the amount of stripping and determining the *burden* (distance the blast holes are back from the quarry face), drill-hole spacing (distance between drill holes measured parallel to the quarry face), drill-hole diameters, drilling speeds, powder factor, bench heights, and number of quarry faces.

Specification Requirements

The specifications should be analyzed to determine whether rock screening, selective quarrying, or wasting part of the quarry's production will be required. The word "analyzed" is used because specification writers seldom state that any of these procedures are necessary. They place the prebid responsibility for this determination on the contractor by specifying permissible rock sizes or by restricting the fines that can be included in the rock fill. Often, to produce fill from local rock deposits that complies with such requirements, it is necessary to perform selective quarrying or rock screening. Also, specifications may include gradations for filter materials so strict that to produce materials that will meet requirements it becomes necessary to separate the material by sizes, stockpile each size, and then reblend the different sizes.

If the rock embankment is the type that has a concrete slab on its upstream slope as a water barrier, the specification for the selectively placed rock facing may be so restrictive that a large amount of work may be required to select the rock.

The specifications for the main fill zone may require that large rocks be placed in its downstream portion and that the smaller rocks be placed in the upstream portion. Often to comply with this stipulation there must be selective loading in the quarry or the rock must be separated on a grizzly.

Exploratory Work Performed by the Owner

The contractor should assemble all the data available from the records of the owner's quarry investigation. This data may be included in the specifications, shown on the plans, or distributed to bidders as supplementary information. Drill-hole cores are often available for inspection at or near the quarry site. Tunnel drifts or surface excavations are usually open for inspection during the contractor's visit to the damsite.

Contractors can seldom secure adequate data for planning the quarry's development from the results of the owner's investigation of the quarry, since these investigations are often limited to drilling a few core holes. A quarry can only be adequately explored by opening up one or more quarry faces with a test blast and by recording drilling speeds, powder consumption, and rock-breakage characteristics during the performance of this exploration. In addition, sufficient core holes should be drilled to prove that the quarry contains the volume of rock required for the embankment's construction. The owner can readily justify the expense of adequately exploring a quarry since contractors will submit lower bids if they are furnished this information.

Often the owner so inadequately explores a quarry that after the contract is awarded, the quarry is found unsuitable or lacking in the required quantities of rock. When this occurs, the contractor must develop a new source of rock, and this will often increase the construction cost of the project. This increased cost is always much more than if the owner had adequately explored the proposed quarry prior to bid advertising.

Contractors' Prebid Investigation of the Quarry

When the owner has inadequately investigated a proposed quarry, contractors are forced to perform their own quarry investigation. Contractors seldom have the time between bid advertising and bid submittal to properly investigate quarries, and since only one contractor will be the low bidder on the job, it is difficult for them to justify the expense of adequate quarry investigations.

Therefore, a contractor's investigation of a proposed quarry is usually limited to surface investigation and performance of a few drilling tests. If contractors do not test-drill the quarry, they may send rock samples to drill manufacturers who will then perform laboratory tests on the drillability of the rock. In addition, contractors often employ consulting structural geologists to report on the suitability of the quarry.

INSPECTION OF DRILL-HOLE
CORES AND GEOLOGICAL LOGS

In most quarry and major rock excavation sites, the owner will drill a number of test holes with a diamond drill bit, which will permit recovery of a solid core. Core drilling is also performed in a similar manner for dam, powerhouse, and other structural foundations. Inspection of the cores and drill logs gives the consulting geologist or drilling and blasting consultant valuable information to be used in developing the parameters for the preparation of the estimate. Inspection of the cores is also an important tool for the estimator in developing the judgment necessary to plan the drilling and blasting program.

In 1964 Don U. Deere[9.1] proposed a quantitative indication of the quality of a rock mass. This method is based upon the extent of core recovery, generally utilizing double-barreled diamond drilling equipment. The Deere method can be utilized to classify rock quality for both open-cut and underground applications.

Deere called this quantitative index of rock mass quality *Rock Quality Designation* (RQD). The RQD is defined as the percentage of core recovered in intact pieces of 4 in (100 mm) or more in length in the total length of the borehole. Cores should be 2 in (50 mm) in diameter or larger.

$$\text{RQD (\%)} = 100 \times \frac{\text{Length of core in pieces} > 4 \text{ in}}{\text{Length of borehole}}$$

Deere developed the following proposed classification of rock quality:

RQD, %	Rock Quality
0–25	Very poor
25–50	Poor
50–75	Fair
75–90	Good
90–100	Very good

RQD is regarded as a quick, inexpensive practical index. Values can be established for core runs of, say, 6 ft (2 m), and since the determination is simple and quick it can be carried out in conjunction with the normal geological logging of the core. With careful handling of the boxed cores, RQD values can be developed later through inspection of the core boxes and the drill log.

The limitations of this method are seen in cases where joints containing thin clay fillings, fault gouge, or weathered material which could result in unstable rock are widely spaced. In these cases core samples may give a misleading value for the RQD.

DEVELOPMENT OF COYOTE QUARRIES

Coyote quarrying is the removal of large quantities of rock from the quarry face by explosives placed in a drift, called a *coyote*, driven parallel to the face. Such quarrying is economical when the proposed quarry is of the sidehill type, has a 75-ft (25-m) or higher face, has a quarry floor sufficient to receive the broken rock (which necessitates a floor width of twice the height of the quarry face), and contains rock that will be well-fragmented when the only breaking force is rock displacement. For example, sound, unfractured rock, i.e., a granite monolith, should not be quarried by coyote blasting, because the rock would break into such large pieces that an excessive amount of secondary drilling and shooting would be required to reduce them to manageable sizes. Coyote blasting is such a specialized field that it should only be planned and performed by experienced personnel; otherwise results may be undesirable.

Coyote quarry development consists of providing a haul road to the bottom of the quarry face, providing a quarry floor to receive the broken rock, excavating the coyote drifts, and exploding the charge in these drifts. Coyote drifts are excavated by driving an adit into and perpendicular to the quarry face for a distance equivalent to 60 to 75 percent of quarry-face height. The adit is driven a shorter distance into the face than the face height so that the relief path from an explosion will be shorter to the quarry face than it is upward, resulting in proper shearing at grade. At right angles to this adit and parallel to the quarry face, cross drifts are then driven in both directions along the length of the quarry face.

Typical dimensions for the adits and cross drifts are 4 ft (1 m) wide by 6 ft (1.8 m) high. They are short and of such small cross section that expenditures for tunneling equipment are held to a minimum by using air legs for drilling and a slusher scraper for muck removal. Typical crew size is three men per shift.

Low-cost explosives such as AN/FO (ammonium nitrate prills mixed with fuel oil) are placed in the cross drifts. The amount of explosive used per cubic yard varies with rock type but often is between 0.75 and 1.75 lb/yd^3 of burden. The farther the adit is driven into the quarry face in relation to face height, the greater must be the powder factor. The explosives are piled in the cross drifts with piles on 25-ft centers and are often detonated with primacord, using two interconnected lines to ensure detonation. The primacord is protected from damage by hanging the cord from the roof of the drift or by placing it on the floor and covering it with sandbags. To provide blowout protection, the pile of explosive in each cross drift adjacent to the main adit is covered with a timber bulkhead, and the remaining length of cross drift between the bulkhead and the adit is stemmed with tunnel muck. The minimum distance for this stemming should be from 12 to 15 ft (3.5 to 4.5 m). As previously mentioned, the explosion will move the rock burden away from the face and deposit it on the quarry floor, which extends out from the quarry face a distance equal to twice the quarry face height.

The volume of rock moved by a coyote blast is bounded in its cross-section area by the rectangle whose height is the quarry-bench height and whose width is the distance from the quarry face to the back of the cross drift. The length of this rock rectangle is the same as the lengths of the cross drifts. A coyote blast will produce such a large volume of broken rock that only a few blasts are required.

The foregoing description is of a simple coyote layout. Coyote layouts should be modified to fit local conditions. Modified layouts may consist of two or more cross drifts extended from one adit, several adits with short cross drifts, a supplementary high coyote level, or a special layout to suit an irregular quarry face.

Figure 9.1 shows a plan and section view of the first blast utilizing almost a million pounds of AN/FO for a coyote quarry for the Ord River project located in the remote Kimberley region of Western Australia. The tunnels were approximately 5 ft (1.5 m) wide and 8 ft (2.4 m) high. Table 9.2 gives the statistics of the blasting operations.

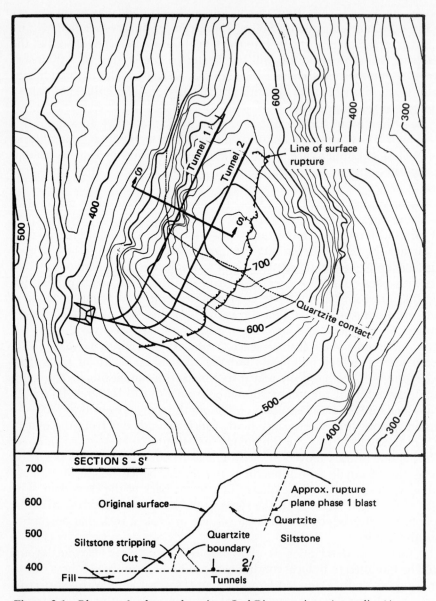

Figure 9.1   Blast no. 1: plan and section, Ord River project, Australia. (American Society of Civil Engineers.)

DEVELOPMENT OF BLAST-HOLE QUARRIES

There are three types of blast-hole quarries: exposed rock sidehill quarries, hillside quarries covered with overburden, and pit-type quarries. All three types of blast-hole quarries are similar in that their development is controlled

TABLE **9.2**   **Blasting Data, Ord River Project**

Tunnel	Explosive Charge (lb)	Volume of Rock (bank yd³)	Shear Factor (lb/ft²)	Powder Factor (lb/yd³)
		First blast		
No. 1	415,000	240,270	10.9	1.73
No. 2	579,000	403,729	13.0	1.43
Total	994,000	643,999		1.54
		Second blast		
No. 3	409,100	191,640	21.6	2.13
No. 4	483,000	280,630	15.0	1.72
No. 5	262,100	153,667	11.0	1.78
Total	1,154,200	625,937		1.84

SOURCE: Lewis L. Oriard and Joseph L. Jordan, "Rockfill Quarry Experience, Ord River, Australia," *Journal of the Construction Division,* ASCE, March 1980, p. 37.

by quarry bench heights, blast-hole diameter, and blast-hole spacing. These controlling factors do not vary with the type of quarry but rather are determined by the characteristics of the rock contained in the quarries.

The remaining quarry development work, i.e., stripping, bench development, and construction of haul and access roads to the quarry faces, varies with the type of quarry.

Bench Heights, Blast-Hole Spacing, and Hole Diameter

The primary objective in blast-hole quarrying is economical production of a pile of broken rock of the proper size and bank height for optimum shovel loading. This is accomplished by selecting a bench height, a hole pattern (a combination of burden, depth, and drill-hole spacing), and a blast-hole diameter that will result in the production of broken rock of the proper size and height for the lowest possible drilling cost. To achieve low drilling cost, a wide drill-hole pattern should be used. To produce small-sized broken rocks, a close drill-hole pattern is necessary. Therefore, quarry planning requires that the proper quarry-shovel size be selected and that the widest drill-hole pattern be used that will produce broken rock that the quarry shovel can handle.

There are no criteria that can be used to establish the drill-hole pattern except past records and experience. However, in theory, broken-rock size in a quarry is dependent on rock breakability. One measure of rock breakability is its hardness. The harder the rock, the harder it is to break. The harder the rock is to break, the closer should be the individual explosive charges in the rock. The closer the individual charges, the closer the drill-hole pattern. Therefore, rock hardness can be used as a gauge for establishing the drill-hole pattern. One measure of rock hardness is Mohs' scale, as given in Table 9.3.

The drill-hole pattern that is selected establishes the blast-hole diameter, drill type, and bench height. The drill-hole pattern establishes blast-hole diameter, since the pounds of explosives required to break a cubic yard of rock does not radically change with rock types. Therefore, as the drill-hole pattern decreases in its dimensions, the diameter of the blast-holes decreases; or, as the drill-hole pattern increases in its dimensions, the diameter of the blast-holes must increase.

Since the drill-hole pattern establishes the diameter of the blast holes, which controls the selection of quarry drills, it also controls the type of quarry drill. And, since the drill-hole pattern establishes drill type, it also limits the maximum height of bench, inasmuch as the different types of quarry drills can economically drill different depths of hole. However, safety requirements prescribe maximum bench heights, dependent upon the nature of the rock.

TABLE 9.3   **Mohs Hardness Scale**

	Origin	Mohs' Hardness Scale	Lb/yd in the Bank	Specific Gravity
Andesite	Igneous	7.2	4,660	2.4–2.9
Basalt	Igneous	7.0	5,080	2.8–3.0
Bauxite	Mineral	2.0	4,290	2.4–2.6
Chalk	Sediment	1.0	3,700	2.4–2.6
Chert	Sediment	6.5	4,320	2.5
Clays	Sediment	1.0	2,700	2.4–2.6
Diabase, dalerite	Igneous	7.8	4,730	2.8
Diorite	Igneous	6.5	5,000	2.8
Dolomite	Sediment	3.7	4,860	2.7
Felsite	Igneous	6.5	4,590	2.65
Flint	Sediment	7.0	4,320	5.5–7.0
Gabbro	Igneous	5.4	4,860	2.8–3.0
Gneiss	Metamorphic	5.2	4,860	2.6–2.9
Granite	Igneous	4.2	4,590	2.6–2.9
Limestone	Sediment	3.3	4,400	2.4–2.9
Marble	Metamorphic	3.0	4,320	2.1–2.9
Marl	Sediment	3.0	3,780	2.2–2.4
Mudstone	Sediment	2.0	2,970	2.4
Porphyry	Igneous	5.5	4,290	2.8
Quartz	Igneous	7.0	4,460	2.65
Quartzite	Metamorphic	7.0	4,320	2.0–2.8
Sandstone	Sediment	3.8	3,920	2.0–2.8
Schists	Metamorphic	5.0	4,660	2.8
Serpentine	Metamorphic	5.0	4,670	2.8
Shale	Sediment	4.0	4,320	2.4–2.8
Slate	Metamorphic	3.0	6,830	4.0
Talc	Sediment	1.0	4,540	2.5–2.8
Traprock	Igneous	7.0	4,870	2.6–3.0

Quarry Drills

There are three main types of quarry drills: percussion drills mounted on air-powered tracks, down-the-hole drills, and rotary drills.

Crawler-mounted percussion drills are used to drill blast holes up to 6 in (15 cm) in diameter. A drill of this type is shown in Fig. 9.2. Air-operated drills are sized by the drill's piston diameter, which should generally be at least 1 in (2.5 cm) larger than the bit. These drills can be secured with or without independent bit rotation. They are used in hard-rock quarries when the rock is to be loaded with relatively small-sized shovels. A close pattern of holes will result in good fragmentation, and secondary drilling and shooting will be minimized. The use of small holes will give better rock fragmentation near the top of the quarry face, since less hole stemming is required and the explosive charge can be brought up higher in the hole. Bench heights often used with this type of drill vary between 25 and 35 ft (8 and 10 m). When it is necessary to break rock into sizes that can be handled by small shovels, a close hole pattern is mandatory. The drilling and explosive-loading cost for this large number of holes will be relatively high, but because secondary drilling and shooting cost will be at a minimum, total drilling and shooting cost will be less than if a wider pattern of holes was used.

Down-the-hole percussion drills are used in hard-rock quarries when the hole pattern can be enlarged, which necessitates the use of blast holes between

Figure 9.2 Crawler-mounted percussion drill with 750-ft³ min compressor. (Ingersoll-Rand Co.)

6 and 9 inches in diameter. Since these drills are larger and more efficient than the crawler-mounted percussion drills, they can drill to greater depths, allowing the use of bench heights of from 35 to 45 ft (10 to 14 m).

Down-the-hole drills follow the bit down the hole, with guidance and support from drill steel suspended from a drill carriage located on the surface of the excavation. The *drill steel* (hollow steel rods connecting the drill to the drill carriage) furnishes support for and conveys air to the drill and rotates the drill and bit, receiving this rotation force from a rotary table mounted on the drill carriage. Since there is a minimum length of steel between the drill and the bit, all the drill's striking force is exerted on the bit. Down-the-hole drills are of simpler construction than drifter drills, since they do not have to furnish a force for bit rotation.

The development of the hydraulic drill powered by either electric motor, air motor, or diesel engine represents a major breakthrough in the development of drilling equipment. Initial appraisal of the use of hydraulic drilling equipment on construction projects is somewhat mixed. Penetration rates in most hard rock can generally be increased utilizing hydraulic drills. However, overall economics continues to be influenced by the first cost of the equipment, the cost of providing utilities and fuel, and the cost of developing the

Figure 9.3 Hydraulic crawler drill with extendable boom. (Ingersoll-Rand Co.)

Figure 9.4 Rotary drill. (Joy Manufacturing Co.)

initial plant. As with all new developments, downtime for the new hydraulic equipment is somewhat of a problem. The estimator will continue to be challenged to adopt the most expedient and economical choice of equipment for any particular project. Figure 9.3 shows an all-hydraulic crawler drill with an extendable boom.

In contrast to percussion and down-the-hole drills, which transmit a primary striking force and only a secondary rotating force to the drill bit, rotary drills use a combination of rotating and thrust forces applied to the drill bit. Rotary drills are used in quarries that contain soft rock that is easily broken, allowing a wide pattern of drill holes and thus requiring drill holes from 6 to 9 in or more in diameter. Since rotary drills can economically drill deep holes, quarry-bench heights can be as high as desired. Figure 9.4 shows this type of drill.

Generally, rotary drilling is used in the softer rock types, and percussion drilling is used in the harder rock types.

Inclined Drilling

There is another explosive distribution procedure that is important in quarry operations. For each blast hole, the explosive should be distributed through the hole in accordance with the work to be done. A quarry face normally forms a slope from the toe of the face to the top of the bench. Therefore, vertical blast holes drilled from the top of the bench have a greater burden of rock to break at the bottom of the hole than at the top. To take care of this change in burden, the greatest concentration of explosives should be at the bottom of the hole and should decrease proportionately up the hole. This gradation of explosive concentrations throughout the hole can be readily accomplished when certain types of slurry are used.

Another method of solving this problem is to incline the blast holes so the bottom of the hole is at the same or a lesser distance from the quarry face as the top of the hole. An additional advantage secured by inclined drilling is that, because of the inclination of the hole, the explosives in the bottom of the hole will have more explosive force reacting toward the quarry face and less wasted in a downward reaction into the quarry floor.

On many rock-fill dams, the fill can be obtained from a cascade spillway excavation consisting of several steps of considerable heights. Where high faces can be safely developed on these cascades, inclined drilling parallel to the inclined face can produce well-broken-up rock at a fraction of the cost of a conventional quarry. One example is on the dumped rock shell of the Hell Hole Dam on the Rubicon River near Auburn, California, where rock developed from the lower cascade portion of the spillway was produced for less than half the cost of either quarried rock or rock obtained from the main spillway excavation in hard granite.

Quarry Stripping, Bench Development, and Access Roads

The best method of determining the amount of stripping, bench development, and access road construction required to develop a blast-hole quarry is to prepare a quarry layout. A layout is helpful in developing the required amount of stripping; the number and extent of the quarry benches, quarry-bench heights, the tonnage of rock available in each bench; and the amount of work necessary to provide haul and access roads. If a quarry layout is not prepared, there will be a tendency to underestimate the cost of quarry development.

The amount of work required to develop a quarry varies with the type of quarry. Exposed rock sidehill quarries require the least amount of quarry development work since stripping is not required. Multiple bench development is often required because, as each quarry bench is excavated back into the hillside, the bench height increases, necessitating the starting of an additional bench. As each additional bench is started, a haul road must be constructed to the bottom of the bench and an access road to the top of the bench.

When hillside rock deposits are located underneath overburden, large areas must be stripped to provide space for the required number of quarry benches. The quarry layout should reflect the amount of stripping required for each ton or cubic yard of quarried rock. Similar to other types of quarries, these benches must be developed and haul and access roads must be provided to each bench.

Pit quarries do not require stripping when the rock is exposed, but more frequently the rock is located beneath overburden, which necessitates a large amount of stripping. Face development for these quarries is costly because the face excavation must start at the surface and must be ramped down until economical face height is achieved. Side drilling is also required to trim up the sides of this pit-type excavation. Generally, only one quarry bench is developed, but if more benches are required, each additional bench must be developed in a similar manner, and haul access roads must be provided to each additional bench.

QUARRY BLASTING

Economical mass-rock excavation and quarry blasting is generally achieved by the utilization of a blasting agent or one of the newer slurry mixtures. However, dynamites continue to be utilized for a variety of applications in and around project areas.

Blasting Agents[9.2]

A blasting agent is a mixture containing both fuel-producing and oxidizing elements. Neither of these elements is considered to be an explosive until mixed together. Blasting agents cannot be detonated by a No. 8 blasting cap when unconfined. The most common blasting agent is AN/FO, a mixture of ammonium nitrate (AN) and fuel oil (FO) in which ammonium nitrate acts as the oxidizer and fuel oil acts as the fuel. AN/FO offers great economy and safety in modern blasting applications. Its cost is substantially less than nitroglycerin explosives, and it is considerably safer to handle. It will produce better fragmentation in many applications because of its high gas-producing properties. AN/FO is generally suitable for coyote blasting and for dry bore holes larger than 2.5 inches in diameter in materials which are conducive to breakage by gaseous expansion. The primary disadvantage of AN/FO is its lack of water resistance and the necessity of properly mixing the materials on bulk-load projects. Explosive trucks can bulk-load AN/FO into dry holes, it can be poured into the hole from premixed 50-lb bags, or it can be utilized in wet holes when purchased in plastic bags.

Slurry Mixtures[9.2]

Slurries, water gels, and emulsions were developed to protect the ammonium nitrate against contamination by water. Slurries are commonly a mixture of an ammonium nitrate base in solution with a combustible fuel, a heat-producing metal, and other ingredients to give a thick, soupy slurry. The slurry contains a gelling agent that solidifies the slurry in the borehole, thus protecting the AN from water. Since the slurry density is greater than water, it will sink to the bottom in wet holes. Slurries can be pumped from tank-type mixing trucks through hoses into the blast holes. Slurries can also be packaged in light plastic cartridges to facilitate transportation and storage. Researchers have also developed cap-sensitive water gels by adding a metallic sensitizer which permits cap sensitivity down to about 20°F (-6°C). Cap-sensitive gelatin, however, must be classified as a high explosive rather than as a blasting agent.

Dynamites[9.2]

Dynamites are cap-sensitive mixtures containing an explosive compound. Dynamites generally, but not always, contain nitroglycerin as the principal explosive. While dynamites are substantially more expensive than the blasting agents and slurries, they remain an important factor in commercial blasting. A major use of dynamites in recent years is in priming and supplementing bulk blasting agents and in controlled blasting operations.

Types of dynamites include straight-nitroglycerin dynamite similar to the original material developed by Alfred Nobel, ammonium dynamites developed to replace straight dynamites, and gelatins and semigelatins.

Straight-nitroglycerin dynamite is made up of nitroglycerin and a carbonaceous material such as sawdust and nitrocellulose (gun cotton). Because of its sensitivity, it is rarely used except for ditching and mudcapping. Ammonium dynamites were developed to replace straight dynamites and use ammonium nitrate as a substitute for some of the liquid nitroglycerins. These dynamites are less sensitive to heat and shock but are also less resistant to water. Gelatin dynamites have as a base a water-resistant gel made by dissolving nitro cotton in nitroglycerin. Higher strength gels are recommended for mudcapping operations. They may detonate at low velocities or fail under high hydrostatic loads. Straight gelatins (without ammonium nitrate) have good water resistance and storage properties. Ammonium gelatins are somewhat lower in cost but are less water-resistant. Semigelatins are designed to combine high water resistance with the lower cost of ammonium gelatins and work well as presplit explosives. Presplit explosives are generally 5/8 to 7/8 inches in diameter and come either in a rigid column (5/8 × 24 in), with couplers to achieve a continuous column, or spaced on a line of detonating cord-like sausages.

Detonators[9.2]

Cap-sensitive explosives are detonated by three types of blasting caps: common caps, millisecond (MS) delays, and standard delays. Detonating cord can be used as a medium to transmit explosive detonations from a blasting cap to the explosive. Common blasting caps are detonated by a fuse. The safety fuse burns at a controlled rate, but it is good practice to time-sample pieces from each coil. Common caps are rarely used today except in unusual circumstances, such as when electric firing is prohibited. Millisecond delays offer the following advantages:

1. They reduce ground vibration.
2. They improve fragmentation.
3. They produce less flyrock.
4. They reduce costs.
5. They reduce overbreak.

Standard delays were the forerunner of the MS delays and are usually utilized for underground work with delays usually set at approximately half-second intervals. The use of millisecond delays in mass-rock excavation reduces the amount of explosive detonated at one given instant and aids fragmentation by permitting the movement of rock at various time intervals. MS delays also reduce flyrock by permitting the rock to move in the desired direction. Millisecond-delay intervals can be obtained from the major manufacturer; the most prevalent delay intervals are spaced apart by 25, 50, and 100 milliseconds.

Primers[9.3]

The primer is that portion of the charge which contains the firing device and serves the purpose of initiating the explosives or blasting agent with which it is in contact. Primers may be a cartridge of dynamite with an electric, common cap, or detonating cord inserted, or they may be non-NG high-detonating-pressure primers specially manufactured to obtain full energy development from non-cap-sensitive products such as AN/FO or slurries.

In quarry blasting the number of blast holes which can be loaded and exploded at one time will vary with the properties of the material, the explosive, and the diameter of the holes. Multiple rows of smaller holes, drilled in a staggered pattern, are often exploded at one time, with different delays used for each row. Large-diameter blast holes are commonly exploded one row at a time. A number of excellent references having comprehensive discussions of quarry drilling and blasting practices are available.[9.2,9.3,9.4]

Secondary Breakage of Quarried Rock

Except in very soft rock quarries, secondary drilling and shooting must be done to break large rocks into sizes small enough for the quarry shovel to handle. This is done by drilling small holes into the rock, loading these holes with an explosive, and then exploding these charges. This procedure is called *block holing*.

Another method of performing secondary breakage, *mudcapping*, is to place explosive charges on the surface of the large rocks, cover the charges with mud, and then explode these charges. This method of performing secondary breakage eliminates the need for jackhammers, steel, bits, and compressed air. Packaged charges of explosives for mudcapping are available.

The required amount of secondary drilling and shooting varies with the quarrying method, with the hardness of the rock, and with the drill-hole spacing. For well-planned quarry work, the cost of secondary drilling and blasting seldom exceeds 20 percent of the primary drilling and blasting cost. If it is neglected, the production achieved by the quarry shovel will decrease, since it must then spend valuable loading time to maneuver the bucket to pick up the large rocks. The speed of the shovel swing will also be reduced, since the large rocks must be balanced on the bucket.

SOURCE OF FILTER MATERIAL

If the rock-fill embankment is designed with an impervious zone for water cutoff, it will also contain filter zones which protect and separate the impervious material from the rock fill. Two sets of two adjacent filter zones are used: fine granular filter zones are located on each side of the impervious zone, and coarser granular filter zones are located between the fine filter zones and the rock fill. Specifications often state that gravel from borrow pits or quarry fines can be used in these zones. Specifications are often misleading in that they imply that this material can be hauled directly from the source without the necessity of processing. This is seldom the case. Often specification restraints are so strict that borrow-pit material must be processed by removing oversized material with a grizzly, by separating the grizzlied material into sizes, and then by reblending the sized material to produce the two types of filters.

If the filter materials are to be produced from gravel pits, the gradation of the pit materials should be examined and flow sheets and processing plant layouts should be prepared showing the required processing equipment. If the filter material is to be produced from a quarry, it may be necessary to provide crushing facilities as well as screening and blending facilities.

PLACING EMBANKMENT FILL

One of two methods is commonly used for placing rock fill into the embankment. The method used influences the layout of the haul roads and the schedules of fill placement.

Placing Rock Fill by the End-Dumping Method

When the available rock fill is sound, of large size, and with a minimum amount of fines, rock embankments are often designed for the end-dumping method of placement of the main rock-fill zone. Specifications designate the percentage of large rock required and the maximum amount of minus 4-in material that will be acceptable. When the end-dumping method is used, the main rock-fill zone is often placed in two lifts, with the bottom lift higher than the top lift. Compaction is achieved by the rock cascading down the lift slope after it has been end-dumped over the edge of the fill. As the rock is dumped, it is struck with high-pressure water jets, which wash all the fines into the previously placed rock fill, permitting all large rocks to have at least three points of contact. Figure 9.5 shows monitors being used for this purpose.

Loads of rock are distributed on the fill so that the larger rocks are placed in the downstream portion of the fill and the smaller ones in the upstream portion. One theory for this is that if there is any leakage through the rock fill, the water will flow through expanding water channels and this will eliminate erosion of the embankment. Water pressure and the volume of water that must be used are defined by the specifications. A typical specification is that the water pressure must be 100 lb/in^2 and the volume of water must be twice that of the rock fill.

The angle of repose of the rock fill establishes the upstream and downstream fill slopes. The design often includes a downstream setback at the top of the first lift.

End-dumping is the most economical method of placing rock fill in the embankment, since only minor amounts of embankment equipment are required. A bulldozer is required to maintain a roadbed on the surface of the

Figure 9.5   Rock washing at Wishon Dam, California. (John W. Stang Corp.)

fill, monitors are required for water jetting, and pumps and water distribution lines are required to furnish water to the monitors.

Placing Rock Fill in Compacted Lifts

The rock fill in an embankment must be placed in shallow compacted lifts when the available rock breaks down into small pieces, when it contains an excess of fines, or when it contains common material. Also, when a rock embankment is designed with an impervious core and the main fill of a rock embankment is placed by the end-dumping method, it is still necessary to place the upstream rock zone and the impervious and filter zones in shallow, compacted layers. In addition, it may be a requirement that a narrow zone of compacted rock be placed on the upstream side of the main rock fill, adjacent to the filter and impervious zones.

Compacted rock fill is placed in lifts that vary between 3 and 10 ft (1 and 3 m) in height. Typical compaction requirements state that each lift must be compacted with four passes of the crawler tracks of a 50,000-lb-minimum-weight tractor.

Selective Rock Placement

If the embankment is designed with a concrete face slab on the upstream slope for a water barrier, the specifications will often require that the foundation for this face slab be constructed by selective placement of a narrow layer of rock on the upstream face of the main rock fill. A typical requirement for selectively placed rock is that each rock must be rectangular-shaped, from ½ to 2½ yd³ in size, and individually selected from quarried material. Chinking of the large rocks may be allowed after they are placed in position.

Placement procedure is to dump these rocks near the upstream edge of the main rock fill. They are then picked up and lowered into position by a crane operating on the surface of the fill. Crane handling can be expedited if each rock is drilled so that it can be handled with a bridle. The drilled holes should be angled toward each other and should be approximately 16 in (40 cm) deep. The bridle consists of a center lifting ring, two 12-in-long iron pins, and cable connecting eyelets on the pins to the center lifting ring. The pins are inserted into the angled holes, and as the crane lifts the bridle, the pull on the cables exerts enough side stress on the pins that they remain in position while the rock is lifted and positioned.

The placement of all other zones in the dams must be keyed to the placement of the selectively placed rock fill. After the first layer of the main rock fill has been placed, placement of the rock facing must commence, since this placement equipment must be supported by the surface of the main rock fill.

Placement of the selectively placed rock is time-consuming, and its placement cost is often more than contractors estimate. Equipment requirements consist of a bulldozer in the quarry to select the rocks, jackhammers for drilling the lifting holes, shovel or crane time in the quarry for loading the rocks, haul trucks, and cranes on the embankment for placing the rocks.

Placement of Impervious Core and Filter Materials

Irrespective of the method used for rock-zone placement, when rock-fill embankments contain impervious cores, the impervious core and the protective-filter zones must be placed in lifts between 6 and 12 in (15 and 30 cm) in height; the moisture content of the impervious material must be brought to close to optimum; and each lift must be compacted in the specified manner. This necessitates that truck access be provided to every embankment elevation.

If the rock-fill embankment has a vertical impervious core, there will be no particular fill-placement scheduling problems. Most rock-fill embankments, however, are designed with sloping impervious cores so that the amount of rock fill required for embankment stability can be kept to a minimum. When the embankment has a sloping impervious core, this core and the filter zones are partially supported by the main rock-fill zone. This necessitates that fill placement start with placement of rock in the main rock-fill zone and that the impervious core and filter-zone placement not start until part of the supporting main rock fill has been placed. Also, placement of the upstream rock zone must follow placement of the impervious fill, since it is partially supported by the impervious fill.

As discussed in Chap. 8, rain and freezing weather will restrict the placement of impervious fill. When impervious borrow pits are too wet, the impervious fill can be dried by draining, blading, harrowing, or by passing the fill material through a pug mill. In extreme cases, it may be necessary to dry the impervious material in mechanical dryers. When the impervious borrow pits are too dry, the water content of the impervious fill can be raised by sprinkling the borrow pits. In some cases, several months of sprinkling may be required.

Placement of the filter zones is relatively simple. The only problem is the dumping and spreading of the material in the narrow zone limits. To prevent materials from mixing, the filter zones are maintained at a slightly different elevation from the impervious zone.

As described in Chap. 8, the impervious- and filter-zone placement equipment will consist of blades or bulldozers for spreading the fill, water wagons, and compactors.

Concrete-Face Rock-Fill Dams

Major progress in the design of the concrete-face rock-fill dam in the past decade has resulted in increased usage of this design type throughout the world, particularly in higher dams. In addition to its excellent static stability, the concrete-face rock-fill dam offers other advantages. Because reservoir pressure is placed upstream of the total mass of the dam in this design, it is particularly attractive in the face of increasingly restrictive seismic criteria.

Construction advantages of concrete facing in favorable situations include construction economy, since the dam acts as its own cofferdam. This design can often be utilized to shorten the construction schedule for impervious-core installation in weather-sensitive areas, since rock-fill placing is independent of rainfall and slip forming of the face is a mechanized operation which proceeds very rapidly.

Leakage is of the order of several cubic feet per second, generally comparable to that of earth-core rock-fill dams. The location of excessive leakage can be detected on the face through utilization of audio equipment or closed-circuit TV. Repair of excessive leakage can be performed underwater through utilization of a fine, silty sand, which is filtered by the crushed rock zone under the concrete face, thus sealing a leaking joint or a crack. Figure 9.6 shows one method of placing a concrete face on a rock-fill dam.

Figure 9.6   Placing a concrete face slab on a rock-fill dam. (Morgen Manufacturing Co.)

HAUL-ROAD LAYOUT

The preparation of a haul-road layout is a necessity for successful embank-ment-construction cost estimating. This layout can be used to compute the quantities of excavation, fill, surfacing, and culverts required for road con-struction. The layout will establish haul distances and haul grades to use in computing truck turnaround cycles. Truck turnaround cycles are used to establish equipment requirements and to estimate the direct cost of fill haulage.

The haul roads should be designed to permit maximum truck speeds and the continuous flow of traffic to and from the embankment. The road layout should show the initial haul roads and how they must be relocated as the embankment is raised in elevation. The number of road-access points re-quired for a rock-fill embankment varies with the method of fill placement and the design of the embankment. This subject was discussed in the preced-ing section. Often, truck access to the embankment between two different haul-road elevations is assured by constructing short lengths of road within the embankment. These haul roads can be graded out when they are no longer needed.

Haul roads should be wide enough to allow haul trucks to overtake and pass each other, so that the speed of all the haul trucks is not controlled by that of the slowest unit. Preferably, the haul-road grades should be under 3 percent, but they should never exceed 7 percent. The roads should be surfaced, main-tained, and watered so that haul trucks can operate at maximum speeds with a minimum of tire wear.

EMBANKMENT SCHEDULING

A schedule for a rock-fill embankment construction must allow time for ac-cess-road construction, for mobilization of plant and equipment, for the con-struction of diversion facilities, and for excavation of the embankment foun-dation. The scheduling of the construction of diversion facilities is discussed in Chap. 4, and the scheduling of the excavation of the dam foundation is discussed in Chap. 5.

To schedule fill placement properly, the *embankment control dates* should be established. These dates are the date that embankment placement can start, the date that maximum placement of fill material can be achieved, the date that maximum fill placement must be reduced (because working areas will be restricted as the embankment is topped out), and the date that the embank-ment must be completed.

Maximum monthly fill-placement requirements can be determined from work quantities and embankment control dates and can be reduced to maxi-mum hourly requirements by selecting the number of days that will be worked per month, the number of shifts that will be worked per day, and the number of hours that will be worked per shift. These choices are discussed under earth-fill embankment scheduling in Chap. 8.

The maximum hourly rate of fill placement divided by the production achieved by each equipment unit establishes the number of operating construction-equipment units that will be required on the project. The number of operating construction-equipment units is also needed for computing the direct cost of constructing the embankment.

The date that fill placement can be started in the embankment is dependent on the topography of the damsite, the design of the embankment, and on specification requirements.

The topography of the damsite controls the start of embankment placement, since rock-fill embankments are normally constructed at sites where the river channel is narrow and abutments are steep. At these sites, foundation excavation can be completed from the top of the abutment down to water level before the stream is diverted. This is often the easiest part of the dam foundation to excavate, and it often can be done in less time than is required to excavate the section of the foundation below streambed level. The portion of the dam foundation that is beneath the stream often cannot be efficiently excavated until the stream is diverted. In most cases, excavation below the stream must be carried to greater depths than is required on the abutments. Considerable time must be scheduled for completing the excavation below the original stream level, since working space will be restricted and pumping will be required.

When a rock embankment is designed for end-dumping of the main rock-fill zone, it is possible to start the placement of rock fill on one abutment before the center of the dam is excavated. If the damsite has steep abutments, only a small amount of rock fill can be placed on this abutment, but a truck-dumping area can be developed. Upon completion of the excavation of the dam foundation, this truck-dumping area will expedite the maximum placement of the dumped rock fill.

When the rock fill is to be placed in compacted layers in the dam embankment, fill placement cannot be started in most cases until the excavation of the entire foundation is completed. Exceptions occur when specifications permit and when the slope of one abutment is so flat that a longitudinal section of fill can be placed on the abutment.

Impervious-zone fill-placement scheduling is dependent on the method of placing rock fill in the dam and whether a vertical or sloping impervious core is to be placed. This subject is discussed under impervious-fill placement in this chapter. The scheduling of impervious fill must be correlated with rock-fill placement. If the embankment is located in an area that has a long rainy season, its rate of placement may control the scheduling of the main rock fill. Impervious-core placement will be slow when it is started because hand compaction will be required and the working area will be restricted. When the impervious zone lengthens out, placement can be done at maximum capacity until the embankment approaches its maximum height. Here production will be again retarded because of lack of working room. Impervious-core placement must be suspended during rainy or freezing weather.

Selective rock placement must also be scheduled to coordinate with the placement of the main rock fill. When embankments contain selectively placed rock fill, its placement often controls the placement of the main rock fill. Also, after the rock-fill layer has been selectively placed, time must be allowed for concrete-slab placement. Figure 9.7 shows a dumped rock-fill dam which is almost complete except for the upstream face material.

EQUIPMENT SELECTION

Upon completion of the construction schedule, the maximum hourly rate of fill placement is used in the computation required to determine the number of operating productive plant and equipment units required to maintain the schedule. To eliminate the chance of making errors in these computations, it is preferable to convert embankment yardage to bank yardage and base all computations on these units.

Determining the Number of Drills

The required number of operating drills is determined by multiplying the scheduled production of rock in bank yards per working hour by the lineal feet of hole that must be drilled to break one bank yard of rock and by

Figure 9.7 Hell Hole Dam, American River project, California. (American River Constructors.)

dividing this product by the lineal feet of hole produced per hour per drill. Many studies have been performed for the purpose of establishing a set of criteria for determining drill-penetration rates for different types of rock. The results of all these studies show that drill-speeds and rock-breakage characteristics can only be determined by drilling and breaking each rock deposit. Therefore, drill-hole diameters, drill-hole spacing, and drilling speeds should be established for each rock quarry from past experience, by performing drilling tests, or by making a laboratory drilling analysis. Field drilling tests or laboratory tests will establish drill-penetration rates.

Drill-penetration rates must be reduced to establish hourly drill production rates. This reduction allows for delays attributable to the operator for time required to move drills, for steel and bit changing, for hole cleaning, for freeing stuck steel, etc. Drill-penetration rates are discussed in detail in Chap. 5. Indications of approximate drilling speeds for different rocks are also given in Chap. 5.

Mohs' scale of hardness (Table 9.3) for the different types of rock will give a rough comparison between the drilling speeds probably required and the breaking characteristics of different rock, but its use is not accurate enough to establish drilling speeds for a cost estimate. For instance, andesite, with a Mohs' hardness of 7.2, is easier ιo drill than basalt, which has a Mohs' hardness of 7.0.

Quarry Loading Equipment

After the rock is broken in a quarry, it must be picked up and loaded into haulage vehicles for transportation to the dam embankment or, if required, to a rock separation plant. The number of operating loading units that will be required is determined by dividing unit production capacity into the total production capacity required.

In most hard-rock quarries, large shovels are used to load out the rock. A typical large quarry shovel is shown in Fig. 9.8. Large shovels must be transported in sections to the job location and then erected. At job completion, they must be dismantled and then transported in sections to their next location. Figure 9.9 shows how the undercarriage of an 8-yd^3 shovel was transported to an isolated location.

One of the world's largest hydraulic shovels is produced in Germany by O&K Orenstein & Koppel, Inc. The 1 million-lb machine is powered by a 2,352-hp plant and can mount buckets ranging in size from 23 to 40 yd^3. Its breakout force when used as a shovel is about 450,000 pounds.[9.5]

For estimating purposes, the number of bank yards that can be loaded by quarry shovels should be established by evaluating their production capacity on previous work. Table 9.4 shows the production rates that should be obtainable with average loading conditions. These production rates should be adjusted for loading conditions that are peculiar to each project.

Figure 9.8 A 15-yd³ quarry shovel. (Marion Power Shovel Co., Inc.)

The production rates shown in Table 9.4 are for optimum bank height. For other than optimum bank heights, these production rates should be reduced in accordance with Table 8.4.

Front-end loaders are used to load out finely broken rock in some quarries. Track-mounted front-end loaders can load broken rock more efficiently than rubber-tired front-end loaders, but maintenance is relatively high and only smaller sizes are available. Rubber-tired front-end loaders are available with a bucket capacity of 15 yd³. When they are used in quarries, chains are often placed over the tires to increase traction and reduce tire wear.

The old burn-and-weld method of changing adapters and hard-facing lips on front-end loaders and shovel buckets has largely given way to the utiliza-

Figure 9.9 Transporting the undercarriage of an 8-yd³ shovel to the Hell Hole Dam, California. (American River Constructors.)

TABLE 9.4   **Production Rates for Quarry Shovels; Average Conditions, 70% Equipment Efficiency**

Shovel-bucket Size (yd³)	Well-broken Rock (bank yd³/h)	Poorly Broken Rock (bank yd³/h)
3½	220	150
4½	300	200
6	370	260
8	490	320
10	600	400
14	800	520

tion of factory-produced points, adapters, and wing-and-lip wear shrouds, which can be changed in a matter of minutes. The utilization of new wear-resistant alloys and flush mounting of wear parts is further minimizing maintenance costs.

Another recent development in equipment maintenance and repair is the utilization of urethane compounds to stop flats, even if tires are cut or punctured. The compounds go into pneumatic tires as a liquid that cures to a solid in 24 hours. The material is very expensive and economic utilization for a particular application depends upon justifying the additional cost to avoid equipment shutdowns due to tire failure.

Other Quarry Construction Equipment

A powder house, a cap house, and a truck are always necessities. Depending on the equipment supplied by the powder company and on the type of explosive used, a truck for bulk loading the explosive into the blast holes may be required. It may also be necessary to supply facilities for bulk storage of explosives.

Secondary drilling and shooting require that the quarry be supplied with jackhammers and portable compressors. If the primary drills do not contain their own source of compressed air, compressors are required. Since electric-driven compressors are more economical to operate than diesel-driven, they should be used where possible and installed at a central point, with pipelines distributing air throughout the quarry. The air required for percussion drills is listed in Chap. 5. The air required for other drills is obtainable from the manufacturers of the drills.

If the quarried rock must be separated into sizes, a separation plant will be required. Separation plants are discussed under quarry development.

Bulldozers will be required for building haul and access roads, for cleaning up around each quarry shovel, and for selective quarrying. If several shovels are used, a rubber-tired bulldozer has the speed to clean up around several shovels.

Quarried-Rock Haulage Equipment

Rear-dump off-highway trucks are used for hauling quarried rock. The trucks should be sized so that they will be loaded by three shovel passes. Although electric-drive rear-dump trucks as large as 200 tons have been used by the mining industry, the largest mechanical drive now in production has a capacity of 130 tons. Figure 9.10 shows a front-end loader dumping into a rear-dump truck.

The number of operating trucks required to maintain any production rate can be found by dividing the required production rate by the truck hourly production capacity. This hourly truck production capacity is determined by multiplying the truck capacity times the number of cycles the truck will make in an hour. Cycle time is composed of loading time, haul time, dumping time, return time, and delay time. An example of the computations required to determine the truck cycle time is given in Chap. 5. When rock-fill embankments are placed in two lifts, only two truck cycles need be computed. When shallow lifts are used, truck cycles are continually changing, and a large number of truck cycles must be computed to arrive at an average cycle. In this second case, a computer can be utilized to determine the average cycle time; its use will eliminate many routine calculations.

Impervious- and Filter-Zone Equipment

The equipment required to load and haul the impervious-zone material is discussed in Chap. 8.

Screening, stockpiling, and reblending equipment may be required to pro-

Figure 9.10   Front-end loader and rear-dump truck. (Terex Corp.)

duce the filter-zone material. This material can be loaded by front-end loaders into haulage equipment. The daily requirements for filter material are so low that any available equipment can be used for this haul. If equipment must be purchased, light, highway-type rear-dump trucks will provide sufficient haulage capacity.

Embankment Equipment

If dumped fill is used in the embankment, the installation of a monitor system may be quite complex. At many sites, the streamflow is not large enough to supply sufficient water during the dry season. Thus it may be necessary to construct downstream storage dams so that the water used on the fill can be collected, stored, and reused. Pumping requirements are controlled by maximum rock-placement rates. Pumps must be capable of lifting the water to the top of the embankment and still furnish 100-lb/in^2 water at this elevation. Banks of vertical high-head pumps are required. Water distribution lines on the embankment must be continually moved and extended. Often a main line is carried up one abutment. Monitors are often mounted on tractors or on air-powered crawler tracks to provide flexibility of movement.

Bulldozers will be needed on the embankment to spread and compact rock in the compacted fill zones and to maintain the road surface on top of any dumped-rock zone. Graders will be required to spread the impervious and filter-zone material. Water wagons will be required for controlling the moisture content of the impervious fill. Compactors will be needed for compacting the impervious fill and filter zones.

Total Embankment Plant and Equipment Requirements

The total plant and equipment requirements are established by increasing the required number of operating units by 25 to 30 percent to compensate for equipment availability. This provides for the equipment units that must be pulled out of operation for servicing or for repairs.

To this fleet of equipment must be added the bulldozers, graders, and water wagons needed for haul-road construction and maintenance and the servicing equipment and facilities. Servicing equipment includes grease trucks, fuel trucks, tire repair trucks, pickups, etc. Servicing facilities include the warehouse, shops, tire shop, rigging lofts, craft buildings, etc. When the amount of servicing equipment and facilities required for rock-embankment construction is determined, the overall project demands should be reviewed so that there will be no duplication of equipment and facilities in any two construction systems.

REFERENCES

9.1 D. U. Deere, "Technical Description of Rock Cores for Engineering Purposes," *Rock Mechanics and Engineering Geology,* vol. 1, no. 1, 1964, p. 17.

9.2 Gary B. Hemphill, *Blasting Operations,* McGraw-Hill, New York, 1981.

9.3 *Blasters' Handbook,* E. I. du Pont de Nemours & Co. (Wilmington, Delaware), 1980.

9.4 John A. Havers and Frank W. Stubbs, Jr., eds., *Handbook of Heavy Construction,* 2d ed., McGraw-Hill, New York, 1971.

9.5 "O & K Shows World's Largest Shovel," *Engineering News-Record,* November 22, 1979, p. 28.

Part

THREE

Tunnels and Underground Structures

Chapter 10 Rock Tunnels

This chapter deals with tunneling in rock. The choice of equipment and methods for tunnel excavation is influenced by many variables: structural characteristics of the ground, length of the tunnel, diameter of the tunnel, grade in the tunnel, and the applicable laws covering the use of diesel engines underground. The discussion of tunnel excavation equipment and methods here includes descriptions of haulage equipment, equipment unique to the drill-and-shoot method of constructing tunnels, excavating equipment, tunnel-boring machines, service equipment and facilities, and experimental methods. Descriptions of tunneling equipment are given in broad terms, since continued equipment development makes any description of equipment rapidly outdated. A tunnel engineer should contact equipment manufacturers and should inspect tunnels under construction to secure more equipment details and to keep up with improvements and new developments in this field.

HAULAGE EQUIPMENT

The type of muck haulage equipment used for any underground excavation will determine the selection of drilling and mucking methods, as the equipment used in drilling, mucking, and hauling must be of the same general type. This makes the determination of the haulage method and equipment the first consideration in the planning of underground construction.

Suitable rubber-tired or rail-supported equipment is available for any size and length of tunnel or underground chamber. The development of small diesel trucks, improvements in ventilation methods, the use of rubber-tired front-end loaders, and the acceptance of the use of diesel engines in underground construction have all resulted in a greater use of rubber-tired equipment. Large rubber-tired front-end loaders are now used for both mucking and hauling in the excavation of short tunnels and in the beginnings of long tunnels of sufficient diameter to provide equipment clearance. Specially built, low-profile front-end loaders, called *load, haul, and dump (LHD) units* have been developed for this use in smaller-diameter tunnels. When LHD or front-end loaders are used in this manner, the only other equipment which is required is drilling equipment. This results in a low excavation cost, as crew size and equipment maintenance are held to a minimum. Except for this special use of front-end loaders, tunnel muck haulage is done with either rail-mounted cars or rubber-tired haul units.

Advantages of Rubber-Tired Haul Units

Rubber-tired equipment can be mobilized very quickly, operates well on grades up to 10 percent, and has restricted operation on grades between 10 and 20 percent. Rubber-tired equipment also requires less capital expense than rail-mounted equipment. For short tunnels, the cost of excavation will be less with rubber-tired haulage than with rail haulage, as smaller crews can be used. Furthermore, rubber-tired equipment furnishes flexibility in excavation operations. It can be readily moved from one heading to another when alternating heading crews are used. When large-diameter tunnels are excavated, it can muck and clean up any width of tunnel. When underground chambers are excavated, it is readily moved from one excavation face to another and it can dispose of the muck at any location.

Disadvantages of Rubber-Tired Haul Units

In small-diameter tunnels, rubber-tired haul units must be of small capacity because of the restricted headroom, and these small units cannot be operated up a steep grade because of the low horsepower. Rubber-tired haul units are wider than rail-mounted equipment of similar capacity, which necessitates more passing room and a wider roadbed. There is more dead weight, less efficiency, and more horsepower required per pay load. Greater ventilation requirements result. On long hauls, this high ventilation load may make the use of rubber-tired equipment impractical. Rubber-tired units also require a well-graded, firm-surface, dry roadbed, which may be impossible to provide in a wet tunnel.

Advantages of Rail Haul Units

On long hauls, trains are more efficient than trucks since they can haul more muck with less horsepower, fewer operators, and less ventilation required. Since rail haulage equipment is compact and relatively narrow, it can operate and pass in small-diameter tunnels; this compactness allows the use of rail equipment of large capacity in relatively small-diameter tunnels. Rail equipment is also more suitable than rubber-tired equipment in wet tunnels, as it can operate on flooded tracks. Furthermore, rail equipment is more adaptable than rubber-tired equipment in converting to concrete placing operations upon completion of the excavation. When rail equipment is operated with cables and a hoist, it can be used on any grade. Additionally, rail-mounted muckers require less maintenance than those used for loading trucks.

Disadvantages of Rail Haul Units

Rail equipment requires a longer period of mobilization than is required for rubber-tired equipment. Excavation of short tunnels is more expensive with rail units than with rubber-tired equipment, as larger crews are required. If rail-mounted muckers are used in large-diameter tunnels, they must muck out one-half the tunnel at a time, or two muckers have to operate abreast of each other, as they are limited in their width of operation. Rail equipment does not give the flexibility that rubber-tired equipment does for the excavation of underground chambers, as rail equipment can only operate where tracks can be installed. Rail equipment operates well on grades up to 2 percent. From 2 to 4 percent it has restricted operations. Above a 4 percent grade, it must be winched up and down the slopes.

Haulage-Equipment Selection

To assist in haulage-equipment selection, types suitable for different diameter tunnels, different haul distances, and different tunnel grades are listed below.

1. For any length or size tunnel, first consideration should be given to tunnel equipment that is owned by the successful bidder. Contractor-owned equipment which has been partially written off on previous work has definite economic advantages. This equipment must be modern, or its increased operating cost may offset the savings in equipment write-off.

2. Tunnels with a grade of over 2 percent and less than 20 percent can be excavated most efficiently with rubber-tired equipment.

3. Tunnels with a grade over 20 percent are excavated with rail equipment that is winched up and down the slope.

4. Short tunnels with a grade of up to 20 percent, a height of over 12 ft (4 m) and a width that prevents the passing of trucks, and the first few hundred feet of long tunnels of similar characteristics can be driven with the least expense by the use of large, articulated, rubber-tired front-end loaders for both mucking and hauling. In this method, crew size can be held to a minimum with reasonable progress. The length of tunnel that can be economically driven in this manner increases as the amount of bucket payload increases. The largest bucket load is secured in well-graded fine material. In ground that produces a fine, well-graded muck pile, heading distances up to 1,500 lin ft (450 m) can be economically driven with one front-end loader, and if a passing niche is excavated and two front-end loaders are used, this distance can be doubled. This method has been successfully used in longer tunnels by hauling in stages. During the first stage the muck is hauled only a short distance, where it is distributed along the tunnel invert. After this stage, supports can be set and drilling can be resumed for the next round. The second stage involves loading and hauling the temporarily stored muck to the portal or shaft while work proceeds at the face.

5. Long tunnels with a grade under 2 percent, a height of under 14 ft (4 m), and heading distances over 2,000 ft (600 m) in length are, in the majority of cases, driven with rail-mounted equipment. This tunnel size does not furnish sufficient headroom for large track-mounted or rubber-tired-mounted muckers to load average-size trucks. Rail equipment is used because large-capacity rail-mounted muckers and cars have lower operating heights. In some cases, particularly in unlined tunnels, overexcavation to accommodate larger equipment may result in an overall reduction in tunneling costs.

6. Tunnels over 14 ft (4 m) in height with heading distances of up to 1 mile (1.6 km) are driven most economically with rubber-tired equipment. The cost of equipping a job with rubber-tired equipment is less than that with rail equipment, and the time required for the plant erection is reduced.

7. Tunnels over 14 ft (4 m) in height, with heading distances over 1 mile (1.6 km) in length, require economic studies to determine the type of equipment that should be used. As the length of the haul increases in a tunnel, the number of trucks necessary to serve the mucker increases. Each additional truck in a tunnel increases the ventilation requirements, and an economic balance is reached where increase in ventilation cost will outweigh the advantage of truck haulage. Since rail-mounted equipment does not require as much horsepower as rubber-tired equipment does to move the same yardage, rail equipment can be used for much longer haul distances with less ventilation than can rubber-tired equipment. Also on long haul distances, one locomotive and one operator can haul as much as several trucks and several truck drivers, which results in a saving in manpower, equipment maintenance, and equipment write-off compared with truck haulage operations.

TRUCK HAULAGE

Gasoline trucks are not allowed in underground construction, since they produce carbon monoxide in their exhaust. Therefore, all underground engines must be diesel, electric, or compressed air, and diesel trucks must be equipped with exhaust scrubbers, which remove most of the engine fumes and particles.

In selecting the type of truck to be used for a particular tunnel, the following criteria should be reviewed:

1. The truck bed must be low enough that the mucker will have sufficient headroom to discharge its loaded bucket into the truck.

2. The truck must not be too wide for the tunnel. In long tunnels, the trucks must be able to pass in the tunnel, or niches must be excavated to provide passing areas.

3. The combined horsepower of the trucks that will be in the tunnel at one time, plus the horsepower of the mucker, must not exceed reasonable ventilation capacity.

A number of manufacturers produce trucks especially designed for underground haulage. These are characterized by a low profile and by an arrangement whereby they can travel speedily and conveniently in either direction, thus avoiding the need to be turned around underground, where space is usually very limited. Such trucks come in a range of sizes from ½ to 10 yd³.

In large tunnels or in large underground chambers where turning around is not a problem, off-highway rear-dump trucks or two-wheeled tractors with rear-dump trailers are useful. Capacities range upward from 8 yd³. Figure 10.1 shows a typical rubber-tired haul unit.

RAIL HAULAGE

In the planning of rail haulage methods and equipment, decisions must be made on the type and size of the muck trains, size and capacity of muck cars, type and capacity of locomotives, train-switching facilities, car-changing equipment, and the size of rail.

Figure 10.1 Dual-control dump truck. (Aveling-Barford, Ltd., and George M. Philpott Co., Inc.)

Muck Trains with Individual Car Loading

In many tunnels, muck cars that are loaded individually by the mucker are used. This provides flexibility in train operation, since the cars can carry as large a piece of rock as the mucker can load and additional trains can be added or released from the operation as desired. When this type of train is used, cars must be continually and rapidly switched to the mucker so it can operate at full capacity. This requires that train-switching facilities be installed in the tunnel and that a fast car-changing method be provided for servicing the mucker. (Equipment and methods that are used for train switching and car changing are described later in this chapter.) With this type of train, passing clearance in the tunnel limits the width of the equipment that can be used. Drawings should be made to check the passing clearances of jumbos, muckers, locomotives, and cars.

In small, unlined tunnels, instead of using small equipment that is able to pass in the tunnel, it may be more economical to use larger equipment and overexcavate areas for passing tracks and for changing cars. In small tunnels one or two trains can usually handle the muck from one excavation round, so train-switching facilities are only required at regular intervals along the tunnel line. However, to change cars at the mucker quickly, the car passer should be kept near the face, which necessitates overexcavation of car-passing niches at 150-ft (45-m) spacing. The cost of the overexcavation necessary for the use of this larger equipment is generally less than the slowdown in operations caused by using smaller equipment. An example of this planned overexcavation is shown by the clearance sections illustrated in Fig. 10.2. When this type of operation is planned, similar cross-sectional clearance drawings should be prepared. It is also necessary to prepare plan drawings showing the length and location of the passing tracks and the length and frequency of the car-changing niches and to check the swing clearance required by the jumbo when it is switched onto a passing track.

Small concrete-lined tunnels are constructed with less cost if equipment is selected that can pass in the tunnel. In constructing these tunnels, any savings resulting from the use of large equipment is less than the extra cost of excavating and then backfilling with concrete the areas where overexcavation for this equipment is required.

Trains Loaded by Conveyor Belt

Conveyor-belt train loading is shown in Fig. 10.3. In this method, car changing at the mucker is not required and train-switching facilities are not necessary if each train has capacity to haul all the muck shot in one round. If the round contains more muck than can be handled by one train, train-switching facilities are required in the tunnel. To use this method, the tunnel must have sufficient diameter to provide clearance for the belt and the belt gantry, and

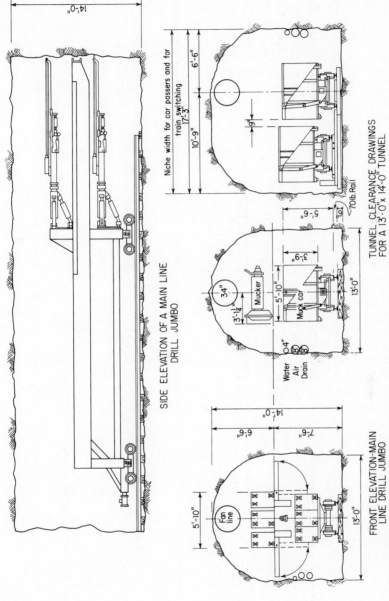

SIDE ELEVATION OF A MAIN LINE DRILL JUMBO

FRONT ELEVATION-MAIN LINE DRILL JUMBO

TUNNEL CLEARANCE DRAWINGS FOR A 13'-0" x 14'-0" TUNNEL

Figure 10.2 Main-line jumbo and clearance diagrams for a small tunnel.

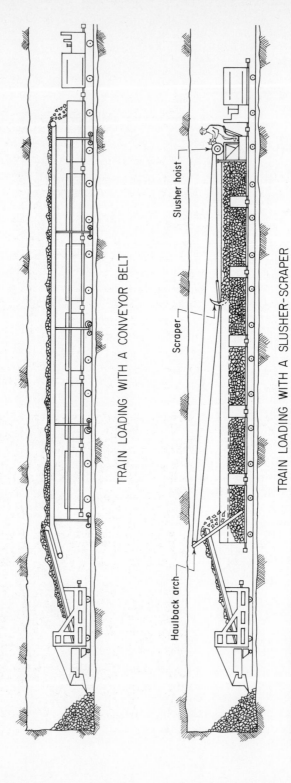

TRAIN LOADING WITH A CONVEYOR BELT

Slusher hoist

Scraper

Haulback arch

TRAIN LOADING WITH A SLUSHER-SCRAPER

Figure 10.3 Methods of loading trains without car changing.

the muck must break fine enough for the belt to handle. These two limitations restrict the use of this type of train loading when the tunnel is excavated in the conventional manner. When the tunnel is excavated with a tunnel-boring machine, muck is removed from the face in small pieces. Since no equipment other than the tunnel-boring machine must have access to the tunnel face, the gantry for the belt causes no equipment interference. Therefore, when tunnels are excavated with boring machines, belt loading of trains is generally used. In large-diameter tunnels driven with boring machines, surge bins may be located at the end of the belt to provide excavation surge storage and to increase the speed of train loading. In larger tunnels it may be feasible to put a tripper on the conveyor belt to load trains alternately on the two sides of the conveyor, thus eliminating the loss of time otherwise required for switching. This system was used on a Robbins tunnel-boring machine on the Washington, D.C., Metro. The trains were stationary during loading, while the tripper moved back and forth along the conveyor.

If belt loading of trains is used when the tunnel is excavated in the conventional manner, a main-line drill jumbo must be used, since a gantry drill jumbo could not pass by the conveyor-belt gantry when the change is made from drilling to mucking. The conveyor-belt gantry must have passing clearance for the mucker and the main-line drill jumbo. This limits the use of this method of train loading to tunnels over 17 ft (5 m) in diameter.

Trains Loaded with Slusher Scraper

As illustrated in Fig. 10.3, the slusher scraper operates in a trough erected on top of and between the cars. The trough is constructed with a closed bottom between the cars and an open bottom over the cars. The lead car in the train is loaded by a mucking machine, and as the muck builds up in this car the machine is pulled back by the scraper to load the other cars in the train. To use this train-loading method, the muck should break fine to permit efficient scraper operation and the scraper should have enough capacity to keep up with the mucker. This train-loading system has been used successfully with all sizes of cars. For example, it was used by Peter Kiewit and Sons' Company in 1959 on the excavation of a tunnel for the Western Pacific Railroad in the Feather River area of California.[10.1] On this project, full-sized railroad gondola cars were adapted for this method of loading. In general, however, it has been most successful with small cars used in small-diameter tunnels where the rock broke into small particles.

Conveyor-Belt Trains

Literature is available concerning trains composed of articulated cars with a conveyor belt forming the bottom of the train. Such cars have side walls, but no end walls. The loading procedure is to load the end car in the train with a mucker; the belt then moves the muck back over the length of the train. The train is unloaded by reversing the belt.

Muck Cars

The most commonly used muck car is the non-self-dumping, side-dump type which has a low bed height and a large muck capacity. This type of car must be dumped with a car dumper as shown in Fig. 10.4.

Another type of muck car is fitted with couplings which can rotate axially in order that the cars can be dumped one or two at a time by a rotary car dumper without need for uncoupling. This arrangement is particularly valuable where muck must be dumped into a hopper, as for removal via a shaft, or for loading trucks at a portal for haulage to a distant spoil-disposal area. Table 10.1 shows the dimensions of typical side-dump muck cars.

The widest cars that can pass each other in a tunnel should be selected in order to reduce the number of car changes required in mucking out a round. The ability of the mucker to spread the load in the car may limit the car's length. Clearance under the mucker discharge point limits the height. If the car is connected to the mucker during the mucking operation, the mucking machine's motive capacity may limit the car's loaded capacity.

Track Gauge and Rail Size

Many contractors standardize on 36-in gauge, as this allows them flexibility in moving the equipment from one job to another. Moreover, because most

TABLE 10.1   Dimensions of Typical Side-Dump Muck Cars

	Capacity of Car			
	8 yd³	9 yd³	10 yd³	12 yd³
Track gauge	36 in	36 in	36 in	42 in
C_L cplg above top of rail*	16 in	16 in	16 in	16 in
Length C_L to C_L cplg*	15 ft 1¼ in	14 ft 4¼ in	15 ft 1¼ in	15 ft 8 in
Wheel base	6 ft 0 in	5 ft 7 in	6 ft 0 in	6 ft 6 in
Length inside	12 ft 0 in	11 ft 3 in	12 ft 0 in	12 ft 11 in
Depth inside	3 ft 9 in	3 ft 9 in	3 ft 9 in	4 ft 0 in
Overall width	4 ft 9 in	5 ft 9 in	6 ft 0 in	6 ft 6 in
Overall height	6 ft 6 in	6 ft 6 in	6 ft 6 in	5 ft 9 in

* C_L cplg = centerline coupling

equipment is 36-in gauge, it has good resale value. Small equipment of narrower gauge may be used. On large tunnels 42-in gauge may be used to provide a more stable track and a wider mucker cleanup width. On some tunnels equipped with specialized slusher trains or other specialized equipment, standard railroad gauge has been used. The weight of rail is determined by the maximum wheel load and the intended tie spacing. Tables 10.2 and 10.3 can be used in determining rail size in a tunnel. The larger the rail used in a tunnel, the more stable the track and the fewer the derails. It is false economy to try to use too light a rail.

Types of Locomotives

Tunnel locomotives are available with diesel engines, battery-powered electric motors, or electric motors receiving energy from a direct-current overhead trolley wire supplemented by batteries.

Battery locomotives were widely used in the past, except where long hauls were required. Later, as a result of improvements in diesel locomotives, battery locomotives were used principally in locations such as compressed-air tunnels, where the use of diesel equipment could not be tolerated. In late years there has been a strong revival in the use of battery locomotives because of more stringent requirements concerning the quality of air in underground operations.

Diesel locomotives are utilized in the majority of tunnels under construction. In comparison with battery and battery-trolley locomotives, they have larger engines and are capable of greater speeds, which results in the use of fewer locomotives and in lower capital expenditures and operating and maintenance costs. Besides eliminating the need for battery maintenance and the use of hazardous trolley wires, diesel locomotives make bonding of the rails and the use of rectifiers unnecessary and reduce the number of transformers required. Since they are independent of any outside power source, they allow more flexibility in operation and more constant use. The disadvantage of the

TABLE 10.2 Recommended Maximum Load, One Wheel, in Pounds*

Weight of Rail* (lb/yd)	Tie Spacing (in)			
	24	30	36	42
8	800	600	500	400
12	1,800	1,300	1,100	1,000
16	2,700	2,200	1,800	1,500
20	3,800	3,100	2,500	2,100
25	4,700	3,800	3,100	2,700
30	6,700	5,400	4,500	3,900
35	8,100	6,400	5,400	4,600
40	9,700	7,700	6,400	5,500
45	11,300	9,100	7,600	6,500
50	13,300	10,600	8,900	7,600
55	15,300	12,300	10,200	8,800
60	17,700	14,100	11,600	10,000

* Standard lengths: 30 ft for weights up to 45 lb/yd.
   33 ft for 50 lb/yd or heavier.

SOURCE: Harold W. Richardson and Robert S. Mayo, *Practical Tunnel Driving*, McGraw-Hill, New York, 1941, p. 111.

diesel locomotive is that it produces fumes which increase the tunnel ventilation requirements. To reduce the amount of fumes, exhaust scrubbers are required on all diesel engines operated underground. Figure 10.5 shows a typical diesel locomotive for underground service, and Table 10.4 lists representative diesel locomotives.

TABLE 10.3 Dimensions and Weights of Light Rail, ASCE Section

Weight of Rail (lb/yd)	Height and Width of Base (in)	Width of Head (in)	Section Modulus	Weight of Rail for 100 ft of Track (lb)	Weight 1 pr. Splice Bars and 4 Bolts (lb)	Size of Bolt (in)	Size of Spike (in)
8	1 9/16	13/16	0.32	533	2.45	3/8 × 1 1/2	3/8 × 2 1/2
12	2	1	0.63	800	4.24	1/2 × 1 3/4	3/8 × 2 1/2
16	2 3/8	1 11/64	1.01	1,067	5.16	1/2 × 1 3/4	3/8 × 3
20	2 5/8	1 11/32	1.43	1,333	5.69	1/2 × 2	3/8 × 3 1/2
25	2 3/4	1 1/2	1.77	1,667	6.56	1/2 × 2 1/4	1/2 × 4
30	3 1/8	1 11/16	2.53	2,000	8.99	5/8 × 2 1/2	1/2 × 4
35	3 5/16	1 3/4	3.02	2,333	9.26	5/8 × 2 1/2	1/2 × 4 1/2
40	3 1/2	1 7/8	3.62	2,667	14.33	3/4 × 3	1/2 × 5
45	3 11/16	2	4.25	3,000	16.71	3/4 × 3	9/16 × 5 1/2
50	3 7/8	2 1/8	4.98	3,333	19.17	3/4 × 3 1/4	9/16 × 5 1/2
55	4 1/16	2 1/4	5.75	3,667	31.81	3/4 × 3 1/2	9/16 × 5 1/2
60	4 1/4	2 3/8	6.62	4,000	35.33	3/4 × 3 1/2	9/16 × 5 1/2

SOURCE: Harold W. Richardson and Robert S. Mayo, *Practical Tunnel Driving*, McGraw-Hill, New York, 1941, p. 111.

Figure 10.5 Diesel locomotive for underground service. (Plymouth Locomotive Works, Inc.)

Locomotive Selection[10.2]

Preliminary selection of a locomotive can be made by determination of the weight of the locomotive required to provide traction, the maximum horsepower required to accelerate the loaded train to a reasonable speed, and the continuous horsepower required to maintain this speed. If long stretches of haulage are on steep grades, careful consideration must also be given to braking requirements. If battery locomotives are being considered, frequency of battery recharging must be determined. In every case, final equipment selection should be governed by the guaranteed characteristics stated on the manufacturer's specification sheets, with due consideration of probable loss of efficiency as the equipment ages.

The tractive effort that can be exerted by a locomotive is a function of

TABLE 10.4   Diesel Locomotives for Underground Service*

Series	Weight (tons)	Track Gauge	Length	Width	Height
TMDR	3–6	18–36	10 ft 6 in	42–48	52
FMD	5–7½	18–56½	11 ft 0 in	42–72	52½
HMD	5–8	18–42	11 ft 0 in	41–54	56
DMD	8–16	23⅝–66	12 ft–14 ft 5 in	46–78	70
JMD	15–25	22½–66	16 ft 0 in	66–75	82
MMD	25–45	30–66	19 ft 8 in	98	104¾

* Various engines, manufactured by Plymouth Locomotive Works, Inc.

locomotive weight and of the coefficient of friction between the steel wheels and the steel rails. Although the coefficient of friction decreases as locomotive speed increases, the decrease is not significant in the ordinary range of tunnel haulage speeds. The coefficient of friction varies from 0.15 for wet, slick track to 0.25 for dry track. The tractive force which a given locomotive can exert is stated as

$$T = W_L F \qquad (10.1)$$

where T = tractive effort, lb
         W_L = locomotive weight, lb
         F = coefficient of friction

The tractive effort required of a locomotive is a function of the weight of the train (including the locomotive) and of the resistance to be overcome. The resistance to be overcome is threefold: rolling resistance, grade resistance, and acceleration resistance.

1. *Rolling resistance* is caused by the deformation of rails and wheels under the weight carried by the wheels. For steel wheels on steel rails, this is approximately 20 lb per ton of train weight.

2. *Grade resistance* is the force which must be overcome in lifting the weight of the train on an uphill grade. On downhill grades, grade resistance is a negative quantity. Grade resistance is approximately 20 lb per ton of train weight per 1 percent of grade.

3. *Acceleration resistance* is the force required to accelerate the train. It is a function of the rate of acceleration. It amounts to 90 lb per ton of train weight per mph per second. Usual acceleration rates in tunnel service are in the range of 0.1 to 0.2 mph per second.

The tractive effort required of a locomotive may be greater to return a string of empty cars upgrade than to move loaded cars downgrade, but it will frequently be necessary in any tunnel to move loaded trains in either direction during switching, train makeup, and other operations, and locomotives should be selected on this basis.

The required tractive effort for a given train during acceleration can be stated as

$$T_T = \frac{(NW_C + W_L)(R_R + R_G + R_A)}{2,000}$$

where T_T = total tractive effort, lb
         N = number of cars
        W_C = gross weight of 1 car, lb
        W_L = weight of locomotive, lb
        R_R = rolling resistance, lb/ton
        R_G = grade resistance, lb/ton
        R_A = acceleration resistance, lb/ton

This is restated for use as

$$T_T = (NW_C{}^1 + W_L{}^1)(20 + 20G + 90A) \qquad (10.2)$$

where N = number of cars
$\quad\quad\quad W_C{}^1$ = gross weight of 1 car, tons
$\quad\quad\quad W_L{}^1$ = locomotive weight, tons
$\quad\quad\quad G$ = grade, percent
$\quad\quad\quad A$ = acceleration, mph

The required tractive effort after acceleration is stated

$$T = (NW_C{}^1 + W_L{}^1)(20 + 20G) \qquad (10.3)$$

The foregoing remarks and formulas pertain only to the weight requirements for locomotives. The second factor in selecting a locomotive, whether diesel or electric, is the peak power requirement for accelerating the train. The formula is derived in many texts from mechanics and is stated as

$$P_P = \frac{T_T S}{375E} \qquad (10.4)$$

where P_P = required peak power, hp
$\quad\quad\quad T_T$ = total tractive effort, lb
$\quad\quad\quad E$ = efficiency of locomotive
$\quad\quad\quad S$ = speed after acceleration, mph

The final factor applicable to all types of locomotives is the sustained power requirement necessary to keep the train moving at a constant speed. This is considered separately because diesel engines and electric motors are not capable of delivering power continuously at peak output. Attempting it will cause early failure of locomotive components.

The required continuous power is stated as

$$P_C = \frac{T_R S}{375E} \qquad (10.5)$$

where P_C = required continuous power, hp
$\quad\quad\quad T_R$ = tractive effort for rolling resistance and grade resistance applied to train weight
$\quad\quad\quad E$ = efficiency of locomotive
$\quad\quad\quad S$ = speed, mph

Equations (10.1) to (10.5) are adequate to determine the power and weight requirements for locomotives. For example, it is required that a locomotive be selected to handle muck trains under the following conditions:

Problem

Tunnel, 16-ft horseshoe, neat line
Pull, 8 ft of tunnel per round
Maximum grade, 0.4 percent adverse
Muck cars, 12 yd³, weighing 13,800 lb empty
Desired acceleration, not less than 0.1 mphps
Tracks damp, but no slippery clay; assume $F = 0.20$
Desired train speed, 10 mph
Rock density, 5200 lb/bank yd³

Solution

Determine the volume of muck from one round:
Neat-line 16-ft horseshoe
Use 17-ft horseshoe; area, 240 ft²
Volume of 8-ft round, 240 × 8 ÷ 27 = 71 yd³ of rock
Volume of muck, 100 percent swell, 142 yd³

Determine the number of cars required and the loaded weight:
Number of 12-yd³ cars required, 142 ÷ 12 = 12 cars
Weight of muck of one car, 12 × 2,600 = 31,200 lb
Weight of one empty car, 13,800 lb
Weight of one loaded car, 45,000 lb = 22.5 tons

First trial, a 15-ton locomotive capable of delivering 225 hp intermittently or 175 hp continuously:
Tractive effort available:

$$T = W_L F \tag{10.1}$$
$$T = 30,000 \times 0.20 = 6,000 \text{ lb}$$

Required tractive effort:

$$T_T = (NW_C{}^1 + W_L{}^1)(20 + 20G + 90A) \tag{10.2}$$
$$T_T = [(12 \times 22.5) + 15][20 + (20 \times 0.4) + (90 \times 0.10)]$$
$$T_T = 10,545 \text{ lb}$$

Compare available tractive effort with required tractive effort: 6,000 lb available, 10,545 lb required. Not acceptable, but proceed.
Determine the peak power requirement, assuming 80 percent locomotive efficiency:

$$P_P = \frac{T_T S}{375 E} \tag{10.4}$$

where $P_P = \dfrac{10,545 \times 10}{375 \times 0.80} = 352 \text{ hp}$

Since the 15-ton locomotive has peak power capacity of only 225 hp, it is apparent that one might operate 6-car trains with 15-ton locomotives or 12-car trains with 25- or 30-ton locomotives. Because the smaller locomotives are more generally useful, assume 15-ton locomotives hauling 6 cars and recompute (10.2) and (10.4).

$$T_T = (NW_C{}^1 + W_L{}^1)(20 + 20G + 90A) \qquad (10.2)$$
$$T_T = [(6 \times 22.5) + 15][20 + (20 \times 0.4) + (90 \times 0.10)]$$
$$T_T = 5,550 \text{ lb} \qquad \text{Acceptable; 6,000 lb available}$$

$$P_P = \frac{T_T S}{375E} \qquad (10.4)$$
$$P_P = \frac{5,550 \times 10}{375 \times 0.80}$$
$$P_P = 185 \text{ hp} \qquad \text{Acceptable; 225 hp available}$$

Determine the tractive effort required after the train has been accelerated:

$$T_R = (NW_C{}^1 + W_L{}^1)(20 + 20G) \qquad (10.3)$$
$$T_R = [(6 \times 22.5) + 15][20 + (20 \times 0.4)]$$
$$T_R = 4,200 \text{ lb}$$

Determine the power requirement after acceleration:

$$P_C = \frac{T_R S}{375E} \qquad (10.5)$$
$$P_C = \frac{4,200 \times 10}{375 \times 0.8}$$
$$P_C = 140 \text{ hp} \qquad \text{Acceptable, since the 15-ton locomotive is rated 175 hp for continuous operation}$$

A 15-ton diesel locomotive, or an electric locomotive of equal capabilities, will be able to meet the stated conditions. However, the 15-ton battery locomotives available are found to be rated at only 90 hp. A 30-ton battery locomotive is available which is rated at 250 hp for continuous duty. It is immediately evident that the heavier locomotive will be capable of the required tractive effort, but it is necessary to check on the horsepower requirements for the heavier train.

The required tractive effort to start and accelerate the train is

$$T_T = (NW_C{}^1 + W_L{}^1)(20 + 20G + 90A) \qquad (10.2)$$
$$T_T = [(6 \times 22.5) + 30)][20 + (20 \times 0.4) + (90 \times 0.10)]$$
$$T_T = 6,105 \text{ lb}$$

The required power to start and accelerate the train is

$$P_P = \frac{T_T S}{375E} \qquad (10.4)$$
$$P_P = \frac{6,105 \times 10}{375 \times 0.80} = 204 \text{ hp} \qquad \text{Acceptable; 250 hp available}$$

Since the battery locomotive can start the train without exceeding the continuous-duty rating, it is evident that it will easily handle the line haul. It is then necessary to consider the frequency of battery recharging. Assume a 3-mile haul and two batteries of 121.6 kWh per battery, or 243 kWh total.

The weight of the loaded train is $(22.5 \times 6) + 30 = 165$ tons
The weight of the empty train is $(6.9 \times 6) + 30 = 72$ tons

Work Performed per Round Trip

Condition	Grade (%)	Weight (tons)	Resistance (lb/ton)	Tractive Effort (lb)	Distance (ft)	Work (ft-lb)
Loaded	+0.4	165	28	4,620	15,840	73,180,800
Empty	−0.4	72	12	864	15,840	13,685,760
Subtotal						86,866,560
25% additional for spotting, switching						21,716,640
Total						108,583,200

The number of kilowatt hours of energy required to perform a given amount of work is derived in standard texts. It can be stated as

$$K = \frac{F}{2,654,155E} \qquad (10.6)$$

where K = kWh
F = ft-lb
E = efficiency

Assuming an overall efficiency of 0.63, the power required per round trip is

$$K = \frac{108,583,200}{2,654,155 \times 0.63} = 65 \text{ kWh}$$

Thus, the 243-kWh battery set provides sufficient power for three round trips. To reduce the time lost in changing batteries, one might consider the use of a direct-current trolley system and using battery power only at the face and in other areas not served by the trolley wire.

TRAIN SWITCHING

To prevent delays in the tunnel-driving operations, it is necessary to have double trackage near the heading for switching trains and also space both for parking the mucking machine when it is not in use and for storage of cars loaded with drill steel, supports, powder, vent pipe, and other supplies. If a main-line jumbo is used, it is also necessary to provide a parking place for it during the mucking cycle.

In tunnels that are too small for trains to pass, this double trackage must be installed at intervals along the tunnel line that are overexcavated to provide the clearance required. As an example of this, refer to Fig. 10.2, which shows the amount of overexcavating required in a 13- by 14-ft tunnel.

In larger tunnels that have sufficient space to enable pieces of equipment to pass each other, portable passing tracks can be used. Two portable passing tracks are needed near the heading; the first one is used for train switching, and the second one is used for storing the mucker, drill jumbo, and cars loaded with supplies. Other portable passing tracks are kept along the tunnel so trains can pass each other on the trip from the portal to and from the heading. These portable passing tracks are called *California switches,* and a detailed description of them is included under the following section on car changing.

Car Changing

When the mucking method used requires that the mucker load individual cars, a fast car-changing method is used for removing the loaded car from the mucker and replacing it with an empty car. This can be done by one of the following five methods of changing cars.

California Switch

The California switch consists of a portable combination of a double siding and switches constructed on a structural mat which is laid over the main track and can be readily moved forward without disturbing the main track. Not only does the California switch allow train switching, but it can be used for switching cars to the mucker. Car changing can be done faster with a California switch than by any other method, since fewer moves are involved. Other methods of car changing require the use of only one locomotive at the face, but the fastest use of the California switch for car changing requires two locomotives.

When cars are changed by a California switch, one switch is located close to the face directly behind the mucker for car changing, and one switch is located a short distance down the tunnel for train switching, the parking of jumbos, muckers, etc. When the switch is used for changing cars, the loaded car is pulled away from the mucker by a locomotive working on one siding, and an empty car is then coupled onto the mucker by a locomotive operating from the other siding. When enough cars are loaded to constitute a train they are hauled away and replaced by a train of empty cars. An illustration of this use of a California switch is shown in Fig. 10.6.

The California switch is built on a structural frame which limits its use to straight sections in the tunnel. The width of the tunnel must be of sufficient size to allow two trains to pass and to allow the California switch to be moved forward as the heading advances. The California switch is a rigid structure, so it must be dismantled to take it around short-radius curves. During this dismantling, moving, and erecting operation, car changing is done by the use of car passers.

The Floor

Floors are used for train switching, car passing, and track laying. They contain mucking tracks and a storage siding for the mucker, provide a smooth, slick mucking surface, and have gantry trackage so that the gantry drill jumbo

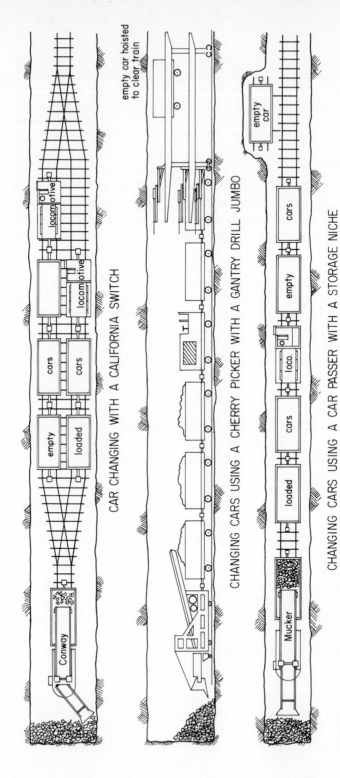

empty car hoisted
to clear train

CAR CHANGING WITH A CALIFORNIA SWITCH

CHANGING CARS USING A CHERRY PICKER WITH A GANTRY DRILL JUMBO

CHANGING CARS USING A CAR PASSER WITH A STORAGE NICHE

Figure 10.6 Car-changing methods.

may be advanced to the tunnel face for drilling and powder loading and then moved to the rear of the floor during the shooting and mucking cycles.

Jacobs Floor.   The Jacobs floor, shown in Figs. 10.7 and 10.8, is suitable for the operation of one mucking machine. It is a three-section California switch, approximately 400 ft (120 m) long, equipped with hydraulic jacks installed where the sections join. To move the floor, the front section is first jacked forward approximately 3 ft (1 m) at a time, with the two rear sections used as dead weight anchors; then the remaining sections are jacked forward individually, with the other sections used as anchors. The floor is maintained in elevation by the depth of cleanup ahead of the front section and the position of the mucker on the first section during its forward movement. When the floor gets below grade, it is raised by placing the mucker in the back of the first section as it is moved. When the floor starts above grade, it is dropped by keeping the mucker in the front of the first section during its moving operation. When the floor gets off line, it is brought back on line by jacking from the tunnel wall. Excavation of the bottom of the tunnel must be watched so no high spots are left, or the floor will ride up the high spot and continue to rise in elevation as the excavation proceeds.

The floor is equipped with one track entering it from the rear and one track

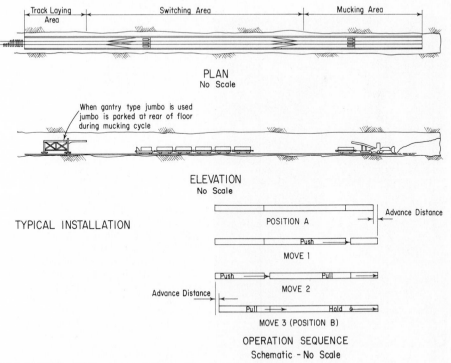

Figure 10.7   Sliding tunnel floor. (Jacobs Associates.)

in the front that is used by the mucker. The remainder of the floor is equipped with four rails, each set on gauge and equipped at the ends with triple switches, so that together they form one main-line track down the center or two passing tracks. The gantry drill jumbo also travels on the two outside rails, so they are run the full length of the floor to permit the jumbo to be stored at the rear during the mucking operation and moved to the face for the drilling operation.

The forward part of the front section where the mucker works is covered with closely spaced rails which provide a smooth, slick bottom for the mucker bucket to slide on while it is working.

The single track at the rear of the last section consists of a supported steel plate on rail gauge, with space under the plate so that standard lengths of rail can be placed under these plates and left in position as the floor is moved forward. Ties can then be slipped under these rails, tie plates placed, and the rail spiked as required.

Standard operation with this floor is to do final cleanup with the mucker and one car. Meanwhile, the jumbo is pushed by a locomotive astride the mucker. After cleanup, the mucker moves the jumbo to its final location, and drilling is started. After drilling and loading the holes, the jumbo is moved to the back of the floor, the round is shot, and the mucker is moved up to the

Figure 10.8 Sliding floor used by Walsh Construction Company for the railroad relocation tunnels near Libby damsite, Montana. (Jacobs Associates.)

muck pile and mucking commenced. The mucker mucks off the slick-plate section of the floor, and the only bottom cleanup is underneath the last round.

Empty cars are stored on one of the passing tracks with one locomotive to shove them to the mucker. Another locomotive pulls the loaded cars from the mucker onto the other passing track. As empty trains are brought in, the locomotives change tracks and the loaded trains are taken out of the tunnel. As mucking is being carried forward, the floor is moved toward the face. When the mucker is cleaning up, the drill jumbo is again moved to the face and operations are repeated.

The advantages of this operation are:

1. Quick car changing is provided so the mucker is not delayed.
2. Neither mucker rails nor gantry rails have to be laid at the heading, which saves time.
3. The jumbo can be moved to the heading while the mucker is cleaning up.
4. The mucker does most of the mucking on a slick plate, which reduces both cleanup and mucking time.
5. The mucker operates from a firm track instead of from an unstable temporary track, which improves mucking operations.
6. The only track laying required is bolting up, placing ties and tie plates, and spiking standard length rails at the rear of the jumbo, which cuts time and labor in the heading.

This floor works best where a gantry jumbo can be used and the bottom width of the tunnel is over three times the gauge of the track. Jacobs floors have been developed for use in tunnels of such widths that two muckers can operate alongside each other. These floors contain three passing tracks which switch into two mucking tracks in the front of the floor and into the main-line track at the rear. This allows each mucker to have one locomotive servicing it during the mucking cycle. These locomotives operate on the two outside tracks, and each outside track is equipped with a car passer. Both car passers shift empty cars to and from the center track. With a separate car-changing arrangement for each mucker, and with each mucker loading a car in less than 4 minutes, the total mucking capacity of the floor is one car in less than 2 minutes.

Navajo Carpet. This is a floor designed and built by Ray Moran for use in small tunnels that have only enough width for two passing tracks. The track arrangement is similar to the track arrangement on a California switch. There is one track in the front of the floor on a slick plate which is used for mucking operations and by a main-line jumbo during the drilling operations. The passing tracks are used for switching the mucker and the main-line jumbo between the drilling and mucking operations. These passing tracks are also used for car changing during the mucking operation. A stub track is used for drill-jumbo and mucker storage; it incorporates the same track-laying facilities as the Jacobs floor, except that rails are fed into the track in the front of the floor rather than in the rear. The floor is fabricated in one piece and jacks itself forward as the heading advances, with jacks reacting against the main-line track.

Car Passers

To maintain production in small-diameter, unlined tunnels, equipment is selected that cannot pass other units except at overexcavated, double-tracked train-switching areas. This requires that car changing be done with car passers, and it is necessary to cut niches in the tunnel wall to allow storage of the empty cars at the car-passing locations. Car passers are also used around short-radius turns in large-diameter tunnels that ordinarily change cars with a California switch. Another use for car passers is on a California switch when two muckers are used as described above. A minor use is in the outside yard area where they are used to store specialty cars such as powder or timber cars, and thus reduce the number of sidings required.

A car passer consists of side-by-side sections of track with short, knife-edge lead tracks. The movable track section is supported on rollers which travel on members perpendicular to the main track and enable the car to be pushed to the side to give clearance for passage of the train. When car passers are used at the heading, the locomotive's position in the train at the start of the mucking cycle is on the tunnel-face side of the empty cars. One car is attached to the mucker. When the first car is loaded, the train travels forward past the car passer to pick up the loaded car attached to the mucker and drops an empty car on the car passer. This empty car on the car passer is then pushed to one side to clear the track. The train in the meantime connects to and removes the loaded car from the mucker and then travels back to the tunnel past the car passer. The empty car is pushed back on the main-line track, and the train moves forward shoving the empty car to connect it to the mucker. At the same time it drops another empty car on the car passer. This process is repeated until the last empty car is connected to the mucker. At this point, the loaded cars are located in the train between the locomotive and the mucker. The loaded train travels out the tunnel past a siding where a trainload of empty cars is waiting. This empty train then moves in to the heading and the process is repeated. This car-changing method is shown in Fig. 10.6.

To maintain a good rate of mucking production, car changing must be done in as short an interval as possible. For this to be accomplished, train travel must be held to a minimum by locating the passing niches at frequent intervals. Preferable spacing of these niches is every 100 ft (30 m); maximum spacing is 200 ft (60 m).

In small-diameter, concrete-lined tunnels the car-passing niches and the overexcavation required for train switching will result in a large amount of overbreak concrete. This added concrete cost may be more than the savings in excavation cost resulting from the use of large equipment. When this is so, the tunnel should be excavated with equipment small enough to pass in the tunnel.

Cherry Pickers

If a tunnel has sufficient diameter that a gantry jumbo can be used, car changing may be done with a cherry picker located on the rear of the drill jumbo. Sufficient hoisting space must be provided in the jumbo to hoist a muck car clear of the train passing beneath it. The jumbo must also be equipped with hoist, cables, and hooks for raising the car. This cherry-picker arrangement on a gantry jumbo is shown in Fig. 10.6. The operation is similar to that of a car passer with the exception that the empty car is hoisted instead of being moved to one side. The cherry picker is an almost obsolete device, but it may still occasionally be useful.

The Grasshopper

This method of car changing has not been used in recent years because of its bulk and the necessity for extra rails. It requires the use of a steel frame about 150 (45 m) ft long, traveling on separate rails set on each side of the main-line track. There are hinged ramps at each end of the framework, which are operated by an air hoist. Tracks are laid up these ramps and over the deck of the framework. Six to eight empty cars are pulled up the rear ramp by a hoist and held on the top deck, and the rear ramp is then raised. After the mucker has loaded a car, the loaded car is pulled away from the mucker by a locomotive to a position clear of the front ramp of the grasshopper. The front ramp is lowered, and an empty car is let down the ramp and coupled onto the mucker.

EQUIPMENT UNIQUE
TO THE DRILL-AND-SHOOT METHOD

Drills

The tunnel-planning engineer has a wide range of choices in drill types and in methods of mounting drills on the drill jumbo. Literature from drill manufacturers can be obtained to gain specific information on the various drills, so only overall concepts will be discussed here.

Most of the drilling in tunnel construction is done with the following drills:

1. Percussion Drill with Rifle-Bar Rotation. This is the most common type of drill and is used in all types of rocks. It is the most satisfactory drill for work in hard rock.

2. Percussion Drill with Separate Positive Method of Drill Rotation. This type of drill is used in soft rock to get greater penetration than the straight percussion drill with rifle-bar rotation can give. When used in hard rock, bit cost may become excessive.

3. Rotary Drill. This drill gives a high penetration in extremely soft rock, such as shale.

4. Auger Drill. This is suitable for use in very soft rock of a hardness similar to coal, such as some volcanic tuffs. Augers are particularly useful in mudstone or claystone, which tend to plug the holes in percussion-type drill bits.

All underground drilling is done wet to reduce the amount of rock dust breathed by the miners. Wet percussion drills are the same as the outside dry percussion drills, except that water rather than air is used for clearing the cuttings out of the hole to prevent dust formation. The use of water has a secondary result in reducing the amount of compressed air required per drill.

In selecting the size and type of drill for short tunnels, consideration should be given to the possibility that the lowest total cost may be the combination of lowest capital cost with a higher direct cost. For short tunnels, capital cost may be reduced, with an increase in direct cost, by selecting light, hand-held drills or air leg-support jackhammers. Although these drills do not obtain as fast a penetration as the larger drills, they use smaller bits, require less capital outlay, and use less air.

In soft rock, a drill should be selected that will furnish good penetration but will also hold the capital cost to a minimum. As an example, the penetration rates of $3\frac{1}{2}$- and $4\frac{1}{2}$-in-size drifters in soft rock are quite similar; the smaller drifter is therefore used as it costs less and uses less air. Rotary drills may be selected as they are less expensive than percussion drills.

In determining the size and type of drill to be used in hard-rock tunneling, selection should be based on the fact that excavation progress is directly proportional to the time required to drill out a round. The fastest penetration in hard rock is obtained by using the largest drill that can be used with standard-size bits without having excessive steel breakage. However, the larger the machine, the greater the steel breakage, the greater the power requirement, and the greater the capital cost. All these factors must be considered in selecting the size of the drill.

Until recent years all percussion drills depended upon compressed air for power. In the past few years drills have been developed which are operated by hydraulic power, with the drills, the drill booms, and the power unit being supplied as a self-contained mobile machine, depending only on an electric power supply. These hydraulic drills are expensive, but they have higher penetration rates than comparable compressed-air drills. Compressed air is still needed for hand tools, cleanup, and other purposes, but the air requirement is greatly reduced.

Drills operated by compressed air are commonly identified by the size of the cylinder bore. Those now used in tunnel driving vary from $1\frac{7}{8}$ to $5\frac{1}{2}$ in. Percussion drills are classified as follows:

1. Sinkers and Jackhammers, Designed Primarily to be Hand-Held. They may also be mounted on a feed leg by an adaptor bracket. These drills are generally classified by weight and vary from a light drill of 30 lb to a heavy drill of 65 lb or more. Feed legs are air-activated, long-pipe jacks which can be used to position the drill and maintain pressure on it during drilling.

2. Feed Legs and Jack Legs, which Consist of Sinker Drills Attached in a Readily Demountable Manner to an Air Feed Leg. They are used for both lateral and overhead drilling. Sizes most generally are:

Bore of air feed cylinder	$1\frac{7}{8}$–$2\frac{5}{8}$ in
Length of feed travel	3–6 ft
Weight of sinker	30–70 lb

3. Drifters. These drills are of the self-rotating type, usually screw-fed by a gear or reciprocating piston motor on a steel or aluminum shell, or chain-fed by a vane or piston motor on a heavy steel channel frame. Sizes most generally used are:

Light	Up to 3-in bore
Medium	$3\frac{1}{2}$–4-in bore
Heavy	$4\frac{1}{2}$-in bore

4. Burn-Hole Drills. These are drifters of $5\frac{1}{2}$-in bore or over, used to drill the large holes required for relievers on a burn-cut pattern of shooting. The number of holes varies from one to three, and the holes are generally 5 inches in diameter. The steel used is generally $1\frac{7}{8}$ in.

5. Stopers. These are drills with an air feed which is usually designed as an integral part of the drill. Stopers are used on up-hole drilling and are classified according to weight, from a light drill of 75 lb to a heavy drill of over 100 lb.

Sinkers or jackhammers can be hand-held on the work platforms on the jumbo. Air legs can be used on these platforms or can be mounted on steel ladders which provide some of the advantages of the jib and drill positioners used with drifter drills.

Drifter drills are mounted on a horizontal support composed of three adjustable arms, called *jibs,* which are bolted to the jumbo. These jibs allow mechanical positioning of the drill at the tunnel face, and jibs are available which will rotate the drill for ease in drilling side holes and lifter holes. Also available are controls for these jibs, called *drill positioners,* which enable the miner to control the position of the drill relative to the tunnel face. The jibs are mounted on the drill jumbo so that one or more drills will be able to drill any spot in the tunnel face. The use of rotating jibs and drill positioners eliminates the need for chuck tenders and thereby reduces the size of the drilling crew.

Hydraulic drills for tunneling use are usually provided in sets of two to four drills driven by a single power unit, either electric or diesel. Each set is carried on a self-propelled mounting, crawler, rail, or rubber-tired. Steel changing is automated, and one driller can readily operate two drills. The equipment can, of course, be mounted on a rail jumbo in order to reach all parts of a large face.

Number of Drills

In determining the number of drills required at the heading, it may generally be assumed that for 12- to 14-ft-diameter standard horseshoe tunnels in granite, one hole will be required per 5 ft^2 of face area.

In smaller-diameter tunnels, the number of holes per square foot will increase to possibly one hole per 2½ ft.^{2} In larger-diameter tunnels the number of holes per square foot will decrease to possibly one hole per 6 or 7 ft.^{2} The use of a burn cut in the smaller-diameter tunnels will reduce the number of holes per square foot. Other factors that influence the number of holes required for a specific diameter tunnel will be the type and general formation of the rock (e.g., heavily jointed, blocky, or massive). For example, tunnels driven through certain granites will require as many as twice the number of holes as those driven through softer rocks. After the number of holes required for any tunnel is determined, it should be compared with the number required on previous tunnels of similar size and with similar rock conditions.

The number of drills is influenced by the fact that each driller will drill from seven to nine holes. The actual number of holes per drill will vary depending upon the rock formations. The type of hole pattern to be used will also determine the type of drills and length of steel changes on the drills.

A common practice in hard rock is to use large-diameter burn-cut holes, which necessitate the use of one large drill per heading. These burn holes furnish *relief* for the explosion (space for the rock to expand when an explosion occurs). This burn cut does away with the requirements for a relief action on the face furnished by the short diamond-cut or V-cut holes and allows each hole to be drilled the full length of the rounds. Furthermore, long feeds can be used on each drifter. Since the feed can be as long as the round, steel changes will not be required for individual holes, which reduces the labor requirements and decreases the actual drilling time but increases the cost of drill steel. The longer the steel, the more the breakage, and the broken steel cannot be reworked for shorter steel changes, as when diamond cuts or V cuts are used. When large drills are used, steel breakage can be reduced by using upgraded hardened steels with traveling centralizers and long feeds.

In closely supported ground, the rounds used are generally much shorter and often a diamond or V cut is used instead of the burn cut. In this case, the drifters should have shorter steel changes, such as a 6-ft change. This will also aid in positioning the drill for the angle holes required for the diamond or V cut.

Drill Jumbos

The tunnel face is drilled with drills suspended from jibs, or booms, and these jibs are mounted on movable frames called *drill jumbos.* Jumbos are equipped with work platforms to give access at a sufficient number of levels to cover the tunnel face, and they contain all facilities required for drilling a round, such as hydraulic pumps, air and water connections to the drills, lights, and equipment for auxiliary face ventilation. Besides being used while drilling out the round, drill jumbos are used as work platforms for loading the holes with explosives, for setting steel and placing timber supports, for drilling and placing rock bolts, and for supporting breast boarding in soft ground. The jumbo may also be equipped with a hoist (a cherry picker) for raising an empty car to a height that clears the train so that cars can be switched during the mucking cycle. The drill jumbo can either ride on rail, be equipped with crawler tracks, or be mounted on a diesel truck. Drill jumbos are of the gantry or main-line type, or in large tunnels may be designed to cover one-half the face.

A gantry jumbo rides on wheels supported near the sides of a tunnel and is designed so that the center work platforms may be collapsed to allow the mucker and haul units to pass to and from the tunnel face through the jumbo. Figures 10.9 and 10.10 illustrate this type of drill jumbo. Rail-mounted gantry

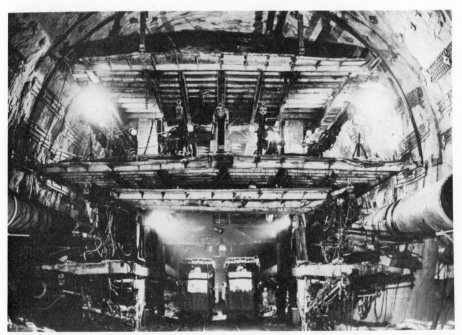

Figure 10.9 Gantry drill jumbo used on Mont Blanc Tunnel, France, (Ingersoll-Rand Co.)

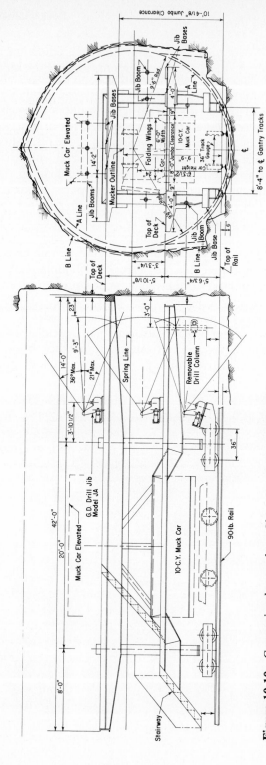

Figure 10.10 Gantry jumbo used on Clear Creek Tunnel, California.

jumbos are supported on special rails laid along each side of the tunnel. While at the face, the jumbo is used for drilling and for loading the holes. Before the charge is exploded, the gantry jumbo is moved back from the face a sufficient distance to protect it from the explosion. It is moved up again to the face for drilling after the mucking cycle.

A main-line jumbo (so called because it rides on the main rail line) is shown in Fig. 10.2. It has collapsible man platforms on the sides that are raised into position for use in the drilling, hole loading, and tunnel-support setting operations. The man platforms are dropped along the jumbo sides to allow space for car passing when it is moved back from the face and parked at a siding during the mucking cycle (which is done to allow access to the face for the mucker and haul units). After the mucking cycle is completed, the mucker must be moved back to a siding before the main-line jumbo can be moved into the drilling position at the face.

Clearance restrictions dictate that only main-line jumbos be used in tunnels that are less than 17 ft (5 m) in diameter. For tunnels over 17 ft (5 m) in diameter, either gantry or main-line jumbos may be used. Some superintendents favor gantry jumbos since they can be moved into the face more quickly than main-line jumbos. They may prefer to set steel, timber, and rock bolts from this type of jumbo. Other superintendents prefer main-line jumbos for any size tunnel, since gantry rail is not required, and they prefer changing cars with a California switch.

Truck drill jumbos are generally of the main-line type with collapsible side work platforms. On short tunnels, the truck jumbos may have fixed platforms which necessitate that they be removed from the tunnel after each drilling cycle to furnish access for the other equipment. On large-diameter tunnels, truck jumbos are often made to cover one-half the tunnel face. Two jumbos are then used side by side during the drilling cycle, and they can be parked end-to-end during the mucking cycle to allow passage of the mucking and hauling equipment. A truck jumbo is shown in Fig. 10.11.

Specification Requirements

Specification requirements for drilling should be checked closely. Some specification writers, because of specific design requirements calling for smooth and sound excavated rock surfaces, call for the *long-hole* technique or for the drilling and blasting procedure known as *smooth, shear, cushion, or periphery blasting*. These methods ensure that a shear plane is established between the periphery holes, which minimizes strain or cracks in the strata beyond the blasting perimeter.

Long-Hole Technique.   Under this method, a minimum heading is driven down the center of the tunnel. Then, at approximately 100-ft (30-m) intervals, the excavation is enlarged to the full diameter of the tunnel. From these enlargements, holes are drilled to the next enlargement on the periphery of

Figure 10.11 Truck-mounted drill jumbo. (Gardner-Denver Co. and George Philpott Co., Inc.)

the tunnel, and also sufficient holes are drilled in the face to break the rock properly. Diamond drills are often used to drill the periphery holes. Another type of drill used successfully with the long-hole drilling method is the drifter drill with independent rotation equipped with drill guides. The holes are loaded and shot; the center excavation provides relief for the explosion. Holes can be shot the full length between the enlargements but are generally plugged and shot in 20-ft (6-m) sections.

Periphery Blasting. This method is similar to the predrilling, presplitting method used on open-cut work. Periphery holes are drilled at the face on approximately 20-in centers and shot with light charges before the remainder of the face is shot. This procedure gives a shear plane which results in a smooth surface at this line.

Mucking Machines

The many different types of mucking machines suitable for tunnel excavation allow the construction planner a wide range in selection of this equipment.

Before a choice is made, state laws should be reviewed to see if there are restrictions on the use of diesel engines underground. The type of muck haulage and the diameter of the tunnel will be the other important factors in mucker selection. The capacity, horsepower, and dimensions of various muckers are presented in Table 10.5. To assist in mucker selection, suitable muckers for various tunnel sizes and methods of muck transportation are described on the following pages.

1. For a truck haulage setup
   a. In small tunnels between 6 and 12 ft (2 and 4 m) in height, Eimco 635 crawler-type loaders may be used for loading small trucks. See Fig. 10.12.
   b. Medium tunnels between 12 and 18 ft (4 and 5.5 m) in height have sufficient headroom for the use of the Eimco 115 crawler-type overshot loader. See Fig. 10.13.
   c. In tunnels over 18 ft (5.5 m) in height, Traxcavators with Libu side-dump buckets or Eimco overshot loaders can be used.
   d. For tunnels over 28 ft (8.5 m) in diameter, large front-end loaders can be used. Also these tunnels have sufficient clearance for the use of a crawler type of power shovel. Depending on local laws, these shovels are equipped with diesel engines or adapted for electric drive.
2. For a rail haulage setup
   a. A rocker type of rail-mounted air-operated overshot loader produced by Eimco and other firms is used in small tunnels.
   b. Eimco 40 H muckers are used in tunnels from 9 to 14 ft (3 to 4 m) in diameter. The 40 H mucker has a bucket which loads a mucker belt. See Fig. 10.14.
   c. Conway muckers are used in tunnels from 10 ft (3 m) in diameter to tunnels with a cleanup width of 23 ft (7 m). When the tunnel has a larger cleanup width than 23 ft (7 m), two Conways are generally used abreast. The Conway mucker consists of an electric-powered shovel loading a short muck belt. Figure 10.15 shows two Conways working abreast in the Mont Blanc Tunnel in France.
   d. Any type of mucker can be used with a slusher train, depending on clearance height. When a crawler-type loader is used with a slusher train and when it is necessary to move this mucker in and out of the tunnel, it is generally transported on a specially built flatcar.
3. Combination excavating and hauling machines
   a. Rubber-tired front-end loaders are used to portal-in long tunnels, to drive adits, and to drive short tunnels. The mucker in this application both loads and hauls muck out of the tunnel or adit.
   b. When headroom is restricted, low-profile, rubber-tired front-end loaders (similar to the Eimco 916) are used to perform the same work as other front-end loaders. Figure 10.16 shows an Eimco 916 used as a load, haul, and dump unit.

TABLE 10.5 Underground Excavating (Mucking) Machines*

	Bucket Size (yd³)	Power	Type of Power	Tramming Width	Relation of Cleanup Width, Operating Height, Tramming Height, and Track Gauge				Belt Width (in)	Mini-mum-Radius Curve (ft)	Operating Weight (lb)	Maximum Car Height	Maximum Car Length
					Cleanup Width	Tramming Height	Operating Height	Minimum Gauge (in)					
						Heights above Top of Rail							
Rail-mounted mucking on rubber belts													
Conway 100-2	1½	1–125 hp / 1–40 hp	Electric	6 ft 2 in	17 ft 0 in	8 ft 2½ in	12 ft 0 in	30	42	25	55,000		
					18 ft 6 in	8 ft 6½ in	12 ft 9 in	30					
					20 ft 0 in	9 ft 3¼ in	13 ft 5 in	36					
					21 ft 6 in	9 ft 10½ in	14 ft 5 in	39⅜					
					22 ft 4 in	9 ft 10½ in	14 ft 8 in	39⅜					
					23 ft 0 in	10 ft 3 in	15 ft 0 in	42					
Conway 100-1	1¼	1–100 hp / 1–30 hp	Electric	6 ft 2 in	16 ft 6 in	8 ft 2 in	11 ft 8½ in	30	38	20	50,800		
					18 ft 0 in	8 ft 7 in	12 ft 6 in	30					
					19 ft 6 in	9 ft 4 in	13 ft 4 in	36					
					21 ft 6 in	9 ft 10 in	14 ft 3 in	39⅜					
					21 ft 6 in	9 ft 10 in	14 ft 3 in	39⅜					
					23 ft 0 in	10 ft 8½ in	15 ft 1 in	42					
Conway 100	1	1–100 hp / 1–30 hp	Electric	5 ft 5½ in	14 ft 10 in	8 ft 0 in	11 ft 4 in	30	38	20	48,500		
					16 ft 0 in	8 ft 3 in	11 ft 7 in	30					
					18 ft 0 in	8 ft 6 in	12 ft 5 in	30					
					19 ft 6 in	9 ft 3 in	13 ft 3 in	36					
					22 ft 0 in	10 ft 2½ in	14 ft 9 in	36					
Conway 75	¾	1–75 hp / 1–25 hp	Electric	5 ft 5½ in	13 ft 0 in	7 ft 6¾ in	9 ft 7 in	24	28		42,000		
					15 ft 0 in	7 ft 6¾ in	10 ft 7¾ in	24					
					17 ft 4 in	8 ft 1 in	11 ft 9 in	30					
					19 ft 0 in	9 ft 1 in	12 ft 11 in	36					
					22 ft 0 in	10 ft 6¾ in	14 ft 14 in	36					

Machine	Capacity†	Power	Type	Width	Length	Dim 3	Dim 4	Grade %	%	Weight (lb)	Dim 5	Dim 6
Conway 50	⅜	1–50 hp / 1–25 hp	Electric	4 ft 3½ in	11 ft 4 in / 13 ft 3 in / 15 ft 0 in	6 ft 6 in / 6 ft 6 in / 7 ft 0 in	8 ft 1 in / 9 ft 1 in / 10 ft 2 in		28	27,000		
Eimco 40 H	½		Air or electric	6 ft 0 in	12 ft	5 ft 10 in / 6 ft 10 in / 6 ft 10 in		24 / 24 / 28		19,000	4 ft 9 in / 7 ft 2 in	9 ft 6 in / 15 ft 0 in
Eimco 40 W	½		Air or electric	6 ft 0 in	15 ft	5 ft 10 in / 6 ft 10 in		28½		19,300		
Rail-mounted mucking into haul unit												
Eimco 12 B	4½–6†		Air	34 in	75–83 in	48–51 in	78½–83 in	15		4,200	46–59 in	
Eimco 25	8–13½†		Air	43⅞ in	96–118 in	52–72 in	91–112 in	18		12,000	52–72 in	
Crawler-mounted mucking into steel flight conveyor belt												
Eimco 635	12½†		Air or electric	6 ft 0 in		4 ft 10 in / 8 ft 8 in	5 ft 0 in / 8 ft 8 in	45		15,000	7 ft	11 ft
Crawler-mounted mucking into haul unit												
Eimco 631	12½†		Air	5 ft 8⅜ in		6 ft 5 in	11 ft 2⅜ in			13,000	8 ft / 3 ft 9 in / 7 ft 2 in	
Eimco 630 E	8–14†		Electric	5 ft 8⅜ in		4 ft 11½ in	6 ft 8 in / 10 ft 7 in			11,300	9 ft 6 in / 11 ft 3 in	
Eimco 105 or 115	1½	143 hp	Diesel	7 ft 8 in		9 ft 6 in.	14 ft 0 in / 17 ft 8 in			42,500		
Caterpillar 977 w/Libu bucket	3	190 hp	Diesel	9 ft 3 in			17 ft 10 in			50,000		
Rubber-tired units												
Eimco 913	3	110 hp	Diesel	8 ft 5 in		7 ft 1 in	12 ft 7 in			28,000		
Eimco 918	8.5	277 hp	Diesel	9 ft 3 in		8 ft 0 in	16 ft 6 in			67,000		
Caterpillar 930	2	100 hp	Diesel	8 ft 0 in		9 ft 4 in	15 ft 8 in			21,200		
Caterpillar 966	4	170 hp	Diesel	9 ft 7 in		9 ft 7 in	17 ft 9 in			37,100		

* Information abstracted from catalogues furnished by the Goodman Division, Westinghouse Air Brake Co.; Eimco Mining Machinery Co.; The Caterpillar Tractor Co.

† (ft³)

307

Figure 10.12 Eimco 635 mucker. (Eimco Mining Machinery Co.)

    c. Self-propelled scrapers are often used in large-diameter tunnels to re-
move the bottom heading. In large-diameter tunnels driven with multiple drifts, these scrapers are used to remove the remainder of the
excavation after the center steel has been set.

Figure 10.13 Eimco 115 mucker. (Eimco Mining Machinery Co.)

Figure 10.14   Eimco 40 H mucker. (Eimco Mining Machinery Co.)

Road Headers

A road header is a machine which excavates by means of a rotating cutter head mounted on a boom designed so that it can pivot up and down and side to side, be extended or retracted. It is thus able to excavate circular, horseshoe, or other tunnel cross-sections. On some road headers the cutter head is mounted so as to rotate in the axis of the boom. On others it is mounted so as to rotate normal to the axis of the boom. All have oscillating muck-gathering arms on an apron at invert level; these arms feed the muck to a short conveyor which discharges at a height convenient for the hauling equipment. Road headers may be mounted on crawlers, may be mounted singly, or may be mounted in groups on a jumbo.

Road headers are very effective tools for the softer rocks. They have also been adapted for use in shields for soft-ground tunnels. Road headers have been used most widely in small horseshoe tunnels, but they have also been successfully applied to excavating small circular tunnels and to excavating larger tunnels by the heading-and-bench method. A typical road header is shown in Fig. 10.17. Figure 10.18 shows excavation for the Dixon Dam diversion tunnel, in which a road header was used for the top heading and a ripper for the bench.

Equipment for Portal Excavation

Usually the planned portal location will be a sloping rock surface covered with soil, and it will be necessary to remove soil and weathered rock to ensure a sound rock surface for the portal. This is most important, because a large proportion of all tunnel failures occur at the portals. The face should be nearly vertical to ensure that there is adequate cover (at least 1.5 diameters) over the tunnel crown.

Rock excavation at the portals is usually performed by drill-and-shoot methods. Presplitting or line drilling is required to ensure that damage to the remaining rock is minimized. Rock saws were used to cut chalk at the portals of the 12 power tunnels for Fort Randall Dam in South Dakota. Rock bolts are installed in areas which exhibit jointing, shearing, or foliation. The rock bolts should be supplemented with wire mesh and shotcrete if the rock is prone to ravel or air-slake. A series of long dowels, grouted or ungrouted, above the future tunnel crown will add stability to that vulnerable area. A canopy is usually erected beyond the face of the portal to divert earth or rock slides. Specialized equipment such as the Gradall shown in Fig. 10.19 is useful for scaling rock at the portals and within large-diameter tunnels.

Figure 10.15 Two Conway 100-1 mucking machines working abreast in the Mont Blanc Tunnel, France. (Goodman Division, Westinghouse Air Brake Co.)

Figure 10.16   Eimco 916 low-profile front-end loader. (Eimco Mining Manufacturing Co.)

TUNNEL-BORING MACHINES

Description

A tunnel-boring machine (TBM) is a large drill. It has two principal parts. One part is stationary during the boring operation. It has grippers to transmit to the tunnel walls the torque and thrust reactions from the rotation and advance of the cutter head, and it is fitted with ways on which the cutter-head mechanism moves forward. The other part is the cutter head, which rotates and forces the cutter bits against the face. The two principal parts are advanced one after another to achieve each advance cycle. As the bits cut into the tunnel face, rim buckets mounted on the cutter wheel pick up the cuttings and discharge them radially to the center and onto a conveyor. This conveyor discharges the muck into cars, into hopper bins, or onto another conveyor. Muck removal from the tunnel is then done by cars, conveyor, or pumping. The most common method of removal is by railcar. To attain high advance rates, TBMs require car loading and hauling capacity sufficient to ensure that no delays occur in this part of the operation. The machine should be furnished with spray nozzles fitted to the front of the cutter head to provide a water-detergent mist for dust control at the face to dampen the cuttings and

Figure 10.17 Typical road header. (A.E.C., Inc.)

to prevent dust at the conveyor discharge points. If a glass-enclosed air-conditioned operator's cab is furnished, the operator's efficiency is increased.

Where the instability of the ground requires their use, shields are used with tunnel-boring machines to maintain the sides of the tunnel until supports can be erected. To aid in handling these supports, cranes and support handling systems are provided, and in order to minimize support-erection time, TBMs are equipped with erector-arm assemblies so that the supports can be placed in position by mechanical means. Most TBMs are fitted with drills for rock-bolting.

Three different classes of bits are used on tunnel-boring machines. In the first class are roller-cutter bits of the type developed for big-hole, vertical-boring machines. These cutters are generally conical in shape, and cutting results from the crushing and pulverizing action of numerous hardened teeth or carbide inserts as the bits roll across the rock face under tremendous pressure. The second class of bits is the disc-type cutter, designed to split the rock instead of grinding it away. These cutters resemble glass cutters, for they have a hard, knife-edged rim, backed up by a husky shank that houses tapered roller bearings. In the third class are simple, solid cutter-tooth bits. Figure 10.20 shows a TBM fitted with disc cutters.

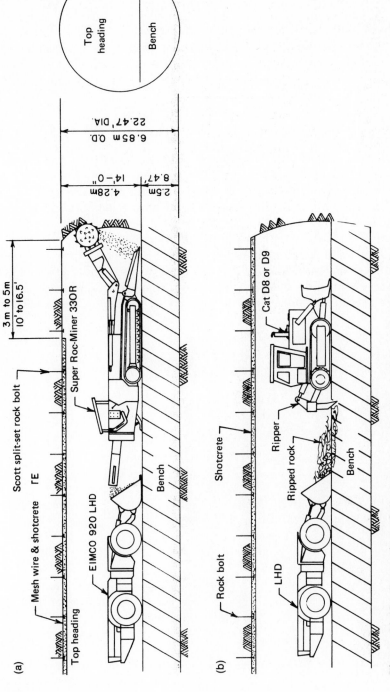

Figure 10.18 Excavation of diversion tunnel, Dixon Dam, Alberta, Canada. (a) Phase 1: excavation of top heading with boom-type continuous road header. (b) Phase 2: ripping of bench with bulldozer. (A.E.C., Inc.)

313

Directional control is achieved by setting the stationary part of the machine to align the ways to the proper line and grade, usually by use of a laser beam and targets mounted on the machine frame.

Each TBM is designed for a particular tunnel and for a particular rock condition. There are, therefore, many variations in design, but a number of machines have had repeated use on several tunnels, with or without modifications.

Applications

Tunnel-boring machines are used in some tunnels because the specifications make such use mandatory, for example, by limiting the amount of shooting that will be permitted in the tunnel. Tunnel-boring machines are also used when conditions indicate that boring will be the most economical tunnel excavation method.

The tunnel-boring machine works on the principle of machining a hole to exact size instead of roughing it out by the brute force of explosives. Some of the advantages of this technique are obvious: The noise and earth shocks of blasting are eliminated; the surrounding rock is left in an undisturbed state; electrically driven machinery produces no fumes or noxious gases that must

Figure 10.19 A G-1000 Gradall. (The Warner and Swazy Co.)

Figure 10.20   Tunnel-boring machine. (The Robbins Co.)

be cleared away; overbreak can be minimized; and since mucking can be carried on continuously under the right rock conditions, progress can exceed that made by any other method. Therefore, in choosing the method of construction of a tunnel in comparatively soft, consistent rock, where the tunnel is long enough to amortize the equipment and when time is available for equipment mobilization, the contractor should give consideration to the use of a tunnel-boring machine, for its use will increase production and permit a savings in overbreak excavation and overbreak concrete.

In determining whether a tunnel-boring machine should be used on any particular tunnel, past records should be consulted. These indicate that the tunnel-boring machine has been successful in rock with a Mohs' scale of less than 6.5. Mohs' scale rates rocks and minerals for resistance to abrasion and lists 10 minerals, arranged in order of increasing hardness:

1.	Talc	**6.**	Feldspar
2.	Gypsum	**7.**	Quartz
3.	Calcite	**8.**	Topaz
4.	Fluorite	**9.**	Corundum
5.	Apatite	**10.**	Diamond

Common rocks encountered in construction have these approximate Mohs' ratings:

Shale	Generally less than 3
Sandstone	Between 3 and 7, depending on cementing agency

Limestone	3
Marble	3
Slate	4–5
Granite	6–7
Schist	6–7
Gneiss	6–7
Quartzite	7

Another condition influencing the application of a tunnel-boring machine is the consistency of the rock. Hard intrusions, faulted formations, crushed ground, squeezing ground, and caving ground have reduced production with these machines. A roof fall occurring in front of the machine often has to be removed by hand, which retards progress.

Tunnel-boring machine manufacturers use the compressive strength of the rock as a measure of whether their machines can economically excavate proposed tunnels. Among different machines, the limit of adaptability varies from the ability to excavate rock of 30,000 to 70,000 lb/in² compressive strength. Since hard-rock boring machines do not vary appreciably in design, a better gauge of the machines' capabilities is obviously needed. Efforts are now being made to develop a standard that will take into account such criteria as the rock hardness, its compressive strength, its abrasiveness, and the measure of frequency of weak bedding planes and of joints. Until these criteria have been considered and applied, the suitability of these machines to specific rock formations is largely a matter of the individual's opinion, based on his experience.

As tunneling machines are improved and new bits developed, this equipment will be used in harder and harder rock. In the future, one of the main opportunities to increase production and lower the cost of tunnel excavation will be provided by the improvement and wider application of the TBM.

Another consideration affecting the decision to use a TBM is the duration of the work. If the use of a TBM will appreciably reduce the duration of the work, there will be substantial savings in those indirect costs which are a function of time.

Unless the contractor already owns a suitable machine, a machine mobilization time of 9 months to a year must be allowed for machine design, construction, and delivery to the job. This time is necessary, since each machine is designed to fit a certain-diameter tunnel in a particular rock formation. Boring machines can only be reused, without rebuilding, for a tunnel of the same diameter and similar rock conditions for which they were designed.

The first few months of operation of the tunnel-boring machine will be a breaking-in period, with slow production. This breaking-in period is necessary because almost every machine manufactured is of different size and characteristics than any other machine and must be adjusted to each job's particular conditions. After the machine is broken in, penetration rates will vary depending on the rock strength and condition, the flow of water encountered, and the amount and degree of faulted, crushed, or caving ground.

Progress in soft consistent rock will be good. Hard intrusions will delay progress owing to reduction in bit life. Roof falls in front of the machine will slow progress, as they often require removal by hand. Overbreak will be reduced since the machine cuts to a straight, smooth line, and depending on the specification requirements, this line may be moved in close to the A line.

In soft consistent rock, the machine often excavates more muck than can be taken away from the machine. In an effort to overcome this, conveyor belts have been used on some jobs to carry the material to the surface. On other jobs, efforts have been made to grind the muck and pump it to the surface with the addition of water to form a slurry. Neither of these methods of muck conveyance eliminates the need for rail or truck haulage for crews and supplies.

Progress in soft rock is very good when the machine is operating, but history on past jobs shows that the downtime is often 50 percent of the available time. Downtime is the result of roof falls, plugging of buckets and conveyors, mechanical and hydraulic repairs, bit changes, delays due to insufficient muck-disposal facilities, and delays chargeable to track and fan installation.

The boring-machine manufacturer will quote penetration rates per hour. The engineers should allow for downtime and then weigh the production rates furnished by the machine manufacturer with average production rates on previous jobs made under similar conditions with similar equipment.

Crew Size

The men required to operate and maintain the tunnel-boring machine replace the men who in conventional tunneling method drill, shoot, and muck at the tunnel heading and those who maintain the equipment necessary for these operations. To compare crew sizes for two different methods, the heading crew plus the powderman, mucker operator, and drill mechanics required by conventional tunneling methods should be counted against the crew required to operate and maintain the tunnel-boring machine. On muck handling, the hauling from the tunnel-boring machine will be a more balanced operation than the hauling from a mucker, so for the same yardage per day, hauling from a tunnel-boring machine will be a more efficient operation.

Specifically, the men required at the heading to operate and maintain a tunnel-boring machine are:

Foreman	Electrician
Mechanics	Muck-hopper operator
Oiler	General laborers
Surveyors	Support crew (if required)
Operator	

This listing does not take into account the tunnel bull gang which lays track and erects ventilation, air, and water pipes or the electricians necessary to extend the power distribution system.

Heading Supplies

Heading supplies required when a boring machine is used include rubber clothing, small tools, and safety supplies for the men. Bits are required for the machine and the cost varies, depending on the hardness of the rock. Other heading supply cost is for power, grease, oil, electrical supplies, mechanical parts, electrical parts, and hydraulic parts.

Tunnel-Driving Equipment

The equipment required for excavating a tunnel with a boring machine should be compared with that required by the conventional tunnel-driving method. The boring machine replaces and eliminates the capital cost of the drill jumbo and drills, powder handling and storage, mucker, a portion of the compressed-air requirements, and the facilities and equipment requisite to the maintenance of this eliminated equipment. Additional expenses inherent in use of the boring machine are the cost of the tunnel machine, a greater power load, and facilities for maintenance of the machine. The remaining equipment necessary for tunnel excavation, such as ventilation, muck haulage, and outside facilities, are required in either case, but occasionally the amount of this equipment will vary.

Tunnels Driven with Tunnel-Boring Machines

Table 10.6 is a partial listing of jobs that have been successfully performed by tunnel-boring machines.

SUPPORTING EQUIPMENT AND FACILITIES

Tunnel Ventilation

The increasing underground use of diesel equipment has resulted in a large ventilation load. The amount of underground ventilation required is almost directly proportional to the amount of diesel horsepower used underground. The larger the tunnel, the more haulage units are required and hence the more horsepower; therefore, as the tunnel length increases, larger pipe and fans are required. State requirements on the number of cubic feet of ventilation air per diesel horsepower used underground is typically 100 ft³/min, plus 200 ft³/min per man. There are additional requirements for air velocity, for air quality, and for ability to reverse the direction of air flow.

TABLE **10.6** **Partial List of Construction Carried Out by TBMs**

Year	Location	Bore Diameter	Rock Type	TBM
1952	Oahe Dam Pierre, S.D.	26 ft 3 in	Soft shale	Robbins
1956	Sewer tunnel Chicago, Ill.	9 ft 0 in	Hard limestone	Robbins
1961	Poatina Hydro Tunnel Tasmania	16 ft 2 in	Sandstone	Robbins
1963	Sewer tunnel Chicago, Ill.	26 ft 0 in	Clay and rock	Calweld
1964	Navajo Tunnel No. 1 Farmington, N.M.	20 ft 0 in	Sandstone	Hughes
1965	Sewer tunnel Philadelphia, Pa.	13 ft 8 in	Mica schist	Jarva
1968	Mine tunnel Wallace, Idaho	9 ft 0 in	Quartzite	Jarva
1970	Water tunnel Vienna, Austria	8 ft 6 in	Limestone	Robbins
1972	Obergesteln Switzerland	12 ft 2 in	Schist	Robbins
1973	Sewer tunnel Ottawa, Ontario	11 ft 2 in	Shale and limestone	Jarva
1974	Sewer tunnel Cleveland, Ohio	13 ft 2 in	Shale	Jarva
1976	Hydro tunnel Split, Yugoslavia	23 ft 5 in	Hard limestone	Robbins
1977	Sewer tunnel Chicago, Ill.	18 ft 2 in	Dolomitic limestone	Robbins
1978	Water tunnel Perry, Ohio	12 ft 2 in	Shale	Jarva
1979	Sewer tunnel Chicago, Ill.	32 ft 3 in	Limestone	Jarva
1979	Sewer tunnel Chicago, Ill.	35 ft 4 in	Dolomitic limestone	Robbins
1980	Subway tunnel New York, N.Y.	22 ft 0 in	Granite-schist	Robbins
1981	Hydro tunnel Fresno, Ca.	24 ft 1 in	Granite	Robbins

The following computations of ventilation requirements are based on the use of axivane fans exhausting air from the tunnel. These axivane fans are the same diameter as the ventilation pipe and can be mounted directly in the pipeline. To assist in the selection of ventilation equipment, the number of cubic feet of air that can be exhausted from a tunnel for different diameters of pipe, different fan spacing, and different fan horsepower is listed in Table 10.7. Table 10.8 gives the reduction in fan horsepower for increases in elevation.

TABLE 10.7 Size of Pipe and Fans for Air Requirements at Sea Level

Air Volume (ft³/min)	Pipe Size (in)	Fan (bhp)	Fan (nominal bhp)	Maximum Spacing (ft) First Fan	Others
80,000	48	117	125	120	720
75,000	48	120	125	320	960
70,000	48	95	125	320	960
65,000	48	100	125	640	1,280
60,000	48	100	100	520	1,120
55,000	45	90	100	720	1,320
50,000	42	75	75	440	960
45,000	42	120	125	1,720	2,240
40,000	42	100	100	2,120	2,680
37,500	38	106	125	1,240	1,680
35,000	38	90	100	1,520	2,000
30,000	38	86	100	2,560	3,000
27,500	36	92	100	2,040	2,480
25,000	36	75	75	2,640	3,080
22,500	34	48	50	1,800	2,200
20,000	32	45	50	1,440	1,800
17,500	32	38	40	1,940	2,300
15,000	30	30	30	1,760	2,080
12,500	26	29	30	1,520	1,800
10,000	24	40	40	3,000	3,240
7,500	24	28	30	4,960	5,200
5,000	18	18	20	1,920	2,120
4,000	18	15	15	3,040	3,240
3,000	18	10	10	4,320	4,480

SOURCE: Joy Manufacturing Co., 283 Wattis Way, South San Francisco, California 94100.

TABLE 10.8 Reduction in Fan Horsepower Requirements with Increase in Elevation

Elevation (ft)	Horsepower Factor
1,000	0.979
2,000	0.944
3,000	0.910
4,000	0.876
5,000	0.845
6,000	0.813
7,000	0.782
8,000	0.755

SOURCE: Joy Manufacturing Co., 283 Wattis Way, South San Francisco, California 94100.

The following example demonstrates the use of Tables 10.7 and 10.8.

Example

Conditions:

Elevation	3,200 ft above sea level
Crew size	24 men
Hauling equipment	3 trucks at 110 hp
Mucker	Front-end loader at 185 hp

Required Air Supply:

Crew	24 at 200 ft³/min	4,800 ft³/min
Trucks	3 × 110 at 100 ft³/min	33,000 ft³/min
Mucker	185 at 100 ft³/min	18,500 ft³/min
Total		56,300 ft³/min

From Table 10.7 this requires a 48-in fan line with 100 hp fans at the portal, 520 ft from the portal, and at 1,120-ft intervals for the remaining length of the tunnel. From Table 10.8 it is apparent that 90-hp motors would suffice, but since these are not a standard size, 100-hp motors will be used.

Vent Pipe

Selection of vent pipe can be made from the following types:

1. Collapsible plastic tubing is used on short tunnels when the capital cost is a large factor in the total cost. With this tubing, ventilation must be a pressure system which blows air into the tunnel.

2. Plastic vent pipe, which is extruded on the job, is a comparatively new product, and its suitability has not as yet been proved.

3. Steel pipe with quick-connecting couplings can be obtained from pipe vendors.

4. Pipe made on the job with airtight clamping rings for quick connections is fabricated with special machines that make pipe of various diameters from coils of steel strip. (See Fig. 10.21, which illustrates the machine making small-diameter pipe.) In order to use a lighter gauge, pipe is made with corrugations which give it extra strength. Fabrication of pipe on the jobsite greatly reduces freight costs, as the charge for hauling steel strip is determined by weight and the charge for hauling fabricated pipe is by volume.

Length of Vent Pipe

In tunnels that have connecting shafts, it is possible to exhaust air through these shafts to save vent pipe. In tunnels that have a shallow cover, it may be economical to drill special vent shafts so that vent lines can exhaust into them, which will reduce the total length of vent pipe required.

Figure 10.21   Machine for fabricating pipe from steel strip. (George M. Philpott Co., Inc.)

Secondary Ventilation System

The end of the main ventilation pipe is kept 200 ft (60 m) from the tunnel face to protect the pipe from blasting damage. This results in a dead-air space between the end of the pipe and the tunnel face. To obtain proper ventilation at the face, small fans are set up on the jumbo with short sections of pipe which extend back past the end of the main vent pipe.

Air Compressors and Piping

Air-compressor capacity should be provided for:

1. Primary drilling
2. Roof bolting
3. Winches
4. Sump pumps powered by air
5. Any air-operated equipment such as air-operated muckers
6. Shops
7. Air leaks

Air pressure required at the tunnel face is 100 lb/in², and any drop in this pressure will decrease the drilling speed and thus decrease production. To maintain the air pressure at the face, air lines and air receivers should be sized so there will be only a small pressure drop in the air system. The pressure at the compressor plant should be set to take care of the pressure drop in the system.

In computing the capacity of the compressor plant, the amount of air required per drill, based on sea-level operation, is as follows.

5½-in-bore burn-cut drill	600 ft³/min per drill
4½-in-bore drifter drill	330 ft³/min per drill
4-in-bore drifter drill	250 ft³/min per drill
3½-in-bore drifter drill	225 ft³/min per drill
3-in drifter drill	150 ft³/min per drill

Increased air consumption caused by increased elevation is given in Table 10.9.

On tunnel jobs where a large number of drills are used, the estimated total consumption of air can be reduced because all the drills will not be working at one time. Tables covering this reduced consumption have been given in various publications. These tables were completed before the development of modern jibs and drill positioners, which reduce nondrilling time, so the reduction factors shown in these tables should be weighted for any particular job.

Stationary electric compressors should be selected, except for short tunnels, since electric power is cheaper than diesel fuel and the maintenance cost for electric compressors is less than for diesel-powered ones. Air handbooks give tables listing the factors to be applied to the air produced by single-stage compressors to get their true production at different elevations. Since the majority of air compressors now used are two-stage compressors, these tables no longer apply. Two-stage compressors are very slightly affected by altitude up to an elevation of 3,300 ft. If they are to be operated above this altitude, the same air production can be maintained by enlarging the first-stage cylinder.

TABLE **10.9 Air Consumption Related to Elevation**

Altitude (ft)	Correction Factor	Altitude (ft)	Correction Factor
0	1.0	7,000	1.26
1,000	1.03	8,000	1.30
2,000	1.07	9,000	1.34
3,000	1.10	10,000	1.39
4,000	1.14	12,000	1.52
5,000	1.17	15,000	1.67
6,000	1.21		

SOURCE: *Rock Drill Data*, Rock Drill Division of Ingersoll-Rand Co. (Ingersoll, New Jersey), 1960.

Example

Compressor and Pipe-Line Selection

If the compressed-air requirements are to be determined for a heading 20,000 ft long at an elevation of 2,000 ft, with the use of one 5½-in-bore burn-cut drill and five 4½-in-bore drifters, the computations are:

Equipment	Nominal, ft³/min	Actual, ft³/min
1 5½-in-bore drill	600	706
5 4½-in-bore drifters at 330	1,650	1,942
Subtotal	2,250	2,648
Shop and miscellaneous consumption	500	589
Total	2,750	3,237
Correction 1.177 × 2,750 =	3,237	

The nominal air requirement was increased by a factor of 1.1 for leakage, and this requirement was increased by a factor of 1.07 for altitude. If 1,100 ft³/min compressors are selected, four should be provided, which includes a spare, so that one unit at a time can be out of service for maintenance.

To determine the size of air line required to provide 2,650 ft³/min at the face, by referring to tables of friction loss in pipe lines,[10.3] the use of 6-in pipe will give a pressure drop of 1.76 lb per 1,000 ft, or 35 lb in 20,000 ft. An 8-in pipe will give a pressure drop of 0.41 lb per 1,000 ft, or 8 lb in 20,000 ft. Therefore, the 8-in pipe will be used.

Water Supply

Water forced through the drill steel is used to wash the cuttings out of the drill holes. Water is also used for wetting down the muck pile to prevent dust while mucking, for invert cleanup, for concrete cleanup, and for grouting. The main water line should be of a size to supply water to a ½-in water-hose connection to each drill. Concrete cleanup requires more water than drilling, so this requirement determines the minimum-size water line. Water lines vary in size from 3 inches in small tunnels to 8 inches in large ones. Couplings should be of the quick-connection type for ease in installing the pipe.

Electric Supply

Blasting Circuits

Single-phase, either 220- or 440-volt, circuits are used for blasting. Using higher voltages decreases misfires in rounds that require a great number of detonators. The blasting circuit should be separate from all other circuits so that fluctuations from motor startups will be at a minimum.

Lighting Circuits

Current safety regulations should be consulted to determine illumination requirements. California at present requires 2 footcandles (fc) at the tunnel invert and 5 fc near the face.

Power Circuit

The main tunnel power transmission lines must be sized to carry power for the muckers (if required), drill-jumbo motors, ventilation fans, pumps, locomotives (if trolley locomotives are used), hydraulic drills (if used), and miscellaneous electrical tools and equipment.

A general procedure for the utility centers is to locate them at specified stations in the tunnel. If these stations are designed at spacings in multiples of 500 or 1,000 ft, there is less waste of parkway cable. Fewer splices will be required and fewer scattered installations will be made. Fan stations, pump stations, transformer stations, and passouts may all be located together, which will give better access for repairs and replacements.

Lighting transformers are normally a 5-kVA single-phase dry type in circuits between utility centers.

The blasting circuit transformer is normally a 5-kVA single-phase dry type on a circuit separate from that for the portable heading center.

In estimating transformer requirements, the lower constant temperature normally encountered underground should not be overlooked. This may give a substantial savings in that a much higher overload for short periods may be possible, which will allow use of a somewhat smaller transformer. This should be checked with the transformer manufacturer.

Light fixtures and sockets must be waterproof. All circuits must be grounded by a ground wire and not by rail or piping.

Drag Cables

When electric-powered muckers or electric-powered hydraulic drills are used, it is necessary to purchase drag cables, which allow both freedom of movement and spacing of the outlets.

Power requirements are quite high in some tunnels; therefore, careful consideration must be given to the total power required, the length of transmission, and the voltage to be used. Power cable may be obtained for underground conditions in voltages ranging from 600 to 6,000 volts. The size of the conductors required for a given power capacity and line losses decrease as the voltage increases. It is therefore advisable to use 2,240-volt or, in longer tunnels, 4,480- or 6,000-volt lines. Either copper or aluminum parkway cables are available; however, in estimating comparative costs, it must be noted that aluminum cable must be two sizes larger than copper cable for equal capacity. For preliminary purposes, until the power load is set, a main power circuit of three-conductor, 350 MCM, 5-kV armored cable should be used.

Underground Transformers

Underground transformers must be either the *dry* type or the *noninflammable* type. Transformers will be located at utility centers spaced through the tunnel and at a portable heading center which is kept as near the tunnel face as practical. The portable heading center will supply power to the mucker, lights at the face, miscellaneous motors, jumbo, and pumps. Transformers should be the portable, noninflammable type.

The transformer power requirements for one utility center are those for the fans, pumps, and lights served from that center. Transformers used should be a portable, three-phase noninflammable type.

Tunnel Drainage

It is standard to allow for a small amount of pumping and drain lines in a tunnel estimate. However, owing to the uncertainty of the exact volume of water to be handled, this portion of the estimate depends on judgment or past experience in similar type of rock. On some wet tunnels drainage may be a major cost factor. If there is any question concerning the amount of water to be handled in any tunnel estimate, it is wise to get advice from a structural geologist.

Drainage lines and pumping are required in flat tunnels or in headings driven down a slope. In the latter case, all water runs to the face, which slows progress and necessitates pumping to keep the lower part of the face free from water.

If a heading is driven up a definite slope, ordinary flows of water can be handled in a ditch alongside the track. When the water flows become excessive, pumps and drainage lines must be installed. On some wet tunnels that are driven with a shallow cover, pumping stations have been installed with discharge pipes extending through holes drilled from the tunnel to the surface. This cuts down on the amount of pipe required and also may save in pump head capacity and power.

Special Vehicles

Besides muck cars, it is necessary to supply the following special cars or their equivalent in truck-mounted equipment if truck haulage is used:

1. Man cars for hauling the tunnel crews inside the tunnel
2. Flat cars for hauling timber, pipe, and miscellaneous supplies
3. Steel car, which carries drill steel and bits in and out of the tunnel
4. Powder cars or prill cars or a combination of the two, complete with prill pots and hoses for loading the holes
5. Vent-line car outfitted with an overhead rack for carrying one section of

vent pipe and with two hydraulic jacks for lifting this vent pipe into place and positioning it

6. Car with built-in cradle for hauling steel sets into the tunnel

Outside Facilities and Equipment

Consideration must be given to provision of various outside facilities and equipment. These include yard grading and trackage to the dump, shops, powder house and concrete-mix plant, and sidings for the storage of spare cars, muckers, and other rail-mounted equipment. If a rail-mounted car dumper is planned, trackage for this is required paralleling the car dump trackage.

A powder magazine and cap house of bullet-proof construction and conforming to the local safety laws are required. A warehouse, electric shop, repair shop, pipe shop, bit and steel shop, and a compressor house are required. These shops necessarily have to be complete with the proper equipment. Shops can be built separately or included under one roof to get more efficient use of personnel. If rubber-tired equipment is used on the job, a tire shop is required.

A separate office is required for the supervisory and engineering personnel. Depending on the location, camp facilities and a mess hall may be required. Also depending on the location, it may be necessary to construct access roads. Besides office, shop, and engineering equipment, it is necessary to provide the following:

1. Pickups
2. Crane for handling lifts, such as a small hydraulic crane, or a front-end loader equipped with a fork lift which furnishes a dual-purpose rig, or a truck crane
3. Bulldozer for shaping muck pile
4. Flatbed truck for miscellaneous haulage
5. Change house

EXPERIMENTAL METHODS

A limited amount of government-funded research has been done on experimental systems of rock excavation. None of these studies has continued after government funding was stopped. Only two have included field trials; these trials tested the use of very high pressure water jets and the use of projectiles fired at the face. Laser beams have been tested in the laboratory. Even more exotic schemes, such as melting rock with nuclear heat sources, are in the discussion stage; none show commercial promise at present.

ADDITIONAL INFORMATION

Other publications may be consulted for additional information.[10.4,10.5,10.6]

REFERENCES

10.1 "King Size Muck Cars Speed Tunnel," *Western Construction Magazine,* April 1959, p. 34.

10.2 *Locomotive Selection Procedure,* Plymouth Locomotive Works, Division of the Fate-Root-Heath Co. (Plymouth, Ohio).

10.3 F. W. O'Neil, ed., "Compressed Air Data," *Compressed Air Magazine,* 1939.

10.4 Robert V. Proctor and Thomas L. White, *Rock Tunneling with Steel Supports,* Commercial Shearing, Inc. (Youngstown, Ohio), 1977.

10.5 George Ziegler and Morris Loshinsky, "Tunneling Under City Streets with Tunnel Boring Machine," *Proceedings of 1981 Rapid Excavation and Tunneling Conferences,* American Institute of Mining, Metallurgical, and Petroleum Engineers, New York, 1981, p. 683.

10.6 C. L. Craig and L. R. Brockman, "Survey of Portal Construction at CEC Projects," in *Underground Rock Chambers,* ASCE, New York, 1971, p. 167.

Chapter 11 Soft-ground and Mixed-face Tunnels

Soft-ground tunneling is the art of tunneling through earth, as opposed to tunneling through rock. Both soft ground and rock are encountered simultaneously in a *mixed-face tunnel*. The crown of a rock tunnel can usually be expected to be self-supporting, or at least to have sufficient stand-up time to enable the miners to install steel sets, rock bolts, or gunite with reasonable safety if support is needed. The face and invert can be expected to be stable, at least in the short term. Soft ground, on the other hand, can be expected to show little or no stand-up time in the short term, and no stand-up time in the long term.

The presence or absence of water is one key factor which determines stand-up time in soft-ground tunnels. The effects of water are generally adverse, weakening the ground and causing it to flow, but moisture in small quantities may improve the stand-up time for sands and other noncohesive materials which run freely if completely dry. The presence of water in a mixed-face situation (usually rock in the invert and soft ground in the crown) is particularly difficult to cope with because the water tends to flow along the contact, transporting material from that zone and thus undermining and inducing collapse of the face.

The cohesiveness of the ground is the second key factor which determines stand-up time in soft-ground tunnels. Admixtures of clay with gravels, sands, or silts tend to make the mass cohesive, which improves stand-up time. Very often the soil particles are found to be cemented into a more or less cohesive mass by infiltration and deposition of relatively insoluble metallic salts. Less frequently soils may be found which are cohesive because of mechanical interlocking of the soil particles. Sands which consist of tightly packed and very angular grains are encountered in some localities; this material is self-supporting in the short term, but is readily excavated with high-pressure water jets. Some sands, as mentioned above, will cohere when damp, but will run when dry. It is relatively easy to control the face of the excavation and to install supports for the crown and ribs in coherent soils, but the converse is true for noncoherent soils.

Soft-ground tunneling is almost invariably conducted with the use of shields to provide continuous support for the tunnel, usually with provisions also to support the face, either intermittently or continuously.

METHODS

Soft-ground tunneling can be accomplished with a variety of methods. The choice of methods will depend upon the presence or absence of water, the nature of the ground, and the diameter, length, and alignment of the tunnel. The length of the tunnel affects the economics of equipment selection; the alignment of the tunnel affects equipment selection in that the ability to steer a shield around a curve is a function of the length of the shield.

The tunneling methods available for soft ground are:

1. Conventional driving in free air
2. Driving with compressed air
3. Driving with slurry-face methods

Again it is pointed out that the common denominator of all these methods is the use of a shield.

The methods and equipment selected for a soft-ground tunnel project depend upon (1) the magnitude of the project; (2) the nature of the ground; and (3) the presence or absence of groundwater.

1. The Magnitude of the Project.   A major project will realize economies by investment of capital for sophisticated equipment with low operating costs. A lesser project may justify equipment with higher operating costs, but with a reduced investment.

2. The Nature of the Ground.   The site may present clay, silt, sand, or gravels, a combination of these soil types, or different soil formations in different parts of the job. Soft ground may consist of materials deposited by wind (*loess*), by slow deposition in still water (*lacustrine or oceanic deposits*), by flowing water (*alluvium*), by torrential flows (*diluvium*), or by glacial deposition (*till*). It will be apparent that boulders will be more common in the diluvium and till than in the other types of deposits. Soft ground may consist of soils formed in situ from the country rock. It is common for some parts of the rock to be more resistant to the weathering process than others; these parts will persist as boulders embedded in soil. This situation is more common near the weathering horizon than it is at higher elevations. Mixed-face situations, typically rock in the invert, passing from rock into soft ground, or passing from soft ground into rock, cause special problems.

3. The Presence or Absence of Groundwater. If there is sufficient free groundwater to transport the ground from the face, methods and equipment must be selected which can cope with this condition. One method is to dewater the ground by the use of wells or wellpoints. Neither is effective for "mopping up" water from the contact between soft ground and rock. Dewatering may be forbidden if lowering the groundwater table is likely to cause settlement of the ground under buildings or if a prolonged period of dewatering is likely to cause deterioration of timber pilings which support them. Groundwater is occasionally removed by electrical osmosis, but this application is rare. Alternative methods are to immobilize the water by freezing or by grouting, but these methods are infrequently used, except locally near shafts. If the free water cannot be removed or immobilized, the most common solution in the past has been the use of compressed-air tunneling. This is still the most common method in the United States, but slurry-face tunneling is superseding compressed-air tunneling in many situations in other nations, particularly in Japan, where slurry-face tunneling has been highly developed. The use of closed-face machines is suitable in some circumstances.

SHIELDS

A shield in its commonest form is basically a steel cylinder, long enough to accommodate crew and equipment and to permit erection of the ground-support system within the tail, thus providing uninterrupted support at all times. The shield is advanced by the use of thrust jacks pushing against the support system within the shield, which in turn transmits the reaction to the previously erected supports behind the shield, and thus to the ground. Figure 11.1 shows such a shield.

The leading edge of the shield is kept continuously against the face in order to minimize loss of ground from outside the zone intended to be excavated. The crown of the shield is often lengthened to form a hood, so that the leading edge will approach the natural angle of repose of the ground being excavated. The leading edge (cutting edge) is reinforced to resist the concentration of forces at that point, and is usually slightly flared. The flare causes overcutting, which helps in steering the shield. Steering is accomplished by varying the pressure applied to the shove jacks, which are mounted on a ring girder within the skin of the shield. The ring girder strengthens the shield and transmits jack thrust to the skin. The number of thrust jacks depends upon the shield diameter and length. The jacks are spaced around the periphery of the shield at an average spacing of approximately 30 to 36 in (75 to 90 cm) center to center. The spacing is usually uniform, but some contractors prefer closer spacing below spring line to provide more thrust there to counter the usual tendency of shields to "dive."

There is a *blade-type shield* that does not transmit the reaction from the force

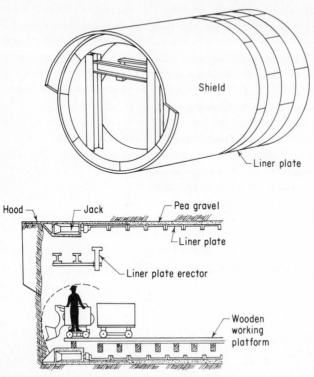

Figure 11.1 Shield operation used in dry, finely particled material.

required to advance the shield to the previously erected liners. It is most often used for horseshoe tunnels. The skin of the shield is composed of relatively narrow blades which are one at a time advanced by hydraulic jacks over the supporting frame. The friction of the ground against the other blades is sufficient to absorb the reaction from the one blade being advanced. A variation of this arrangement is called a *forepoling shield.* It is a complete cylinder, but is fitted with blades on the crown. These blades can be jacked forward for ground support. Most shields are fitted with breast-board jacks which can be extended to hold breast boarding in place in order to control running ground, exposing only a part of the face at a time. (See Fig. 11.2.)

It is apparent that if the support system is to be erected within the shield, the shield will necessarily have a greater outside diameter than the lining system. It is less apparent that additional space must be provided so that the lining system can be correctly located in line and grade even though the shield has deviated slightly from perfect alignment. This additional space is also necessary to accommodate curved alignment of the lining system within the straight shield, even though a few shields are articulated to help in this respect. The tail of the shield is often made of thinner plate to provide flexibility and to reduce the overall diameter and the size of the annular void. If sufficient space is not provided, the shield will become "iron-bound" and cannot be steered. It will be apparent that the space provided between the exterior of the assembled ring and the exterior of the shield will leave an annular void which must be promptly filled to prevent inward movement of the surrounding ground. This is accomplished by gravel packing and grouting the gravel, or by grouting. Steering a shield, which is rectangular in section parallel to its axis, around a curve creates a "shadow" of overexcavation which enlarges the annular void. This occurs on both horizontal and vertical curves, and it occurs almost continually, but to a lesser degree, from routine steering corrections, even on tangent alignment. (See Fig. 11.3.)

The foregoing paragraphs describe features common to most shields. Other features of shields vary with the tunneling methods and will be described with the discussions of those methods.

CONVENTIONAL TUNNELING IN FREE AIR

Conventional tunneling in free air is the usual method of choice if it is physically and economically feasible. The most common determinant is the presence or absence of water.

Figure 11.2 A blade-type shield. (Westfalia Lunen.)

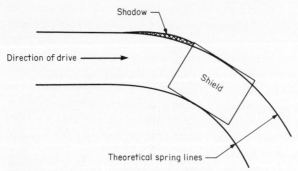

Figure 11.3 The shadow produced by a shield being driven around a horizontal curve.

The presence of water does not necessarily bar the use of free-air methods. The water may be removed by the use of wellpoints (for shallow tunnels) or by deep wells. Other methods, such as causing water migration by electrical currents, have been used on rare occasions. The water may be immobilized by grouting or by freezing the ground, but the cost of these measures is usually so high as to prohibit their use except for short distances, such as in areas adjacent to shafts, or in special circumstances, such as when tunneling under structures susceptible to damage from ground subsidence. If dewatering is to be considered, it is necessary to investigate the possibility of damage to surface structures by subsidence induced by groundwater removal. In some areas timber piles may fail because of rapid deterioration when exposed to air after long submersion. Either case may require that expensive underpinning be provided if dewatering is the method of choice. Ground freezing is sometimes a hazard to surface structures; it can prompt the formation of ice lenses, which cause heaving, or can result in subsidence during thawing because of restructuring of the soil from the freezing process.

The absence of free water does not necessarily clear the way for free-air tunneling. Some clays are so plastic and have such a high pore pressure that compressed air or other methods are necessary to hold the face. One method which is occasionally used in these circumstances is to "shove blind," as shown in Fig. 11.4. The front of the shield is completely closed, except for one or more apertures. When force is applied by the shove jacks and the shield advances, the highly plastic clay is extruded through these apertures and is removed by the conveyor belt as the tunnel progresses, as shown in Fig 11.5.

A more common method of soft-ground tunneling in free air is to excavate the face with hand tools, a front-end loader, or a mucking machine. With hand tools, such as air-operated clay spades, workmen loosen the face, dropping the muck onto a conveyor which loads the muck cars. In a large-diameter tunnel the crew will work from platforms at several levels. There are usually provisions for breast boarding the entire face in case the ground is unstable,

Figure 11.4 Shield for "shoving blind." (Tekken Construction Co., Ltd.)

and exposing only a part of the face at any one time. It is also usual to extend the top of the shield to form a hood, so that most of the ground above the natural angle of repose is supported.

If a mucker is used, the operation is very similar to the method used for hand excavation, loading onto a conveyor or directly into muck cars. If a front-end loader is used, there is usually no need for a muck conveyor. The

Figure 11.5 Shoving blind in soft plastic clay. (Tekken Construction Co., Ltd.).

loader either loads directly into the hauling equipment, or carries the muck directly to the portal or shaft.

Digger shields use a power-operated digger, usually resembling a backhoe, to excavate the face, dropping the muck onto a conveyor for loading into muck cars. Road-header machines are similar, except that the excavation is performed by a rotating cutter head, mounted on a boom to enable it to reach all parts of the face. Some manufacturers mount the cutter head with its axis parallel to the axis of the boom. Others favor a transverse mounting.

Tunnel-boring machines have been widely used for soft ground. The rotating cutter head may be completely open, or it may be closed except for slots for muck passage. Retractable overcutters are sometimes provided to improve steerability. The rotating cutter head is dressed with an array of cutter teeth, usually drag bits. The muck is picked up by buckets on the periphery of the wheel and is then dropped on a muck conveyor which loads the muck cars. There has been some development of methods to transport the muck as a slurry to be pumped out of the tunnel, or to transport the muck by conveyors. Neither method has gained much acceptance, probably because of the fact that neither method eliminates the need for transportation for men and supplies.

Support systems for soft-ground tunnels may consist of prefabricated liners (described in Chap. 12), with or without secondary linings, or of steel ribs with timber or other lagging. Ribs and timber lagging are regarded as temporary support, usually completed with cast-in-place concrete lining for permanent support, as shown in Fig. 11.6. Ribs should be jacked tightly against the perimeter of the excavation. This requires that fillers ("dutchmen") be used to close the gaps which result from expanding the ribs. Ribs and lagging are used only under favorable ground conditions, and are not suitable for running ground.

COMPRESSED-AIR TUNNELING

Compressed-air tunneling is a technique for tunneling below the water table, where it would be otherwise impossible to support the face. The principle of the method is that the air pressure in the tunnel is raised to counterbalance the sum of the hydrostatic and earth pressures at the face. This is necessarily a compromise, since the air pressure is approximately the same at the crown as it is at the invert, while the opposing pressures increase from crown to invert, at a rate approximately 0.5 lb/in^2 per foot of depth. If the air pressure balances the opposing pressure at the crown, there is danger of a "run" at the invert; if the pressure is balanced at the invert, there is danger of a "blowout" at the crown. The danger from this compromise increases with tunnel diameter. It is essential that there be at least two diameters of cover to ensure against a blowout. Runs are serious matters, and may fill the tunnel, but blowouts can

Figure 11.6 Steel ribs and timber lagging on the Berkeley Hills Tunnel. (Bay Area Rapid Transit District.)

be catastrophic. The sudden outburst of air has been known to carry away workmen, and the sudden loss of pressure will cause an immediate inflow of ground. The usual method of coping with the matter of insufficient cover is to deposit a temporary berm of earth on the ground surface above the tunnel location, but this is often an unacceptable solution, particularly if the tunnel route is under a busy street or a busy waterway.

The effect of air pressure in impervious soils is simply that of a counterbalancing force. The effect of air pressure on pervious soils includes support, but it also drives groundwater away from the face, thus stabilizing the ground. The drying effect may be a nuisance in clean sands, if these sands tend to run when dry.

Air pressure in a tunnel is supplied by low-pressure compressors or blowers on the surface. Pressure in the tunnel is maintained by air locks installed in the tunnel, or occasionally in the shaft serving the tunnel. The air must be free of oil or noxious gases. It is essential that there be standby blowers and alternate power sources available in the event of failure of the primary blowers or power sources. The presence of compressed air increases the oxygen supply and thereby increases fire hazards.

No internal combustion engines are permitted underground in compressed-air tunnels, and none should be located on the surface where the exhausts can be picked up by the intakes for the air compressors. Underground haulage for compressed-air tunnels is universally by electric-battery locomotives. Other underground equipment is powered by electricity, compressed air, or hydraulic systems. Shields for compressed-air tunnels may have any of the features used in other shields for soft-ground tunneling.

It will be apparent that compressed-air tunneling is expensive and that the expense compounds as depth increases. Equipment costs are high, since the contractor must supply blowers, backup power supplies, sophisticated systems to regulate air pressure, muck locks, man locks, battery locomotives and battery chargers, and special safety equipment such as medical locks, emergency walkways and locks in case of flooding, emergency breathing apparatus, air-quality monitoring equipment, etc.

Labor costs are high. Additional men, such as lock tenders, are needed. The basic labor scale for compressed-air tunneling is higher than the scale for free-air tunneling. The full shift is paid for, although the crews actually work fewer hours. More time is devoted to emergency drills than is required for free-air tunneling.

Table 11.1 gives typical working times under compressed air.

Other costs are high. All workmen must have pre-employment physical examinations and frequent reexaminations to be sure that they can work safely or continue to work safely under compressed air. Arrangements must be made for instant availability of specially trained physicians and medical facilities specially suited for examining and treating compressed-air workers.

SLURRY-FACE TUNNELING

Slurry-face tunneling is a technique developed in the past few years. It has been developed and used on a few projects by British and German contrac-

TABLE **11.1   Allowable Working Time in Compressed Air***

Pressure (lb/in²)	Allowable Working Time (hours:minutes)	Required Decompression Time (hours:minutes)
14 or less	6:00	0:06
Over 14, through 18	6:00	1:03
Over 18, through 22	4:00	1:08
Over 22, through 26	4:00	1:44
Over 26, through 32	4:00	2:43
Over 32, through 38	3:00	2:58
Over 38, through 44	2:00	2:34

* Allowable working times are from a typical union contract; the applicable safety codes generally permit longer working hours. See *Tunnel Saftety Orders* and *Compressed-Air Safety Orders*, both issued by the State of California, Division of Industrial Safety, Sacramento.

tors, with methods which require the use of bentonite. It has been used widely by Japanese contractors, who have developed methods which do not require the use of bentonite.

The slurry-face tunneling machine consists of a shield fitted with a pressure bulkhead at the face, with a rotating cutterhead in front of that bulkhead. A slurry, consisting of local clays suspended in water, is pumped through the bulkhead at a pressure slightly in excess of the sum of the hydrostatic pressure and the earth pressure of the ground being penetrated. The rotating cutter head produces muck which is suspended in the slurry, which is then pumped to the surface where the muck is recovered, and the slurry is recirculated, after adjusting its specific gravity and viscosity as may be necessary. Japanese engineers have developed highly sophisticated control systems to regulate the entire process.

The big advantage of slurry-face tunneling is that it permits the crew to work in free air, with the face well supported, under conditions which would otherwise require the use of compressed-air methods. It is also an advantage that the slurry, which has a specific gravity which approximates that of the ground plus groundwater, has the same pressure distribution as the ground plus groundwater from crown to invert. This is not true of face support by compressed air, where the air pressure is essentially the same at the crown and at the invert, thus leading to danger of a "blow" at the crown or a "run" below spring line, depending upon the compromise pressure adopted by the tunnelers. It is for this reason that slurry-face tunneling can be safely accomplished in situations where there is not sufficient cover for compressed-air tunneling or where dewatering the ground is not practical. Slurry-face tunneling works have been successfully completed under the sea at depths in excess of 100 ft (30 m) below sea level. This is approximately the maximum depth for compressed-air tunneling, which is conducted at that depth only at great expense and at great risk to the health and safety of the crews.

Equipment for slurry-face tunneling is expensive. It is, however, a cost-effective method in soft ground with a high water table where there are few or no obstacles such as boulders or pilings. The method is effective in clays, silts, sands, and gravels. Experiments with adaptations to cope with obstacles, including mixed-face situations and the presence of boulders, are underway. These adaptations have been successful, but further development continues. Figure 11.7 is a schematic representation of the slurry-face system. Figure 11.8 is a photograph of a typical slurry-face shield.

Earth Pressure Balanced Shield Tunneling[11.1]

The earth pressure balanced shield consists of a closed-face shield with a rotating cutter head. Muck is discharged through the cutter head via a screw conveyor into muck cars. This method is competitive with slurry-face tunneling methods in most types of soils, but it has not yet been used where hydro-

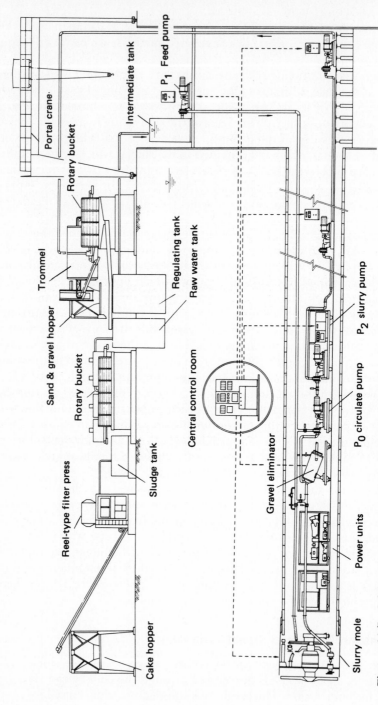

Figure 11.7 Schematic representation of the slurry-face tunneling method. (Tekken Construction Co., Ltd.)

Portal crane

Intermediate tank

P_1 Feed pump

Rotary bucket

Trommel

Sand & gravel hopper

Regulating tank

Raw water tank

Rotary bucket

Central control room

Sludge tank

Reel-type filter press

Gravel eliminator

P_2 slurry pump

P_0 circulate pump

Power units

Cake hopper

Slurry mole

340

Figure 11.8 Typical slurry-face shield. (Tekken Construction Co., Ltd.)

static pressures are high. Equipment costs for this system are lower than equipment costs for slurry-face tunneling. This system, which was developed in Japan, was used with great success on the N-2 sewer tunnel in San Francisco, which set new records for both Japanese and American soft-ground tunneling. A variation of this machine, called the *water pressure balanced shield*, uses water pressure at the face and slurry transportation for the muck.

MIXED-FACE TUNNELING

The most common situation in mixed-face tunneling is the transition from tunneling in earth overburden to tunneling in rock. The interface between earth and rock may be steep, which implies a short length of mixed face, or it may be relatively flat, which implies a longer length of mixed-face work. If there are to be shafts or crossover structures, it is very desirable to locate these structures so as to separate the rock tunneling from the soft-ground tunneling, thus eliminating or minimizing the length of mixed-face tunneling and simplifying the equipment changes which may be required. Another common situation in mixed-face tunneling is to encounter rock in the invert of a tunnel which is otherwise completely in soft ground.

Since the equipment required for the soft-ground section will usually be different from the equipment required for the rock section, it is often convenient to drive from both ends toward the interface, using the appropriate rock tunneling and soft-ground tunneling methods, and abandoning the shell of the shield where the two methods meet.

If the mixed-face situation is relatively short, it is often feasible to grout or freeze the overburden and excavate the mixed-face length by rock-tunneling methods. This alternative is more favorable if the depth of overburden and other considerations permit the freezing or grouting to be completed from the surface of the ground before tunnel driving reaches the mixed-face zone. If the freezing or grouting must be performed from the tunnel, the delays incident to that work tend to make this alternative less favorable.

If the rock is soft it may be feasible to continue driving by soft-ground methods, attacking the rock with paving breakers or with very light use of explosives. Even where the rock is hard, this is the usual response to intermittent encounters with rock in the invert of a tunnel which is otherwise in soft ground.

If the mixed-face section is approached from rock it may in some cases be possible to excavate all but a thin shell of rock, providing temporary support as necessary, and then complete the rock excavation as a part of the soft-ground work. This method was used in a tunnel in Atlanta.[11.2]

If the mixed-face section is approached from the soft ground it may be possible to hand-mine the overburden, providing support by ribs and lagging or by ribs and spiling. The ribs will rest on a rock berm outside the excavation limits for the rock excavation. This method was used in a tunnel in Washington, D.C.[11.3]

PIPE-JACKING

Pipe-jacking is a method of tunneling, usually in soft ground, which may or may not utilize a shield, with the shove jacks located in the starting shaft. The jacks shove the string of pipe and the shield (if used) ahead as excavation progresses. The shield is very short, since the shove jacks are at the shaft. The trailing end of the shield is articulated and fitted with short-stroke steering jacks. The trailing end is actually a shove ring, profiled to match the profile of the leading edge of pipe, which is usually precast concrete. As the pipe is advanced, additional pipe sections are added at the starting shaft.

Two factors limit the length of pipe which can be successfully advanced. One is skin friction. This is usually minimized by providing holes in the pipe through which a lubricant, usually bentonite, can be introduced. The second limiting factor is the capacity of the pipe to withstand the axial loads from the shove jacks. This loading is at a maximum at the starting shaft and increases as the length of the pipe string increases. This factor is countered by increasing the wall thickness of the pipe above the thickness required for the permanent radial loads on the pipe. Pipe strings are usually readily jacked to distances of

300 to 400 ft (90 to 120 m). Longer pipe-jacking projects have been success-fully completed using intermediate jacking stations.

Excavation is usually accomplished by hand methods, for there is little working room in the usual range of pipe diameters, which is from 4 to 6 ft. The power for the muck car may be manpower or may be a tugger at the starting shaft. Slurry-face pipe-jacking eliminates the need for such strenuous labor.

CUT-AND-COVER TUNNELS

Cut-and-cover operations may require that the sides of the excavation be supported by the methods used for the support of shaft excavations, as de-scribed in Chap. 13. This is particularly true if the work is to be performed under city streets, where the use of sloped sides on the excavations cannot be tolerated. Cut-and-cover operations in the streets must often be decked over so that traffic can flow more or less normally. The deck system is usually integrated with the support of excavation system, using caps on top of the soldier piles to support the deck beams. The deck beams are in turn used to support temporarily the numerous utility lines which are uncovered in the course of excavation. Excavation by backhoe is usually preferable to ex-cavation by clamshell in the upper few feet, particularly in the zone in which utility lines are encountered. Particular care must be exercised in dealing with gas lines, steam lines, and high-voltage electric lines. Gas lines should be temporarily rerouted during construction to avoid the danger of a leak under street decking, as a gas leak in such a confined area might lead to a catastrophic explosion. Figure 11.9 shows a completed cut-and-cover subway tunnel.

The Milan method of cut-and-cover construction involves construction of structural walls cast in slurry-filled trenches, followed by excavation to the bottom of the structural roof. The roof is poured on the excavated sur-face, the space above the roof is backfilled, the street surface is restored, and traffic reverts to normal. The remaining excavation below the roof slab is performed by underground excavation methods, and the structure is com-pleted. This method was developed in Italy to reduce the time period during which cut-and-cover subway construction interfered with normal use of the streets.

ADDITIONAL INFORMATION

Manufacturers' literature and current publications offer additional informa-tion about soft-ground and mixed-face tunneling practices.[11.4,11.5,11.6,11.7]

Figure 11.9   Completed cut-and-cover subway tunnel. (Bay Area Rapid Transit District.)

REFERENCES

11.1 Shigeo Kurosawa, "Earth Pressure Balanced Shield Tunneling," *Journal of the Construction Division,* ASCE, December 1981, p. 609.

11.2 Thomas R. Kuesel and Harvey W. Parker, "MARTA's Broad Street Tunnels," *Proceedings of 1981 Rapid Excavation and Tunneling Conference,* American Institute of Mining, Metallurgical, and Petroleum Engineers, New York, 1981, p. 393.

11.3 Vinton A. Garbesi, "Integrated Mixed-Face Mining and Support Systems, WMATA contract No. IKDO11," *Proceedings of 1979 Rapid Excavation and Tunneling Conference,* American Institute of Mining, Metallurgical, and Petroleum Engineers, New York, 1979, p. 194.

11.4 Robert V. Proctor and Thomas L. White, *Earth Tunneling With Steel Supports,* Commercial Shearing, Inc. (Youngstown, Ohio), 1977.

11.5 H. Takahasi and H. Yamazaki, "Slurry Shield Method in Japan," *Proceedings of 1976 Rapid Excavation and Tunneling Conference,* American Institute of Mining, Metallurgical, and Petroleum Engineers, New York, 1976, p. 261.

11.6 Takeshi Watnabe and Hironobu Yamazaki, "Giant Size Slurry Shield Is a Success in Tokyo," *Tunnels and Tunnelling,* January/February 1981, p. 13.

11.7 Mike Richardson and John Scraby, " 'Earthworm' System Will Threaten Conventional Pipe Jacking," *Tunnels and Tunnelling,* April 1981, p. 29.

Chapter 12 Tunnel Linings

The purpose of this chapter is to describe the constuction methods and equipment necessary for the installation of permanent linings in tunnels. Transportation and utility distribution facilities, compressors, shops, and other outside facilities involved in the excavation operation and used to service the lining operations are described in Chap. 10. Descriptions of equipment included in this chapter are limited to the additional equipment that is necessary for and unique to tunnel lining operations.

Although the majority of tunnels constructed have permanent lining installed before they are placed in service, some tunnels are left unlined, e.g., those constructed in sound, hard rock which are to be used for the conveyance of water under a low hydraulic head. If short sections of these bare-rock tunnels require steel supports, these sections are lined with concrete to protect the steel sets and to assist in providing permanent support. Some unlined hydraulic tunnels have paved inverts to prevent uneven erosion of the invert; such uncontrolled erosion may cause dams and eddies and restrict the flow. Other unlined hydraulic tunnels have paved inverts to facilitate motor vehicle operation when tunnel inspections are indicated in later years. If access tunnels to a powerhouse or other underground facility pass through suitable rock, they may also be left unlined.

In permanently lined tunnels, the type of lining used varies with the competency of the ground and the purpose of the tunnel. Poured-in-place concrete lining is used for hydro, highway, and subway tunnels. Tunnel linings composed of precast-concrete sections or heavy cast-iron or steel liner plates are commonly used in subway tunnels located in areas where incompetent ground is encountered. These liners, installed within the tail of the excavation shield, restrict ground subsidence and minimize building foundation underpinning requirements. Shotcrete lining is used for all types of tunnels driven in incompetent material because of its ability to provide both temporary and permanent support. Other linings used for special conditions are precast-concrete or steel pipe, where the space between the lining and the tunnel surface is backfilled with concrete.

CAST-IN-PLACE CONCRETE LININGS

Terminology

The concrete lining of a tunnel is divided into two general areas. *Invert concrete* comprises approximately the bottom 90° of the concrete lining of a circular tunnel, the bottom and a short distance up each side wall in a horseshoe tunnel. This "wall distance" in a horseshoe tunnel must be of sufficient height to anchor the arch forms. *Arch concrete* is simply the remainder of the concrete in the tunnel's lining (i.e., the top of the arch and the rest of the sides). Some concrete lining operations include the pouring of *curbs*. Curbs are continuous concrete walls poured along each side of the tunnel. Specifications often require that the surface of these pours be set back from the finished tunnel surface. If so, all finished surfaces in the tunnel are made on either the invert or arch pour. Because of limitations of the placing equipment, *tunnel concrete* should have more plasticity than concrete ordinarily placed in open forms. To achieve this plasticity and workability, this concrete has a higher cement and sand content and the maximum size aggregrate is usually under 2½ in. Gravel is preferred to crushed rock as an aggregate. Before the concrete is placed, forms must be erected in the excavated tunnel. There are several types of forms that can be used in this operation. *Telescopic forms* are either full circle or arch forms which can be stripped, collapsed, and moved through the center of other forms previously erected and which can then be erected in front of the concrete pour. *Nontelescopic forms* are constructed so that they cannot be moved through the center of other similar forms and are used only when noncontinuous pours are made. They are less costly than telescopic forms. *Bulkhead forms* are forms placed between the arch forms and the tunnel rock to give a vertical construction joint at the end of any pour. A *form traveler,* a traveling jumbo equipped with hydraulic cylinders, is used to strip, collapse, transport, and erect the full circle or arch forms.

Concrete is placed in "pours." In a *full-circle pour,* the concrete is placed completely around the tunnel in one operation. A *continuous pour* is concrete placement that continues 24 hours a day for the entire workweek. Bulkhead forms are installed at the end of the workweek only, after the pour is stopped. When a continuous pour is made, only one construction joint is formed per week. *Bulkhead pours,* with bulkhead forms and vertical construction joints, are made at definite intervals along the tunnel line. They are made when specifications limit the length of pours or when nontelescopic forms are used; the result is a cyclical operation of concrete placing. (This cycle consists in concrete placing, form stripping, form setting, and the resumption of concrete placing.)

The concrete placement operation requires specialized equipment. A *concrete placer* is a pump or air gun used to force concrete through a pipe embedded in the fresh concrete being placed between the form and the tunnel surface. That length of pipe that transports the concrete from the placer to pour area is called the *slick line*. The slick line must be embedded in the fresh concrete to prevent segregation of the concrete as it is discharged from the pipe. The traveling frame that supports the slick line is known as the *pipe jumbo*.

Another traveling frame, used in tunnels with a large diameter, provides a working platform for the finishers. This frame is called the *finish jumbo*. From its platform the finishers repair any flaws, such as honeycombed areas, in the fresh concrete after the forms are removed and apply curing compounds if required.

After the concrete is placed, low-pressure grout is placed between the concrete lining and the excavated tunnel surface. The purpose of the grouting is to fill any voids left after the concrete-placing operation.

Concrete Lining Sequence

If a tunnel is to be lined with concrete, the construction planner must first decide whether the tunnel lining operation is to be done concurrently with the excavation or whether lining is to be installed upon completion of the excavation. Concrete lining and tunnel excavation are done at the same time when specifications make it mandatory or when tunnels are driven in mud, soft ground, or sand, which requires that the lining be placed as close to the heading as possible. Concurrent lining and excavation operations are sometimes used in subways constructed in material other than competent rock, in a downtown area where any ground subsidence would effect building settlement. When concreting and excavation operations are performed at the same time, the resulting interference between the two operations slows the progress and increases the cost of both. Two crews are needed if excavation and lining are carried on simultaneously. Therefore, when concurrent operations are necessary, the engineer should make allowances for this increased cost and the increased construction time required.

The preferred method of placing concrete lining is to complete the tunnel excavation before lining operations are started. Completing excavation before lining is begun provides the work crews with total access to the tunnel and allows management to give full attention to the operation under way.

Concrete-Placing Methods
for Completely Lined Tunnels

When a tunnel is to be completely lined with concrete, it is placed by one of five methods. Descriptions of these methods follow.

1. Full-Circle Method.   The full-circle method involves the placement of the complete circumferential concrete lining in one operation. This method is used when specification requirements or job conditions make its use mandatory or when restricted concrete placement makes its use economical. Either full-circle forms or arch forms with a screeded invert may be used. The most difficult aspect of this placement method is the work required to keep the forms on line and grade because of the tendency of the plastic concrete to raise the forms. Most other concrete-placing methods take advantage of previously poured concrete that can be used both as a line and grade reference point and as an anchorage for the arch forms. In the full-circle method, the forms must be set to line and grade by the use of bottom jacks. Flotation can be prevented by blocking from the tunnel roof. This blocking is removed as the pour advances.

The full-circle method of concrete placement may be the most economical method when production is not critical; for concreting can be done by a small crew. Cleanup is done directly ahead of the pour, and the length of tunnel which can be completed each day is often determined by the length of tunnel that can be cleaned up in one shift. Working procedure is to assign one shift to cleaning up, one shift to moving forms, and one shift to pouring. With this method of operation, all labor is worked as efficiently as possible, with crew sizes restricted to the number required for each task, on each particular shift.

When full-circle pours are made, the design of the form traveler presents problems, since the surface on which it must travel will vary from concreted invert, to the top surface of the invert forms, and finally to the surface of the unpaved invert. To travel on these three surfaces and support the forms in a level condition, jacks must be provided over both the rear and front wheels. A preferred form-traveler design, for use with telescopic forms, is one designed to travel only on rails mounted on top of the invert forms. In order to restrict the form traveler to this area, cantilever beams, capable of supporting a length of forms, must be provided on the rear and front of the form traveler. Facilities for sliding the forms from the rear cantilever over and under the jumbo to the front cantilever must also be incorporated in the form traveler. Form-moving procedure consists in backing the form traveler up to the last section of forms and raising the invert form by the rear cantilever arm. This form is then moved forward so that it is supported by the front cantilever arm. The form traveler is then moved forward in the tunnel, and the invert form is set on jacks by the front cantilever arm. The traveler then goes back and picks up the arch form with the rear cantilever arm. The arch form is then slid forward so that it is over the main section of the form traveler. The traveler then moves forward and sets it in place. The arch forms can be held in position by blocking from the tunnel roof. As the concrete is placed around the form, the blocking is removed.

In large-diameter tunnels when full-circle forms are used, the invert section of the full-circle forms should be designed so it can be lifted from the invert shortly after the pour is made in order that air bubbles can be troweled out and any concrete deficiencies can be corrected. Concrete deficiencies may occur in the invert section of large-diameter tunnels because of the long horizontal distance the concrete must flow across the invert to fill the form properly.

2. Invert Pour, Then Arch Pour Method. Invert pour and then arch pour is the method of concrete placement in most common use. It does not have the problem of form referencing and form flotation encountered in the full-circle method, since the arch forms can be firmly secured to the previously placed invert concrete. This method has a cost savings over the curb, invert, and arch method because the lining is poured in two operations instead of three.

In this method, the invert is poured first by using a slip-form screed to form the invert. Usually the invert is poured for the full length of the tunnel, or for a reach from one access to another, before starting arch concrete. Arch pours may be continuous, using telescoping forms, or may be in lengths which may be completed in one shift. Telescoping forms are economical only for long tunnels because of the high initial investment. With this placement method and the proper selection of equipment, rapid, continuous pouring of both invert and arch can be made.

3. Arch Pour, Then Invert Pour Method. If specifications permit, invert cleanup and the placement of invert concrete can be done concurrently by the "retreat" method of invert placement. With this invert-cleanup method, if low-pressure grouting and final cleanup are done with rubber-tired equipment, it will not be necessary to relay track, provided the arch concrete is poured before the invert concrete. However, since there is no previously placed concrete that can be used for arch-form anchorage, there will be added cost and time required to anchor the arch forms and hold them to line and grade. The added time and cost required for form anchorage often surpass the savings that result from not relaying the track. For this reason, this method of concrete placing is seldom used.

4. Curb, Invert, and Arch Method. The curb, invert, and arch was the method most popular in the past because of the simplicity of having the curb concrete in place to provide a reference and a firm anchorage for both the invert and the arch pours. This method is becoming less popular because it is more expensive than the invert and arch method. When the curb, invert, and arch method is used, there is a tendency to overexcavate to allow for the curb pour. This overexcavation increases cost by increasing the nonpay yardage of both the excavation and the concrete per lineal foot of tunnel. Another reason for increased cost is that the concrete lining is placed in three operations instead of two and invert cleanup must be done in two operations instead of one.

With this method, the procedure is to clean up under the curb pour, place curb forms, and then pour the curbs. To save time, these curbs can be placed while excavation is being performed without too much interference with excavation. After the curbs are poured, the remainder of the invert is then cleaned, and with the curbs for anchorage and grade reference, the invert is placed. The arch pour is then made in the same manner as in the invert and arch method.

When the concrete lining must be placed concurrently with excavation, this is the method most commonly used. It has the advantage, under these conditions, of giving less interference with the excavation operations than other methods, while affording constant access to the tunnel heading during all concreting operations. Such access can be maintained during the curb and arch pour by using California switches. Similarly, when the invert is cleaned and the concrete is placed on this prepared area, the main-line travel to the excavation heading can be preserved. The curbs, previously placed, provide a simple method of supporting a bridge necessary for main-line access in the clear of the concreting operation.

5. Curb, Arch, and Invert Method.   The curb, arch, and invert method is used for the same purpose, and under the same conditions as the arch and invert method. As with that method and under the right conditions, track relaying will not be necessary. This saving of track relaying can be accomplished when concrete and invert cleanup are done concurrently, when low-pressure grouting and final cleanup are carried out with rubber-tired equipment, and when arch is poured before invert. The difference between this method and method 3 is that the arch forms can be anchored to the curbs; however, additional expense is involved in curb placement.

Concrete-Placing Methods for Partially Lined Tunnels

When specifications require only a paved invert, the invert is placed by methods similar to those used in the invert, arch lining method, which will be described in detail below. On tunnels where cleanup is required to bare rock, the large amount of nonpay overbreak concrete in the invert will affect cost, production, and the capacity of the equipment required.

If a tunnel has only a few areas where concrete lining is required, bulkhead pours are made with nontelescopic forms. These pours can be made while excavation is being performed by using the curb, invert, and arch method. Placing time and cost are high, relative to the amount of concrete poured, because the operations are intermittent. Transition forms are required at the start and stop of each concrete section, and because of limited yardage, the equipment write-off will be high. Concrete placement in the arch section is done by pumping methods or by use of *pots*. These pots are large-capacity air guns used to transport the concrete into the tunnel, where they are connected to the slick line and air supply in order to shoot the concrete into the forms. Pots serve a dual purpose since they replace both the concrete agitators and the air guns.

The Invert, Arch Concrete-Placing Method

The invert, arch method is the only concrete-placement method that will be described in detail. If this method is understood, the requisite changes in procedure for using other placement methods can be readily made. For descriptive purposes here, the invert, arch placement method is separated into the several elements of work or costs. The descriptions generally pertain to cast-in-place linings for rock tunnels, but the methods used for cast-in-place linings for soft-ground tunnels are essentially the same.

Removal of Fan Line

Often, after the tunnel is holed through and there is an uninterrupted passage of air from one tunnel portal to the other, the natural flow of air through the tunnel will satisfy ventilation requirements. If this natural ventilation is not adequate, ventilation requirements can be met by placing a bulkhead equipped with exhaust fans over one end of the tunnel. These exhaust fans will then draw fresh air through the entire length of the tunnel. Automatic opening and closing doors can be installed in the bulkhead to provide access for men and equipment. Shafts are convenient locations for ventilation fans.

Since ventilation requirements can thus be met by other means and because the large-diameter vent pipe will interfere with the concreting operations, the first step in preparing the concrete lining is the removal of this vent pipe. The pipe can be uncoupled and lowered by cranes and the vent-pipe car by which it was placed. But because it will be removed much faster than it was placed, the vent car will not provide sufficient transporting equipment. To provide sufficient pipe-hauling units, special cradles can be mounted on some of the haul units that were used for muck disposal in the excavation operation.

Arch Cleanup—Resetting
Steel Sets, Retimbering, Removing Tights

First, surveyors run templates down the tunnel line, marking all sets that encroach on the minimum-clearance line and all rock that projects past the minimum-excavation line. The sets that encroach must be reset, the excavation tights removed, and the lagging and blocking removed, rearranged, or replaced to conform to the specification requirements. No special equipment is needed for this operation, for it can be performed with some of the equipment used for tunnel excavation.

The concrete specifications should be checked for any special requirement affecting this work. For example, some specifications state that lagging and blocking can only cover up to 50 percent of the surface area to be concreted.

Invert Cleanup

Specification requirements influence invert cleanup more than any other step in the lining procedure. This is due to wide differences among specifications on the type of tunnel cleanup required.

When specifications do not require cleanup to bare rock, invert cleanup can be combined with invert concrete placing, which eliminates a separate invert cleanup operation. Concurrent invert cleanup and concrete-placing operations result in a savings to the owner. (Many designers believe that this procedure produces as competent a concrete lining as any constructed under the bare-rock cleanup requirement.) Under this type of specification, the only cleanup required is the removal of loose material that is unsatisfactory for backfill, the leveling of the remainder to tunnel grade, and the removal of track. Cleanup operations required by this type of specification can be combined with the "retreat" method of placing invert concrete (see below).

A separate cleanup operation completed before the invert pouring can begin is necessary when the specifications require that all loose material be removed from the tunnel invert and that the invert surface be free of all water. Because of the additional time and man-hours required for such cleanup, the added amount of excavated material to be handled, and the increase in overbreak concrete, this type of cleanup amounts to a large percentage of the concrete cost. Cleanup operations are started at one end of the tunnel and are carried through to the other. As the cleanup progresses, the track must be lifted, and track materials hauled out of the tunnel or hung on the tunnel sides. All muck must be removed and the rock surface cleaned with air and water jets. Provisions must be made for surface drainage and all surface water must be removed.

Under bare-rock cleanup requirements, some contractors have obtained a cost savings, after rock cleanup is completed, by installing drainage systems in preference to hand mopping the surface water (required to keep the invert dry ahead of the concrete pour). The type of drainage system used is a wellpoint system of drains, which is covered with gravel to the concrete pay line. Plastic is placed over the gravel before the invert concrete is placed. Suction is maintained on the drain pipes while the invert concrete is placed and cured. After the invert concrete has strengthened, the gravel under the plastic is grouted.

Much of the equipment used for the tunnel cleanup is available from that used for tunnel driving, such as the muckers, muck cars, locomotives, drills, and air facilities. This equipment may be supplemented by the following specialized units:

Tractor-mounted front-end loaders

Gradalls or backhoes for removing the tunnel muck, used in conjunction with air and water jets for final cleanup of the tunnel

Conveyor belts for transporting the muck from the excavating equipment to rail cars or rubber-tired hauling units

Placement of Reinforcing Steel

Reinforcing steel is often installed by a subcontractor. In this case, the prime contractor transports the invert steel into the tunnel for the subcontractor before the invert cleanup is started, and the steel is hung from the sides of the tunnel until the invert cleanup is completed. The subcontractor then places the steel in the invert ahead of the concrete placement. Reinforcing steel for the arch is taken into the tunnel after the invert pour is made and is placed prior to the arch concrete.

In some pressure tunnels with low cover, the designers often call for heavy mats of reinforcing bars of large diameter, such as #18 bars. This reinforcement causes unique problems and increases the cost of tunnel concrete. If the invert is poured separately from the arch, the steel must be spliced above the invert pour and the bars must be lapped or coupled together. Specifications should be checked to see if the steel in the laps is a pay quantity. When the arch is poured, the reinforcing steel may interfere with retraction of the slick line as the pour advances, which will require that either intermittent pours be placed with the slick line located above the reinforcing steel or the slick line be lodged in a dimple in the tunnel arch. If a dimple is used, the designers may allow this extra concrete in the arch to remain; in other instances the dimple must be removed and the concrete ground to a polished surface. The vibration and flow of the concrete also becomes more of a problem in a reinforced tunnel, and this factor should be reflected in the estimated cost.

Figure 12.1 shows reinforcing steel in place along with the invert screed setup ready for invert concrete.

Panning and Care of Water

If the tunnel has water inflow in the crown and sides, it may be necessary to guide this water down the sides of the tunnel to discharge pipes placed in the bottom. Such guidance is accomplished with light-gauge metal or metal foil placed between the supports or pinned to the rock. The drainage system in the bottom may have to be as extensive as a modified wellpoint system, with pumps exhausting the bottom drainage system to the main discharge pipe. Main drainage pipe of large diameter should be installed along the sides of the tunnel so that all water will be removed. This drainage system must be maintained until the completion of the arch concrete. The bottom drainage system can then be grouted or, in some cases, left as a permanent drainage system with flap valves on the exit pipes. The main drainage pipes should then be removed from the tunnel.

Placement of Invert Concrete

There are two main methods of placing invert concrete. One is the "advance" method, the other the "retreat" method. The following description of the two methods of placing invert concrete is for a tunnel excavated with track-

Figure 12.1 Reinforcing steel and invert screed setup. (Bay Area Rapid Transit District.)

mounted equipment. If the tunnel is driven with rubber-tired equipment, the procedure is the same, except there will not be any track to take up and relay.

Advance Method.   The advance method of placing invert concrete, schematically shown by Fig. 12.2, because it is more complicated than the retreat method, is normally only used when specifications require that the invert surface be cleaned up to bare rock and be free of water. With the advance method, the cleanup operations are done well in advance of the invert placing operation. If a tunnel has two open ends, or portals, the cleanup is started at one portal. Track, muck, and other material from the cleanup operation are hauled out the other end of the tunnel, with concreting operations following the cleanup. In a tunnel which has only one portal, cleanup is started at the blind end and carried back to the portal before the concrete invert is started.

The key piece of equipment with the advance method of placing concrete is an invert traveling bridge supported from the sides of the tunnel and of such length that the concrete at the rear of the bridge has sufficient strength to support the railroad track. A cross section taken through an invert bridge at the screed location is depicted by Fig. 12.3. The bridge must be suspended high enough so that space is available beneath it for finishing the invert concrete and for relaying the track. One method of support and travel for this bridge is by rails that run on wheels located on each side of the tunnel. The

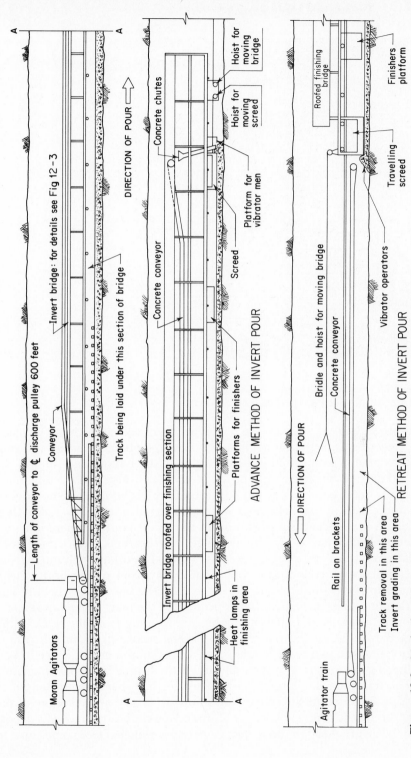

Figure 12.2 Methods of invert pour.

Moran Agitators

Length of conveyor to ℄ discharge pulley 600 feet

Conveyor

Invert bridge: for details see Fig 12-3

Track being laid under this section of bridge

DIRECTION OF POUR

A — A

Concrete chutes

Hoist for moving bridge

Hoist for moving screed

Concrete conveyor

Platform for vibrator men

Screed

Platforms for finishers

Invert bridge roofed over finishing section

Heat lamps in finishing area

ADVANCE METHOD OF INVERT POUR

A — A

Roofed finishing bridge

Finishers platform

Hoist for moving bridge

Travelling screed

Bridle and hoist for moving bridge

Concrete conveyor

Vibrator operators

DIRECTION OF POUR

Rail on brackets

Track removal in this area

Invert grading in this area

Agitator train

RETREAT METHOD OF INVERT POUR

355

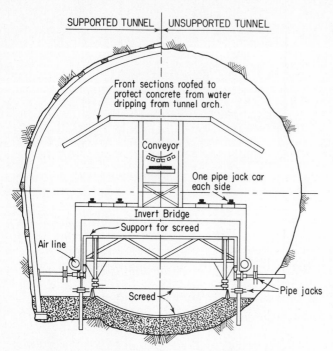

SUPPORTED TUNNEL | UNSUPPORTED TUNNEL

Front sections roofed to
protect concrete from water
dripping from tunnel arch.

Conveyor

One pipe jack car
each side

Invert Bridge

Support for screed

Air line

Screed

Pipe jacks

Figure 12.3 Section through invert bridge and invert
screed for the advance method of invert pour.

wheels are supported and kept on line either by brackets which are attached
to the steel sets or, if the tunnel is unsupported, by brackets which are supported by pipe jacks inserted in holes drilled in the bottom and sides of the
tunnel. If desired, the wheels can be installed in the bridge and the rail on the
brackets. If this type of bridge support is used, rail is hung from the sides of
the tunnel when the track is removed for invert cleanup. The rail is then
installed on the brackets shortly ahead of the invert pouring operation.

After the bridge passes, the rail is taken off the brackets and used for track
installation on the invert. The brackets are then taken off and rehung in front
of the bridge. A traveling screed is mounted in front of the bridge, with
independent travel on duplicate wheels so it can be more easily held to line
and grade and so its movement is independent of the main bridge movement.

A conveyor belt is mounted on the bridge, running from the track in the
rear of the bridge to the front of the bridge, where it terminates over a
hopper connected to a three-barrel chute. Concrete placement is performed
by delivering the concrete to the rear end of the bridge by trains of agitator
cars. The trains are discharged onto a conveyor belt that runs the length of
the bridge and deposits the concrete in a hopper. A gate at the bottom of the
hopper directs the concrete into one of three pipe chutes by which it is deposited on the invert ahead of the screed. Concrete is switched from pipe chute to

pipe chute to deposit it evenly across the invert. Concrete is vibrated from platforms in front of the screed so that it is forced under the traveling screed. As the screed advances, the bridge advances, and the whole operation moves down the tunnel.

Concrete finishing is performed under the bridge with heat lamps to hasten the curing time. Fast curing is necessary with the advance method because all traffic to the invert-placing operation must take place over recently placed concrete. Under the rear of the bridge, first paper, for concrete protection, then longitudinal stringers, ties, and rail are laid on the concrete so that as the bridge advances, there is always rail connection to the bridge. The paper is placed to aid in curing as well as to protect the concrete. The longitudinal stringers are laid under the ties to give firm support to the ties on the curved surface of the invert. Ties and rail are brought in from the portal as required, and the ones taken up during cleanup operation are reused. In a two-portal tunnel where cleanup and concrete are carried on at the same time, it is generally necessary to truck the rails and ties from the cleanup portal to the concreting portal. If track material was hung on the tunnel sides during the cleanup operation, it is only necessary to reclaim this material.

On some tunnels, the specification requirements are so strict that mopping up water is more expensive than installing a wellpoint system, covering the aggregate with plastic, and grouting this aggregate after the concrete invert is placed. If this method of invert drainage is used, the aggregate is brought in by concrete agitators, unloaded, and transported in the same manner as the concrete. However, the aggregate is discharged ahead of the concrete pour by a special belt provided for this purpose.

Hence, special equipment required for the advance method includes:

Bridge—complete with guides, conveyor belts or pump, brackets, power drum for moving, power, heat lamps, etc.
Invert screed—complete with guides and power drum for moving
Concrete vibrators
Agitator trains

Retreat Method. The retreat method of concrete invert placing, as shown in Fig. 12.2, is the most common method of placing invert concrete. Cleanup of the invert is part of the concreting cycle, and it is not necessary to route traffic over recently placed concrete.

The first operation is to excavate all excess tunnel muck from both sides of the track with the mucking machine. The invert pour is then started at one end of the tunnel and progresses toward the other. This invert pouring method uses a bridge and a screed, but the bridge can be very light, since it is only used to provide the concrete invert with protection from water dripping from the roof of the tunnel and to furnish support for the concrete finishers.

Concreting operations are started backing up from one end of the tunnel where lengths of rail, sufficient to support the bridge, are removed from the track and then fastened to brackets along the sides of the tunnel. The rails must be set to line and grade since they are used for guides and support for the screed and the bridge. The ties are then removed and hauled outside the tunnel, the tunnel invert is graded or cleaned, and the bridge and screed are erected. Concrete is transported into the tunnel by agitator trains and discharged onto a conveyor belt which deposits it in front of the screed. Vibrators then consolidate the concrete as the screed is pulled forward. As the screed and bridge advance with the concrete pour, other sections of rail are taken up and placed on the sides, and the invert is prepared for concrete. Mechanized screed units can be used which perform the same function as the vibrator operators; this reduces the manpower required but increases the equipment cost.

Behind the bridge, paper is laid on the invert, stringers and ties are hauled into the tunnel and laid on the paper, and rail is taken off the brackets and relaid on the ties. The brackets that supported the rail are taken off the sides of the tunnel and carried forward to be rehung. Finishing and curing are performed under the roofed bridge. With this method of invert placement, the invert cleanup and invert pour are carried on alternately.

The special equipment required for the retreat method includes:

Bridge—riding on rails supported on the sides of the tunnel or steel sets (when available); designed to support a roof and to carry the finishers; sometimes equipped with heat lamps to hasten the curing time

Traveling screed—riding on the same rails as the bridge

Short conveyor belt—used for transporting concrete from the agitator trains to the screed (a concrete pump can be utilized as an alternative)

Agitator trains

Placement of Arch Concrete

The following is a description of a method in common use for the placement of arch concrete. This method is schematically shown in Fig. 12.4. A length of telescopic arch forms, sufficient to maintain a continuous pour, is erected at one end of the tunnel. These forms are fastened to the invert concrete by the use of anchor bolts installed in threaded inserts placed in the invert concrete when it was poured or installed in holes drilled later in the invert concrete. If the tunnel is too short to justify the cost of telescoping forms, or if it is not feasible to pour continuously, an appropriate length, perhaps 50 or 100 ft (15 or 30 m), of forms is purchased without the telescoping feature. The only difference in using these forms is the necessity for constructing a bulkhead at the end of each pouring cycle. Some specifications permit a sloping end instead of a formed joint at the ends of pours, which results in far lower costs for the owner and for the contractor. Concrete for either system is transported into the tunnel by agitator trains or by mixer trucks and is then discharged onto a short transfer belt. This transfer belt elevates the concrete and

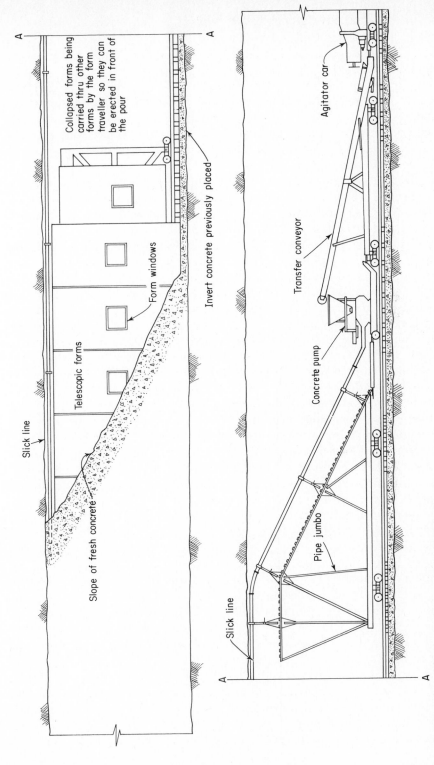

Collapsed forms being carried thru other forms by the form traveller so they can be erected in front of the pour

A —— A

Slick line

Telescopic forms

Form windows

Slope of fresh concrete

Invert concrete previously placed

Agitator car

Transfer conveyor

Concrete pump

Pipe jumbo

Slick line

A —— A

Figure 12.4 Method of pouring arch concrete.

deposits it in a hopper located above the concrete pump. The concrete is then batch-fed from the hopper into a concrete pump which forces the concrete into the crown of the arch and in back of the forms through the slick-line pipe. This pipe is supported by a pipe jumbo, and after the pour is started, the discharge end of the pipe is kept embedded in the fresh concrete. Embedding the end of the pipe is necessary to prevent segregation of the concrete. The concrete placed in the arch flows down the sides of the tunnel and connects with the invert concrete.

As the pour is made, the belt, pump, and pipe are moved backward at the same speed at which the concrete fills the forms. When the concrete has set sufficiently behind individual lengths of forms, these forms are stripped, collapsed, and carried through the other forms. The collapsed forms are then erected in front of the forms still in position. This stripping, collapsing, transporting, and erecting of the forms is done with a form traveler jumbo. Concrete haul, concrete placing, and form moving are synchronized so that the arch concrete can be placed continuously (except for shutdowns over the weekend). At the end of the week, bulkhead pours, which form a vertical construction joint, are made.

On small tunnels, only one slick line is used; it discharges into the crown of the arch. On large tunnels more discharge pipes, located at various points in the tunnel circumference, are used, and they are connected in rotation to the placer. These additional pipes reduce the distance that the fresh concrete must flow around the sides of the tunnel. On tunnels that have a large volume of concrete per lineal foot of tunnel, two or more pumps may be used, and the mixing and transporting equipment must be sized for their capacity. In such cases, it is common practice to use one or two concrete pumps, plus one air placer. The air equipment is used at the crown, where it is most difficult to fill against the irregular rock surface. The air placer has a very effective "slugging" action which improves placement there. When placement rates, equipment capacity, and cost are estimated, the amount of nonpay overbreak concrete must be taken into consideration.

The arch forms can be either telescopic or nontelescopic. Telescopic forms are used for continuous pouring, and enough forms should be provided for 16 hours of concrete placing, taking into account the front slope of the concrete inside the form. Nontelescopic forms are used for short pours, and the form covers the length of tunnel to be poured at one time. After curing, the forms are stripped, moved, and erected for the next pour. Forms are moved by a form traveler riding on railroad track or on rubber tires and containing jacks for collapsing and setting the forms.

Forms should be provided with windows through which concrete placing can be watched and the concrete vibrated. On large tunnels, concrete distribution pipes may be carried down the inside of the forms to these windows. They should also be equipped with form vibrators.

Figure 12.5 shows tunnel arch forms previously used for the tunnel con-

Figure 12.5 Tunnel arch forms with outside forms for cut-and-cover construction. (Economy Forms Corp.)

crete and now erected with an outside form for cut-and-cover construction beyond the tunnel portal.

There are three types of concrete placers: concrete pumps, air guns, and pots. Concrete pumps force the concrete through the discharge pipe by mechanical means. Pumps are used to pump concrete long distances, since they can pump concrete approximately 1,200 lin ft (365 m), while a gun operates best at a distance of 200 ft (60 m). Finally, pumps are used for placing short pours since it is easier to start and finish pours with a pump than with an air gun.

Pneumatic placers or air guns force batches of concrete through the discharge pipe by the use of air pressure. Their placement capacity exceeds that obtained with pumps. Operating and maintenance cost of air guns is less than that of pumps because they have fewer moving parts. Space occupied by this placer is smaller than that necessary for a pump. Most pneumatic placers operate on the same principles as the Press Weld, an all-air-operated concrete gun shown in Fig. 12.6. These units are large steel containers equipped with a discharge nozzle and a sliding door mounted on top of the container. The unit is charged with batches of concrete by opening the sliding door and dropping concrete into the container. The sliding door is then closed, air pressure is applied, and the concrete is forced through the nozzle. The unit requires a single operator. This man operates a quick-acting valve controlling air to the back of the container and another quick-opening valve controlling

Figure 12.6   Press Weld placer. (George M. Philpott Co., Inc.)

air to the nozzle. When both valves are opened together, concrete starts out of the container through the nozzle in a steady stream. As concrete travels through the 6-in discharge pipe and reaches the forms, a quick drop in pressure results. Both valves are then closed and the machine is ready to recharge and place another batch of concrete. An upper automatic charging gate is almost always used with Press Weld units. This is a small hopper which mounts on top of the Press Weld unit and holds one charge of concrete. As the sliding door opens on the Press Weld, a sliding gate in the bottom of the charge gate opens and drops a charge of concrete into the Press Weld. The upper automatic charge gate is necessary for the most efficient operation of Press Weld units.

Listed below are dimensions and capacities of various sizes of Press Weld placers:

Size (yd³)	Height ft	Height in	Width ft	Width in	Length ft	Length in	Approximate Hourly Capacity (yd³)
1⅛	4	0	3	8	9	0	60–100
¾	3	6	3	1	9	0	45–60
½	3	0	2	8	9	0	34–45

Pots are constructed so that they will hold up to 8 yd³ of concrete. They are charged with concrete outside the tunnel and then travel on rail to the point in the tunnel where the concrete is being poured. At this point they are connected to a concrete discharge pipe and the air supply. Thus, pots act as air guns, shooting the concrete into the forms. Pots are used in small tunnels that have restricted working area or in large tunnels which have intermittent pours. The advantage of using pots is that they replace the agitators, transfer conveyor, hopper, and placer. They are disadvantageous in that the amount of concrete placed per day with this operation is comparatively small unless a large number of pots is purchased.

A summary of the special arch concrete equipment would include:

Arch forms
Form traveler
Slick line
Slick-line jumbo to support the pipe
Concrete placer
Concrete-charging hopper
Concrete-transfer conveyor belt
Agitator trains
Form vibrators
Hand-held vibrators

Low-Pressure Grouting

After the concrete is placed in the tunnel, any voids left between the concrete lining and the tunnel surface must be filled with low-pressure grout. Specifications should be checked to see if this is a pay item, or if the contractor must include the cost of this operation in the concrete pay item. Grout pipes are installed in the lining as it is poured. These pipes must be cleaned out and connected to the grout pump. Grout is then pumped as required.

Equipment essential to the grouting operation includes:

A grout pump serviced with air, water, and electricity and equipped with a grout mixer to feed the pump (a grouting setup is shown in Fig. 12.7)
Equipment to haul cement and sand to the grout mixer

Final Cleanup

Final cleanup involves the removal of track, pumps, pipe, electrical facilities, and all other items of plants and equipment. The operation further entails being certain that all loose material has been removed from the tunnel after grouting operations, and includes the cleaning and patching of all concrete.

CAST-IN-PLACE SECONDARY LININGS

It is sometimes required that secondary linings be poured in soft-ground tunnels, or rock tunnels in very weak rock, where the primary lining is steel, cast-iron, or precast-concrete liner segments. This requirement may arise

Figure 12.7   Grouting equipment. (Gardner-Denver Co. and George M. Philpott Co., Inc.)

from the need to provide a smoother surface to reduce friction losses in sewer or water tunnels, to provide increased strength in a pressure tunnel, to provide increased resistance to chemical attack in a sewer tunnel, or to provide a level invert in transportation tunnels. Secondary linings are almost universally reinforced. The methods and equipment are generally identical to those used for cast-in-place linings for rock tunnels, although full-circle pours are far more common for secondary linings.

SHOTCRETE LININGS

Shotcrete is a term used to describe pneumatically placed concrete. It differs from gunite in that a coarse aggregate is used in shotcrete. Shotcrete lining has been used successfully to provide temporary support for tunnels during excavation and, when later increased in thickness, to provide the permanent lining. It is most economical when it can be used in this dual capacity.

As a temporary support, a thickness of approximately 2 in (5 cm) of shotcrete is applied to the excavated surface immediately after the round has been exploded and the surface has been scaled. Thickness depends on the diameter of the tunnel and should be established by field test for any tunnel job. To furnish this temporary support, a high-early-strength shotcrete is of greatest importance. This can be secured by the use of a hardening-acceleration admixture. The shotcrete placed in the Tehachapi tunnels in California included a German Tricosal admixture; the amount used was from 5 to 7 percent of the weight of the cement. Since this admixture reacts quickly with the cement, it is necessary to apply the shotcrete in the dry-process manner, i.e., by mixing the ingredients and water at the nozzle. If high early strength is not required, wet-process shotcrete may be satisfactory. This 2-in (5-cm) lining of shotcrete is later increased to 6 or 8 in (15 or 20 cm) to serve as a permanent lining. These additional applications of shotcrete can be delayed so they will not interfere with the tunnel excavation.

Temporary support by shotcrete is dependent on its interaction with the rock surface. A shotcrete layer applied immediately to a new rock face acts as a tough surface, by means of which a rock of minor strength is transformed into a stable one. The shotcrete absorbs the tangential stresses which build up to a peak close to the surface of a cavity after it is opened up. As a result of the close interaction between the shotcrete and rock, the rock retains almost all of its original strength and is able to supply arch action. The effective thickness of the zone of arch action is in this way increased to a multiple of that of the shotcrete. In this way tensile stresses due to bending are diminished, and compressive stresses are easily absorbed into the surrounding rock. The zone of arch action can be increased with rock bolting. Mixes containing steel fibers have been used on a limited basis to improve shotcrete strength.

A typical mix design for one cubic yard of shotcrete might be:

Cement	660 lb
3/4-in minus stone	1,300 lb
Sand	1,790 lb
Accelerator	90 lb
Water-cement ratio	4 : 1

The quantities of materials which must be batched to produce one cubic yard of pay shotcrete greatly exceed the theoretical requirements. The first reason is that in order to place a minimum 2-in (5-cm) thickness (for example) on a rough rock surface, it may be necessary to place an average thickness of 3 or 4 in (8 or 10 cm). The second reason is the loss from rebound of materials from walls and crowns of the shafts or tunnels. The combined effect may mean that several times the theoretical quantities are actually required.

Shotcrete also prevents rock disintegration, which always starts by the opening of a minute surface fissure. Since shotcrete prevents this initial movement, the rock behind the shotcrete remains stable.

The use of shotcrete as a temporary and permanent lining has had wide application in countries that have a low labor cost permitting more delay in the excavation cycle. Because of high labor cost in the United States, the use of shotcrete here is not so advantageous, unless its application can be made without slowing the mucking cycle. Slowing the mucking cycle can be prevented if there is enough room in the heading to allow a shotcrete jumbo to be used to support the nozzlemen without interfering with the mucking operation. If mucking is delayed until the shotcrete is applied, the extra excavation cost plus the cost of the shotcrete lining has to be balanced against the cost of the tunnel supports and concrete lining that would otherwise be used, to determine the economical procedure for any given tunnel. The equipment used in the application of shotcrete will deliver approximately 10 yd^3 per hour through the nozzle. Equipment needed for shotcrete application consists of a gun, material hose, air and water hoses, nozzle, and sometimes a water pump. Jumbos are used to suspend the nozzlemen over the muck pile. These can be simple platforms suspended on drill jibs which are mounted on rubber-tired, self-propelled units, or more elaborate jumbos as necessary. A supply of compressed air is required, and equipment is needed to supply aggregate, sand, and cement to the gun. Figure 12.8 shows the typical equipment used to supply shotcrete.

Figure 12.8 Typical equipment used to supply shotcrete. (Reed International, Inc.)

CONCRETE-SERVICING EQUIPMENT

Compressed Air

More compressed air may be used for one concrete setup than for one excavation setup, if air-placement methods are used, but since most jobs have several excavation headings and only one concrete setup, the air provided for excavating the tunnel generally takes care of the concreting. This may not be the case when only one heading is driven, and additional air may then be required for concreting. Air required for concreting is used for:

1. Drilling out tights: allow about 100 ft^3/min per drill.
2. Air and water cleanup: allow 1,000 ft^3/min for each ¾-in nozzle.
3. Air-operated sump pumps: allow about 40 ft^3/min for every 100 gal/min pumped.
4. Air-operated concrete placers: for Press Weld placers it takes an average of 800 ft^3 of compressed air at 100-lb pressure to deliver 1 yd^3 of concrete 100 ft and lift it 20 ft. For greater distances, add 1 ft^3 of air for each additional foot of horizontal pipe and 5 ft^3 for each foot of additional lift.
5. Air-operated vibrators: allow 25 ft^3/min each for light-duty vibrators, 80 ft^3/min each for medium-duty vibrators.
6. Miscellaneous small tools and air leaks: allow 200 ft^3/min.

Aggregate, Concrete-Batch, and Concrete-Refrigeration Plants

The aggregate and refrigeration plants, if required, as well as the concrete-batch plant, are in most cases located near the tunnel portal. Since these plants include the same equipment as can be used on any type of construction, a description of equipment will not be included. However, capacity of these plants must be balanced with the placing equipment.

Concrete Haulage

If the tunnel muck is removed from the tunnel with rubber-tired equipment, diesel-driven agitator trucks are generally used to haul the concrete from the batch plant into the tunnel.

If the tunnel muck is removed from the tunnel with rail-mounted equipment, rail-mounted agitator trains are used to transport concrete into the tunnel. These agitators are of two types, standard and tubular agitators. Standard agitators are located on cars so that they will be able to side-dump onto a conveyor belt. This conveyor belt is located along the tunnel wall and is used for transporting the concrete to the concrete placer. Tubular agitators discharge into each other. The front agitator discharges directly onto a belt which conveys the concrete to the concrete placer. With tubular agitators, a full trainload of concrete can be discharged at one point. This is very important in a small tunnel that does not have clearance for a conveyor-belt installation along the side of the train.

PREFABRICATED TUNNEL LINERS

Prefabricated linings are manufactured from concrete, steel, or cast iron, including malleable iron. All are manufactured in segments which are joined in the tunnel to form rings. The segment-to-segment and ring-to-ring connections are most often bolted, but may be (in the case of precast concrete) unbolted knuckle joints. Sealing between segments may consist of gaskets of rubber or similar material cemented to the flanges, usually with caulking grooves in case of leakage. Sealing for metal segments may consist of machining the faces, which then tend to rust together. Caulking grooves are always provided with this system. Bolt holes are sealed with plastic grommets under steel washers on both sides of the joint. Tapered rings are provided for use on curves and for use in maintaining alignment. *Taper* refers to nonparallel ends.

Grout holes are provided in each segment so that the annular space between the liners and the ground may be filled after passage of the shield. Each segment must have some arrangement to enable the erector to grasp and manipulate it into place. Rings are commonly 30 to 48 in long (75 to 120 cm). The shorter lengths are selected for tunnels with curved alignments in order to reduce shield length and thus to improve steerability. The number of segments in a ring varies with tunnel diameter; one key segment, which has a shorter arc length than the typical segment, is usually included. Most specifications require that the longitudinal joints be staggered. Liners are designed for:

1. Permanent exterior loading resulting from earth pressure.
2. Permanent interior loading if the tunnel is to be a pressure tunnel.
3. Temporary axial loading resulting from advancing the shield by jack pressure on the liners. This pressure is sometimes distributed by providing a jacking ring between the jack pads and the liners.

Experimental work is being done on slip-forming tunnel linings, on segments which can be assembled as a helix, on wedge-shaped segments which permit the rings to be expanded as they leave the shield, and on other improvements in current soft-ground tunnel lining technology, but few of the improvements appear to be sufficiently developed to show promise of general use in the immediate future.

Precast concrete may be reinforced or unreinforced. Reinforcing can be conventional or prestress. Experimental precast-concrete liners have used fiber (usually steel fiber) reinforcing as a replacement for or a supplement to conventional reinforcing. Precast-concrete tunnel liners are almost universally provided with gasket-type sealants, often with two sealing zones in each plane. Liners for smaller diameters are usually plain; liners for larger diameters are usually coffered to reduce weight.

Steel tunnel liners are pressed from plate or fabricated by welding. All are assembled by bolting. The simple pressed plates are commonly corrugated during the pressing operation. The flanges are necessarily curved at the corners, which causes difficulty in sealing out water. The liners have very little resistance to axial loads unless stiffeners are added. For these reasons, the pressed liners are seldom used except for short hand-excavated tunnels where water problems are not significant.

Tunnel liners fabricated by welding are most often built from plate but are occasionally fabricated from rolled sections, such as large channels. They may be made watertight by machining the abutting surfaces or by applying sealing strips. Tee sections are commonly used between the flanges to resist jacking forces. It is necessary to provide high-quality coatings to prevent rusting, and cathodic protection and electrical bonding are often specified. Fabricated steel liners as shown in Fig. 12.9 are relatively light in weight and are easy to handle and assemble.

Cast-iron or cast-malleable-iron liners were at one time the standard prefabricated liners, but there are at present only a few suppliers. Cast-metal liners have excellent resistance to corrosion. They may be made watertight by

Figure 12.9 Typical fabricated steel liners. (Bay Area Rapid Transit District.)

machining the edges or by using sealants. Cast-metal liners are usually coffered to save weight and are usually specified to be coated with a high-quality paint system. Cathodic protection will be required if there are stray electrical currents, such as those often found in subway tunnels.

It is important in all uses of prefabricated liners that the annular space between the liner and the excavation be promptly and completely filled. This prevents ground subsidence, which may be disastrous for structures on the surface, and ensures uniform distribution of soil loads on the liners. The packing is placed through grout holes which are provided in the segments. Sand-cement grouts are the commonest packing. Pea-gravel packing, with the interstices filled with neat cement grout, is also commonly used. There have been some experimental uses of foamed concrete, but this material has not been generally adopted. It is important that the exterior surfaces of liner segments be as smooth as possible in order to reduce wear on the tail seals of the shield. Tail seals between the shield skin and the erected liners are easily damaged, and repair and replacement are very difficult. One Japanese contractor has developed a wire-brush type of seal which is very effective and very durable.

Chapter 13 Shafts

Shaft construction is the excavation and lining of a near-vertical tunnel. It differs from other types of underground construction because excavation is performed either downward, from the excavation surface that also supports the men and equipment, or upward, from platforms suspended underneath the face of the excavation. This type of excavation requires the use of special shaft equipment not used in other types of underground excavation. It is preferable to use experienced shaft-excavation personnel for this type of work; if they are unavailable, allowances must be made for job training.

TYPES OF SHAFTS

Shafts may be classified by the type of ground in which they are constructed. There are shafts in overburden or soft ground and shafts in rock. Most shafts in rock will be started in overburden. If there is only a slight depth of overburden, it may be removed so that all shaft construction is in rock. If the overburden depth is significant, shaft construction will be started with methods appropriate for soft ground and continued with methods appropriate for rock. The rock may be so completely weathered near its surface that it requires soft-ground construction methods.

Shafts may be further classified as to whether or not there is access to the bottom of the shaft at the time that shaft construction is started. If there is no such access, the shaft must be "sunk blind," meaning that all operations must be performed from the surface. This is frequently the case in mine shafts, missile silos, and underground storage of oil or natural gas. If there is access to the bottom, the shaft may be constructed by shaft raising or by slashing down from a pilot hole. Either of these methods will facilitate the work because the muck can then fall to the tunnel below instead of being lifted to the top of the shaft. These methods are not suitable for shaft excavation in soft ground, but may be used after the part of the shaft which is in overburden has been excavated by other methods.

Shafts incidental to other underground construction are usually advertised in conjunction with tunnel, subway, or underground powerhouse contracts. Examples of this type of shaft are ventilation shafts for subways, vehicular tunnels, or powerhouses; gate shafts, inlet shafts, or surge shafts for hydro tunnels; and access shafts, penstock shafts, or elevator shafts for underground powerhouses. Since access is usually available to either the top or bottom of these shafts, they can be constructed by large-hole drilling, sinking from the surface, raising from the bottom, or any combination of these methods.

Shafts necessary to service tunnel driving are constructed by the tunnel contractor. These shafts are used when tunnel driving cannot be performed through the tunnel portal and the topography prevents the construction of short construction adits. Such shafts are also utilized on long tunnel projects to divide the tunnel into shorter distances that can be excavated from each heading. Since access is not available to the bottom of these shafts, they must be constructed by sinking from the surface. Their large diameter makes the excavation of these shafts unsuitable for large-hole drilling methods.

SHAFT CONSTRUCTION METHODS IN ROCK

Five main methods are employed in the construction of shafts: sinking the shaft from the surface, raising from the bottom of the shaft, large-hole drilling, pilot-shaft enlargement methods, and long-hole raising methods.

Shaft Sinking

Limitations.   Shafts are sunk when access is not available to the bottom of the shaft or when the shaft passes through incompetent material which would cause a hazard in shaft-raising operations. Production is limited in this process by two conditions: the restricted working area in the shaft and the necessity that all movement in the shaft be vertical, with slow hoisting methods.

The only space available for men to work in the shaft is on the surface that they are excavating or on work platforms. If work platforms are used, these platforms must be suspended by cables anchored to the sides of the shaft or by cables from the hoisting equipment. This restricted working space limits the number of men, the number of drills, and the size of the equipment that can be used in the shaft.

The hoisting capacity provided for shaft sinking limits the weight of the individual pieces of equipment that can be used in the shaft. Hoisting capacity further limits the speed of mucking, since the rope speed controls the number of buckets of muck that can be hoisted out of the shaft per unit of time.

Equipment and Procedures.   All methods of shaft sinking use drilling, mucking, and hoisting equipment. Some of the differences among the methods of shaft construction lie in the type of equipment used in these three operations and other differences lie in whether working platforms are used and, if used, in the platform arrangement. A discussion of the basic equipment and procedures, common to all shaft sinking operations, will precede a detailed discussion of the operations themselves.

1. Hoisting Service. Progress in shaft sinking is controlled by the main hoist since the length of the mucking cycle is dependent on the speed of muck removal. Hoisting service in shallow shaft construction can be provided for a short distance, approximately 100 ft (30 m), by crawler crane or truck crane. The limit on this type of hoisting service is the crane's hoisting-cable-drum capacity. When this capacity is exceeded, a headframe and hoist are required.

When a headframe is used, it must be strong enough to resist the stresses caused by the line pull and tall enough that the longest lift can clear the shaft opening. The main sheave must be of sufficient diameter to prevent excess wear on the main hoisting cable. If the shaft is to be equipped with a permanent headframe and hoist, these may also be made available for shaft construction. If the shaft is not to be so equipped, a sinking headframe and hoist are required. Skid-mounted portable sinking headframes and hoists are available from equipment manufacturers.

The main hoist should have a cable drum of sufficient diameter and capacity to hold all the cable required to reach the shaft bottom. The hoist motor must be of sufficient size to furnish the line pull necessary to lift the heaviest load and a rope speed adequate to prevent the hoisting of the muck buckets from delaying the mucking cycle.

In addition to the main hoist, two or three smaller, air-operated hoists are used. Cables from these hoists pass over secondary sheaves, and these cables are used to move and control muckers, platforms, forms, etc.

2. Other Surface Facilities. A variety of additional surface facilities are needed at the construction site. A list of these facilities would include a shaft door, ventilation equipment, a water supply, power supply or distribution, and concrete production facilities. For the proper disposal of muck, it is essential to provide a muck bin equipped with automatic shaft-bucket dumping facilities, and truck-loading gates. Provision should also be made for a hoist house, a change house, a shop and warehouse, a place for powder storage, a cap house, and office facilities. Finally, it is often necessary to construct a concrete collar around the shaft opening, and this possibility should not be overlooked.

3. Drilling. Drilling in the shaft can be done with hand-held drills, with drills mounted on a collapsible shaft drill jumbo, or in large shafts, with track drills. A shaft drill jumbo has arms like umbrella ribs that are used to support either air-leg jackhammers or jib-mounted drifter drills. The jumbo frame is equipped with an index for orientation of the arms so that the drill pattern can be mechanically set. When the shaft drill jumbo is located on the excavation face and the umbrella-like top deck of the jumbo is expanded, it can be plumbed and held rigidly in position by jacks which react against the sides of the shaft. Figure 13.1 shows this type of drill jumbo.

4. Mucking.   Shaft mucking is done by hand, with a clamshell, by the use of mechanical muckers such as the Cryderman or the Riddell, or by a crawler type of front-end loader. If a clamshell is used, remote controls for its operation are available for use in the bottom of the shaft. The Cryderman mechanical mucker is air-operated. It has a sliding boom which activates a positive-action grab bucket that picks up material and deposits it into skips. The mucker is supported and controlled from the work deck of a platform. Cryderman muckers are available in sizes that have $\frac{1}{2}$- and $\frac{3}{16}$-yd^3 bucket capacity. The manufacturer rates the production capacity of the $\frac{1}{2}$-yd^3 machine at 40 to 50 ft^3/min and the $\frac{3}{16}$ yd^3 machine at 20 to 30 ft^3/min. The Riddell mucker is a movable bridge which supports and controls a cable-operated clamshell. The bridge rides on rails laid on a work platform. Both the Cryderman and the Riddell mucker platforms contain openings for muck-skip passage. Crawler front-end loaders are used in shafts which have a diameter in excess of 17 ft (5 m). They vary in size depending on the diameter of the shaft. The smallest unit in common use is the Eimco 625 B with a 21-ft^3 bucket capacity. Front-end loaders with a greater capacity are available up to the size

Figure 13.1   Collapsible shaft drill jumbo. (Shaft and Development Machines, Inc., Salt Lake City, Utah.)

of the Caterpillar 983 B, which has a 4½-yd^3 bucket. Muck skips are required for hoisting the muck out of the shaft. These vary in size depending on the shaft hoisting capacity and the shaft diameter. Free-swinging muck skips are used on shallow shafts. On deeper shafts, the muck skips are equipped with crossheads and have safety dogs on either side for engaging guides in the event of a cable failure. The crossheads and bucket ride on cable guides or fixed guides installed on the shaft sides. If muck bins are used, the skips are equipped with dumping rollers that engage the dumping scrolls in the headframe, which results in automatic dumping of the skip into the muck bins. Shallow shafts are excavated with one muck skip, but as the shaft deepens, two or more muck skips are used to expedite the mucking cycle. While one is being loaded, the other or others are hoisted, dumped, and returned.

Safety codes require that men not work below a free-swinging bucket.

5. Work Platforms.   The use of work platforms and the number of decks required on these work platforms depend on the method of shaft construction and on the depth of the shaft. In shallow shafts that are completely excavated before the concrete lining is placed, work decks are not required during the shaft excavation, and only a simple work deck is necessary during the concrete-placing operations. In deep shaft construction, when concrete is placed directly behind the excavation and work is done by the cycle type of operation (to be later discussed in detail), additional work platforms are required. These may be separate platforms that are removed from the shaft when the changeover from excavation to concrete is made, or they may be multideck platforms which remain in the shaft and are used for both excavation and concrete. If multideck platforms are used, the bottom deck supports and contains the mucker controls and the upper decks are used for form moving and for concrete pour. The platforms contain openings for the passage of muck skips and drill jumbos.

In deep shaft construction, when concrete is placed directly behind the excavation and work is done by the concurrent type of operations (to be discussed in detail later), multideck platforms are required which are not removed from the shaft. In South Africa, where shafts are sunk with greater rapidity than in any other country, five-platform jumbos are used. Passageways are provided through the platform for the muck skips. Since all drilling is done by hand-held drills, the bottom platform contains ladders for access to the excavation face and is equipped with piping manifolds for servicing the drills. The bottom platform also suspends the mechanical clamshell mucker and contains the mucker controls. The second platform is used for setting the curb, or blast ring (see below), and contains hydraulic jacks which react against the shaft sides for stabilization of the platform. The third platform is movable and is used as a work deck for setting forms and pouring concrete. The fourth platform contains blocks for movement of the total platform. The fifth, or top, platform is used for adjusting the suspension, stabilizing the platforms, and as a work platform for setting eyebolts in the concrete lining. These eyebolts are used for platform suspension during the greater part of its use. When movement of the multideck platform is required, it must be moved with the main hoist cables.[13.1]

6. Placement of Concrete.   Concrete lining operations within the shaft require facilities for transporting the concrete to a central container on the concrete work deck and for then distributing it behind the forms. Concrete can be lowered and discharged into the central container by concrete buckets or by a vertical slick line. Concrete can then be distributed around the inside of the forms with hoses or swing spouts. If the shaft excavation is completed before concreting is started, concreting is done from the bottom of the shaft upward. Slip forms are often used under these conditions. If the shaft lining follows along with the excavation, concreting is done downward in progressive pours, and a heavy bottom bulkhead form called a *curb,* or *blast, ring* is installed at the bottom of every pour. Forms are then erected to connect this blast ring to the previously placed concrete. Ventilation pipes, compressed air pipes, drain lines, water lines, and electrical service lines are required in the shaft. When concreting follows excavation, these temporary lines are frequently embedded in the concrete lining. Providing multilift pumping for proper drainage in deep shafts necessitates the excavation of booster-pump stations in the sides of the shaft, at approximately 200-ft (60-m) intervals.

When excavation must pass through a water-bearing rock stratum, drilling and grouting equipment is required to grout ahead of the excavation in order to cut off water flow into the shaft. Freezing has also been utilized for this purpose.

7. Shaft Supports.   When shafts are constructed in incompetent material, supports are required to restrain the sides of the excavation. Supports used may be steel ring beams, wood sets, blocking between beams or sets and the excavated surface, solid-steel or wood lagging, shotcrete, or rock bolts.

Methods of Shaft Sinking.   As suggested above, there are different methods of sinking shafts. These methods are three: completing excavation before concreting is started; employing an alternating method of operation; or excavating and concreting at the same time.

1. Completing Excavation before Concreting.   In shallow shafts and some medium-depth shafts passing through competent material, the shafts are constructed by completing the shaft excavation before the concrete placement is started. One advantage of this method is that better progress can be obtained by performing only one type of operation at a time. Another advantage is that the concrete lining can be placed by starting at the shaft bottom, with slip-form pouring to the surface. The disadvantage of this construction method is that there is more exposure to rock falls, and if incompetent rock or water is encountered, temporary bracing and construction drainage become both extensive and costly.

This method of shaft excavation is a drill, shoot, muck, and support type of operation. First the drilling equipment and personnel are lowered to the face of excavation. If hand-held drills are used, 5-ft (1.5-m) lifts are excavated. If drilling is done with a drill jumbo, the lifts are increased to approximately 10 ft (3 m). When a drill jumbo is used, it is lowered to the bottom of the shaft where the center leg is placed on the exact center of the shaft. The jumbo is then plumbed and anchored by jacking against the sides of the shaft. The drill pattern is then drilled out, the drilling equipment is removed from the shaft, and the holes are loaded with explosives. After removing all personnel from the shaft, the holes are exploded. The shaft is ventilated until all fumes are removed, and then the mucking equipment is lowered into the shaft. In shafts of less than 17 ft (5 m) in diameter, mechanical shaft muckers like Cryder-mans may be used. In larger shafts, mucking may be done with front-end loaders. In shafts with very large diameters, when the crawler type of mucker is used, only half of the shaft area may be drilled and shot at a time, while the mucker is placed on the other half and covered to prevent damage from the shot rock. On shallow shafts, when the hoisting time is of short duration, only one muck bucket may be used. On deeper shafts, two or more are used. In this case, while one is loaded, the other or others are hoisted out of the shaft and dumped. Upon completion of the mucking cycle, supports are placed and the mucking equipment is removed from the shaft. Then the drilling equipment is again lowered and the cycle repeated.

Upon completion of all excavation, the shaft is checked for tights, and, these being removed, the concrete operations are started. Forms are installed in the bottom of the shaft, and concrete is placed from a work platform suspended from the shaft sides. Concrete may be lowered to the work platform in buckets or by use of a vertical standpipe. The concrete is then distributed from the central container by either swing spouts or flexible hoses. Slip forms may be used.

2. Excavating and Concreting Alternately.   In the alternating method of shaft construction, after three rounds, or approximately 30 ft (10 m), of shaft excavation are completed, excavation is stopped, and an equal distance of the shaft is concreted. This method is preferred in incompetent rock, where temporary support for the shaft walls would present a problem. The cycle method is also preferred in all deep shafts since the hazard of rock falls is eliminated. This method is widely used in the United States for deep shafts, since crew size can be held to a minimum by using the same crew to excavate and concrete. Separate equipment units may be used for excavation and for concreting. This necessitates that each type of equipment be removed from the shaft before the next type is lowered. Other setups use a three-deck jumbo with the mucker supported and controlled from the bottom deck and the other decks used for concreting. If hand-held drills are used, they can be carried on the bottom deck. If a shaft drill jumbo is used, it can be raised and lowered through the muck-skip passages.

In this method of shaft sinking, the third round exploded is not mucked out. Instead, the broken rock is used as an additional working surface in the concreting operations. The other working surfaces for concreting may be a separate two-deck platform or a combination multideck platform used for both excavation and concrete. First, the curb ring, or as it is often called, the blast ring, is stripped from the previous pour and anchored in position at the bottom of the new pour. This curb ring supports the bottom of the pour and also protects the concrete from blast damage when excavation is resumed. The remainder of the forms are then stripped from the previous pour and erected in position. Concrete is lowered in the shaft to a central container by buckets or a vertical pipe. From this central container, concrete is deposited behind the forms by a swing spout or by flexible hoses. Upon completion of the concrete pour, the excavation cycle is resumed by mucking out the broken rocks left from the last round and then drilling the next round.

3. Excavating and Concreting Concurrently.   The concurrent method of shaft sinking requires that excavation and concreting be performed at the same time with separate crews. Production by the crews and the crew size are balanced so that the concrete lining is always maintained at the same distance from the excavated face. Faster progress is obtained with the concurrent method of shaft construction than with the alternating method, but a larger crew is required, there is more interference between the two operations in the shaft, and a more complicated equipment setup is required. The procedure for both excavating and concreting operations is the same as in the alternating method, except that both are done at the same time. This method definitely requires a multideck work platform.

Using the concurrent method of shaft sinking, a progress rate of 8 lin ft per shift on a 24-ft-diameter shaft has been made in South Africa. Similar shaft-sinking methods and equipment have been used in the United States, but the African production rates have never been equalled. This high production rate in South Africa may be attributed in part to the availability of low-cost labor which permits use of large crews. The shaft-sinking platforms used in Africa were previously described under shaft-sinking equipment and procedures. The construction procedure used in Africa is to lower access ladders from the bottom platform to the bottom itself. After the bottom is cleaned, the shaft centerline is marked, the drill-hole pattern laid out, and the round drilled out by hand-held drills. The holes are then loaded, the platform is hoisted clear of the shot, and the personnel are removed from the shaft. The round is shot, the shaft is ventilated, and the platform is again lowered to a mucking position. The majority of the muck is loaded into skips with a mechanical mucker and the final cleanup done by hand mucking. The excavation process is then repeated.

Concreting operations are carried on at the same time and are timed so that three shifts of work will result in the same lineal-foot production as is accomplished in shaft excavation. The first shift strips the pour ring, or curb ring, and prepares it for lowering. Then all protruding concrete is chipped away, and the various utility lines, such as compressed air, water, and electrical service, are extended. The second shift lowers and sets the curb ring, lowers and erects the forms, and starts the concrete pour. On the third shift, the remainder of the forms are erected, and the concrete pour is completed. The top forms are placed last to facilitate the concreting operations.[13.1,13.2]

Shaft Raising

The most economical method of constructing small shafts or pilot shafts for large shafts is the raising method, provided access is available to the bottom of the shaft or that it is feasible to delay shaft excavation until such access is available. However, shaft raising should not be planned unless the shaft passes through competent material, for a raising operation in incompetent rock is a hazardous one. Shaft raising is economical because mucking time is eliminated; the shot rock falls directly away from the excavation face. Another cost advantage in shaft raising is the fact that air, water, and power services can usually be taken off the tunnel-driving service lines, which eliminates the duplication of these services. Some of the disadvantages of this method are that all excavation must be done overhead from platforms suspended beneath the rock surface and that the movement of men, equipment, and materials for shaft raising is slower and more complicated than that in a shaft sinking. Shaft raising can be done in a variety of ways, and new methods are constantly being developed. The three most successful methods to date of raising small shafts are described here.

The Manway and Muckway Method. The manway and muckway method is shown in Fig. 13.2. As the shaft excavation is carried upward from the bottom of the shaft, the excavated space is timbered off into two compartments. One compartment is used as a muck chute, and the other is outfitted with platforms and ladders and used to provide access to the face. On pilot shafts, the overall dimensions are held to a minimum of approximately 6 by 8 ft (2 by 2.5 m). If it is permissible to obstruct the area under the shaft, the muck chute can be left open at the bottom and the muck allowed to drop on the tunnel floor. Since only a small amount of muck is produced, a day's production can then be loaded out, on one shift a day, with a front-end loader.

If it is necessary to keep the area under the shaft clear, the shaft muck can be retained in the chute by hydraulically operated gates located at the bottom of the muck chute. Then, if a siding for muck cars is installed underneath these gates, car loading is easily and rapidly accomplished. If the shaft is over a small-diameter tunnel, it may be necessary to widen the tunnel at this point to furnish room for the siding.

Excavating procedure requires that miners stand on a platform covering the top of the manway and drill out the round with stoper drills. After loading the holes, they climb down the manway and explode the shot. Ventilation

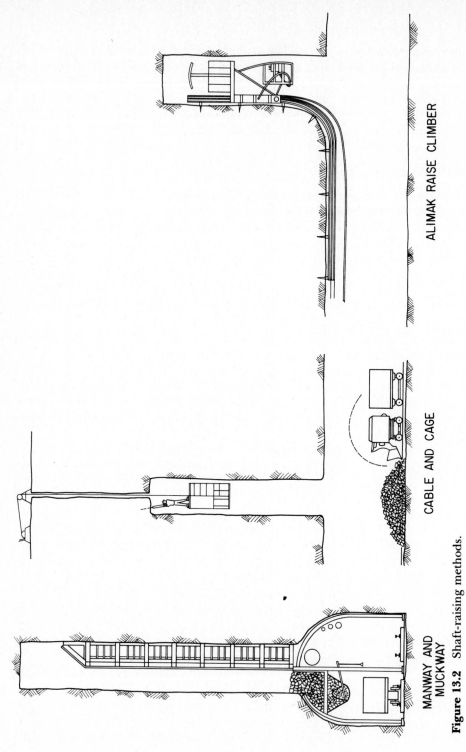

MANWAY AND MUCKWAY

CABLE AND CAGE

ALIMAK RAISE CLIMBER

Figure 13.2 Shaft-raising methods.

removes the powder fumes, and they climb back up the manway to clean the muck off the platform and scale the sides and the new rock face. Next, the timbering is extended so that the miners can drill the next round. This process is then repeated.

On small, inclined shafts, where the angle of inclination is great enough to cause the muck to slide down the shaft, the shaft is divided into upper and lower compartments by heavy timbers spanning the shaft and pinned to the wall surface on each side. The space under the timbers supplies a muck chute. Ladders are constructed above and on the dividers to provide access to the face. In order to furnish mechanical transportation to the face, a track is built on these dividers and a small car is moved up and down this track by an air hoist located near the face.

On inclined shafts where the angle of inclination is such that the muck will not slide on the rock surface, generally less than 40 percent, slick plates are installed on the bottom wall to decrease the friction. The slick plates should be spaced at intervals that will reduce friction enough for the muck to maintain motion in the chute. If the shaft is so inclined that the muck will not slide even when the entire shaft bottom is covered with plate, a shaft division is not used, slick plates are not installed, and instead, the muck is pulled down the slope with a slusher scraper.

The Cable and Cage Method.   A sketch of the cable and cage shaft-raising method is shown in Fig. 13.2. When shafts are raised by this method, a hole approximately 6 in (15 cm) in diameter is drilled from the surface down the centerline of the shaft. A hoist is erected over this hole, and a hoisting cable is passed through it. A man cage is attached to the cable underneath the shaft. The miners are then hoisted in the cage to the shaft face. There, they stand on the cage roof while scaling the face and drilling out the round. The cage is then lowered, explosives and powder are picked up, and the cage is again raised so that the men can load the holes with explosives. Again the cage is lowered to the bottom where it is this time unhooked from the cable and moved back a safe distance to be clear of the falling rock. In the meantime the cable is hoisted up the hole clear of the shot. The charge is exploded and the shot rock falls down the raise. After ventilation has removed the fumes from the raise, the cable is lowered and the cage is moved under the cable and reattached to it. The cage and men are hoisted to the excavation face, and the process is repeated. As the men drill out the next round, the muck at the bottom of the raise can be removed.

The Alimak Raise-Climber Method.   The Alimak raise climber consists of a working platform and a man-cage elevator that travels on a rail fastened to the side of the shaft on a vertical shaft or the top side of the shaft on an inclined shaft. Different-sized Alimaks can be purchased with cage capacities varying from one to six men. The rail that the cage rides on is a pin-rack track fastened to four pipes which provide the face with ventilation, compressed air, water, and electrical service. The rail weighs approximately 33 lb/ft and comes in short sections. As the excavation advances, the rail can be extended by bolting these sections onto the end of the rail and onto the shaft sides. The excavation procedure itself, with this method, is quite similar to that used in the cage and cable method. A schematic layout of the Alimak raise-climber method is shown in Fig. 13.2.

Shafts Excavated by Large-Hole Drilling[13.3]

When access has not been available to the bottom of the shaft, large-hole drilling methods have only been successful on shafts up to 8 ft (2.5 m) in diameter. On shafts larger than this, the removal of cuttings from the hole has presented a difficult problem.

When access has been available to the bottom of the shaft, larger shafts have been drilled. This operation entails drilling a pilot hole of approximately 12 in (30 cm) in diameter, the length of the shaft. The main function of this hole is to furnish a method of disposal for the cuttings from the reaming operations. This hole can be reamed out to the required diameter in progressive steps. Two methods are used in the reaming operation. One is to ream downward, and the other is to take the reamer to the bottom of the shaft and, with hoisting equipment located over the pilot hole, to ream upward.

When a large shaft is to be excavated by the pilot-shaft-enlargement method, large-hole drilling is often used to furnish a pilot shaft of 60-in (1.5-m) size.

Shafts Excavated by the Pilot-Shaft-Enlargement Method

The pilot-shaft-enlargement method of excavation is used for large-diameter shafts when access is available to the bottom of the shaft. First, a small-diameter pilot shaft, located on the centerline of the main shaft, is raised to the surface. The shaft-raising method may be any of those previously described. Depending on the type of rock encountered, it may be more economical to drill a large-diameter hole from the surface for the full shaft length than to construct a raise. After the pilot shaft is finished, it is enlarged to the full shaft diameter, with the pilot shaft or drilled hole used as a means of muck disposal. This is the most economical method of large-shaft excavation, for it utilizes the force of gravity to provide a fast and efficient method of muck disposal.

This shaft excavation method was used to construct the surge shaft located over the Angeles Tunnel in southern California. A schematic layout of this operation is shown in Fig. 13.3. After the top heading of the tunnel was driven and a 6-ft pilot shaft was drilled, excavation for shaft enlargement was started from the surface. This rock was soft enough to be loosened by ripping and then shoved down the hole with bulldozers. At the bottom of the shaft it was loaded into rear-dump haulers by a belt loader. The sides of the shaft

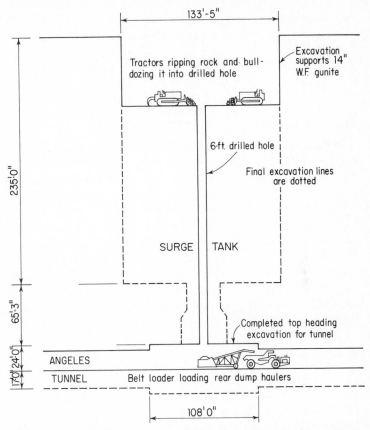

Figure 13.3 Method of excavating the surge tank for the Angeles Tunnel in southern California.

were supported with 14-in WF (wide-flange) ring beams in various spacing, with gunite applied between the rings.

In some cases the rock must be drilled and shot. In this situation, air-propelled drill units or hand-held jackhammers can be used for the drilling for the shaft enlargement.

On shafts where diameter limits the use of standard-size tractors, small crawler muckers can be used. When shafts are of such small diameter that there is insufficient area to use mucking machines, the muck from the shaft enlargement operations must be pushed down the pilot shaft by hand, or in some instances, high-pressure water can be used to wash the muck down the hole. If there is not sufficient space at the bottom of the shaft to operate a belt loader or when the rock size is not suitable for its use, other means of loading out the shot rock at the bottom must be planned. One method is to allow the muck to fall onto the tunnel floor where it can be loaded out with front-end loaders. Or if the flow of traffic through the tunnel must be maintained, the pilot shaft can be used as a muck pocket. The haul units can then be loaded from this muck pocket by means of gates installed at its intersection with the tunnel.

Some shafts must be constructed through a top layer of such incompetent material that pilot-shaft raising through this material is too hazardous an operation to undertake. In these cases the full diameter of the shaft must be sunk from the surface until material is encountered of such a nature that pilot shafts can be excavated.

Special Equipment Needed for Pilot-Shaft-Enlargement Method. When using the pilot-shaft-enlargement method of shaft construction, certain special equipment is required. To begin with, pilot-shaft-raising equipment is necessary. If required, drills for drilling out the shaft enlargement must be provided. Special mucking equipment is needed, including the crawler type of bulldozers (if the shaft is of sufficient diameter to warrant their use), or high-pressure water. Provision must also be made for muck loading and hauling equipment at the bottom of the shaft. Hoisting service must be furnished to the shaft enlargement operations by either a hoist or a crane. A man elevator is essential to that same operation. Finally, utility services such as compressed air, water, and electricity plus ventilation fans and piping must be supplied to the shaft enlargement operations.

The Long-Hole Method of Shaft Construction

In the long-hole method of shaft construction a small pilot shaft is raised the full length of the main shaft. At approximately 100-ft (30-m) intervals this pilot shaft is enlarged to the full shaft diameter. Then, from one enlarged section to the next, holes are drilled on the main shaft's perimeter and in the center so that the remaining rock sections can be shot and removed. In progressive sections from the bottom, these holes are loaded with explosives and the rock is shot down the raise. One of the most critical problems in this method of shaft construction is keeping the shaft perimeter holes on line from one enlarged section to another.

This method was used for the excavation of the spillway shafts for Hungry Horse and Monticello dams. It has not been used recently, because less expensive shaft construction methods have been developed.

SHAFT FACILITIES FOR
SERVICING TUNNEL DRIVING

Some tunnels may be excavated more economically by driving portions of the tunnel from shafts. These shafts may be included in the contract documents as pay items or may be constructed by the contractor with their cost spread to tunnel excavation. In tunnels where there is a long distance to be driven between two headings, construction time and travel time can be reduced by furnishing additional access to the tunnel by a shaft located midway between the two headings. Tunnel headings can then be driven in each direction from this shaft, which reduces individual heading lengths. When surface elevations permit, the best method of providing additional tunnel access is with an adit. When the topography is such that an adit must be driven a long distance, a shaft sunk down to the tunnel line may be preferable. These shafts must be sized and equipped to service the tunnel heading operations. Shafts are also used to service tunnel driving operations when it is impossible to use the tunnel portals for construction purposes.

When shafts are required to service tunnel driving operations and plans call for permanent shafts in suitable locations, these permanent shafts may be altered by the contractor for construction use. Upon completion of tunnel construction, the shaft must then be changed back to serve its original purpose. Sometimes the design engineers will design the permanent shafts in such a way that they may be used for both tunnel construction and permanent use. The subsequent conversion of one to the other is made with a minimum of changes.

Shafts used for servicing tunnel driving are sunk from the surface by the same methods as those described above under Methods of Shaft Sinking. When the shaft excavation exceeds 100 ft (30 m), the headframe and hoist required for tunneling service are often erected to aid in the sinking operations. If they are not available, however, a sinking headframe and hoist are used.

Shaft equipment necessary to handle servicing of the tunnel driving operations varies with the depth of the shaft and the size of the tunnel. On shallow shafts, in order to save the capital and operating expense of erecting and operating headframes, contractors are currently handling the muck removal and servicing of tunnel driving with large portable cranes. Either these cranes lift the loaded muck cars or the muck car bodies from the bottom of the shaft, or the muck cars are dumped into a muck skip located in a pocket at the bottom of the shaft and this skip is hoisted to the surface.

On deeper shafts, headframes, complete with hoist and muck bins, are used. The shaft size must be sufficient to provide space for two main compartments, an emergency access ladder, and utility services for each tunnel heading. One of the main compartments is for a counterbalanced muck skip, equipped with an automatic trip which dumps the buckets into the muck bins. The other compartment is for an elevator that lowers and raises the men, tunnel equipment, and materials required for construction of the tunnel.

The utility services needed at each tunnel heading are the ventilation pipes, compressed-air pipes, water pipes, and electric cables. The shaft is excavated deeper than the tunnel invert to provide a muck pocket and space for skip travel and loading. Muck cars from the tunnel heading are dumped into the muck pocket. By using gates, this muck is fed by gravity into the muck skip. The skip is then hoisted to the top of the shaft and engages guides in the headframe which cause the skip to dump the muck into the muck bins. From these bins the muck can be hauled away by trucks. The hoist must have drum capacity for the total length of hoisting cable required and a rope speed sufficient to dispose of tunnel muck as fast as it is excavated. Two drum hoists are used, one drum for the muck skip, and one drum for the elevator. Headframes used are of two types; one type has the hoist mounted on top of the headframe, while the other type has the hoist mounted to one side of it. In the latter case, the headframe must be braced in this direction to resist the side stresses. The other facilities on the surface surrounding the shaft are similar to those used at the tunnel portal location.

Figure 13.4 is a photograph of the headframe and muck bin that was installed over a 300-ft shaft for the Eucumbene-Tumut Tunnel for the Snowy Mountains Hydroelectric Authority in Australia. This shaft serviced two 21-ft-diameter tunnel headings. The length of tunnel driven from this shaft was 4 miles in one direction and 5 miles in the other direction. After tunnel construction was complete, the shaft was converted to its permanent use, which was a water intake for the tunnel from the stream flowing by the shaft. Because of the heavy stress, the headframe was of substantial construction. Muck was hoisted from a muck pocket in the bottom of the shaft by a 12-yd^3 counterbalanced muck skip which dumped automatically into the muck bins.

Figure 13.4   Headframe, muck bins, hoist house, shops and office, and junction shaft for the Eucumbene-Tumut Tunnel, Australia.

Muck was hauled away from the bins and disposed of by trucks. Weights of steel in the headframe and bin were:

Headframe	253,000 lb
Muck bin	132,000 lb
Rail used in the bin for reinforcing	105,000 lb
Total	490,000 lb

The hoist for this shaft was a two-drum hoist with 60-in grooved drums, air-activated automatic safety brakes, block friction clutches, and cable-payout indicator. The hoist was driven by a 300-hp 380-volt 3-phase 50-cycle motor which produced a line speed of 50 to 300 ft/min. The line pull on the 1⅜-in-diameter cable was approximately 32,000 lb. The capacity of the shaft was sufficient so its use did not restrict tunnel mucking operations, and progress of 80 ft per day was often made in each tunnel heading.

At the bottom of the shaft, a muck bin was excavated below the tunnel grade. Cars were unloaded into this bin by a rotary car dumper. The muck skip extended past the muck bin and was loaded by hydraulically operated gates which controlled the flow of the material from the bin to the skip.

SHAFT CONSTRUCTION METHODS IN SOFT GROUND

Shaft excavation in soft ground requires that the sides of the excavation be supported. The support system may be completed before excavation is started, may be built during the course of excavation, or may be partially constructed before excavation is started and completed during the course of excavation. Most specifications for public works require that the support system and the sequence of installation be approved by a registered structural engineer.

Support Systems

Support systems which can be completed before excavation is started must rely upon their own geometry instead of internal bracing or external tieback for stability, or must be so shallow that the support system can be safely cantilevered out of undisturbed ground below the planned excavation. An example of the former case is the tremie concrete wall, with or without soldier piles. If the design requires no additional support during the course of excavation, the hole is limited in area and is probably circular in horizontal section.

The support system can be partially constructed before excavation is started. One such system uses sheet piling. Internal bracing, consisting of ring beams for a circular shaft, or walers and struts for a rectangular shaft, is installed level by level as the excavation proceeds. It may be feasible to avoid interference from struts by securing the walers through the use of tiebacks. *Tiebacks* consist of rods inserted in holes which are drilled into the surrounding soil and grouted. After the grout has attained sufficient strength, jacks are used to stress the rods against the walers, which in turn support the sheet piling and the adjacent ground. If the shaft is to be constructed where artesian water is a problem, it may be excavated in the wet. After the bottom is sealed with a tremie concrete slab heavy enough to resist the artesian head, the shaft is dewatered, with the support system being installed as the dewatering progresses.

Another system in which the support is partially complete before excavation is started consists of soldier piles and lagging. Any geometry is feasible with this system. The soldier piles are driven or inserted in prebored holes before excavation is started. Timber or precast-concrete lagging is installed as the excavation progresses, with levels of walers with struts or tiebacks, or with ring beams installed level by level per design. The lagging may be inserted between the flanges of the soldier piles and wedged back against the earth or may be attached to the exposed faces of the soldier piles by welded studs, by bolting through the flanges, or by the use of hardware which grips the flanges. If the ground tends to run, it will be necessary to caulk the interstices of the lagging with rags, excelsior, or grouting. The soldier piles should be set back a few inches to allow for probable deviations in pile installation. If the lagging is designed to be attached to the face of soldier piles, it can be moved back of the flanges if piles are found to be too close to the neat line of the structure. The converse is true if lagging is intended to be inserted behind the flanges, but piles are found to be back too far from neat line. Either case will reduce the quantity of nonpay concrete required if concrete is to be poured directly against the support structure.

If ring beams and vertical lagging are used, the support system is entirely constructed during the course of excavation. Steel lagging or timber lagging may be used, or liner plates may be fitted between the ring beams. If there is sufficient standup time, the ground may be supported during the course of excavation by erecting rings of steel liner plate, precast-concrete segments, or cast-iron segments.

If the shaft in overburden is to be continued into rock, it is prudent to set the support system back from the perimeter of the proposed rock excavation. This will ensure that overbreak in the rock excavation will not undermine the support system for the overburden. In such cases, the setback is usually sufficient to permit forming the exterior of the shaft in overburden, in order to conserve concrete.

Excavation

The clamshell bucket is the most common tool for shaft excavation in overburden. If the shaft is large the use of the clamshell bucket will probably be supplemented by the use of a front-end loader or a small bulldozer to move material to the clamshell operation. If the spoils are to be removed by truck, the efficiency of the operation will be improved by mucking into an elevated hopper from which trucks are loaded as they arrive.

Large-Hole Drilling

Large-diameter augers are frequently used for the construction of shafts in overburden. Many missile silos have been constructed by this method. If the ground has excellent standup time and if it is not necessary for men to enter the hole, the drilling may be accomplished without ground support until a liner is inserted and the walls are poured. More often it is necessary to provide a steel casing to follow the drill. The hole is overexcavated by retractable reamers to permit auger withdrawal and to facilitate the downward movement of the casing.

Placement of Concrete

In placing concrete for lining shafts in overburden which continue into rock, the same inside forms and methods are used in the overburden as in the rock section of the shaft, but the addition of outside forms will ordinarily be required. Concrete for shafts wholly in rock is most often placed against the support system, using inside forms only. The specifications may require that waterproofing, often bentonite panels, be applied to the support system, and that a membrane, usually plastic sheeting, be placed under the floor slab. If the bottom is soft, it may be desirable to put down a "mud mat," a thin slab, generally 4 to 6 in (10 to 15 cm) thick, of lean concrete. This provides a support for the waterproofing, which is then less liable to be punctured, and provides an unyielding support for the chairs which will support the reinforcing steel at the proper elevation.

Caissons

It is sometimes necessary to construct shafts in completely saturated clays, silts, sands, or gravels. The locations may be on land or in water. The shafts may be permanent structures or may be temporary structures, serving, for instance, as cofferdams within which bridge piers will be constructed. These conditions may dictate that shafts be constructed by sinking caissons.

Caissons consist of prefabricated shaft walls which are sunk more or less continuously as excavation proceeds. The sinking force is gravity, aided by water jets, by lubrication, by vibration, or by impact. The caisson walls are most frequently concrete, and the lower edges are fitted with continuous steel cutting edges. Caissons may be open or closed, and are most often circular or rectangular, but may be any required shape.

Open caissons consist only of walls and cutting edges. Open caissons may be completely constructed before the sinking operation is started, but are commonly constructed in sections. Sections are added from time to time as excavation proceeds. Open caissons may be sunk by excavation in the dry or in the wet. The dry method is not feasible if the nature of the soils and the hydrostatic head indicate any possibility of a "boil." In this case, not only would the caisson be filled with soil, but the loss of ground might cause the caisson to tip. Excavation in the wet avoids most of this danger, since the water within the caisson tends to counterbalance the forces which cause a "boil." Open caissons sunk by excavation in the wet must be bottomed-out in an impervious stratum or must be closed with a tremie concrete seal before they can be safely dewatered. The thickness of the tremie seal needs to be approximately 40 to 50 percent of the hydrostatic head to be safe against uplift.

Closed caissons are sometimes referred to as *pneumatic caissons,* since compressed air is used in them to counterbalance hydrostatic pressure. A closed caisson is like an open caisson except that the top is closed and is fitted with man locks and muck locks. It is not ordinarily feasible to stack up sections as the work progresses, but this is not impossible. The required air pressure increases as the caisson goes deeper. The safety requirements are the same as for compressed-air tunneling and for shaft work.

The most common tool for caisson excavation is the clamshell bucket, but other tools, including dredge pumps, are occasionally selected, depending upon the caisson size and type and upon the nature of the ground. Excavation must be performed with great care to ensure that the caisson remains plumb at all times. Steering a caisson is even more of an art than is steering a tunnel shield.

ADDITIONAL INFORMATION

Other publications may be consulted for additional information.[13.4,13.5]

REFERENCES

13.1 "South Africans Shatter Soviet Shaft Mark," *Engineering and Mining Journal,* June 1960, p. 128.

13.2 D. M. McGillivray, "High Speed Shaft Sinking Techniques in South Africa," *Proceedings of the 1979 Rapid Excavation and Tunneling Conference,* American Institute of Mining, Metallurgical, and Petroleum Engineers, New York, 1979, p. 1197.

13.3 George E. Kemnitz, "Big-hole Drilling Methods are Used for Mine Shafts," *Oil and Gas Journal,* January 9, 1967.

13.4 Scott F. Andersen, "The Helms Underground Pumped Storage Project Shaft Development," *Proceedings of the 1981 Rapid Excavation and Tunneling Conference,* American Institute of Mining, Metallurgical, and Petroleum Engineers, New York, 1981, p. 887.

13.5 James E. Friant, "Blind Shaft Construction New Equipment Update," *Proceedings of the 1979 Rapid Excavation and Tunneling Conference,* American Institute of Mining, Metallurgical, and Petroleum Engineers, New York, 1979, p. 1247.

Other Heavy Construction Applications

Chapter 14 Piling, Caissons, Dredging, and Marine Construction

Piling, caissons, dredging, and marine construction are treated as a group because they are so often interrelated that all or several of these construction activities may be required in one contract. A number of contractors specialize in this group of activities.

PILINGS AND CAISSONS

Piles and caissons may be required to support new structures, or they may be required as underpinning to provide additional support for existing structures. Sheet piling, soldier piles and lagging, or tremie concrete walls may be constructed in order to support deep excavations. Sheet piling may be used to enclose areas such as those to be filled and used for marine terminals or for cofferdams.

Pilings

The materials commonly used for piling include steel, precast concrete, and timber. Steel piling may be H section or may be interlocking sheet piling. Precast-concrete piling is usually prestressed and is usually square in cross section. It is comparatively easy to make fully effective splices by welding steel pilings, but it is difficult to do so with timber pilings. Precast-concrete pilings are sometimes spliced using dowels and epoxy cements. If it is at all possible, pilings should be purchased in full lengths or spliced before driving, because making splices in partially driven piles is time-consuming and expensive. Pilings are usually purchased in lengths greater than the anticipated final lengths in order to assure that no splicing will be required, even when an occasional pile must be driven deeper than was anticipated. The excess lengths are later cut off.

Piles may be driven or may be placed in predrilled holes. Setting piles in the predrilled holes may be necessary where absolute accuracy in alignment must be assured or where the noise and vibration caused by pile drivers is objectionable. Piles are sometimes set in predrilled holes and then driven an additional few feet. If piles are set in predrilled holes without additional driving, it is often specified that the holes be drilled deep enough to permit placing a concrete pad to support the pile and that the drilled hole be filled with lean concrete after the pile is in place.

Pile drivers may be *reciprocating* or may be *vibratory*. The oldest form of pile driver (except for the simple drop hammer) is the reciprocating machine driven by steam or compressed air. In the single-acting type the hammer is raised by power and falls under the force of gravity. In the double-acting type power is applied on both the upstroke and on the downstroke. Diesel pile drivers use the force of gravity for the downstroke, and at the same time compress a volume of air between the piston and the cylinder. A charge of diesel fuel is then injected. This detonates to provide power for the upstroke. Reciprocating hammers are available in a wide range of sizes and are satisfactory for all types of ground. Vibrating pile drivers grasp the top of the piling while an eccentric mechanism induces strong vibrations. The vibration plus the weight of the machine causes the pile to sink. Vibratory hammers are not effective in all ground conditions.

All pile drivers operate in leads which guide the hammer. If only plumb piles are to be driven, simple hanging leads will suffice. If batter piles are required, there must be provision for aligning the leads to the correct batter. The leads also guide an anvil which transmits the hammer blow to the piling. Anvils are fabricated to fit the top of the pile being driven in order to avoid battering the pile out of shape. Most pile drivers are mounted on crawler cranes, but other mountings are occasionally used, particularly in marine work.

Sheet pilings are usually driven with templates to ensure that they are started in good alignment. Circular templates are used for driving sheet-pile cells, which are often used for cofferdams and are sometimes used for permanent structures, usually in connection with marine works.

Test piles are usually required if the piles to be driven are to support structures. Most specifications provide for extra payment for load tests on piles and permit the engineer to require such tests whenever he deems it necessary. Measurements of settlement under load are taken over a period of several days.

If the piles are to be set in predrilled holes, the drill is most often a continuous auger mounted on a crane. The auger is often rigged so that lean concrete may be pumped into the hole as the auger is withdrawn. The auger may be mounted on the pile driver leads if the piles are to be driven after they are set in the predrilled holes.

Steel H piles, referred to as *soldier piles,* are often used in conjunction with timber, precast-concrete, or steel lagging to support excavations. The soldier piles are set from 3 to 8 ft (1 to 2.5 m) apart, and the spaces between the piles are closed by the lagging, which is placed as the excavation progresses in a tight or loose pattern, depending on the nature of the soil. Lagging may be wedged between the flanges of the piles or may be attached to the exposed faces of the flanges by the use of welded studs or of clamps which grip the pile flanges. Soldier pile and lagging walls permit water to seep into the excavation; since this prevents a hydrostatic head from developing behind the wall, it is frequently a very desirable attribute of this system.

All except the most shallow systems require bracing to resist earth pressure. The bracing system will be installed at one or several elevations as the excavation progresses. Each level of bracing should be designed to clear a stage of the permanent structure and to be removed when some element of the permanent structure can resist the earth loads. Most specifications require that the design be prepared by a registered structural engineer. The bracing may be within the excavation, in which case it will probably consist of walers bearing against the piles, with the loads carried from the walers to struts or to rakers which bear on earth. If external bracing is used it will usually consist of tiebacks, which are tension members drilled and grouted into the adjacent soil. The topmost level of bracing may consist of wire-rope ties to deadman anchors set back from the edge of the excavation.

Tremie concrete walls are often used for support of excavation if the support system is to be watertight. These walls are constructed full depth before the excavation, usually by excavating a trench kept full of bentonite slurry to prevent the earth walls from collapsing. The completed trench is filled with tremie concrete, which displaces the bentonite slurry as the concrete is placed. Variations of this system are common; usually excavating and concreting the wall take place panel by panel between previously placed steel or drilled-in-place concrete soldier piles. Steel H piles are readily connected to a bracing system. It is common practice to utilize the soldier pile and tremie concrete (SPTC) walls as the permanent structural wall of the excavation. In these cases the SPTC wall design is furnished as a part of the bid documents.

Hollow steel shells are often used for pilings. These may be cylindrical or tapered. The cylindrical shells may be of a constant diameter or may consist of a series of diameters increasing by steps from bottom to top. The shells are driven with steel mandrels. After the mandrel is withdrawn the shell is filled with concrete. Figure 14.1 shows Step-Taper* piles being driven in Singapore.

* Registered trademark, Raymond International, Inc.

Figure 14.1   Driving Step-Taper piles in Singapore. (Raymond International, Inc.)

Caissons

Caissons are sometimes referred to as *drilled-in-place concrete piles.* A caisson consists of a hole drilled in the earth and filled with concrete, usually unreinforced except for a group of dowels at the top to tie into other parts of the structure foundation. The bottom of the hole is sometimes belled to increase the bearing area. If the ground tends to cave, it may be necessary to use a steel casing or to fill the hole with bentonite slurry to provide support. The steel casing is usually withdrawn as the hole is filled with concrete, but it may occasionally be left in place. If bentonite slurry is used, it is displaced as concrete is introduced through a tremie pipe. Drills may be continuous flight augers or bucket augers, or may be reverse circulation drills similar to those used for well drilling.

UNDERPINNING

Underpinning may be required to support bridges or buildings which are settling or which may be threatened by settlement from planned tunneling or excavation operations. The simplest method of underpinning is to construct concrete piers under existing footings. This work is performed in stages, each stage consisting of excavating under a part of the footing and constructing a new pier under that part until the whole of the footing is supported by a new and deeper foundation.

In underpinning operations, it is very often found that conditions as encountered are substantially different from those anticipated. Because it is often impossible to locate foundation plans for old buildings, it is usually prudent, and sometimes essential, that some exploratory work be performed in advance of actual underpinning operations.

If the underpinning must be carried to depths greater than would be feasible by extending the piers, the underpinning may consist of steel shells jacked down from access pits under the existing footings. The shells are purchased in short lengths and coupled by welding as the work progresses. The soil is removed from the shells during the jacking process and the shells are later filled with concrete. The final step is to wedge the new work tightly against the existing footing and to fill the access pit with concrete.

It is occasionally necessary to transfer the load on a footing to new foundations in different locations. This is usually accomplished by constructing caissons in the new locations and bridging between them with needle beams, which in turn carry the load which is to be transferred. Needle beams may be steel or concrete. Steel needle beams are usually encased in concrete.

There are other underpinning methods, such as grout injection and patented slender piles installed through the existing foundation, which are sometimes more useful or less costly than the conventional methods.

Because many underpinning jobs require a long lead time before the other work can begin, the underpinning may be performed under separate contract in advance of the contract for the main work. Even where separate contracts are not the case, most underpinning is performed by companies which specialize in this work and which have developed many ingenious tools and methods to reduce the cost of performance.

DREDGING

Dredging is underwater excavation. Most dredging involves removal of common materials, but removal of rock is often required. The dredging plant and equipment may consist of standard excavating equipment, or it may consist of equipment specially designed and built for this purpose. Floating plant must be serviced by tugs, tenders, and crew boats.

Even moderately heavy seas will seriously interfere with the operation of floating plant. For this reason, dredging in the surf zone, or in other areas where heavy seas may be expected and where water depth will permit, is often performed from platforms elevated above the wave crests. The platforms may be trestles or may be movable towers which rest upon the bottom. Dredging in the surf zone, and perhaps elsewhere, may be frustrated by continuous refilling of the excavation by materials transported by the turbulent water. If the dredging is for the purpose of laying an underwater pipeline, it is often necessary to build a cofferdam to keep the trench open until the pipe is laid and backfilled. Backfill in such areas is usually protected by an armor of heavy rock.

Standard Excavation Equipment

Clamshell and orange-peel excavators, draglines, and backhoes are often employed for dredging. They may work off the ground or from trestles, or they may be chained down on barges to serve as floating plant. Wheels or crawlers may or may not be removed. If employed as floating plant, barges are provided to receive the spoils. Most such equipment is diesel-powered. This type of equipment is particularly useful for rock excavation. Production rates will be lower than for excavation in the dry.

Underwater drilling and shooting is usually performed using standard equipment modified as necessary for this service. Drills are converted to use water instead of air to clean cuttings from the holes. Drilling and loading are usually performed through a lightweight casing to facilitate locating the holes. The drill barge is usually fitted with spuds to enable it to be held rigidly in position. Powder factors for underwater blasting are from two to four times as high as for work in the dry. Explosives used must be those that are effective when wet.

Bucket-Line and Bucket-Wheel Dredges

Bucket-line dredges consist of a continuous chain of buckets supported on a heavy steel boom or "ladder," together with a power source to pull the buckets and thus perform the excavation. The spoils may be retained in hoppers on the dredge or conveyed to barges. The ladder can be raised or lowered, but cannot be swung except by swinging the dredge. This is accomplished by pivoting around one spud, using the swing winches to take in the line to an anchor set out by a work boat for this purpose. A bucket-wheel excavator is similar, except that the buckets are mounted on the periphery of a wheel which can be raised or lowered. The lips of the buckets of either type of dredge are usually fitted with teeth to break up the formation being dredged.

Suction Dredges

Suction dredges are floating plant. Excavation is accomplished by a pump which draws a stream of water through a pipe which reaches to the bottom. The stream of water picks up the material to be excavated and carries it in suspension to the point of disposal. Unless the materials to be dredged are loose, it is necessary to mount a cutter head and a power source to rotate the cutter head on the ladder which supports the suction pipe. This cutter head is used to break up the formation to be dredged into fragments which can be pumped. The suspended materials will vary from 15 to 20 percent by volume, depending on density, of the total flow through the pump. Suction dredges are generally designated by the discharge pipe diameters, usually ranging from 8 to 36 in.

The suction dredge may be of the hopper type, where spoils are retained on board until the hoppers are filled. In this case the water containing the spoils is pumped to the hoppers. The spoils settle, and the water flows over a weir and over the side. When the hoppers are full, dredging stops, the ladders are raised, and the dredge moves to the disposal area, where the hopper gates are opened and the spoils fall to the bottom. The big advantages offered by hopper dredges are the fact that there is minimum interference with other water traffic, the spoils can be readily transported many miles for disposal, and the unit is entirely self-contained. The big disadvantage is the fact that a large part of the time is devoted to transportation instead of to dredging. This can be obviated by pumping into bottom-dump barges tied alongside, but this requires towing service.

Suction dredges may be of the pipeline type, pumping through floating pipelines from the dredge to the disposal area. Interference with other water traffic may be minimized by submerging the pipeline in areas where it would otherwise obstruct shipping channels. The dredge is held in place with anchors set by its tenders, and the area to be dredged is covered by swinging the dredge about one of the spuds set at the stern. When it reaches the limit of swing in one direction, the spud about which it has been swinging is raised, the spud on the other side is lowered, and the swing is started in the opposite direction. Thus the dredge walks one swing cycle after another in the direction of advance.

If the spoils are to be pumped great distances it will be necessary to employ booster pumps at intervals in the discharge line. The discharge line is supported on floats, a few of which are anchored. The discharge pipe sections are rotated from time to time to distribute abrasion around the entire inner surfaces. Rubber linings will prolong the life of dredge-pipe sections. Dredge-pipe sections are fitted with simple couplings. Special fittings, such as swivels, will occasionally be needed.

Dredging quantities are measured by "before-and-after" surveys. These are usually performed by echo sounding, corrected for water-level fluctuations caused by tides or variations in riverflows. The corrections are derived from the tape of a continuously recording water-stage indicator installed in the vicinity of the work.

The larger suction dredges are extremely expensive and require a substantial number of auxiliaries. They are usually electric and may require a power supply of 25,000 kW or more. Figure 14.2 shows a modern suction dredge.

Only experienced personnel can be entrusted with planning and managing major dredging operations. The dredging plant must be continuously employed if possible; this means that operations must be scheduled for months or years in advance. Unless a company is regularly engaged in major dredging projects, dredging work is usually not sought; if it is incidental to other work, it is usually subcontracted to companies which specialize in this field.

Figure 14.3 shows a pipeline dredge at work. It has just completed a swing to the right, pivoting around the starboard spud. The port spud will now be lowered, and the starboard spud will be raised to start a swing to the left. Solid lines show completed dredging; dashed lines show the dredging which will be

Figure 14.2 A modern oceangoing dredge. (Richard B. Costain, Ltd.)

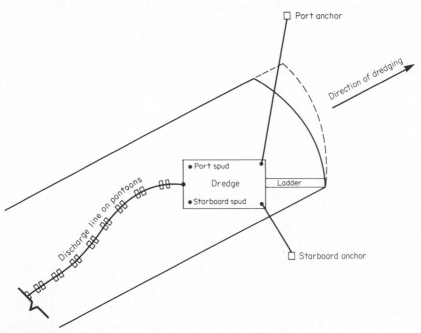

Figure 14.3 Pipeline dredge at work.

accomplished on the swing to the left. The swing action is the result of taking in one anchor line while paying out on the other. The anchors must be reset periodically by an anchor barge. Note that the discharge line has slack in it to permit dredge movement. It has a swivel connection where it leaves the hull of the dredge. Additional pipe sections, mounted on pontoons, are towed behind or alongside the dredge to be added as needed.

MARINE CONSTRUCTION

Marine construction is generally understood to refer to building major structures in situations where work must be performed from floating plant, or by progressively working outward from the shore. It is construction which is performed in water. Typical marine construction projects involve wharves, breakwaters, drydocks, bridge foundations in deep water, offshore drilling platforms, production platforms, submarine pipelines, sunken tubes for transportation, and dredging, which is treated elsewhere in this chapter.

Most of the activities for marine work are the same as for land work, but must be performed where water levels may fluctuate daily with the tides, or seasonally, and where tidal or river currents complicate the work. Marine work hours may vary from day to day, as dictated by the progression of the tides. Special quick-setting concrete may be required for construction in the intertidal range.

Major pieces of equipment for a full-service marine construction firm include derrick barges and heavy-lift crane ships. Figure 14.4 shows the 600 ft-

Figure 14.4   2,000-ton crane ship. (Raymond International, Inc.)

long, 116-ft-wide crane ship *Sirius* at work installing components of a platform. The *Sirius* has a 2,000-ton lifting capacity over the stern and a 1,600-ton
revolving lift capacity.

Marine construction may be facilitated by, or may require, off-site prefabri-
cation of elements to be incorporated in the final structure. It is often feasible,
for instance, to precast concrete deck sections or to fabricate steel bridge
spans on shore or in drydock, and then bring these elements by barge to the
site for erection. On relatively low structures it may be possible to transport
such sections on elevated pedestals on the barges, positioning the barges at
high tide, and using the falling tide to lower them into position. Where this is
feasible, it may be possible to position extemely heavy elements without the
need for on-site lifting equipment.

For extremely heavy lifts, such as long bridge spans, transportation by
barge and lifting by jacks is often feasible. A number of companies specialize
in designing, furnishing, and operating this type of lifting gear. It is prudent
to employ specialists where this service is required.

An increasing number of bridges are being constructed from precast deck

Figure 14.5 Schematic installation of sunken tubes.
(Bay Area Rapid Transit District.)

Figure 14.6 Sunken tube positioning, Fort McHenry Tunnel, Baltimore, Maryland. (Raymond International, Inc.)

elements cantilevered out from the piers, and joined by post-tensioning rods or strands and epoxy cements.

Sunken tubes for highway and other tunnels are constructed in segments, often in the range of 325 ft (100 m) long, prefabricated in drydocks.* The tubes may be concrete or may be steel shells lined and ballasted with concrete. The segments are completed or partially completed to a stage where they have only a small positive buoyancy when the ends are bulkheaded, and are then towed into position over a previously dredged trench. Each segment is in turn sunk by filling external ballast pockets and joined to the previously placed segment. The final joining usually requires welding and always requires additional concrete placement. After the segments are in position and are watertight, the tube is backfilled. The backfill will be protected with stone armor in areas where currents might move the backfill or in areas where a ship might drop anchor in an emergency.

Figure 14.5 shows schematically the methods utilized for installing sunken tubes. Figure 14.6 shows an actual installation in process on the Fort McHenry Tunnel project in Baltimore, Maryland.

* A list of 60 sunken-tube projects in all parts of the world, together with their respective dimensions, materials, and cross sections, is found in an article by D. R. Culverwell, "World List of Immersed Tubes," in *Tunnels and Tunnelling*, September 1981.

ESTIMATING

Only a contractor regularly engaged in the specialized work discussed in this chapter is likely to have employees capable of making proper estimates of the cost of such work. Therefore the contractor not regularly engaged in this type of work should employ experienced estimators, engage consultants, plan to subcontract such work, or seek an experienced contractor with whom to joint venture such work.

Chapter 15 Concrete and Steel Structures, Nuclear Plants, and Powerhouses

This chapter discusses the civil engineering aspects of the construction of such structures as bridges, locks, sewerage plants, water-treatment plants, and other buildings incidental to heavy construction. Also included here is a section on the special requirements for construction of nuclear power plants. The special requirements for powerhouse construction and for equipment installation for hydroelectric developments are also considered.

GENERAL BUILDING CONSTRUCTION METHODS

Construction methods used for buildings associated with heavy construction projects are quite similar in foundation, concrete, and structural steel applications. These aspects of general building construction are treated here through a review of functional requirements, including those for foundation preparation; formwork; shoring; concrete supply; concrete placement; slab finishing; construction joint preparation; formed surface finishing; curing; reinforcing steel; precast, prestressed and post-tensioned concrete; miscellaneous embedments; and structural steel erection.

Foundation Preparation

Some bid schedules provide a payment item for this work, but there will still be a need to clean up loose bits of wood, paper, or other debris and to mop up or control water. This will probably be performed by the pour crew, and provision should be made in the estimate for the costs which will be incurred.

If there is no separate payment item for foundation preparation, it will involve more than merely picking up loose trash, and a special crew may be required to perform the work. On soil foundations, preparation will probably involve fine-grading, recompaction of disturbed surfaces, control of water, and dampening of soil if it is so dry that it will absorb too much moisture from the concrete. In the case of rock foundations most specifications require the removal of all loose or drummy rock. Cleanup of rock foundations requires the use of such hand tools as picks, shovels, bars, and paving breakers. Small backhoes are often useful for cleanup and for loading out the spoils.

Foundation cleanup is measured in square feet, square yards, or square meters. The cost of cleanup of rock surfaces should be estimated separately from the cost of cleanup of earth surfaces.

Formwork

Concrete structures require forms to define the shape of the structure and to retain the concrete in that shape until the concrete has hardened. The faces of forms may be of lumber, plywood, plastic and fiberglass, hardboard, cardboard impregnated with wax or plastic, or steel. Plywood forms are often faced with plastic. Aluminum faces are usually forbidden by specifications. The supporting framework may be made of lumber or steel or occasionally aluminum. Special form linings may be specified for architectural effects. Cardboard forms are used almost entirely to form cylindrical columns or cylindrical voids. Plastic and fiberglass forms are used most frequently for forming cylindrical columns, coffered soffits, or special shapes. A common rule of thumb is that lumber or plywood forms fabricated on site will be less expensive than steel or fiberglass forms, unless at least 15 reuses can be foreseen for the latter types. Cardboard forms are strictly single-use, but are very often the least expensive forms on the job.

Job-built forms may be built in place or may be fabricated in a shop. Built-in-place forms are most often used where they must be fitted to very irregular surfaces, as starting off rock. Other forms can be more economically produced in the carpenter shop, using such power equipment as radial saws, band saws, table saws, planers, and drill presses.

Except in the case of built-in-place forms, the crew used for erecting and stripping formwork will be different from the crew employed in form fabrication. The erect-and-strip crew will need the part-time services of a truck and a crane.

The direct cost of built-in-place forms will include labor, lumber and plywood, form oil, form hardware, and the costs of small tools such as power saws and drills.

The direct cost of form fabrication will include labor, maintenance of carpenter shop equipment, and form materials, including lumber, plywood, and such hardware as nails, screws, and bolts. Labor costs and labor-related costs will vary in proportion to the complexity of the formwork.

The direct cost of erecting and stripping formwork will include labor, equipment operating invoice, and form oil and form hardware. It will include cleaning and repairs for reusable forms. Form hardware consists of the metalwork necessary to align the forms and to resist the pressure of the plastic concrete. Form hardware may consist of one-use ties with reusable fittings, one-use rods with reusable she-bolts and fittings, or reusable taper ties. The taper ties are not suitable for concrete which must be watertight. Most specifications require that no hardware be left closer than 1½ in (4 cm) from the finished surfaces. Therefore, the breaking plane for snap ties must be held back, and a removable conical insert used to provide space to bend the tie until it snaps. The cone also serves to space the forms. Forms should be designed with cleanout panels at the bottom to facilitate last-minute cleanup.

Forms are designed on the basis of the maximum pressure that will be exerted by the plastic concrete, with due allowance for shock loads and for the effects of vibrating the concrete. The pressure will vary with lift height, placing rate, and placing temperature. Several handbooks are available which are invaluable for form design.

Forms should be designed to assure safety to the workmen who erect and strip them and the workmen who will be engaged in pouring concrete. Scaffolds supported by the forms should conform in all respects to the standards promulgated by safety authorities.

Formwork is measured by the square foot or by the square meter. Form fabrication or form purchase or rental costs should be estimated separately from erect-and-strip costs. Both categories should be subdivided to reflect requirements for off-rock, build-in-place, flat surface, cylindrical surface, conical and bellmouth surface, and warped surface conditions. Some contractors further subdivide these costs for estimating and cost-accounting purposes.

The work of estimating formwork costs can be expedited by using a well-arranged tabulation of the different classes of formwork for each bid item. Plywood and lumber costs are functions of the areas of forms to be fabricated. Hardware costs are functions of the total set-and-strip areas. Form fabrication costs and form erect-and-strip costs are estimated separately, using the total man-hours and supply costs for each operation. Setting screeds, installing chamfer strips or keys, and similar operations which do not increase the formed area are not included in the total areas because they are considered to be incidentals, but man-hours and supply costs are included.

The estimator should consult the specifications about the time at which forms can be stripped, since this will affect the cycle time for reuse of forms. A separate formwork tabulation is prepared for each bid item which involves formwork, as shown in Table 15.1. Figure 15.1 shows formwork used for a pier cap.

TABLE 15.1   Estimate of Form Costs for Mass-Concrete Rio Bonito Spillway

Classification	Setting and Stripping				Fabrication			Plywood		Lumber		Hardware	
	ft²	ft²/MH*	MH*	Uses	ft²	ft²/MH*	MH*	ft²/ft²	Cost at $1.00/ft²	FBM†/ft²	Cost at $0.50/FBM†	Cost/ft²	Cost at $1.00/ft²
Contraction joints													
Off-rock	2,520	2	1,260	1	Build-in-place			1.2	$ 3,024	2	$ 2,520	$ 0.60	$ 1,512
Standard panels	20,085	7	2,869		Steel							0.60	12,051
Fillers	6,695	4	1,674	3	2,232	6	372	1.1	2,455	3	3,348	0.60	4,017
Downstream face													
Off-rock	1,320	2	660		Build-in-place			1.2	1,584	2	1,320	0.60	792
Standard panels to el. 1,100 ft	7,464	7	1,066		Steel							0.60	4,478
Lift at 1,100 ft	1,860	4	465		Steel							0.60	1,116
Above el. 1,100 ft	3,472	7	496		Steel							0.60	2,083
Upstream face													
Off-rock	1,240	2	620		Build-in-place			1.2	1,488	2	1,240	0.60	744
Standard panels	13,964	7	1,995		Steel							0.60	8,378
Gallery													
Drainage trench	496	2.5	198	1	496	4	124	1.2	595	1.5	372	0.20	99
Walls & roof	4,464	3	1,488	2	2,232	5	446	1.1	2,455	4	4,464	0.20	893
Total form area	63,580	4.97	12,791		4,960	5.27	942		$11,601		$13,264		$36,163
incidentals													
Filler at lip	1980 lin ft	10	198		Lump Sum					1.5	496	0.10	99
Screeds	L.S.		124				28		182		248	0.10	198
Install water stop	430 lin ft	12	36									0.10	43
Keys 2" × 6"	1224 lin ft	12	102		1300 lin ft	24	54			1	650	0.10	122
Chamfer strips	6840 lin ft	40	171		7000 lin ft	200	35			0.1	350	0.02	137
TOTAL	63,580	4.74	13,422		4,960	4.68	1,059		$11,783		$15,008		$36,762

* Man-hours.

† Feet board measure.

NOTE: The purchase of steel forms is included in plant and equipment cost.

Figure 15.1 Formwork for a pier cap. (Economy Forms Corp.)

Shoring

Concrete decks and beams must be supported until the concrete has attained sufficient strength to be safely self-supporting. The most common types of shoring are tubular metal shoring and timber shoring. These are most valuable for large areas, such as floor or roof slabs. Shoring with steel beams supported on piers or walls or with separate columns or bents below the pouring level is a technique used where conventional shoring is uneconomical or where it might obstruct the area under the work, such as when building an overpass over an existing highway or railroad.

Many states require that the shoring installation be designed by a registered structural engineer. The plan should include the sequence and timing of shoring removal, the reshoring which may be necessary if rapid construction is required on a multilevel structure, and all other requirements necessary to assure the safety of the workmen and of the general public.

There is no generally satisfactory means of estimating shoring costs. Some contractors measure the volume between the soffit and the supporting surface below. Some contractors combine the estimated cost of shoring with the estimated cost of erecting and stripping soffit forms.

Concrete Supply

The cost of concrete will appear in the estimate as permanent materials if concrete is purchased from a supplier. It will appear under the estimated costs of labor, repair labor, equipment-operating invoice, small tools and supplies, and permanent materials if the concrete is produced on site. A separate and subsidiary estimate will be required if concrete aggregates are produced on site.

The costs of operating concrete-production facilities are in part a function of the volume of concrete to be produced and are in part a function of time, since during periods of production the batch plant crew which will be required to produce 2,000 yd³ (1,500 m³) in a month will be the same crew which will be required when production is 10,000 yd³ (7,500 m³) per month.

There are four measures of the volume of concrete for a given payment item. One is the *bid schedule quantity*, which may be a lump sum. The next is the *neat-line quantity* per the bidder's takeoff. The next is *the volume to be placed*, which is equal to the neat-line quantity plus overbreak quantity. The last is the *batched quantity*, which is equal to the placed quantity plus waste. Waste results from spillage, from faulty batching, from delays which cause rejection of the concrete due to partial setting, and to overorder on the last order for a particular pour. The contractor must bear all the costs of concrete used for filling avoidable overbreak and for wasted concrete, but revenue will be limited to the neat-line quantity. It is therefore necessary to distribute the cost of the total volume expected to be batched over the neat-line volume. If cement is to be paid for as a separate payment item, the cost of cement for nonpay concrete and the cost of cement handling losses must be distributed in the same way.

Concrete Placement

No concrete placement should be started before a pour checkout sheet has been approved by the inspector and the foremen of the carpenters, electricians, pipe fitters, iron workers, and other craftsmen and by the party chief for the survey crew.

The usual placing methods for concrete include placing from buckets, using cranes or cableways, pumping, conveying, or directly discharging from the hauling vehicle, commonly called *tailgating*. If the concrete is to be discharged from an elevation substantially above the point of deposition, it should be placed through an "elephant trunk" to avoid segregation. If concrete is to be deposited underwater, it must be placed through a tremie to avoid segregation and to avoid washing the cement out of the mix. Elephant trunks and tremies are usually steel tubes fitted with a hopper at the top. Rubber tubes are often used for elephant trunks or are attached to the lower end so that they can be flexed to assist in concrete distribution. Concrete should never be moved by the use of vibrators, as this will cause segregation. Tremies should always be immersed in the pour and should be kept full; a sufficient number of tremies should be provided to ensure that the concrete need not flow very far to cover the pour area. Specifications for concrete to be tremie-placed require that the mixes be extra rich in cement to compensate for losses to the water. All concrete except tremie concrete should be vibrated to ensure that the forms are filled and that no rock pockets or voids develop.

Concrete should be deposited in layers from 18 to 24 in (45 to 60 cm) deep to avoid concrete flow and resulting segregation. If the area is large, a pouring pattern should be developed to ensure that layers above or adjacent to previously placed concrete are placed before the earlier placements have had time to set. The vibrators should pass through the interface between the underlying concrete and the concrete being placed to ensure that the two layers are properly consolidated. Placing the vibrator against the reinforcing steel will minimize air pockets which may otherwise be trapped under the horizontal bars. Extra care should be exercised adjacent to anchor bolts or other embedded metals.

Placing concrete from a bucket is a simple operation, but it does require crane or cableway time in competition with other needs for hook service. In order to reduce the demand for hook service, it may be prudent to schedule concrete placement with buckets to the night shift or shifts, thus freeing the cranes and cableways for other duties during the day shift.

Concrete buckets are available in sizes ranging from ½ to 8 yd³. Sixteen-cubic-yard, two-compartment buckets have occasionally been used. Most buckets are fitted with bails, but buckets intended to be refilled without being detached from the hook may omit the bail and be supported by wire-rope pendants from a spreader bar. Most buckets are circular, but some are square. Laydown buckets are available. These lie flat for filling and hang vertically after being picked up. The discharge gate is located at the side. The laydown bucket is useful where concrete must be discharged into an opening in a vertical form.

Buckets may remain hooked to the crane or cableway and be refilled from the vehicle hauling the concrete, or the crane or cableway may land an empty bucket and pick up a full one. In the latter case an automatic hook will reduce the hook time. An automatic hook consists of sister-hooks powered by a compressed gas, usually nitrogen. A trigger opens or closes the sister-hooks whenever the trigger is lowered to touch the bail of the bucket. Operation is safe because the sister-hooks cannot open when the bail of the bucket is engaged. If buckets of concrete are hauled, the hauling vehicle must have one empty slot to receive the just-emptied bucket. Another slot will be left empty when the last full bucket is removed and the hauling vehicle departs.

Concrete pumps have been developed which are capable of high delivery rates, even if the concrete must be delivered to elevations several hundred feet above the pump. Concrete pumping has the further advantage that it requires only minimum hook service. The principal disadvantage of the pumping method is that setup time and cleanup time will substantially exceed the time requirements for bucket placement. Concrete pumps are often mounted on trucks with a boom on which the pump line is mounted. Beyond the boom a rubber pump line serves as an elephant trunk. Use of this type of machine greatly reduces setup time for pours within its reach.

Concrete conveyors have most of the advantages and disadvantages of concrete pumps, but since the efficiency of concrete conveying is a function of slope, conveyors are less suitable where concrete must be deposited at elevations substantially higher than the elevation where conveying starts. Conveyors can be arranged to travel axially or laterally or to swing in order to cover the placement area, and are frequently operated in series. The final conveyor is fitted with a rubber elephant trunk to assist in distribution. Figure 15.2 shows concrete being placed with portable conveyors.

Tailgating is the least expensive way to place concrete. The trucks are equipped with chutes which can be rotated and raised or lowered to reach the work area. The trucks may discharge directly to the placing area or may discharge to an elephant trunk if they can readily reach locations above the placing area. This last is most common in construction of cut-and-cover structures in cities, where trucks can operate on temporary decking over the work area.

Safety measures in concrete placement include avoidance, insofar as possible, of situations where buckets may fall on or swing against the workmen or

Figure 15.2   Placing concrete with portable conveyor belts. (Morgen Manufacturing Co.)

the forms, the use of proper guards on moving parts of pumps or conveyors, and proper grounding of electrical equipment. Men working with concrete should wear hand and eye protection against cement burns.

Concrete placement is measured by the cubic yard or cubic meter of concrete placed. Various types of placements, such as mass concrete and wall concrete, are estimated separately, even if not segregated in the schedule of payment items.

Slab Finishing

Unformed surfaces are usually horizontal, or nearly so, and may simply be screeded off to the required grade, given a float finish, a broomed finish, or a trowel finish. Nonslip granules, concrete hardeners, waterproofing admixtures, or coloring agents may be incorporated in the surface during finishing. Other special finishes are occasionally specified.

If a float finish or a broomed finish is required, it follows the screeding operation. If a trowel finish is required, it follows the floating operation. Machines are available for each of these operations, and are used if the areas to be finished are large or if use of machines is required by the specifications.

Finishing is measured by the square foot or square meter. Each class of finish should be estimated separately.

Construction Joint Preparation

Construction joints must be treated to ensure good bond between elements of a monolithic structure. The treatment will include scrupulous cleaning and roughening of the concrete and may include the use of bonding agents.

Proper cleaning of concrete will remove all laitance, grease, oil, mud, and loose debris, and will leave a roughened surface. The methods available include wet or dry sandblasting, air-water blasting, high-pressure water blasting, and the use of chemicals on the surface to delay surface set, followed by water blasting. Specifications may require scabbling (chipping) to roughen surfaces where machinery is to be set on mortar beds. Proper timing is essential if the surfaces are to be properly prepared. If the work is attempted too soon, too much concrete will be removed, and the coarse aggregate may be undercut. If the work is started too late, the less costly methods may not be effective and sandblasting may be required. Wet sandblasting will usually be preferred for vertical surfaces and for small areas.

Proper bond between pours requires that the older concrete be thoroughly moist before the new pour is started. The specifications may require that a thin layer of mortar be thoroughly broomed into the older horizontal surface as the first activity in a pour. Some specifications may require that an epoxy or similar bonding agent be used on vertical surfaces and, less often, on horizontal surfaces.

Construction joint preparation is usually measured by the square foot or square meter. Various parts of a structure will employ different treatments for construction joints, and the costs of each class of treatment should be estimated separately.

Formed Surface Finishing

Most specifications state that surfaces against which additional concrete, back-fill, or embankment is to be placed will require only that rock pockets be repaired and that holes left by removal of form ties be packed with nonshrink mortar. Permanently exposed surfaces must usually have a higher degree of finish. Areas not used by the public may require only repair of rock pockets, packing of form tie holes, and removal of fins. Areas used by the public may additionally require such treatments as sack-rubbed, ground, sandblasted, or bush-hammered finishes. Concrete walls and ceilings which are to be painted should be spackled. Spackling involves filling all the minute air holes which will invariably be found on such surfaces with mortar in order to achieve a smooth surface and hence a good paint job.

Finishing is measured by the square foot or square meter. Each class of finish should be estimated separately.

Curing

Concrete should be kept moist and should be protected from freezing or from excessively high temperatures until the cement is completely hydrated and the concrete has reached its design strength. Most specifications prescribe a minimum curing period.

The specifications may require that certain surfaces be water-cured by sprinkling, by flooding, by the use of curing blankets kept continuously wet, or by covering with sand kept continuously wet. Certain surfaces may be cured by covering them with a plastic film or by spraying them with a curing compound. Most curing compounds leave a film of wax. Some curing compounds are designed so that the film will weather away under the influence of sunlight and rain.

Protection from temperature extremes can be afforded by insulating forms, shading surfaces, protecting surfaces by insulating mats, or by applying heat. Heating will be more effective if the area can be enclosed. Proper measures for fire safety are essential where heating is employed.

Curing is usually measured by the area involved. A separate estimate should be made for each curing method used.

Reinforcing Steel

On most jobs the reinforcing steel is purchased from suppliers, with the price covering the costs of detailing, preparation of placing plans, cutting, bending, and delivery. The steel may be placed by the contractor, or the placing may be subcontracted. The subcontractor may on occasion furnish and place the steel.

Placing accessories are usually purchased separately by the contractor or, if the placing is subcontracted, by the subcontractor. These include tie wire, support steel, chairs for supporting steel on soffit forms, and dobies. *Dobies* are mortar briquettes with wires embedded, which are tied to the bars to space them away from forms or to support the bars.

If welded connections or electrical bonds are required, due allowance must be made for the extra costs which this requirement will entail.

Reinforcing steel is measured by the pound or by the kilogram. The weight of bars or wire-mesh fabric required for laps at the approved splice locations is also measured for payment.

Precast, Prestressed, and Post-Tensioned Concrete

Precast concrete members may be unreinforced, conventionally reinforced, prestressed, or post-tensioned. The precast members may be purchased from a supplier or may be cast on the jobsite. Typical precast items which are neither prestressed or post-tensioned are manhole sections and wall panels. Wall panels may be cast away from the work area and lifted into place or may be cast in the work area, tilted up to their final positions, and secured there by welding or bolting. Generally, in the case of wall panels the stresses imposed during erection will exceed the stresses in service, and the panels must be designed with this in mind.

Prestressed members are invariably precast and are usually purchased from an outside supplier because the heavy buttresses needed to maintain the prestress until the concrete has reached sufficient strength are too expensive to construct unless they can be used over a long period of time. Typical prestressed members are concrete pilings, small beams, wall panels, and hatch covers.

Post-tensioned members may be precast or they may be cast in place. Typical examples include concrete girders and concrete floor slabs. Post-tensioning through cast-in-place concrete may also be employed to secure trunnions for tainter gates, to resist hydrostatic uplift in the slabs of a spillway chute or stilling basin, or to resist overturning forces in the case of a concrete gravity dam or spillway. Unless a contractor is experienced in the work and has all the necessary equipment, he may purchase the precast members and subcontract the prestressing operation. Most precasting companies have forms for the commonly used girder sections and can deliver the girders ready for erection at a far lower cost than the costs of construction at jobsite.

In recent years a number of major bridges have been constructed of precast, conventionally reinforced box girder segments joined by post-tensioning. Several have been built by erecting segments in opposite directions from each pier in order to maintain balance. Each segment is post-tensioned to the previously erected segments, and the partial spans function as cantilevers until they are joined at midspan.

Miscellaneous Embedments

Most concrete structures require water stops, keyways, expansion joints, contraction joints, and embedded metals, such as anchor bolts, pipe sleeves, and hatchway frames. Many of these items are temporarily attached to or supported by formwork. They are therefore usually installed by the carpenter crews. The gratings, hatch covers, and other metalwork which is not embedded is later installed by ironworkers. It is relatively simple to segregate the costs of permanent materials and the costs incurred by the ironworker crews, but it is difficult to determine the costs of setting embedded items, since these costs, despite efforts to the contrary, are usually mixed in with the costs of formwork. The estimator should endeavor to make spot checks on the job to enable him to estimate the man-hours required to install these miscellaneous items.

Structural Steel Erection

The steel structures most often encountered in heavy construction are bridge girders, building frames, such as frames for the superstructures of powerhouses or pump stations, switchyard structures, and steel penstocks, tunnel liners, or pump discharge lines.

Most structural steel is erected by ironworkers, but switchyard steel is usually erected by linemen, who also erect takeoff towers, transmission line towers, and other facilities pertaining to the transmission of electricity on aboveground outdoor structures. Carpenters usually install anchor bolts and other items which are embedded in concrete and which will later anchor the structural steel. Boilermakers usually install penstocks and pump discharge lines.

The greatest variable in the cost of erecting structural steel is in most cases the location and type of field connections. The location of connections will determine the need for temporary support and the weights of the members. The type of connection will determine the special skills and the time which will be required to complete each connection.

In determining the locations of field connections, if the specifications permit any latitude, the fabricator and the erector should work together. Connections are the most costly element in fabrication cost; so minimizing the number of connections reduces both fabrication and erection costs. The size of members will, however, be limited by transportation constraints and must be balanced against the lifting capacities which will be required during erection. The size may also determine whether temporary support will be required.

Field connections may be bolted, riveted, or welded, although field riveting is very seldom seen in modern structures. Turned pins or bolts may also be used in some connections in some structures. Bolted connections will in general require a greater number of pieces, such as splice plates, and a greater weight of metal than is the case for welded connections. This is particularly true for connections, or those parts of connections, which are in tension, because the bolt holes greatly reduce the net cross section of the metal in the basic member. Bolted connections are, however, quickly and uniformly installed to specified torques and readily inspected. Welded connections require more time, as well as the use of certified welders, and are not as easily inspected. Inspection of welds requires x-ray or sonic examination and expert interpretation of the photographs or recordings produced by these methods. Welded connections may require preheating and stress relieving.

The critical element in erection is to ensure that all members are properly aligned before the connections are completed. In bolted structures the bolts should not be completely tightened until the structure is checked to be certain that members are plumb or level, as the case may be. In welded structures the initial connections are made with temporary bolting. In either case it is exceedingly costly to correct misalignments after the work has progressed to other parts of the structure.

Steel structures common to heavy construction are most often erected by crawler cranes or truck cranes working from the ground or by tower cranes within or adjacent to the structure. Other lifting equipment often used on high-rise buildings is not often used in the types of structures encountered in heavy construction, but may occasionally be employed. These include guy derricks or stiffleg derricks, sometimes located on the structure being erected, jib cranes, and basket booms.

Small penstocks are usually shipped to the job as "cans," which require only that successive cans be joined in the field. Larger penstocks are usually shipped as plates cut to size, and rolled to the correct curvature, with the edges prepared for field welding. The plates are then welded into cans in the field fabricating shop and then moved into position and joined. The specifications may require stress relieving by heating, particularly in the case of complex pieces such as bifurcations. Special carriers have been developed for transporting penstocks and tunnel liners. These are wheeled bridges which can pass one set of wheels through the can and later raise that set of wheels to enter the previously installed can.

Bolted structures require the use of torque wrenches, usually air-operated, and compressors for air supply. Welded structures will, of course, require welding machines. There should be a good supply of wire rope and turnbuckles to be used in plumbing the structure and holding it in position until the field connections are completed.

Safety in steel erection involves all the usual protection needed when welding or bolting. Temporary or permanent decks should keep up with the work in order to limit the distance that workers can fall, edges and openings must be enclosed by railings, and men working above the decks must be required to use safety belts and safety lines. Good scaffolding should be installed where feasible and should be frequently inspected; this applies particularly to hanging scaffolds. Work in exposed locations should cease during periods of high winds, snow, sleet, or rain.

Structural steel is usually purchased fully fabricated, and erection is frequently subcontracted. A furnish-and-erect subcontract should obtain the optimum balance between shop costs and field costs if the specifications allow any latitude in the location of connections. Such latitude is most common in the fabrication and erection of penstocks, tunnel liners, and pump discharge lines.

NUCLEAR POWER PLANT CONSTRUCTION

Because of their large scope, stringent quality assurance requirements, and complexity, nuclear power plants present unique construction requirements. The following section describes some of the unique construction requirements which affect nuclear projects, along with innovative techniques which have been developed for their fulfillment.*

Concrete Construction

Several special requirements make concrete construction on nuclear projects very different from that used in conventional building. Although industry codes and standards are used as the basic requirements, they are supplemented by more stringent technical specifications and quality assurance requirements. In addition, the structural design criteria for nuclear projects result in massive structures, with uncommonly high reinforcing steel densities and unique configurations.

Because of the high levels of demand for concrete, in strict compliance with specification requirements, dedicated concrete batching plants are generally provided for nuclear projects. These plants are usually designed for 150 to 250 yd³ (115 to 190 m³) per hour production. Special monitoring and control capabilities are frequently incorporated to allow consistent production of high-quality concrete. Several materials and fresh concrete properties, such as gradation, slump, air content, and temperature are monitored at a frequency significantly greater than that for conventional construction. Maintenance of a low rejection rate requires sophisticated plant equipment, personnel training, and consistent monitoring and analysis of trend data. Because of the rigid enforcement of technical requirements, which are more stringent than con-

---

* This section is based in part upon information prepared by C. B. Tatum, project superintendent, Ebasco Services, Inc., Elma, Washington.

ventional practice, concrete supply costs on nuclear projects exceed those on commercial plants.

Some of the unique aspects of concrete structures on nuclear projects are indicated on Fig. 15.3. The high reinforcing-steel density is apparent. In addition, several other types of embedments are used to a degree substantially in excess of conventional construction. These include support plates, electrical conduit, penetrations, piping, and other items required to support plant structural, mechanical, and electrical features. The composite construction technique found on many nuclear projects is also illustrated. Under this approach, structural-steel floor framing is provided to support steel decking for use as stay-in-place forms. This significantly reduces shoring requirements and allows easier access for mechanical and electrical work below completed slabs. The structural steel also provides attachment points for pipe hangers, cable tray supports, HVAC duct supports, and other plant features. As a result of the increased congestion and more complex configuration of many concrete placements on nuclear projects, preparation for placements is considerably more involved. The reinforcement and embedment configurations frequently dictate a specific installation sequence for each embedded item.

Figure 15.3 Concrete nuclear plant construction. (Washington Public Power Supply System.)

With the precise location tolerances specified for many types of embedments, additional layout and checkout are required. Unusual wall shapes and projecting embedments frequently require special formwork approaches. Each of these conditions highlights the need for detailed analysis of the plans and specifications.

Concrete placement presents special requirements on nuclear plants. Stringent requirements for systematic vibration, free-fall limitations, and restrictions of horizontal movement result in the need for strict control of placement operations. When combined with frequent testing, strict slump control, and high levels of congestion, these factors make concrete placement on nuclear projects substantially different from and more expensive than placement for conventional construction. Recent trends have been toward extensive use of pumping, frequently with specialized placing booms. These differences must be recognized in estimating the man-hours for placing, finishing, and concrete repair.

Slip-Form Construction of Shield Building

The use of innovative progressive techniques in heavy concrete work has proved cost-effective on nuclear projects. Slip-form construction of the heavily reinforced concrete shield building is one example of such an innovation. Using this method, shield buildings have been completed in less than 1 month on several projects, and at substantial cost savings as compared with conventional forming techniques.

In the slip-form approach, the concrete is placed between shallow forms, which are raised by pneumatic or hydraulic jacks at an average rate of approximately 10 in (25 cm) per hour. The forms typically include three work decks. Interior and exterior support assemblies are required to provide adequate formwork rigidity. Reinforcing steel is installed from the main or upper work platforms. Concrete placement on the main platform generally is by pumping, with finishing performed from the lower work platform below the forms. Large lift cranes are generally required to supply reinforcing steel, embedments, and other materials to the appropriate working platform. The slip-form operation generally proceeds without planned stoppage until completion; however, planned stopping points and cold joints may be used if required.

Containment Vessel Erection

On nuclear plants, the major components of the nuclear steam supply system are located within a containment designed to withstand diverse accident conditions. Several variations of this structure have been used, including heavily reinforced concrete, post-tensioned concrete, and freestanding steel vessels. The freestanding steel design presents opportunities for construction innovation and use of specialized technology to lessen erection man-hours and shorten schedules. Each of the vessel designs must be erected, tested, and documented in accordance with the strict quality assurance requirements of the American Society of Mechanical Engineers' Boiler and Pressure Vessel Code.

The inside view of a containment vessel is shown in Fig. 15.4; the vessel is of the freestanding steel design, with an inside diameter of 150 ft and an overall height of approximately 270 ft. The nominal wall thickness is $2\frac{5}{16}$ in, with special reinforced sections up to 6 inches in thickness. Over 10,000 ft of field welding was required during field erection, after which the bottom head and shell were postweld heat-treated in a single operation. This heat treatment for stress relief was not required on the roof or dome because the hemispherical design allowed the use of a plate thickness less than $1\frac{1}{2}$ in.

A schedule savings of 2 months on one site was obtained by allowing the start of bottom head erection immediately following completion of the slip-formed shield wall. Under this approach, the bottom knuckle plates, support frames, and a lift crane for their handling were prestaged inside the circular foundation wall prior to the slip forming. After slip forming of the shield wall

Figure 15.4 Inside view of containment vessel. (Washington Public Power Supply System.)

was completed, bottom knuckle erection, using the prepositioned lift crane, proceeded in parallel with erection of the guy derrick which was used for the remainder of the containment. This parallel work effort is an example of the type of innovations which can significantly benefit nuclear construction if the potential is realized and the approaches are adopted in early planning.

Both shell erection and welding provide opportunity for construction innovation. At one site, the erector designed and fabricated a circular beam assembly with jacks to raise and lower a temporary cover for containment erection. This cover and beam served the triple functions of weather protection, staging support, and thermal barrier for the heat-treatment operation following erection and welding. Installed following erection of the bottom head, this cover was suspended from the concrete shield wall and jacked up the 159 ft of shell height using hydraulic jacks. As a shell course (approximately 11 ft) was completed, the cover was raised. Two levels of staging were suspended below the cover to provide access for shell fitup and welding. This approach resulted in several important benefits to shell erection and welding. Weather protection was provided for both interior and exterior shell welding, through the use of drop panels supported from the cover. The movable staging provided access for welding both the shell seams and attachments such as support plates and penetrations. Figure 15.5, an overhead view of the site, shows the cover suspended from the finished shield building wall.

The large quantities of welding on steel containment vessels of this design allow the use of automatic welding with important benefits. Using current technology, welds following the root pass can be made simultaneously from each side. This automatic welding equipment, working at high weld-metal deposition rates, can produce welds which consistently meet radiographic examination requirements.

On the vessel shown in Fig. 15.4, three automatic welding machines were used simultaneously. Approximately 6 working days were required to complete each 500-ft-long girth weld. The layout of seams in the vessel design allowed the work to proceed in a spiral manner, with erection of the next shell course above beginning prior to completion of girth welding below. Using this automatic welding technology, a low rejection rate was attained, despite the requirement for 100 percent radiographic examination.

Figure 15.6 shows an alternative design for a nuclear power plant; the containment vessel is shown in place and heavy concrete structures are seen to be underway.

Heavy Rigging Activities

The transport and setting of nuclear steam supply system (NSSS) components, some of which weigh in excess of 1,000 tons, present a key challenge to nuclear plant constructors. Building configurations frequently restrict the

Figure 15.5 Plan view of nuclear power plant, Satsup, Washington. (Washington Public Power Supply System.)

access for lifting equipment and require innovative approaches for vessel setting. Overall project benefits can result if the specialized equipment necessary to handle the NSSS components is also used to increase the efficiency of other construction activities.

Several alternative approaches have been used to transport NSSS components to the project site and set them onto foundations. The most frequently used methods include hauling by crawler or rubber-tired transporters and setting by lifting frames and special configurations of permanent plant equipment or crawler cranes. Figure 15.7 illustrates the use of a specially designed crawler crane which has handled heavy lifts at a number of nuclear projects.

The setting of the large reactor vessel and steam generators is a major milestone at a nuclear project. Early studies must be performed to evaluate the cost and schedule trade-offs of alternative rigging equipment and methods. If these studies indicate that overall advantages result from setting the large components in the shortest possible time frame, equipment such as that shown in Fig. 15.7 is one option. This approach allows setting of the components with the minimum impact on collateral work.

Figure 15.6 Aerial view of William H. Zimmer nuclear power station in Moscow, Ohio. (Cincinnati Gas and Electric Co.)

On the Washington Public Power Supply System project in Satsup, Washington, the 450-ton reactor vessel and two 750-ton steam generators were set on three successive weekends. Because of the location of other structures around the reactor building, the lift radii were 190 ft and 155 ft, respectively. The crane used a 400-ft boom length for the reactor vessel set and a 340-ft length for the steam generator. The setting operations involved upending (using special lifting fixtures), raising, swinging (using the crane's rear crawlers), and lowering into position. Each of the operations, from initial hoisting to turnover for final set, was completed in less than a single shift. Important cost and schedule savings were realized by the use of this method for NSSS vessel setting.

Several additional project benefits resulted from the use of this specialized lifting equipment. The use of yard preassembly and installation of other large components was increased because of the greater lift capacity available. The containment dome and attached piping was fully completed on the ground at a substantial savings in erection man-hours. Essentially all of the reactor building equipment was set in place over the top, avoiding the multiple handling operations required for setting through the construction hatch. These diverse applications of the rigging equipment necessary for setting NSSS components illustrate the potential for construction innovation present on nuclear projects.

POWERHOUSE CONSTRUCTION

This section describes the planning, scheduling, and selection of construction equipment for the construction of powerhouses associated with hydroelectric projects. Concrete and structural aspects are covered earlier in this chapter and in Chaps. 6 and 7. Additional methods for structural work and equipment installation methods are covered in this section.

Stage Construction of Powerhouses

Most powerhouses are designed so that they can be constructed in two stages, as shown in Fig. 15.8. The first-stage construction is often the only powerhouse work included with a dam contract. First-stage powerhouse construction will vary in scope, but it generally consists of the erection of the basic powerhouse structure, including the main permanent crane and the draft-tube gates. First-stage construction does not include the concrete that encases

Figure 15.7 Containment dome placement at the Perry nuclear power plant unit no. 1 near Cleveland, Ohio. (Newport News Industrial Corporation, The Cleveland Electrical Illuminating Co.)

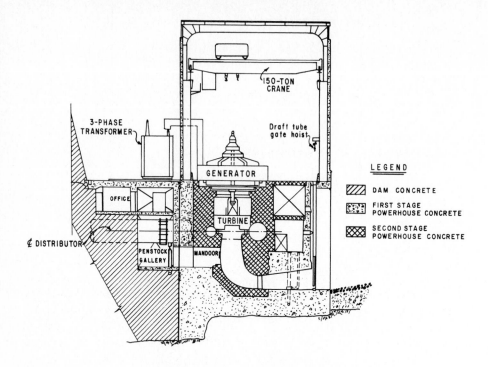

TRANSVERSE SECTION THRU
¢ GENERATING UNIT

Figure 15.8 First- and second-stage concrete construction for the generating bays of a hydro power plant.

the embedded turbine parts. Thus concrete placement is not delayed by the installation of turbine parts, and first-stage construction is expedited.

Second-stage construction consists of the fabrication, installation, and concrete encasement of the draft-tube liners, the scroll cases, and the turbine embedments; the placement of the remainder of the second-stage concrete; the installation of the turbines; the assembly and installation of the generators; the installation of the transformers; and the installation and wiring of all major powerhouse equipment. This second-stage construction may be included with the main dam contract, but often it is advertised as a separate powerhouse completion contract.

Some powerhouses are designed so that all concrete is placed at one time. When this type of powerhouse is constructed, the draft-tube, pit, and scroll-case liners and the turbine embedded parts must be fabricated and installed prior to the placement of the surrounding concrete. This design eliminates the formwork required between the first- and second-stage concrete and also eliminates minor amounts of concrete. Concrete placement is more difficult, however, since it must be suspended while the embedded liners, scroll cases, and embedded turbine parts are being installed.

Often powerhouses are constructed with extra generating bays that provide space for future installations of power-generating facilities. When this is done, only first-stage construction is completed in these extra, or "skeleton," bays.

The foregoing paragraphs describe the power-generating bays of a powerhouse. In addition to generating bays, powerhouses contain two other bays or equivalent areas along the length of the powerhouse. One bay or area is required for assembling and repairing generators and other powerhouse equipment. The other bay or area contains the powerhouse control equipment and provides office space.

Powerhouse Excavation

Foundation excavation for powerhouses and dams is described in Chap. 5. Special attention should be given to the amount of presplitting that may be required to maintain the sides of the powerhouse excavation. Many designers prepare drawings that depict excavation pay lines for the finished foundation surface of the powerhouse as a series of humps, valleys, and trenches. When this is done, the cost of excavation and foundation preparation is greatly increased, and the quantity takeoff should reflect this situation. When the drawings establish excavation and concrete pay lines, the specifications frequently require that the contractor place backfill concrete in any area excavated beyond these pay lines, without receiving reimbursement for this work. Since it is economically impossible to excavate any foundation exactly to prescribed lines, the takeoff quantities should show the estimated amount of overbreak so that the estimator may charge the cost of performing this additional excavation and the cost of placing the overbreak concrete against the appropriate pay quantities.

Penstock Construction

Penstocks are required to convey water from the reservoir to the turbines when powerhouses are located adjacent to the downstream toe of the dam or are located a distance from the dam. When the powerhouse is located at the downstream toe of the dam, the penstock is embedded in the dam concrete. When the powerhouse location is at a distance from the dam, open penstock runs are installed between the dam and the powerhouse. Penstocks of this type are shown in Fig. 15.9.

Penstock construction consists of aligning, fitting, and welding curved plates into lengths of penstocks at an assembly yard at the jobsite; transporting these assembled sections to the job locations; placing these sections on the foundations; joining these penstock lengths by the use of Dresser couplings or by welding; anchoring the penstocks to the foundations by ring girders bolted

Figure 15.9   Constructing Guri Dam powerhouse, Venezuela.

to concrete foundations or by concrete encasement; and sandblasting and painting the exposed surfaces.

A penstock assembly yard may be outfitted with concrete footings to furnish a level support while the curved plates are aligned and fabricated. Assembly-yard equipment consists of cranes, plate-alignment tools, welding machines, weld-testing equipment, and heat-treatment equipment. *Spiders* (interior supports) are installed in each section of penstock to maintain its shape. These spiders are left in the penstock sections until the sections have been joined together in their final location and anchored to their foundation or encased in concrete.

Heavy-duty transport trucks or truck tractors pulling special trailers are required to transport penstock sections from the assembly yard to the placing cranes. If the project is equipped with cableways of a capacity sufficient to handle the penstock sections and if the cableways can supply hook coverage of the penstock area, they can be utilized to place the penstock sections on their foundation. Otherwise, guyed derricks or other types of cranes are required.

When penstock sections are placed in position, the ends must be in alignment and spaced the correct distance apart for welding or for Dresser coupling installation. This final alignment is often done by welding bolt anchors on the inside and close to the edge of each penstock section. Each penstock section is then pulled into alignment by adjusting bolts that span the openings between penstock sections and between adjacent bolt anchors. Kickplates are also often welded to the edge of each penstock section to force the penstock edges into alignment. After the sections have been aligned and joined together by welding or with Dresser couplings, they are anchored into place, the spiders are removed, and the penstocks are sandblasted and painted. Where the penstock joins the scroll case, a blocked-out concrete section is left. This blocked-out concrete is poured after the connection is made.

Powerhouse Concrete Placement

When powerhouse construction is limited to first-stage construction, the placement of this first-stage concrete presents no special problems. When the contract work includes the placement of second-stage concrete or when the powerhouse design does not provide for first- and second-stage construction, special concrete-placement procedures are necessary to prevent misalignment, distortion, and uplift of the embedded liners, scroll cases, and embedded turbine parts. These special placement procedures include: placing concrete in shallow lifts and maintaining a minimum height differential between adjacent pours to reduce the fluid pressure of the freshly poured concrete; dividing encasing lifts into a group of pours and controlling the time interval between pours to reduce concrete shrinkage stresses on the embedded steel; eliminating the bracing or attachment of concrete forms to the embedded parts; preventing contact of concrete buckets or vibrators with embedded parts or supporting members; prohibiting the use of large vibrators and the dumping or chuting of concrete from high distances into the pour; restricting the rate of concrete placement; and using cold-water sprays on all internal surfaces of embedded parts during the concrete pouring and curing period to dissipate the heat that is produced by hydration of the cement in the concrete.

Storage of Powerhouse Equipment

If the contract work includes the complete construction of the powerhouse, the powerhouse equipment must be received and stored in suitable building and storage areas.

Specifications may delineate storage requirements for the powerhouse equipment, but storage often remains as the contractor's responsibility. Specifications may require that all critical items be stored in buildings where the temperature and humidity are accurately controlled. If storage requirements are not covered by the specifications, the requirements recommended by the equipment manufacturer may be used as a guide to determine the extent and type of the required facilities.

All powerhouse equipment should be protected from weather damage, dirt, corrosion, and distortion. Whether buildings are required, whether plastic covering is sufficient, what extent of heating is required, etc., depend upon the protection provided by the manufacturer when the equipment was crated and upon site weather conditions.

Electrical equipment and delicate equipment should be stored in a dry building. Governing equipment should be stored in a clean, dry, well-protected place. In cold or damp weather, space heaters should be used to heat the storage areas. If equipment susceptible to damage by dust or moisture is installed for a considerable period of time before operation, it should be enclosed with clean tarpaulins or plastic to prevent ingress of dirt, dust, and moisture during construction. Space heaters and moisture absorbents may be required within the enclosure.

Turbine parts that control the turbine final alignment and running clearance are machined to a fine tolerance. During storage, these parts should be well-blocked to prevent distortion of the machined surfaces. Runners, head covers, bottom rings, wearing rings, and the like should be stored with their axes vertical to maintain roundness. Shafts, runners, and other attached equipment should be supported, so that their weight is uniformly distributed to prevent any deformation. Supports should not be placed under bearing surfaces. The shaft should be stored so that it has no direct contact with wood supports. The shaft should be rotated 180° periodically and lifted so that additional rust-preventive material can be applied to prevent pitting at the points of support.

INSTALLATION OF POWERHOUSE EQUIPMENT

First-stage construction of a powerhouse does not present any construction problems in regard to the installation of permanent materials and equipment that have not already been discussed under concrete dam construction. If the project work involves the construction of the second-stage work in the powerhouse or the construction of a powerhouse without first- and second-stage construction, specific problems are encountered during the installation and testing of powerhouse equipment and the fabrication, alignment, and encasement of the embedded parts.

The installation of powerhouse equipment and the embedded plate-steel liners is such specialized work that often the prime contractor will prefer to subcontract this work. To prevent scope-of-work disputes by subcontractors, it is preferable to subcontract this work to contractors who perform both mechanical and electrical work and who specialize in installing powerhouse equipment. If the prime contractor undertakes the installation of powerhouse equipment, he should make maximum use of manufacturers' field representatives and his construction superintendents should be experienced in this field. There should be close cooperation between the manufacturers' representatives and the contractor's supervisors, and they should agree on erection procedures before work is started. Adequate space must be available for turbine subassemblies and for the assembly of the generators and other equipment. Since space for this work is restricted, the equipment for one generating bay is completely assembled and is moved into position before the assembly of the equipment for the next generating bay is started.

Following are descriptions of the procedures required to install the main embedded parts and equipment for one generating bay of a powerhouse. The first description is of a generating bay where the first-stage concrete has already been placed and a Kaplan-type turbine is to be installed. (The main parts of a Kaplan turbine are shown in Fig. 15.10.)

There are work variances from this description when concrete is not placed in two stages or when other types of turbines are installed. The order of work and work procedures is an additional variable dependent upon the installers. The installations of Kaplan and Francis turbines are quite similar. If the powerhouse contains a Pelton turbine, however, installation will be different. Therefore, a brief description of the installation of a Pelton turbine will follow that of the Kaplan turbine.

These installations are so complex that cost estimates for this work should only be performed by engineers who have had experience in this field. The purpose of these descriptions is only to provide the background required to schedule powerhouse construction, based on the assumption that the prime contractor will subcontract the equipment installations. Therefore, the following descriptions should not be used as a basis for estimating installation costs.

Installation of Kaplan-Type Turbine Equipment

The embedded parts for a turbine must be set with fine accuracy since the final leveling and alignment of the turbine operating parts depend on their correct elevation, concentricity, and trueness. Before any placement of embedded parts is undertaken, the proper grade and centerlines for each powerhouse unit should be accurately determined and referenced.

The machined surfaces of all parts of the powerhouse equipment must be cleaned of rust, paint, or protective coating, and burrs and nicks must be removed with a fine emery cloth before assembly. This often requires months of man-labor.

The first embedded part to be installed is the *draft-tube pier nose*. When the draft-tube liner extends past the dividing pier in the draft tubes, it forms a pants leg, and the pier nose is a part of the pants leg. When this liner stops short of the pier nose, draft-tube pier noses are required. The installation of these pier noses does not present any particular problem. Pier noses must be received at an early date, however, so that their installation will not delay concrete placement.

The next embedded part is the *draft-tube liner*. The draft-tube liner is located below and connected to the discharge ring. Its length varies with the

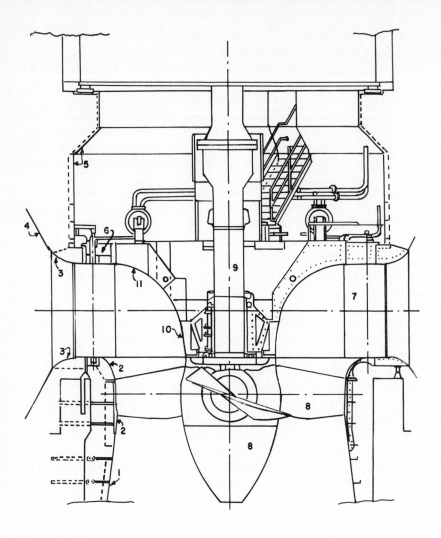

Figure 15.10   Parts of a Kaplan turbine.

powerhouse design. If the turbines are small or if the draft-tube liners are short, the liners can be installed in one piece. Otherwise, they are installed in sections; the sections are then aligned and welded together. One installation method is to place spiders in the prefabricated liner sections; they are then lifted and set on adjustable jacks mounted on concrete pedestals poured with the first-stage concrete. The liner sections are connected to steel rods fastened to anchors placed in the first-stage concrete and equipped with turnbuckles. The liner sections can then be positioned and aligned by adjusting the jacks and tightening the turnbuckles on the anchored rods. The rods prevent the fresh concrete from floating and moving the draft-tube liner. Welding is then done; the welds are tested, and depending on specification requirements, heat treatment may be performed. The concrete encasement can then be poured as described under concrete placement, and after it has set, the spiders can be removed.

After the draft-tube liner has been partly encased in concrete, the *discharge ring* can be installed. Another name for the discharge ring is the *throat ring*. The bottom of the discharge ring is attached to the draft-tube liner and it extends to the *stay ring*. It contains the bushings for the lower stems of the wicket gates. It may come in two sections, depending on the particular design. If it is in two sections, the section next to the stay ring, which contains the stem bushings for the wicket gates, is called the *bottom ring,* and the remainder is called the *discharge ring.* Also, the discharge ring may have a removable section. Installation of the discharge ring is done by using jacks and turnbuckle-equipped anchor rods to bring it into alignment. It must be positioned accurately because the final leveling and alignment of the turbine depend on an exact setting of the discharge ring. After alignment, it is connected to the draft-tube liner and the concrete that encases it can be poured.

After the discharge ring has been partly encased in concrete, the stay ring, also known as the *speed ring,* is installed with the inner surface of its bottom ring flush with the top of the discharge ring. The stay ring consists of a top and bottom ring separated by wicket vanes. Its purpose is to guide water into the wicket-gate openings. The bottom ring spans the distance from the discharge ring to the scroll case. The *top ring* fits between the scroll case and the pit liner. The top ring is also connected to the outer turbine cover (a ring). The stay ring can be assembled on the powerhouse floor and then set in place, or it can be assembled in place. Like the other embedded parts, it is positioned on jacks and aligned with rods equipped with turnbuckles. It must be set to exact elevation and alignment since it also controls the turbine position. The stay ring is bolted to the discharge ring, and their mating surfaces must match. If there are any surface discrepancies, these must be removed by grinding.

The *scroll-case liner,* also known as the *spiral-case liner,* is next in the order of installation. This is a steel-plate liner for the water passage that partly encircles the turbine pit and supplies water to the turbine. A penstock section connects the scroll case to the turbine shutoff valve. The scroll cases may also be connected to a discharge pressure-relief valve located so that it can discharge at the downstream face of the powerhouse. For Kaplan and Francis turbines, the scroll cases are of welded construction. For impulse turbines, scroll cases are bolted construction to prevent distortion and consequent misalignment of the nozzles. Scroll-case liners are assembled, fabricated, and encased in concrete with the same type of procedure as that used for the draft-tube liners. Jacks and anchor rods equipped with turnbuckles are used in alignment. After they are aligned, welding, weld testing, and heat treatment can be performed and their concrete encasement can be poured. The exterior surfaces of some scroll-case liners may be covered with cork before the encasing concrete is poured.

The pit liner is then erected next and fastened to the upper ring of the stay ring. The *pit liner* is a steel plate that lines the pit between the scroll case and the turbine gallery. It is the last major embedded part and is placed with the same procedure as that used for the draft-tube liner. It is set back along one wall to form a shelf for support of the servomotors. When a Pelton turbine is installed, the liner comparable to the draft-tube liner is commonly called a pit liner, though its proper name is a *wheel pit liner*.

The remaining pieces of major equipment are nonembedded. The first nonembedded item to be set is the *outer head cover,* which is a ring that fastens to the upper ring of the stay ring and contains bushings for the top stems of the wicket gates. It also is equipped with removable cover plates which allow removal of the wicket gates. Its other purpose is to provide the upper support for the wicket gates and to support the intermediate turbine cover, which in turn supports the inner head cover, the runner, and the turbine shaft.

If the wicket gates have not been previously installed, they should be positioned at this time. They are installed between the discharge or bottom ring and the outer head cover. Their purpose is to control the water flow past the turbine runner.

The *turbine assembly,* consisting of the *runner, shaft, inner head cover,* and *intermediate head cover,* is next installed. These parts can be assembled before being lifted into place. Figure 15.11 shows a bridge crane setting a Kaplan wheel and an inner head cover at the McNary Dam powerhouse on the Columbia River. The intermediate head cover rests on the outer head cover and supports the inner head cover. The inner head cover contains the shaft bearing and seals, and its bottom edge is equipped with a renewable bronze wearing ring to take the upward thrust of the runner. Final positioning of the turbine moving parts is not done until the turbine shaft has been coupled to the generator shaft. Lubricating, cooling, and control piping can be connected at this stage of erection.

The *gate-operating ring* can then be mounted on top of the intermediate head cover. It is connected to and controls the opening of the wicket gates. In turn, it is connected to the servomotors located in a shelf provided in the turbine pit.

In the meantime, second-stage concrete placement can be completed and the installation of the generator can commence. Figure 15.12 shows the generator parts whose installation is described in this chapter. Anchor bolts can be installed and plates can be set on the concrete and grouted to support the stator, the bottom-generator frame, and the rotor brakes. The stator on large generators comes in sections. These sections are erected and assembled on the sole plates; the stator plates are stacked and all coils are installed. A minimum time of 1 month is required for this work. The lower bracket arms for the generator and the generator shaft also can be erected in place. The lower bracket arms support the generator shaft and contain the shaft bearings. The rotor is assembled and the stator sections are connected on the rotor-erection stand in the assembly bay. All windings are connected. After the stator work is completed, the rotor can be placed on and connected to the shaft. Figure 15.13 shows a bridge crane moving a generator rotor at the McNary Dam powerhouse. Next, the generator shaft is connected to the turbine shaft. This is usually done by reaming and grinding all holes to fit the bolts. After this connection is made, final adjustments are made to the generator and turbine bearings to be sure that they are in perfect alignment. Ladders, platforms,

Figure 15.11 Bridge crane moving a Kaplan wheel and inner head cover, McNary Dam powerhouse, Columbia River. (General Electric Co.)

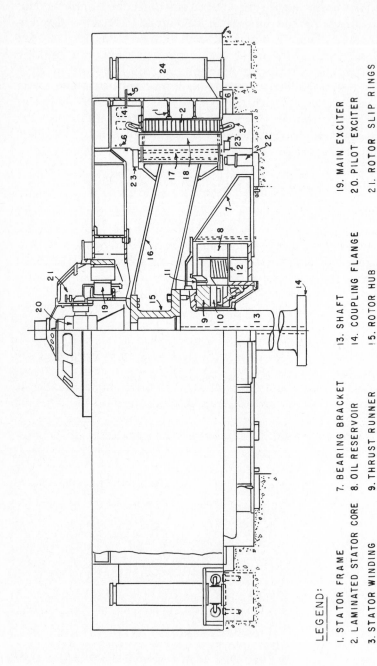

LEGEND:

1. STATOR FRAME
2. LAMINATED STATOR CORE
3. STATOR WINDING
4. STATOR END CONNECTIONS
5. STATOR LEADS
6. FIRE EXTINGUISHER PIPING

7. BEARING BRACKET
8. OIL RESERVOIR
9. THRUST RUNNER
10. THRUST-BEARING SHOES
11. GUIDE-BEARING SHOES
12. BEARING-OIL-COOLING COILS

13. SHAFT
14. COUPLING FLANGE
15. ROTOR HUB
16. ROTOR SPIDER
17. LAMINATED FIELD CORE
18. LAMINATED FIELD POLES

19. MAIN EXCITER
20. PILOT EXCITER
21. ROTOR SLIP RINGS
22. BRAKE AND JACK
23. ROTOR FANS
24. AIR COOLER

Figure 15.12 Parts of an electric generator.

438

Figure 15.13 Bridge cranes moving a generator rotor, McNary Dam powerhouse. (General Electric Co.)

stairways, oil piping, cooling piping, fire-protection piping, and electrical connections can be installed. The upper generator bracket, the exciters, the air housing, and the generator coolers can be placed, controls connected, and painting completed.

The governor should be installed last, because it should not be installed and then be exposed to other construction. The governor is mounted on the floor adjacent to the generator. Piping and control wiring can then be connected. Figure 15.14 shows the generator floor of the McNary Dam powerhouse.

Transformer installation consists of unloading and positioning the transformers; installing their wheels, heat exchangers, and bushings; hooking up the cooling and oil piping; filling them with oil; and connecting them to the generator and to the lines leading to the switchyard.

Installation of Pelton-Type Turbine Equipment

The installation of a Pelton-type turbine is quite different from the installation of a Kaplan turbine. For a Pelton turbine, the major work is the placement, alignment, anchoring, and bolting of the scroll case; the installation of the needle valves and nozzles; and the turbine encasement. Bolted construction is used to prevent misalignment of the turbine nozzles. After these parts are bolted and aligned, they are encased in concrete. Passageways to the needle valves and chambers behind the needle valves must be constructed to permit removal of the needle valve and nozzle when repairs are necessary.

The installation of these turbine parts is illustrated by three progress photo-

Figure 15.14   Generator floor, McNary Dam powerhouse. (General Electric Co.)

graphs taken during the construction of Middle Fork powerhouse on the American River, California. Figure 15.15, taken during the installation of the bolted scroll cases, shows the nozzles and the pit liner. Figure 15.16 was taken after the scroll cases and pit liners had been installed in concrete and while concrete formwork and reinforcing steel were being placed for the generator supports. The penstock and the penstock bifurcation are visible at the left of the photograph. Figure 15.17 was taken while the generators were being installed.

TESTING POWERHOUSE EQUIPMENT

Testing, starting up the units, and bringing them on the line are done under the direction of the equipment suppliers' representatives. Since the owner will operate the completed powerhouse, it is to the owner's advantage to supply the testing crews and have them trained by the equipment representatives during the testing period. Since this work is handled by the equipment suppliers and by the owner, its substance will not be discussed. For scheduling purposes, testing and start-up take approximately 2 months per unit, of which 1 month is required for drying out the generator during slow-speed operation. In special cases, the contractor must provide operating crews to the equipment suppliers' representatives. When this is the case, the cost of supplying these crews must be included in the project cost.

Powerhouse-Construction System

The powerhouse-construction system is established by determining the work quantities, scheduling the construction of the powerhouse, and selecting the required construction equipment. Powerhouse construction is so rigidly controlled by the specified requirements for equipment installation that construction systems vary only in the use of different equipment units for concrete placement and the handling of the powerhouse equipment and embedded materials.

Quantity Takeoffs

Similar to the procedures required for planning the construction of all other types of heavy construction, quantity takeoffs of the work required for powerhouse construction furnish the basic data for planning and estimating its construction. If the powerhouse contains multiple power units, the work should be scheduled and the quantity takeoffs based on constructing the units in a stepped manner. Stepped construction means that each unit is started, built, and finished with approximately a 2-month interval between units. This

Figure 15.15 Installation of the bolted scroll case, nozzles, and pit liner for a Pelton-type turbine, Middle Fork powerhouse, American River, California. (American River Constructors.)

Figure 15.16   Installation of concrete formwork and reinforcing steel for the support of a generator, Middle Fork powerhouse, American River, California. (American River Constructors.)

allows better utilization of trained crews and maximum reuse of forms, since forms can be progressively used on each power-generating unit.

Decisions must be made on the extent to which powerhouse work will be subcontracted before quantity takeoffs are started. Most prime contractors subcontract the electrical, piping, heating, plumbing, and air conditioning items; the installation of reinforcing steel, turbines, generators, transformers, governing and associated equipment; the erection of gates, stop logs, and much of the architectural work. If this work is to be subcontracted, the quantity takeoffs need only cover excavation, backfill, concrete forms, concrete pay quantities, overbreak concrete, cement grout, floor finishes, wall finishes structural steel framing, and miscellaneous embedded metalwork, such as embedded frames, guides, and anchor bolts. This miscellaneous embedded metal is installed by the prime contractor, since it is closely tied to concrete placement.

Special attention should be given to the forms for the galleries and passageways, block-out forms, draft-tube forms, and the scroll-case forms for low-head powerhouses where steel-lined scroll cases are not used. Form cost increases when large amounts of embedded metal project through the forms. The forms for the draft tubes are expanding curved forms that must be set to strict alignment tolerances. As a result, they are costly to build and require a long erection time. When powerhouses contain multiple units and when the powerhouse pours are designed to permit their use, steel forms are used in preference to job-built wooden forms. When concrete-finished scroll cases are used, this forming is also expensive and often difficult. In most cases steel forms cannot be used for forming scroll cases, since, after stripping, the forms must be dismantled into small enough pieces to go through the wicket-gate openings. This requirement also makes the reuse of wooden draft-tube forms impractical. If the forms for all water passageways in a powerhouse are not constructed with smooth surfaces, a large amount of hand grinding and finishing will be required.

As the quantity of subcontract work is reduced, the quantity of takeoff work is increased. Piping, liquid storage, liquid pumping, and electrical installations are quite extensive in a powerhouse and detailed takeoffs must be prepared for estimating this work. As an example of work complexity in a powerhouse, separate piping systems must be installed for raw-water supply, potable-water supply and treatment, sewage, drainage, cooling water, compressed air, piezometer installations, the turbine shaft seal, the bearing thermometer, governor oil supply, fuel supply to the standby generator, CO_2 fire protection, generator and transformer fire protection, lubrication oil supply, and transformer oil supply. Figure 15.18 shows a view of the turbine-generator erection for a nuclear power station.

Takeoffs alone cannot indicate the extent of work involved in the installation of embedded turbine parts and the erection of turbines, generators, transformers, governing equipment, gates, etc. However, takeoffs can indicate the weights to be lifted and the amount of welding required. Estimating

Figure 15.17 Installation of a generator, Middle Fork powerhouse, American River, California. (American River Constructors.)

Figure 15.18 Turbine-generator erection in process at the William H. Zimmer nuclear power station, Moscow, Ohio. (Cincinnati Gas and Electric Co.)

the cost of installing these items and determining the time required for their installation should be based upon past job records or on estimating experience.

SCHEDULING POWERHOUSE CONSTRUCTION

The scheduling of first-stage construction work for a powerhouse consists of determining the time required to make each concrete pour, the time required for erection of the powerhouse superstructure, and the time required to erect and test the cranes. The total time required is dependent on work complexity and size of the powerhouse. The first powerhouse unit often can be completed in 8 months from the start of concrete placement. Reuse of forms and utilization of construction equipment and trained crews are accomplished more efficiently if powerhouse units are constructed in a stepped manner. For stepped construction, an additional 2 months of construction time are required for each additional generating bay in the powerhouse.

Powerhouse scheduling is more complex when a completed powerhouse is to be constructed. If every pour is made on schedule, approximately 12 to 14 months from the start of concrete placement is required to place all the concrete for the first powerhouse generating bay, erect the crane rails, and complete the crane erection. Two more months are required to complete the concrete pour in each additional powerhouse generating bay. After the powerhouse permanent cranes are in operation, approximately 4 additional months are required to set the first turbine, the first generator, and the first governor and transformer and to complete the piping and electrical work. Again there will be a 2-month lag from unit to unit when multiple-unit powerhouses are constructed.

To determine accurately the time required for the placement of powerhouse concrete, a detailed pour schedule must be prepared. The powerhouse generating bays will control job completion since concrete placement will not be delayed by major equipment installation in the control and assembly bays. The pouring schedule should show the number of pours, the pouring sequence, and the periods when pouring must be suspended for the erection of the draft-tube forms and the installation of the embedded steel liners and turbine parts. Powerhouse scheduling requires that time be allowed on the schedule for placing special forms and for installing pipes, electrical conduit, and embedded turbine parts. Concrete placement will also be controlled by the dates that embedded materials will be received at the jobsite. The first embedded items to be installed are the draft-tube pier noses; therefore, the pier noses must be delivered to the job at an early date. Following this, the draft-tube liner, scroll-case liner, stay and discharge rings, and pit liner must be received in time to make the pour schedule.

It is preferable to have the same individual who made the concrete takeoff prepare the powerhouse-construction schedule since he will have knowledge of the amount and complexity of the required forms.

CONSTRUCTION EQUIPMENT SELECTION

If the powerhouse is an integral part of the dam, the powerhouse concrete is often placed in the same manner and with the same equipment as the mass concrete. When the powerhouse is located as a separate unit at the downstream toe of the dam, a powerhouse concrete-placing method and placing equipment are selected that will not interfere with dam concrete placement, since the placement of mass concrete is of prime importance on any dam project.

When the dam concrete is placed with a cableway, this separation of powerhouse concrete-placement equipment is of special importance. A large cableway has such a high hourly operation cost that low concrete-placement cost can only be obtained when the cableway is being used to its maximum capacity. The cableway maximum capacity can only be achieved during the placement of mass concrete. Placing powerhouse concrete with a cableway is slow because most of the powerhouse pours are small and placement is delayed because accurate bucket spotting and bucket stability are required to allow concrete to be deposited in narrow form openings. The pendulum action or drift of a bucket suspended from a cableway and the jump that occurs when the bucket is discharged do not particularly delay the placement of mass concrete, but on powerhouse pours these cableway characteristics result in many pouring delays because the concrete must be discharged from the bucket into restricted areas and it is necessary to protect protruding reinforcement steel and other embedded items from bucket damage. If cableway-tower runways must be extended to provide cableway coverage to the powerhouse, this will also increase the cost of using a cableway to pour the concrete in a powerhouse.

In consideration of these factors, the cableways used for massive concrete placement are seldom used for powerhouse pours; instead, powerhouse concrete is usually placed with separate equipment. Occasionally, a smaller-capacity cableway is used. This was done at Hungry Horse Dam on the Flathead River near Kalispell, Montana, where powerhouse concrete was placed with a 4-yd^3-capacity cableway with its traveling tower located on a separate runway from the three 8-yd^3-capacity cableways used for placement in the dam concrete. In other instances, the main dam cableways are only used to transport the concrete buckets to the powerhouse site, and final placement of the concrete is done with other equipment. This does not tie up a main cableway exclusively for powerhouse pouring and permits more accurate concrete-bucket handling for powerhouse pours. More often concrete buckets are transported from the mix plant to the powerhouse site by trucks and then handled with other equipment. Since the powerhouse is often located at the bottom of the canyon, requiring steep road grades between the mix plant and the powerhouse, trucks are used instead of rail equipment for hauling concrete.

If the dam concrete is placed with a trestle-crane placement system, powerhouse concrete placement can often be done with a similar type of crane added to the system for this purpose. If the placement of the dam concrete necessitates the use of a trestle, it is sometimes possible to locate this trestle so that it can service both the dam and the powerhouse, or it may be possible to provide trestle service to the powerhouse by constructing a branch trestle curving off the main trestle.

When the concrete in the powerhouse is to be placed with a separate system, the selected system should be one that will result in low concrete placement cost and will also provide hook service to handle most of the lifts required in its construction. If only first-stage powerhouse construction is required, hook coverage is only needed for form moving, for handling reinforcing steel, for placing other embedded materials, for erection of any structural-steel framework, and for crane erection. If the work consists of constructing a completed powerhouse, a large amount of additional hook time is required. Hook time is needed for handling the draft-tube liners, discharge rings, stay rings, scroll-case liners, pit liners, wicket gates, stop logs, water-control gates, pumps, control motors, transformers, turbines, and generators. Some of these lifts will be beyond the capacity of one crane. When the powerhouse is equipped with a permanent crane, this crane can be used to handle the heavy lifts required for setting the turbines and generators and to handle the transformers.

When cranes are provided for powerhouse concrete placement and for hook service, it is necessary to check the required boom length, lifting capacity, and gantry height before their size can be determined. This information should be plotted on a drawing containing the plan and sections of the powerhouse. This drawing can also be used to determine whether a trestle will be required. It is usually necessary to try cranes with different boom lengths and gantry heights. If a trestle is required, the drawing will show its location and elevation. It is desirable to keep the trestle outside the powerhouse area, preferably upstream. For some powerhouses, however, hook coverage cannot be provided except by placing a trestle on the center line of the powerhouse.

Gantry cranes provide a fast and efficient method of pouring concrete and give good hook coverage for powerhouse construction. Gantry cranes provide bucket stability and can accurately spot and maintain the bucket position. They have sufficient boom and hook capacity to handle all but the heaviest of lifts. The crane tracks can be located on the ground or on a trestle. The choice between the two depends on the results of the previously described hook-coverage study.

Self-propelled crawler or truck cranes also provide concrete coverage and hook capacity. Whether they should operate from ground level or from a trestle is also determinable by the hook-coverage study. As crane capacities keep increasing, the need for trestles keeps decreasing. Because the undercarriage of crawler cranes is not supported in as rigid a manner as that of rail-mounted gantry cranes, crawler cranes will not provide equal concrete-bucket stability. The advantages of the use of crawler cranes when compared with gantry cranes are that the purchase price and erection cost are less; a higher percentage of the purchase price will be recovered when they are sold because of their greater versatility; and they are more mobile, which allows them to be used on other project work. The boom of one of these crawler cranes is shown on the left side of Fig. 15.9, the Guri Dam powerhouse. This photograph is deceptive since the steep excavated rock surface that is shown in the left of the photograph stops short of the generating bays. Excavation in front of the generating bays permitted water to flow freely from the draft tubes. Two crawler-type cranes were operating on this excavated surface when the construction of the powerhouse was started; later they were operated from a timber trestle placed on this excavated area. These cranes handled the concrete buckets for most of the powerhouse-concrete pours.

Tower cranes can be used for placing concrete in small powerhouses. They have a limited hook capacity, so that they can only handle small concrete buckets and make light lifts. For heavy lifts, other types of cranes such as guyed derricks must be provided. The guyed derrick shown in Fig. 15.9 of the Guri Dam powerhouse was used to handle the heavy lifts during the construction of the powerhouse structure.

Stiffleg cranes have been used for placing concrete in powerhouses. They are so slow that it is hard to keep the pour "alive" while it is being placed. Because they are anchored to one position, several stifflegs may be required to furnish complete hook coverage for a powerhouse. This is an older type of construction equipment, and its use has been largely replaced by modern self-propelled cranes. The two stiffleg cranes shown in Fig. 15.9 of the Guri Dam powerhouse were so slow in operation that they were mostly used for moving forms and setting embedded parts and materials.

The use of conveyor belts for making powerhouse pours should increase as specifications are revised to permit their use. They can only be used for concrete placement, however, and hook coverage for the powerhouse must be provided by material-handling cranes. Conveyors are a fast and economical method of making the small pours in a powerhouse structure.

If the powerhouse must be completely enclosed before second-stage concrete is started, special equipment will be required for second-stage concrete placement. When specifications permit or when its use is made mandatory, pumped concrete or intrusion-prepacked concrete is used for second-stage powerhouse concrete. Second-stage concrete placement can be accurately and evenly controlled by either method. The intrusion-prepacked method will reduce the lifting force against the scroll-case liners. Permanent powerhouse cranes may be available for placing this concrete, but because they are so slow, they are unsuitable for this work.

Chapter 16 Pavements, Pipelines, and Rail Networks

Most of the basic planning and estimating requirements of the fundamentals of heavy construction have been discussed in other chapters. This chapter will concentrate on the aspects unique to pavements, pipelines, and rail networks.

HIGHWAY, AIRPORT, AND CANAL PAVING

Most support aspects of paving construction are similar to the basic excavation, drilling and blasting, and embankment installation procedures discussed in other chapters. However, the introduction of the slip-form paver has resulted in significant improvements in productivity over older methods of building concrete highways, runways, and parking aprons. The utilization of reclaimed aggregates has proved economical in many locations, and improvements in asphaltic concrete practices have also contributed to increased productivity. Increasing attention is being focussed upon the replacement, maintenance, and improvement of existing highways in the United States and other developed countries, while in the developing countries the demand for new highway construction remains high.

Portland-Cement Concrete Pavements[16.1]

Bulk earthwork is first completed to rough tolerances as discussed in earlier chapters. Final subgrade preparation for new pavement is performed utilizing a multipurpose combination trimmer-spreader which trims the subgrade,

449

spreads the subbase materials, and trims the compacted subbase to prescribed tolerances. Basic components include a rotary cutter, a screw spreader which rotates in both directions, and moldboard templates. Tolerance control is achieved to ⅛ in through an automatic level-control system. Equipment for preparing the subgrade for maintenance and repair operations is dependent upon the size of the project.

Concrete Supply

Central mix plants are generally favored for large new highway paving projects. Highly portable automated batching and mixing equipment can be relocated and set up in 1 day or less at the new location. Central-mixed concrete is hauled to the work location utilizing special side- or rear-dumps, agitating or nonagitating bathtub-type trucks, ordinary dump trucks, or in some cases agitating mixers. This method has largely replaced the use of batch plants, compartmentalized batch haul trucks, and dual- or triple-drum pavers located at the pour location.

Forming and Placing

Slip-form paving is one of the major developments bringing improvement in productivity in paving construction. Slip-form pavers spread, level, vibrate, consolidate, and finish the pavement to the specified tolerances with a minimum of handwork. Side forms may range to 50 ft (15 m) in length and slide forward with the machine, leaving the edges of the consolidated concrete pavement strip unsupported. Slip-form pavers are availabe for standard paving-lane widths for highway construction and for up to 36- or 48-ft (10- to 15-m) widths for airport pavements and expressways. Production rates of up to 1 mile (1.6 km) per day have been achieved with portland-cement concrete spreads on the interstate highway system in the United States. Of course, conventional side forms continue to be utilized where long strips suitable for slip forming are not practical. Figure 16.1 shows a typical slip-form paver.

A special slip-forming machine has been developed to extrude curbs, gutters, and highway barrier walls. Production is reported to average 1,000 lin ft (300 m) per day on 4- to 5-ft (1- to 1.5-m) barrier walls on one project, compared with 200 ft (60 m) per day using a carpenter and labor crew with steel forms in the conventional manner. Slip forming has also been successfully applied to bicycle paths up to 12 ft (3.5 m) wide and to median barrier walls utilizing 1 yd³ of concrete for 3.0 to 3.4 lin ft (0.9 to 1 m) of wall.[16.2]

A special slip-form machine has also been designed to allow paving on several different slope angles for canal construction, as well as on a flat configuration for the canal invert.

Bituminous Pavements[16.3]

Flexible pavements consist of a number of courses of treated or untreated materials placed over a subgrade and compacted foundation. Courses can include a subbase, base, leveling, binder, and surface. The quality of the materials usually increases with each successive course above the subgrade.

Prime, Seal, Macadam, and Road Mix Applications

With some exceptions, this type of pavement will not generally be the final riding surface for high-type highway construction. A *tack coat* is a thin application of bituminous material to an old road surface in order to supply a bond to the new surface. Cutbacks, tars, or emulsions are used for this application, utilizing a distributor truck at coverages up to 0.1 gal/yd.[2]

The *prime coat* is applied to an absorbent base to act as a bonding agent and to seal the joint between the new pavement and the base. The prime coat is applied through the distributor at a rate of 0.25 to 0.50 gal/yd.[2]

Figure 16.1 Typical slip-form paver in operation. (Crawford, Murphy, and Tilly, Consulting Engineers, Springfield, Illinois.)

Seal coats consist of thin bituminous-aggregate courses applied to existing bases or surfaces of any type. Their application consists of sweeping, priming, and curing, followed by binder application by distributor, aggregate application, and rolling and broom dragging to compact the finished surface.

Penetration-macadam roads can be 4 in (10 cm) and more in thickness and can generally handle heavier loads and may be used for base courses. The *macadam road* is laid by sweeping, applying large size aggregate, rolling, and applying binder by distributor; a second application of somewhat smaller aggregate, a second rolling, and again a binder application can be followed by a third or even fourth course, depending upon the specified requirements.

Paving with *road mix* involves mixing the aggregate and binder on the road. Aggregate and binder are blended together using blade graders, mechanical mixers, or traveling plants. Installation involves sweeping, priming and curing, applying aggregate, aerating and drying, mixing, spreading and leveling the mix, and compacting with a steel or pneumatic-tired roller.

Central Hot-Mix Bituminous Concrete

Central-mixing plant design is in many ways similar to concrete plants as previously described. Plants can be portable, semiportable, or stationary, depending upon the economics of a particular application. Aggregate drying is one of the most critical features of a hot-mix plant.

Asphaltic concrete is generally mixed in a central mix plant, hauled to the site in rear-dump trucks, and spread and compacted by a spreading and finishing machine. The finishing machine must spread and compact the materials evenly, leaving a finished surface ready for final compaction by suitable steel rolling equipment. A typical finisher consists of a tractor unit and a floating screed unit. The tractor unit includes a receiving hopper, feeders, distributing augers or screws, power plant, transmission, controls, and operator seat. The screed unit includes tamper or vibrating units, thickness and crown controls, screed heater, and screed plate.

Reclaiming, Recycling, and Repair

Reclaiming and recycling of both bituminous and portland-cement concrete is becoming increasingly economical in urban locations. During 1981 the State of Wisconsin repaved about 400 miles (640 km) of two-lane highway with reprocessed material, about 75 percent of the total program. The contractor found that existing batch-mix plants could handle only about 50 percent reclaimed material, while a drum mixer handled 70 percent or more.[16.4]

Recent tests by the Southwest Research Institute, San Antonio, Texas, show that a 100 percent sulfur binder will make a blacktop mixture with stability equal to a conventional asphaltic mix. A compressive strength of 2,500 lb/in^2 was developed after a curing period of 24 hours.[16.5]

One of the world's largest highway recycling projects is the rebuilding of the Edens Expressway in Chicago, Illinois. On the first half of the project, completed in late 1979, the contractor recycled 97,700 yd³ (75,000 m³) of concrete, or about 200,000 tons. The old reinforced concrete was taken to a central crushing plant equipped with two primary jaw crushers, a magnet and mechanical picker to remove reinforcing steel, and a triple-roll secondary crusher.[16.6]

Many of the early portland-cement concrete interstate, primary, and secondary highways are beginning to deteriorate. Typical repairs are exemplified by the overlay of 11.5 miles (18.5 km) of Virginia's I-95 with 4.5 in (11.5 cm) of asphalt. The overlay consisted of two 1.5-in-thick lifts of intermediate course and a final 1.5-in-thick asphalt-concrete surface course.[16.7]

PIPELINES

Pipelines for economical transportation of oil, gas, coal, and other materials are receiving renewed interest in the United States and around the world. The currently planned 745-mile (1,200-km), \$30-billion Alaska segment of the 4,800-mile (7,700-km) Alaska-Canada natural gas pipeline may be the most expensive single construction project in American history.[16.8] This section will review some of the features peculiar to pipeline construction. Pipeline construction is highly specialized, however, and only a summary presentation is included here; additional information is available in the *Handbook of Heavy Construction*.[16.9]

Trenching Excavation Equipment

Trenching excavation is a fundamental factor in successful pipeline construction. Where conditions are favorable, large wheel-type trenching machines are generally the most economical. Ladder-type machines have certain advantages for many projects, and backhoes continue to perform specialty work on almost every pipeline project.

New trenching machines have been developed which now make trenching practical in rock, frozen ground, and in other conditions under which trenching was not previously practical. Large, automatically controlled trenchers simultaneously cut and trim the trench bottom, and excavate and trim the trench sides to required slopes. Curved buckets dig round-bottomed ditches which exactly fit the pipe shape, thus reducing placement, bedding, and backfilling costs. Laser-controlled machines produce finished line and grade to 0.10-in tolerances. On one project a medium-sized trencher equipped with weighted buckets, rock teeth, larger-than-standard drive-shaft bearings and heavier counterweights achieved horizontal production rates of up to 100 ft (30 m) per hour in cuts up to 10 ft (3 m) in soft limestone and shale. Another laser-controlled machine equipped with special hard carbide-tipped teeth averaged 150 lin ft (45 m) per day in a 27-ft (8-m)-deep cut through soapstone. In good soil with no crosslines, trenchers can average 1,500 to 2,000 lin ft (450 to 600 m) per shift of 36-in-wide trench up to 10 ft (3 m) in depth.[16.10]

Utility Pipe Installation[16.9]

Shoring requirements are very important, especially where laborers must work in the trench. In hard or compact soil, all trenches over 5 ft (1.5 m) deep should either be shored or side slopes should be laid back on a 1 : 1 slope or other safe angle of repose. The *Construction Safety Orders* of the State of California give standards for wood shoring, which is becoming obsolete in most installations. The use of aluminum vertical shoring with horizontal trench jacks or braces is one of the most effective methods, since all workers can stay on the surface during installation. Shoring can also be accomplished through the use of a trench shield attached to a ladder trencher. The length of the shield must be sufficient for the installation of a full length of pipe. A system for backhoe excavation utilizes a similar shield for pipe installation; the shield follows the backhoe and serves as a protective cage for the workers. The backhoe operator uses the bucket to advance the shield.

Manual pipe installation is usually economical when sections are 20 ft (6 m) in length and weigh under 150 lb (70 kg). Heavier pipe lengths require equipment to lift the pipe and lower it in place. A swing crane mounted on a small track-laying tractor has proved economical for intermediate pipe sections, and side-boom tractors are used for larger sized pipe. These pieces of equipment usually require only one operator. Production rates of 2,000 ft (600 m) per shift can be maintained in good ground in open areas requiring little or no shoring.

Testing is important for water mains as well as for water testing of pipelines containing other materials. When possible all joints should be left exposed. The liner should be filled to operating pressure and allowed to stand for 1 to 2 days. The line should be walked and each exposed joint checked when pressure is first applied and again after the test pressure has been applied. Test sections should never terminate against old valves or connections to other lines; temporary bulkheads are usually used. Specifications often provide for a maximum amount of leakage per inch of pipe diameter per day at a given pressure. The amount of water pumped into the line to maintain pressure over the time period is therefore a convenient method of measuring the leakage.

Transmission Pipelines[16.9]

Modern transmission pipelines are envisioned in such major projects as the Trans-Alaska pipeline for the transmission of crude oil, the proposed Alaska-Canada pipeline for the transmission of natural gas, and the 1,378-mile

(2,218-km) project in the midwestern United States proposed by Energy Transportation Systems, Inc., San Francisco, for the transmission of coal slurry.

Planning and estimating major transmission pipelines requiring huge sums of money is a mammoth task. A project such as the Trans-Alaska pipeline, which is subject to intense regulatory scrutiny, harsh, unpredictable weather and ground conditions, and an imported work force, is often constructed on some sort of combination reimbursable and fixed-cost basis, with incentives for the achievement of superior performance. Here ingenuity and skill on the part of the estimator is very important in first obtaining the contract through presentation of the best technical plan and in developing a workable program for construction which will ensure execution of the plan in accordance with predetermined cost and schedule goals. Factors which must be taken into account in the initial planning of projects which feature both unit-price and cost reimbursements include the size and location of the project; the type of terrain; the weather conditions; the logistics of supply of materials and equipment; the supply, skill, and militancy of the work force; the special problems associated with right-of-way considerations; and the extent of federal, state, and local regulations.

Transmission line spreads are generally made up of a number of functional operations, all advancing at approximately the same installation rate. Figure 16.2 shows an organizational chart for a typical pipeline spread.

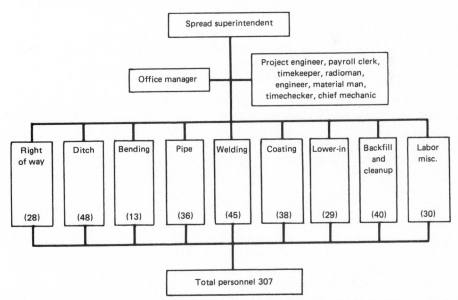

Figure 16.2 Organizational chart for typical pipeline spread. (John Havers and Frank Stubbs, Jr., *Handbook of Heavy Construction*, 2d ed., McGraw-Hill, New York, 1971.)

The Trans-Alaska pipeline was called the outstanding civil engineering achievement of 1978 by the ASCE.[16.11] Half of the pipeline was constructed across 420 miles (672 km) of ice-rich permafrost on special elevated pipe foundations when outside temperatures were as low as −80°F (−62°C). Another 380 miles (610 km) of the 48-in pipe is buried. Here special precautions were necessary because of the compressive stresses built up as a result of the constrained thermal growth caused by the pipeline through which oil at 145°F (63°C) flows. Thirteen thousand pieces of construction equipment were used on the project to complete the 1.2 million bbl/d pipeline. Critical to the program were the welding of 100,000 40-ft (12-m) sections of 48-in-diameter pipe with 100,000 circumferential welds, 100 percent x-ray examination, and hydrostatic testing to 125 percent of operating pressure. Delays in the commencement of construction because of environmental lawsuits, native claims, and regulatory delays are estimated to be responsible for half of the $8-billion-dollar cost.

RAIL NETWORKS

Renewed interest in light rail systems is being experienced as a part of worldwide construction of urban rapid-transportation systems. Massive projects are bringing the advantages of modern railroad technology to the developing countries. The construction of both conventional cross-country railroad networks and urban rapid-transit systems involves the use of applications, including excavation, tunneling, and heavy concrete and steel construction, covered elsewhere in this book. A general description of some of the current and evolving applications is included here, since actual track installation is normally handled by specialty contractors. Technical data regarding railroad engineering practices can be found in the *Civil Engineering Handbook*.[16.12]

Light Rail Systems

Much research has been conducted in recent years on the design of rail and roadbeds for high-speed trains. On the Metropolitan Atlanta Rapid Transit Authority System in Georgia,[16.13] continuous-welded rail was placed in lengths 1,440 ft (440 m) long. Trackways were designed to 4 ft 8½ in gauge, which is standard in the United States. Surface lines were placed on heavy prestressed concrete ties on rock ballast. On concrete slab sections in subways and elevated structures the rails rest on elastomeric and steel sandwich cushions to reduce noise and vibration. Trains will operate on 750-volt direct current picked up from a continuous-contact "third rail." An electronically controlled rail-mounted machine was used to set the rails to a tolerance of ⅛ inch in 33 feet. The contractor also developed a unique machine for handling the 8½-ft-long, 525-lb ties, placing them four at a time on the ballast.

Conventional Rail Systems

Major conventional rail systems are either underway or being planned by many of the oil-rich developing countries. A Brazilian contractor is building a new rail line in Iraq from Baghdad west to Akashot near the Syrian border. Other Arab states are pursuing plans to link a number of countries in the Persian Gulf with a rail network.[16.14]

An ambitious project in Gabon, Africa, is designed to link the Atlantic Ocean port of Owendo with Franceville using a single-track line. Government authorities have currently slated $550 million, or one-third of the 1980–1982 budget, to speed completion by 1986 or 1987. Construction costs are expected to run about $4.7 million per mile, or $2 billion for the overall project. Current estimates are about twice the original estimate. Major problems have been attributable in part to annual rainfalls of up to 120 in (300 cm).[16.15]

REFERENCES

16.1 "Portland-Cement Concrete Pavements," sect. 25B in John A. Havers and Frank W. Stubbs, Jr., *Handbook of Heavy Construction*, 2d ed., McGraw-Hill, New York, 1971.

16.2 "Slipforming Successfully Used for Unusual Application," *Rocky Mountain Constructor*, May 15, 1981, p. 14.

16.3 "Bituminous Pavements," sect. 25A in Havers and Stubbs, op. cit.

16.4 "Wisconsin Sets a Hot Recycling Pace," *Engineering News-Record*, July 16, 1981, p. 65.

16.5 "Revolution in Paving Counters Rising Cost," *Engineering News-Record*, March 8, 1979, p. 22.

16.6 "Urban Expressway Rebuilt on Recycled Concrete Base," *Engineering News-Record*, Nov. 29, 1979.

16.7 Information supplied by the Asphalt Institute, College Park, Maryland.

16.8 "Investors to Decide Gas Line's Fate," *Engineering News-Record*, June 4, 1981, p. 11.

16.9 "Pipelines," sect. 26 in Havers and Stubbs, op. cit.

16.10 "There's a Lot Happening in Underground Excavation Around California," *California Builder and Engineer*, May 14, 1979, p. 77.

16.11 "Trans-Alaska Pipeline," *Civil Engineering*, June 1978.

16.12 Leonard Urquhart, ed., *Civil Engineering Handbook*, 4th ed., McGraw-Hill, New York, 1959.

16.13 "Engineering the MARTA System," *Georgia Professional Engineer*, vol. XXII, no. 4.

16.14 "Gulf Nations Seek Rail Link," *Engineering News-Record*, September 17, 1981, p. 51.

16.15 "$2-Billion Rail Job Back on Track," *Engineering News-Record*, February 5, 1981, p. 25.

Estimating Considerations

Chapter 17 Estimating Procedure

The preceding chapters have discussed the construction methods and equipment used in various types of heavy construction. A working knowledge of both methods and equipment is essential to cost estimating. This chapter discusses estimating procedure. The procedures described here will be applied in Chaps. 18 and 19 to a fictitious hydroelectric project. All three chapters will be better understood if frequent reference is made from one to the other.

The basic fact is that an effective estimating procedure is analytical in nature. The work to be performed is broken into its elements, and each element is then analyzed to determine the materials, manpower, and equipment necessary for its accomplishment. A good example of this is the procedure for estimating the cost of a concrete structure. The cost elements usually consist of foundation cleanup, formwork (subdivided into form fabrication or purchase, and form erection and stripping), concrete production or purchase, concrete haul and placement, wet finish, hard finish, and construction joint preparation. The need for such subdivision of costs is apparent from the fact that if a given volume of concrete is used in a thin slab on grade, very little formwork is required, but there is an extensive surface area requiring wet finish (floating and perhaps troweling). The same volume of concrete, if used in a thin wall, requires a substantial area of formwork, but little or no wet finish. There are, of course, some operations, such as loading and hauling earth or rock, where both elements are considered together to ensure a proper balance between loading capacity and hauling capacity.

It will be noted that the estimators produce estimates of costs, not estimates of bid prices. These costs are later marked up to bid prices. It should also be apparent that good estimating is essential if the contractor is to survive. If the estimates are too low the contractor loses money. If the estimates are too high, the contractor spends money pursuing work, but is never the successful bidder.

TYPES OF ESTIMATES

Estimates may be prepared either by the longhand method, or by the computer method. The longhand method requires that the estimator write down all explanations and computations and that all calculations be performed on a calculator. By the computer method, scheduled rates of production, crew sizes, wage rates, and other basic factors are fed into a computer which then computes the cost and turns out a typed tabulated estimate.

Both of these methods have advantages and disadvantages. The primary advantage of a computer estimate is that the computer does not make arithmetical mistakes and tabulates the information in such a manner that details can be readily obtained. The main advantage of the longhand method of estimating is that the estimator, in spending more time on all the details and calculations, has more opportunity to change and refine his assumptions and judgment factors.

The estimator who uses a computer must thoroughly understand the procedure used in a longhand estimate. He must furthermore compile all basic information in detail before programming the computer, so that the computer run is a purely mechanical operation. Any change in assumptions or facts after the start of the computer run requires that the estimate be run through the computer again.

The longhand method of estimating is used in the example shown in this book because it better illustrates heavy-construction estimating procedures. The bulk of the time spent in preparing an estimate is devoted to selecting equipment and estimating progress and crew size. All of these things must be determined before either a longhand or computer estimate is made.

ESTIMATE FORMS

There are many methods of setting up an estimate, and advantages may be presented for any method. Some estimators estimate only direct labor and job supplies, such as powder and exploders, for the individual bid items. They then place all maintenance labor, service labor, supervisorial labor, equipment, plant, and repair parts in a general account. These are then prorated to the bid items.

This method is not recommended, for the greater the amount of expense that is prorated, the less accurate is the estimated cost of the individual items. Since an estimate is a collection of individual items of cost, the more accurate the estimated cost of each item, the more accurate will be the complete estimate.

An estimating format should be designed to provide ease in estimating and to act as a guide for cost accounting. If cost accounts follow the estimate, or the estimate follows the cost-accounting method, it is easier for the job supervisor to compare the actual cost with the estimated cost and exercise closer job control. Keeping estimates and cost reports on the same form furnishes job records of the cost of individual items that can be directly compared with similar items in future estimates.

The estimate format used here has the following advantages:

1. It follows good accounting procedure in classifying expenses, capital accounts, and overhead.
2. It places expense that can be reduced with good job control in the direct cost division and furnishes unit costs of individual items, giving ready comparison with actual job costs.
3. If cost accounting follows the estimate form, good cost records will be available for bidding on future work.
4. It makes all details of the estimate readily available to compare with estimates prepared by joint venture partners.
5. It allows the estimate to be prepared in a logical progression of steps and provides an arithmetical check on most of the estimated cost.

This recommended form of estimate is divided into the following categories of cost:

1. Direct Cost. This category comprises those costs which can clearly be identified with particular bid items. Direct cost is the part of the total cost that can vary with good supervision and job control and is the part most difficult to estimate. Thus, good job records are invaluable in furnishing a guide to bid work. Direct costs are commonly subdivided as follows:

a. *Direct labor,* which includes the cost of fringe benefits, payroll taxes, and those insurance premiums which are a function of labor hours or labor costs. Direct labor includes foremen but excludes higher levels of supervisors.
b. *Repair labor,* on the same basis as direct labor, but segregated in order to ascertain the total cost of equipment operation. If mechanics are included in the crews, as is the custom for tunnel estimating, there will be no separate category of repair labor.
c. *Equipment operating invoice,* which includes all operating costs for equipment, exclusive of labor costs. Operating cost does not include ownership cost or depreciation. It does include repair parts, tires, etc.
d. *Supplies,* which includes form lumber, explosives, and other expendable items. It may also include the costs of short-term equipment rentals. It should include the cost of the small tools which will be used in the performance of the work. Small tools are usually defined as those which have a unit value of $1,000 or less. Examples are vibrators, cutting and burning

outfits, small power tools, and hand tools furnished by the contractor. The cost of small tools is usually estimated as a percentage of labor costs.

e. *Permanent materials,* which includes items such as concrete, structural steel, or equipment supplied by the prime contractor and incorporated in the finished work. The nature and quality of these materials are specified in the bid documents.

f. *Subcontract costs,* which includes the costs of furnishing and installing permanent materials supplied by subcontractors, or the costs of having a subcontractor install permanent materials supplied by the prime contractor or by the owner. Some subcontracts may include only labor.

The greatest differences among contractors in the matter of what constitutes direct costs is in the area of equipment costs. Many contractors subdivide direct costs further than the method used in Chaps. 18 and 19, providing separate categories for repair labor and for equipment operating invoice. Some contractors further subdivide repair labor into field repairs and overhaul, and subdivide equipment operating invoice into fuel, lubricants, tires and tubes, repair parts, etc. Some contractors carry operating costs for general-purpose machines such as cranes, flatbed trucks, pickup trucks, and maintenance vehicles as indirect costs. Some contractors operate separate equipment owning subsidiaries or divisions, and charge equipment ownership costs to direct costs as "company rentals." Company rentals make good sense for jobs of short duration or jobs which are not joint ventures.

2. Plant and Equipment Cost.   This category includes purchases and long-term rentals, including sales taxes, plus freight charges, plus setup costs, less salvage value. It thus includes some labor costs. A form is included in Chap. 18 for tabulating these costs. Rules cannot be formulated to work in all cases when making a distinction between plant and equipment. Therefore, since each estimator's distinction between the two classifications will differ, the two are added together and the sum is used to compare estimates at a prebid estimators' meeting. Most contractors use the following definitions when making the distinction between plant and equipment: *Plant cost* is the cost of facilities necessary to construct the project which do not have a salvage value at the end of the job, and it includes work performed on equipment which will not increase its salvage value. *Equipment cost* is defined as the cost of machinery and facilities necessary to construct the project that will have a salvage value at the completion of the project. When estimating the net cost of plant and equipment chargeable to the job, the salvage value should be deducted from the total cost. The deducted salvage value should be the net salvage value, after allowance has been made for the cost of moving out, cleanup, and storing and selling the equipment.

If equipment is rented for the project, the rental cost of the bare, nonmaintained, nonoperated equipment should be included as a separate item in the plant and equipment cost. If this cost is placed in the direct cost division it will distort the unit direct cost items when recorded cost is used to estimate future work.

The net cost of plant and equipment is determined at the start of the work when methods and equipment are decided upon. After the method of construction is determined, this cost should not vary to any great degree from the original estimate during the construction of the project, since it is not dependent on job control.

3. Indirect Cost. This category includes all cost not charged to the other divisions of cost. It includes project supervision, job engineering, office payroll and expense, insurance, property taxes, special taxes, and bond premiums. Indirect cost should not include sales tax, since this tax should be added to the cost of plant and equipment, permanent materials, and supplies. The amount of indirect cost is dependent upon the efficiency of the engineering and office staff, the complexity of cost records required, and the number of reports that must be submitted.

4. Camp Cost. If a project is in an isolated area, it may be necessary to install camp facilities and operate them for the life of the contract. If a prefabricated camp is planned and a caterer used, quotations for these costs can be secured before bid submission. Thus, this cost of the operation can be well determined before bid submission. The major cost variable will then be the number of men who will use the camp.

5. Escalation. Most estimates are made by using wage rates in force at the time of bid submission. The escalation item covers the cost of anticipated wage increases occurring during the life of the project. This cost is dependent upon future wage rates and fringe benefits that will be negotiated by the union, plus any anticipated increases in workers' compensation insurance and payroll taxes. If it is anticipated that during the contract period there will be a rise in material and supply prices, escalation on these items should also be computed.

The use of this estimating format is exemplified in Chap. 18, where the estimate for the fictitious Sierra Tunnel is drafted, based upon the categories of cost outlined above.

After an estimate is completed, the estimator should evaluate the following expenses and recommend to his principals that proper consideration should be given to them when job markup is determined.

6. Interest Expense. This expense is the interest on the contractor's capital required to finance the work. To determine the capital and interest expense properly, it is necessary for the estimator to prepare a cash-flow forecast for the job.

7. Contingencies. On work where there is an exposure that is indeterminate and cannot be estimated, the probable maximum cost to the contractor for this exposure should be determined and explained to the principals. The principals can then determine to what extent the job will be subject to any contingencies and can determine markup accordingly.

STEPS IN PREPARATION OF THE ESTIMATE

The steps taken in estimating heavy construction are interrelated, so judgment factors used in one step may affect decisions made in another step. This makes it impossible to prepare the estimate in the exact order presented below and means that there must be a continual review of all preceding steps as each additional step is completed. In following these steps in estimating, the engineer should always keep in mind the fact that the controlling and largest item in the cost of construction is labor and that labor cost varies with the construction progress, construction methods, and type of equipment. So that comparisons can be readily made with the estimating procedure described in this chapter and with the contractor's estimate presented in Chaps. 18 and 19, the following steps have been numbered. The same numbers are used for estimating computations in Chaps. 18 and 19.

     I. Studying plans and specifications
    II. Inspecting the jobsite
   III. Reviewing geological reports
   IV. Taking off quantities
    V. Requesting quotations on permanent materials and subcontracts
   VI. Computing labor rates
  VII. Tabulating equipment operating costs
 VIII. Transmitting estimating instructions
   IX. Selecting construction methods
    X. Selecting plant and equipment
   XI. Preparing a construction schedule
  XII. Estimating direct costs
 XIII. Tabulating direct costs
 XIV. Estimating plant and equipment costs
  XV. Estimating indirect costs
 XVI. Estimating camp costs
XVII. Estimating escalation
XVIII. Tabulating total costs
 XIX. Analyzing cash flow
  XX. In-house review of the estimate

I. STUDYING PLANS AND SPECIFICATIONS

The first step in estimating the cost of any type of construction is to become thoroughly familiar with the plans and specifications in order to grasp the overall job picture and to understand the specific conditions and work requirements stipulated by the owner. When the plans and specifications are reviewed, consideration should be given to the fact that they are the owner's and that they are written for the benefit of the owner and will be interpreted and enforced by the owner's representative.

Specifications and bid advertising announcements should be checked for any requirements regarding prequalification of the contractor. Prequalification is required by some governmental agencies and private owners. If it is, fulfillment of this requirement should be expedited. Prequalification requirements vary from proof of financial responsibility to the listing of job experience in similar types of work. Contractors' licensing requirements should also be reviewed; this point is not mentioned in the specifications but is covered by local laws. When receiving joint venture bids, some states classify bids as nonresponsive if they are not presented in numbered bid forms checked out to the joint venture, and will rule out bid forms checked out to individual members of the joint venture.

The general and special conditions of the specifications should be carefully examined. Special attention should be paid here to determine whether they furnish any relief to the contractor in contract price, and whether extension of contract time is provided in the event of unforeseen underground or other physical conditions, acts of God, strikes, war, weather, etc. Contract completion dates and liquidated-damage provisions should be understood, and the bond and insurance provisions should be reviewed.

Payment provisions and bid items should be reviewed to determine whether provisions have been made for reimbursements to the contractor for mobilization, whether the owner will withhold a percentage from each progress payment until job completion (known as *contract retention*), when and how progress payments will be made, and whether payment is on a unit price or lump sum basis.

Most bid schedules contain both unit-price bid items and lump-sum bid items. The bid documents will usually state that the estimated quantities tabulated in the bid schedule are to be used only for comparison of the proposals from the several bidders, and that payments will be made at the unit prices in the proposal for the quantities of work which are actually required. Some bid documents provide for adjustments in unit prices in the event that the estimated quantities differ significantly from the actual quantities. This provision may apply to all bid items or to certain specified bid items, or it may apply only to the contract as a whole, in which case the criterion is contract price as bid versus contract price as constructed. It should be noted that certain items when bid on a lineal foot basis actually behave as lump sum items, since the number of lineal feet is fixed. Tunnel work is a particularly good example, since the length of the tunnel is always very well defined before bidding. On a lineal-foot basis of payment, the quantity of steel supports, the excavation delay for supported ground, the quantity of lagging and blocking, and the volume of contact grouting constitute some of the risks passed on to the contractor, and the contractor should recognize this when he determines the markup.

The plans and specifications should be examined closely for any requirements that may change the standard method of construction and for any limitations restricting the amount of work that can be done from any area. Specific points to review are limitations on blasting; the maximum size of aggregates in the concrete; restrictions on the placing of concrete; the types of concrete forms required; the curing time, i.e., the length of time before forms can be stripped; and the type of rock cleanup required. Restrictions on any of the above will affect the contractor's cost and should be reflected in the estimate.

The plans and specifications should be examined to determine if there are any special costs connected with construction. These costs may result from the location of the job or special construction considerations. Typical examples of these special cost items are given below.

A diversion tunnel may be located with its portals below the streambed elevation. To keep the portals dry, it may be necessary to construct cofferdams. At other types of locations, unexcavated plugs may be left adjacent to the streambed. Transportation to and from the tunnel portal will be complicated because of the presence of these cofferdams or plugs. If the plug method is used, the plug must be removed upon job completion. In diversion tunnels that are to contain a permanent concrete plug, special excavation of anchorages is required. Diversion-tunnel construction may require the use of high-pressure consolidation grouting at the tunnel portals and at the permanent plug location. Occasionally diversion tunnels must be holed through into a body of water. Extra work and extra costs occur in this operation, such as excavating a bottom pocket in the tunnel to receive the rock when the final round is shot.

A project may be located in such a remote area that special off-site cost is required for labor recruiting, purchasing, and expediting. At remote locations larger stocks of spare parts, materials, and supplies are needed. Initial transportation as well as daily transportation from hiring hall to jobsite may be required for job personnel. Travel-time payments to and from the job might be a requirement. Labor turnover will increase. There will be added material handling and hauling cost. It may be desirable to provide helicopter or airplane service to the job area, including landing and servicing facilities at the jobsite. A project may be located in an area where heavy snowfalls occur. The equipment must be winterized, and snow removal may be necessary. The weather may be so extreme that there will be a loss in job efficiency.

If a project is located in a state or national park or a national forest, there will be added cost of waste disposal, as preservation of trees and stream pollution are rigidly controlled in these areas.

Work in military installations may cause additional costs owing to security requirements. Such costs may be for security clearance, special transportation of men because of private-car restrictions, and additional travel-time payments.

When projects are located in populated areas, there are many additional cost items. Blasting may be restricted to daylight hours, and restrictions may be applied to the amount of powder that can be exploded at one time. Spoils of excavation must usually be hauled a long distance to the disposal area, and this haulage may be restricted to specified hours. Construction will present special problems because of the congested area and because utilities may have to be moved or altered. The space available for a contractor's work area may be limited. Noise must be suppressed, which will increase the cost of compressor installations. Ground settlement will be a problem and expenditures may be necessary for foundation underpinning and street repair. The work may be located in a filled area where piling, foundations, and even sunken ships may be encountered. Employee parking will be a critical and often an added cost.

If the work is located in a foreign country, many special costs will be incurred. Passports, shots, health clearance, and air transportation are required for each American employee. Turnover of American personnel is quite high on foreign work. Premium wage rates are paid for overseas duty. Board and lodging are usually supplied without cost to the single American employees. The American employees are reimbursed for any local income taxes which exceed the United States rates. The local labor must be trained to use American equipment and American methods. Job logistics are a problem, and if large inventories of supplies, materials, and spare parts are not maintained, air freight will be used for emergency orders. Local operators may be hard on equipment and tires. Maintenance cost of equipment may increase greatly if local personnel operate the equipment and if local mechanics are inefficient. Local equipment operators may not be the equivalent of their American counterparts in production rates. Job inspection may be less standardized. Local corporate taxes may reduce profits. Import duties may apply to the importation of equipment, materials, supplies, and commissary items. Camp cost, including mess hall and commissary cost, may be high. It may be desirable to provide a purchasing and expediting office either in the United States or in a European country. Travel expenses to and from the site by the contractor's American executives, engineers, and consultants will be incurred. It may be necessary to hire local lawyers and engineers to help in job negotiations. Finally, at job completion the sale of the plant and equipment may present considerable difficulties. Construction equipment has been shipped back to the United States by American contractors from other countries upon job completion.

Before bidding, the contractor should bring to the owner's attention any discrepancy between the plans and specifications, mistakes in the specifications, or any instance where the owner's intention needs clarification. Verbal clarification of the plans and specifications by the owner or the owner's representative does not hold the owner to these opinions or interpretations; this clarification must be obtained in writing. In all cases the specifications place the primary responsibility of interpretation on the contractor, so it is important that they be thoroughly understood by the estimator.

One method of becoming familiar with the plans and specifications is to prepare an abstract as described in Chaps. 18 and 19. This abstract is also useful to have on the job inspection trip and to furnish to management to use for quick job review purposes.

II. INSPECTING THE JOBSITE

The job inspection should be made after the plans and specifications have been studied so that the inspection team will have a concept of the job and some grasp of the problems resulting from specification requirements. On this inspection trip, attention should be paid to the points involving special problems as well as to an overall job review. Most contractors use a standard checklist to ensure that important matters are not overlooked during the course of the inspection.

Particular points that should be considered during the job inspection are outlined below.

1. The surrounding area should be examined for towns and other places having capacity to house the construction force. This is necessary before a decision can be made concerning camp requirements.

2. Access to the job should be investigated to ascertain what difficulties will be encountered when bringing in equipment, supplies, and materials and to determine whether access roads need to be constructed or improved. Weather conditions should be checked to determine their influence on job access and construction progress.

3. The availability of power and water should be examined so that the cost of the facilities to the job can be estimated.

4. All work areas should be examined and photographed.

5. All areas suitable for plant installations and as spoil disposal areas should be examined and photographed.

6. Gravel pits, borrow areas, and quarry sites should be examined and photographed. If royalties must be paid for the materials, the costs should be ascertained.

7. All rock outcrops and all drill cores should be examined and photographed. If a consulting geologist is to be employed, it is desirable that he accompany the inspection team in order that his observations about site conditions may be fully appreciated.

8. If in situ drilling tests can be arranged, this should be done. If this is not feasible, rock samples should be procured for laboratory drillability tests.

9. Geological reports by the owner or by the federal or state government not included with the bid documents should be read, and adequate notes made. If written copies of the reports are not available, it may be possible to tape record them.

10. Labor rates, crafts employed, and availability and suitability of labor vary from area to area. Therefore, labor should be investigated and local labor agreements obtained. These agreements will cover, for the specific area, craft classifications, existing wage rates, any future wage rates that have been negotiated, hours of work, overtime rates, fringe benefits, subsistence rates and areas where paid, travel time, and hiring rules.

11. Sources of supply of ready-mixed concrete, concrete aggregates, cement, explosives, lumber, plywood, etc., in nearby communities should be investigated. The inspection team should procure telephone directories for all nearby communities.

12. Particular attention should be paid to prospects for obtaining minority subcontractors, suppliers, and workers. Efforts to arrange their participation should be well documented.

13. The inspection team should prepare a comprehensive report of their observations, complete with carefully identified photographs. The consulting geologist should deliver his report as early as possible.

III. REVIEWING GEOLOGICAL REPORTS

If the project involves tunneling or extensive excavations, it is prudent to employ a geotechnical engineer to work with and advise the estimators about the geology of the worksite. He should study all available geological reports and maps, examine the drill cores and jar samples from the investigative drilling program, and examine test excavations and outcrops of rock. Rock samples should be obtained for laboratory analysis of drillability. The geotechnical engineer should prepare a report which describes the nature, stratification, jointing, and faulting of the rock, the presence of water, and the effects of all these factors on the choices of methods and the progress of the work. In soft-ground work, he should forecast probable loads on tunnels and retained excavations. The report should include observations about similar work in adjacent areas if similar conditions are likely to prevail.

IV. TAKING OFF QUANTITIES

The quantity takeoffs should be started as soon as the plans and specifications have been received. As soon as any portion of the quantity survey is completed, this information should be distributed to the estimators who are preparing the construction schedule and those who are estimating the direct-cost items. All quantity takeoffs must be completed before direct-cost estimating can be finished. It is very desirable that each estimator participate in the takeoff of the bid items with which he will be concerned, or that he closely supervise the people who make this takeoff.

Prior to discussion of quantity takeoffs, it is necessary to properly define three terms: takeoff quantities, bid quantities, and final pay quantities. Each of these quantities may differ from the others for any item of work.

Takeoff quantities are quantities taken off the bidding plans and specifications by the estimator and represent to the contractor the most accurate estimated quantities of work to be performed.

Bid quantities are quantities printed on the bid comparison sheets. These quantities are furnished by the owner's engineer and are used for comparison of bids. The sum of the extensions of all bid quantities, each multiplied by the unit bid price, will equal the total bid submitted by each contractor.

Final pay quantities are the pay quantities of materials that are handled during construction of the project. The sum of the extension of all final pay quantities, each multiplied by the appropriate unit bid price, will equal the total payment to the contractor.

Accurate quantity takeoffs of all work quantities are required for the preparation of estimates for the following reasons:

1. To Establish Work Quantities for Lump-Sum Bid Items. These quantities must be determined so that they can be used for construction cost estimating. The specification writers often show bid quantities as a lump sum when the work required to construct the item is of a composite nature requiring the performance of many different tasks. Or, they may show bid quantities as a lump sum for work items that are of such a nature that work quantities are difficult to establish. Use of the lump-sum bid quantity, in this latter case, is to relieve the specification writers of responsibility and to force the contractor to accept the responsibility.

2. To Determine Work Quantities That Form Part of the Contractor's Cost but That Are Not Listed in the Bid Comparison Sheets. Depending upon the extent of the work items included in the bid schedule, quantity takeoffs are required to determine: the pounds of powder consumed per cubic yard of excavation, the nonpay cubic yardage of excavation in road construction, the estimated yards of overbreak excavation, the square footage of concrete forms, the form ratio, the square footage of concrete cleanup and finish, the tonnage of aggregates, the lineal footage of welding, the quantity of nonpay overburden in a borrow pit, etc.

3. To Check Bid Quantities. If takeoff quantities vary greatly from the bid quantities, this will influence the method of spreading the cost of the plant, equipment, indirect cost, escalation, and markup to the bid items to arrive at unit bid prices. The spread of these costs to bid items whose bid quantities are greater than takeoff quantities should be held to a minimum to permit the contractor to recover this cost.

4. To Determine Unit Cost More Accurately. An estimator may estimate the total direct cost of performing some bid item work and then establish unit cost by dividing this total cost by work quantities. If takeoff quantities are used as the divisor, the resulting unit cost will be the most accurate cost obtainable, since takeoff quantities are considered by the contractor to be more accurate than bid quantities.

V. REQUESTING QUOTATIONS ON PERMANENT MATERIALS AND SUBCONTRACTS

In preparing an estimate, little time or effort should be spent on estimating the cost of permanent materials or subcontracts. (An exception is the case of work to be performed so late in a long-term contract that no sources can be expected to submit firm proposals at bid time.) Usually these costs are put in the estimate using past prices as a guide. If the work is to be bid by a joint venture, the sponsoring partner often sends out these prices to be plugged into each partner's estimate, so that at the estimators' meeting, there will not be any time wasted in comparison of costs on these items. An example of the use of plugged-in prices is given in Chap. 18. Quotations on these items are received from suppliers and subcontractors shortly before the time for bid submission. At that time adjustments are made in the estimate and bid for the difference between these quotation prices and the prices used in the estimate. The procedure used in making these adjustments is described in Chap. 20.

The estimator should advertise for prices on permanent materials and subcontracts as soon as possible after securing the bidding documents so that suppliers and subcontractors are informed that the prime contractor is interested in securing their quotations.

Permanent materials may include cement, concrete aggregate, concrete, steel tunnel supports, turbines, generators, pumps, etc.

Subcontracting the work gives the advantage of fixing the cost of the work before bid submission, but it has the disadvantage of transferring direct control of the work from the prime contractor to the subcontractor. This can cause grave difficulty if the subcontractor's work is essential to day-to-day progress on the principal features of the project. For this reason, subcontracting is generally limited to specialty work or to work on features not directly affecting progress on the contract as a whole. Some work may be elective, in that subcontract proposals will be compared with the estimate of cost of performance by the prime contractor. Work commonly subcontracted includes drilling and grouting, electrical work, mechanical work, placing reinforcing steel, and spoil disposal for work performed in urban areas.

VI. COMPUTING LABOR RATES

Estimates are usually prepared by several engineers functioning as a team. In order that all participants use the same labor rates, these computations are usually one of the first tabulations prepared for any estimate.

The starting point for determining labor costs is to obtain copies of all pertinent labor agreements. These agreements will state the basic hourly wages, overtime rates, fringe benefits, and working rules for each class of labor. It is also necessary to know the federal and state payroll tax rates and the rates for workers' compensation and public liability and property damage (PL&PD) insurance. Premiums for these types of insurance are paid on the basis of labor hours or labor costs and hence are readily and properly included here.

To prepare the labor cost tabulation it is necessary to calculate for each class of labor the cost of hours worked at straight-time rates, at time-and-a-half rates, and at double-time rates. The elements of this cost are:

1. Basic hourly wages.
2. Overtime factors to be applied to the basic hourly wages.
3. Fringe benefits, such as payments to health and welfare funds, pension funds, apprenticeship training funds, etc. These payments are usually based on the number of hours worked, irrespective of whether the hours are straight-time or premium-time hours.
4. Subsistence or travel pay. These are usually paid (if applicable) on the basis of days worked, irrespective of hours. There may also be a one-time travel allowance for new hires.
5. Workers' compensation and PL&PD insurance are usually paid on the basis of percentages of the total wages paid, exclusive of overtime premiums and of payments for fringe benefits and subsistence.
6. State and federal taxes for social security and unemployment insurance are assessed on the basis of total wages paid, exclusive of payments for fringe benefits. These payments are not required after reaching a fixed annual level for each individual, but since labor turnover is generally quite high, this cutoff point is often disregarded in computing labor costs.

It is standard practice for many companies to build into the tabulated rates an allowance for unscheduled or incidental overtime work.

VII. TABULATING EQUIPMENT OPERATING COSTS

A tabulation of equipment operating cost should be started for a project as soon as the wage rates applicable to the project have been established, but it cannot be completed until all equipment selections have been made. Most estimating teams will designate one member to be responsible for the preparation of this schedule. The estimator can prepare a partial tabulation covering basic construction equipment, e.g.: front-end loaders, rear-dump trucks, drills, air compressors. This partial tabulation is distributed to each member of the estimating team; then, as more types of construction equipment are selected, they are added to the tabulation, and these revisions are distributed.

Continued revisions must be made until all direct cost items have been estimated and all plant and equipment units have been selected.

Equipment operating cost constitutes another basic cost tabulation that should be established and used by each member of the estimating team so that all work will have the same cost basis. Tabulation of equipment operating cost consists of listing each construction plant and equipment unit, and estimating each unit's average operating cost per hour, excluding ownership costs. The average operating cost includes maintenance labor; consumption of fuel, oil, and grease; cost of repair parts; and cost of tire repair and replacements. These costs are estimated by adjusting past equipment operating-cost records for changes in wage rates and for changes in supplies and tire prices. Most contractors exclude operating labor, but a few others include it. This inclusion or exclusion of operating labor results from the use of different cost-accounting procedures.

VIII. TRANSMITTING ESTIMATING INSTRUCTIONS

In the heavy construction industry, when the work is to be bid by a joint venture, it is the practice for the sponsoring company's chief estimator to inform all partners of the date, time, and location of the prebid estimators' and principals' meeting, and to send them estimating instructions and estimating data. Estimating instructions and data are distributed to simplify the comparison of estimates at the prebid estimators' meeting, which will free meeting time for comparisons of cost differences resulting from the use of different construction systems and evaluations of labor productivity.

These instructions should be sent to each partner as soon as the applicable wage rates and payroll burdens are available and after plugged prices have been established. Distribution should be made of most or all of the following data:

1. **Estimate Comparison Forms.**
2. **Descriptions of the Cost Divisions Shown on the Estimate Comparison Forms.** Estimating instructions should clarify sufficient detail to permit meaningful comparisons.
3. **Wage Rates, Fringe Benefits, and Payroll Burdens.** An easy method of transmitting this information is to distribute the wage rate tabulations described under estimating step VI.
4. **Insurance Rates.** The rates for insuring the contractor's construction equipment and plant facilities, for providing builder's risk insurance, and for securing any special insurance that is required during construction of the project should be included in the estimating instructions.
5. **Tax and Assessment Rates.** The tax and assessment rates for personal property taxes should be determined, as should the percentage of sales tax

and of any business tax. This information should all be included in the estimating instructions.

6. Bond Rates. These rates are normally supplied by the sponsoring company.

7. Power Cost and the Up-and-Down Charges. The power cost and the "up-and-down" charges for the installation of any required substation must be included. This information may have been established during the field-investigation trip or it can be secured from the power company that services the area.

8. Plugged or Bogie Prices. Estimating instructions should tabulate the plugged or bogie prices for permanent materials and subcontracted work.

9. List of Available Equipment. A descriptive list and the acquisition cost of any suitable available construction equipment owned by the sponsoring partner should accompany estimating instructions.

10. Request for Equipment Lists. All partners should be requested to distribute a similar listing and pricing of any suitable construction equipment they have available for use on the project.

11. Geological or Materials Engineering Reports. Any geological or materials engineering reports on the project site, quarry site, or borrow areas should accompany the instructions. If a consultant has been retained to examine any of these areas, a copy of his report should be distributed to each partner.

IX. SELECTING CONSTRUCTION METHODS

There are some jobs where the choice of construction methods is dictated by factors which leave little room for alternatives. An example might be a high concrete-arch dam on a small river in a narrow gorge in a remote area. The river diversion scheme will almost certainly require a diversion tunnel, and the terrain will almost certainly require the use of a cableway or possibly more than one cableway. Concrete aggregates will be produced from the closest available source, and concrete will be batched in a plant at jobsite.

Other jobs may involve more choices. An example is a long rock tunnel which might be driven by drill-and-shoot methods or by the use of a tunnel-boring machine. Another example is a concrete gravity dam in a wide valley near a city where ready-mixed concrete is readily available. There will then be several options for river diversion, hook service, and concrete supply. If there is adequate time and manpower for bid preparation it will be desirable to prepare separate estimates for each of the most attractive alternatives. This may entail only comparison of direct costs and plant and equipment costs, or it may further require preparation of separate schedules and estimates of indirect costs for each combination of alternatives under consideration. If there is not adequate time or manpower to prepare the alternative cost estimates, the estimating team will necessarily rely upon past experience to select the construction methods which probably best fit the work in hand, and proceed on this basis.

X. SELECTING PLANT AND EQUIPMENT

The basic choice of equipment will be dictated by the choice of construction methods. It may be that the bidder owns equipment which is suitable for some construction methods, and that his interest in the project is limited to the opportunity to utilize that equipment.

The construction methods which the bidder proposes to use may determine the type but not the capacity of the units of equipment. For instance, it may have been decided that a quantity of excavation and embankment is to be performed using motor scrapers. If the scraper spread is not already owned by the bidder, the estimator then has a choice of self-loading or pusher-loaded machines available in a wide range of sizes. The factors which will influence the choice of equipment in the scraper spread will be the volume to be moved, the time available for performance, labor rates (large machines usually result in low unit-labor costs), delivery time for the preferred equipment, equipment prices, and the market for resale of the preferred equipment when it is no longer needed.

XI. PREPARING A CONSTRUCTION SCHEDULE

The preparation of the construction schedule should be started simultaneously with the review of the plans and specifications. During this review, the bid date, anticipated award date, start of construction, completion date, climatic conditions, riverflow information, periods when concrete placement will be restricted because of climatic conditions, dates when different stages can be started, and dates when owner-supplied permanent equipment and material will be received should be plotted on the schedule.

Construction scheduling must be correlated with the selection of the types of construction equipment and plant units and with the selection of construction methods. The construction schedule is developed from the quantities of work to be performed and the rate at which the work is to be performed. If the available time is constrained by other factors, the schedule will demonstrate the sizes of crews and the number of equipment spreads which will be required, the need to work multiple shifts, or the need to work more than 5 days per week.

Scheduling construction is a continuing operation, since the schedule must be changed as new or additional work concepts are developed. For example, during direct cost estimating (estimating step XII), it may develop that changes or additions must be made to the construction equipment, which may necessitate changes in the construction schedule. Or it may be found that cost can be reduced if the schedule is rearranged. In this case it may be necessary to completely reschedule the work. For these reasons, construction scheduling is never finalized prior to estimate completion.

A typical bar-chart construction schedule is shown under estimating step XI for the sample estimate in Chap. 18.

XII. ESTIMATING DIRECT COSTS

A direct cost estimate is prepared for each bid item in the bid schedule. Costs are estimated for the takeoff quantity. After the costs for each bid item have been estimated, the total is adjusted to the bid quantity. This procedure is not necessary if all costs vary linearly with quantities, but is essential if there is a cost, such as the cost of a special form, which is fixed, regardless of quantities. In this case the actual cost will be recovered, even though the bid quantity is in error.

The starting point for estimating direct costs is the tentative decision, based on the preliminary schedule, about working hours. If it is apparent that shifts longer than 8 hours or more than 5 workdays per week are required, the work will be performed as scheduled overtime at the appropriate wage rates. If multiple shifts are required, and if there is a shift differential, either in wage rates or working hours, the appropriate wage rates are used. If no determination is possible at this time, it is common practice to estimate on the basis of a 40-hour week, and to make an adjustment when a decision about working hours can be made.

One bid item may include several subdivisions of direct costs. The costs of furnishing and erecting the steel frame for a powerhouse superstructure may be a good example. The costs will include purchasing the fabricated steel, installing the anchor bolts, unloading, sorting, erecting dry-packing base plates, and field painting. Each of these subdivisions will be estimated separately and then summarized.

The anchor bolts will probably be installed by a carpenter crew. The estimator will state the makeup of the crew—for instance, a foreman, six carpenters, two laborers, a crane oiler-driver, a flatbed driver, and one-quarter of the time of a crane operator—and will then compute the labor cost per hour. He will then state the production rate, for instance, 15 anchor bolts per crew hour. If 192 anchor bolts are to be installed, this will require 12.8 crew hours, which he will probably round off to 13 crew hours. The cost of direct labor will then be 13 times the hourly crew cost. The costs of repair labor and equipment operating invoice will be for one-quarter of 13 hours, or 3 hours for the crane and flatbed truck, 13 hours for the foreman's pickup, and perhaps 6 hours for a welding machine. The cost of supplies will be 192 times the estimated cost of the supplies necessary to space and hold one bolt in position until concrete is poured. The costs of direct labor, repair labor, equipment operating invoice, and supplies will be extended and tabulated on the estimating form.

An ironworker crew will unload the steel, shake it out, set masonry plates, erect and plumb the frame, and dry-pack the masonry plates. The estimator will proceed in the same manner as he did for the carpenter crew, establishing a crew size and makeup and production rates for each element of the ironworker activities, and extending and tabulating the costs.

The estimator will then turn to the cost of permanent materials, using the plug price for the structural steel. The plug price may be a unit price, for instance, $1.40 per pound, or it may be a lump sum price. The cost will be entered into the permanent materials column of the estimating form. He will then estimate the cost of dry packing materials, and enter that cost.

The estimator will then estimate the cost of field painting, assumed to be performed by subcontract, using the appropriate plug price, and enter this cost in the subcontract column.

Finally the estimator will add the direct labor and repair labor columns. With this as a basis, he will calculate the cost of small tools, perhaps 5 percent of labor costs, and enter this cost in the supplies column. He will then complete the addition of all columns and check that the sum of the column totals agrees with the sum of the totals for all horizontal lines. Finally he will adjust the costs, column by column, if the bid quantity differs from the takeoff quantity. If the bid quantity is a lump sum there will be no adjustment of the takeoff quantity.

It is essential that the estimator be familiar with the work assignments for different crafts if the work is to be performed by members of the building trades unions. Custom has dictated different assignments in different parts of the nation. The estimator should also realize that no workman has his tools in hand for 60 minutes per hour and should factor his production rates accordingly. It is reasonable to expect 45 to 50 minutes of productive work per hour in good weather, and fewer minutes per hour for outdoor work during inclement weather.

XIII. TABULATING DIRECT COSTS

As each estimator completes his work on the direct cost of each bid item, he delivers the totals to the lead estimator, who will have these costs entered column by column on the "spread sheets." Usually the spread sheets will be arranged so that there are subtotals for each group of related items (such as earthwork, concrete work, metalwork, mechanical work, and electrical work) to correspond with the format of the joint-venture comparison sheets.

XIV. ESTIMATING PLANT AND EQUIPMENT COSTS

The *gross plant and equipment cost* is the cost of providing the construction equipment and erecting the plant facilities that are used for constructing the project. The cost chargeable to each project is the *net plant and equipment cost,* or the *depreciation,* which is the gross cost less the net salvage value. The *net salvage value* is the estimated sales price of the plant and equipment at the time of project completion, less dismantling, shipping, storage, and selling expense.

The distinction made between construction plant and construction equipment was given earlier in this chapter. Since these definitions are extremely broad and since different definitions are used by many construction companies, plant and equipment costs are often added together and the sum used for the purpose of comparisons at joint venture meetings.

The performance of estimating step X, the selection of the construction plant and equipment, will establish a list of the major construction equipment and plant units required for the project and a list of all the quantities of work required to erect the plant facilities, construct the roads and parking areas, and build the service facilities.

Costing the construction plant and equipment can begin as soon as the first equipment is selected. Prices should be solicited from equipment manufacturers so that cost will be based on the latest prices. If possible, it should be determined if the quotations received from the suppliers were for estimating purposes, and whether they can be reduced by a percentage to arrive at actual *buy out* prices, or prices paid at the time of equipment purchase.

The plant and equipment cost should be tabulated on a form that includes identifying descriptions, required number of units, weights, sales tax, freight charges from factory to delivery point, transportation costs from receiving point to jobsite, labor costs for assembling equipment units and erecting plant facilities, invoice consisting of plant materials and consumable supplies, used equipment purchase prices, new equipment purchase prices, total costs, net salvage values, and net job cost. A typical tabulation of this kind is shown in Chap. 19.

Plant and equipment estimating cannot be completed until all direct cost items have been estimated, since, during the performance of this work, it may develop that different types of equipment or additional equipment units and plant facilities may be required, or a more economical construction system may be adopted. Also, direct cost estimating may indicate a need for a change in construction scheduling, which in turn may change plant and equipment requirements.

Depending on each construction company's accounting procedure, the purchase of sedans, pickups, office equipment, surveying equipment, and first-aid equipment is either charged to plant and equipment cost or to indirect expense. Accounting practice also varies in regard to how rock-quarry development cost is handled. This cost can be carried as part of the construction plant and equipment cost, or it can be placed in a quarry clearing account and cleared out of this account at an applicable rate per cubic yard of rock produced.

Both the number of years that the equipment will be in service and the number of hours it will work should be considered when setting equipment salvage value. The maximum equipment life of construction equipment such as trucks, tractors, and scrapers is often set at 5 years, or 10,000 hours of service. The usable life of large equipment units such as mixing plants, aggregate equipment, cableways, gantry cranes, and large shovels is comparatively much longer, but it is dependent on whether there will be construction projects that require this type of equipment at the time of project completion.

Several publications are available to aid in determining the salvage value of secondhand construction equipment. One of these is the *Green Guide,* published by Equipment Guide Book Company in Palo Alto, California. If allowances are not made in the salvage value for the cost of dismantling the equipment, moving it out, general job cleanup, and projected cost of storage until resale, a separate item containing this cost should be added to the estimate.

If part of the construction work is done with nonoperated, nonmaintained rental equipment, the cost for renting this equipment should be tabulated on a separate schedule and shown as a separate sum in the estimate summary. Some estimators include this equipment rental in direct cost, but this has the disadvantage of distorting direct unit cost so that it will be unsuitable for use in future estimating. Other estimators mix this rental cost with plant and equipment cost, but this does not follow good accounting practice since rental cost is not a capital expense, and if handled in this manner, distorts the relationship of plant and equipment acquisition cost to salvage value. This may also confuse company management when they review this relationship during markup determination.

XV. ESTIMATING INDIRECT COSTS

The estimate of indirect cost must be based on the total construction time, the length of time that each subsystem will be operated (e.g., excavation subsystem), the complexity of the project, the number of workers that will be employed, the number of shifts that will be worked, the number of days worked per week, the extent of the work that will be subcontracted, the contract volume, the application of computers, the efficiency of the engineering and office staff, and the complexity of the cost records and reports that must be submitted to the owner and to the contractor's main office. The majority of the foregoing information can be obtained from the construction schedule and other sections of the estimate.

The indirect cost estimate can be started as soon as the construction schedule is established, but it cannot be completed until the contract volume is available for bond rate calculating, and the value of the plant and equipment is available for computations of insurance costs.

The following are listings and brief descriptions of the construction cost items that belong in indirect cost. These costs are separated by accounts to provide a checklist so that items of expense will not be neglected. Many construction companies use printed estimate forms listing all these items; the forms serve as checklists and may be more extensive than this list.

1. Job Management.   This includes the salaries and payroll burdens of the project manager, his staff, and secretaries.

2. Job Supervision.   The wages, salaries, and payroll burdens of the general, shift, and major craft superintendents, and of their staffs and secretaries are placed in this account. The personnel charged to this account should be held to a minimum.

3. Engineering Labor.   The wages, salaries, and payroll burdens for the project engineer, office engineers, draftsmen, field engineers, and surveyors should be charged to this account.

4. Office Labor.   This category includes the wages, salaries, and payroll burdens of the administrative manager, accountants, payroll clerks, timekeepers, personnel manager, and other clerical labor.

5. Purchasing and Warehouse Labor.   The salaries, wages, and payroll burdens of the purchasing agent, chief warehouseman, warehousemen, and if a service truck is used, driver wages should all be charged to this account.

6. Safety and First-Aid Labor.   The salaries and payroll burden of the safety engineer and any required nurses are placed in this account. The specification requirements determine if it is possible for the safety engineer to have other duties. This will determine whether all or only part of his time is chargeable to this account.

7. Janitorial and Security Labor.   The wages and payroll burdens of janitors and job watchmen are chargeable to this account.

8. Operation of Pickups and Sedans.   This account includes the cost of operation and maintenance of automobiles and pickups that are required on the project, except for any portion of these costs which are included in direct costs.

9. Office Supplies.   The cost of postage, stationery, office forms, and all other types of office supplies are charged to this account.

10. Computer Cost.   This account covers the cost of renting computer time from the home office or from an outside agency.

11. Telephone, Telegraph, and Cable.   This account covers telephone and telegraph charges. Most estimators do not allow sufficient funds to cover this item of cost.

12. Office Rent.   This account is only used when applicable.

13. Heat, Light, and Power for the Office.

14. Engineering Supplies.   The cost of engineering supplies for the office and field, including surveying supplies, comprises this account.

15. Office Equipment.   The cost of office equipment may be charged either to this account or to plant and equipment. This should be the net cost to the job of the office equipment provided. The net cost is first cost less salvage, or the rental cost of this equipment.

16. Engineering Equipment.   This is another cost item which may be carried either in indirect cost or in plant and equipment. It is the net cost to the job of engineering equipment.

17. First-Aid and Safety Supplies.   This account should receive the cost of first-aid and safety supplies. On some jobs, the industrial compensation insurance company may provide first-aid services, and as a result, it is not a separate item of expense to the contractor but rather is included in the workers' compensation rates.

18. Entertainment and Travel Expense.   The cost of entertaining the client and other visitors should be included in this account. This cost also covers the expense connected with any traveling done by supervisory personnel.

19. Blueprints, Copies, and Photographs.   This account covers the cost of making prints and copies and taking job pictures.

20. Outside Consultants.   This account provides funds for hiring outside legal or engineering help to assist on the job or to help present claims.

21. Legal Cost.   This covers the cost of any legal help required by the project.

22. Audit Cost.   This covers the cost of yearly audits performed by outside auditing firms.

23. Testing-Laboratory Expense.   This cost account is used, when applicable, to accumulate charges for the testing of concrete or other items performed by independent laboratories.

24. Job Fencing.   Many work areas are fenced in preference to hiring watchmen and guards. Fencing could also be listed under plant costs.

25. Mobilization of Personnel.   This account covers the cost of moving key personnel and their dependents and possessions to and from the jobsite.

26. Licenses and Fees.   This account covers the cost of licenses and fees required by the project.

27. Sign Cost.   This is intended to cover the cost of erecting both the owner's and the contractor's signs. For example, the Army Engineers Department always requires the contractor to erect a sign designating that it is an Army Engineers' job.

28. Drinking Water.   The cost of distributing iced drinking water to the construction personnel should be charged to this account.

29. Chemical Toilets.   The rent of portable toilets is charged to this account.

30. Miscellaneous Cost.   Any item of cost not defined above is included here.

31. Insurance.   This account covers the cost of insuring the plant and equipment, builder's risk insurance, vehicle insurance, individual bonding, and any special insurance required. A special insurance cost occurs when construction is carried near railroad tracks. In this case, the railroad often requires bonds or insurance to be furnished by the contractor.

32. Taxes.   The cost incurred by the assessment of county and city property taxes on the contractor's plant and equipment is covered by this account. Some states have an additional business tax and other special taxes. Sales tax should not be included in indirect cost, but should be added to the cost of plant and equipment and to the direct cost of materials and supplies.

33. Bond Premium.   In the majority of jobs, the contractor is required to post bonds to guarantee the successful completion of the job and to guarantee payment for all material and labor utilized for the job. Typical bond premium rates appear in Table 17.1.

The tabulated bond premiums are based on a contract period of under 24 months. If the contract covers a longer period of time, a charge by the bonding companies of 1 percent per month of the total premium is added to these rates. Bond premiums are computed on the total estimated contract price and are adjusted to reflect the final contract price at completion.

TABLE 17.1   Typical Bond Premium Rates

| | Bond Premium | |
Contract Price	Association Group of Bonding Companies	Nonassociation Group of Bonding Companies
First $500,000	$12.00/1,000	$7.50/1,000
Next 2,000,000	7.25/1,000	5.25/1,000
Next 2,500,000	5.75/1,000	4.50/1,000
Next 2,500,000	5.25/1,000	4.00/1,000
Over 7,500,000	4.80/1,000	4.00/1,000

XVI. ESTIMATING CAMP COSTS

There are two types of construction camps, married camps and single-status camps, which are used for different purposes. Married quarters are provided to house the contractor's key supervisory, office, and engineering personnel. Single-status camps are provided to house the remainder of the working force. At very remote areas, both types of facilities are required. Since contractors often lose money on camp operations, single-status camps are only provided when their use is a necessity. On the other hand, contractors often provide married-quarter camps on nonremote locations in order to attract desirable personnel and to enable supervisory personnel to live near the job-site and be available when any construction problems arise.

The simplest method of providing married quarters is to purchase house trailers. These are subsequently rented at nominal rates to the key personnel. Camp size will vary from one unit for the project superintendent to enough units to house every key man. In addition to the trailer cost, the cost of constructing the camp includes clearing the site, grading, providing water-supply facilities, providing a sewage-disposal system, and furnishing and distributing electricity. At remote areas, either school facilities and instruction or the furnishing of daily transporation from campsites to schools for the children of the camp occupants may be required.

The facilities provided for a single-status camp consist of sleeping quarters, washrooms, mess hall, and may include a clubhouse. Local regulations should be checked to determine the number of men that are permitted to occupy one room and the square footage of room area that is required for each man. Locked storage for personal articles and clothes is a necessity. It is now possible to secure quotations on prefabricated single-status camps, completely erected, including mess hall, clubhouse, sewage disposal, water supply, heating, and electrical distribution. The prime contractor must furnish the graded camp location to the camp subcontractor. The advantages in using this type of camp are that it has good salvage value at the end of the project and the cost of construction can be definitely determined before bid submission.

Single-status camp operation is often subcontracted to a caterer who will contract for a price per man-day to supply meals and run the camp. Prices for catering service can be secured before bidding and this establishes the operating loss before bid submission. When a caterer is used, the camp operating cost will be the difference between the caterer's per diem charges and the per diem revenue received from the men for staying in the camp.

The field investigation will determine whether a married-quarters camp and/or a single-status camp are required. If camps are required, then quotations should be requested for house trailers, for the furnishing and erection of single-status trailer camps, and for the operation of these camps. Firm quotations cannot be received for furnishing and operating single-status camps until the estimated occupancy can be determined. Since the prime contractor often provides camp space for subcontractors' personnel, their demands on the camp facilities must also be established. In addition, some project specifications require that the prime contractor provide camp facilities for the owner's engineers.

The total number of employees and subcontractors' personnel cannot be estimated until the direct cost, the plant and equipment cost, and the indirect cost have been estimated. When these tasks are completed, the number of construction men for each time period can be determined by using the construction schedule and the estimated labor cost to make a labor-time distribution. This labor distribution is then divided by the average man-day cost to determine the prime contractor's work force. Each subcontractor's labor cost must be estimated and his labor cost distributed so that his total work force can be determined. This total work force can be reduced by the number of men that will be housed in married quarters. A judgment must be made to determine what percentage of this remaining work force will wish to stay in a single-status camp. This is a difficult number to establish, since construction personnel often prefer to drive long distances to work rather than live in single-status camps. To this total, then, must be added any owner's engineering personnel that must be housed.

As soon as this estimated camp occupancy has been established, it should be transmitted to the subcontractor who is quoting on the camp construction and to the caterers, so that they can firm up their quotations. The caterers' firm quote will be a cost per man-day that varies with the camp occupancy. The prime contractor's estimator must base his estimate of camp operation on an estimated occupancy and use the applicable man-day quote from the caterer. This can never be accurately estimated, so for any project there will always be differences between estimated cost and actual cost for camp operation.

XVII. ESTIMATING ESCALATION

Escalation computations cannot begin until all of the preceding estimating tasks have been completed, since these computations are based on the total project labor cost and permanent materials cost.

Estimates are based on wage rates and material prices applicable at the time of bidding. To determine the true cost of constructing the project, the increase in cost that occurs during its construction must be determined. These cost increases are called *escalation*. The basic data required for escalation computations are: the time of year that labor rates will increase; the labor expenditures that will be made in each time period, starting at the time when labor rates increase; and the percentages that labor cost will increase for each period. The escalation cost is computed for each period by multiplying the unescalated labor cost by the percentage of increase for that period. The sum of these increases will equal the total labor escalation. A typical labor escalation computation is shown in the Denny Dam estimate in Chap. 19.

The percentages of labor cost increases are established by reviewing labor contracts to determine when the labor increase will occur and through what date future wage increases have been negotiated. For the periods that are not covered by existing labor contracts, it is necessary to estimate the anticipated labor increase. The trend of labor increases in the area for each craft should be reviewed, and future wage increase predictions should reflect this past information; predictions should be influenced as well by overall wage increases throughout the country. The total project labor can be distributed over yearly periods by using the construction schedule to determine what work will be performed in each period.

Material and subcontract escalation cost is often very minor since many quotations will be firm for the life of the project. It cannot, however, be determined prior to receipt if the material and subcontract quotations will be firm. These final quotations are not received until shortly before bid submittal time. Therefore, escalation costs must be determined at this late date. These computations are performed using the same procedure as that used for labor escalation.

XVIII. TABULATING TOTAL COSTS

After all construction costs have been estimated, they should be summarized in a form that is consistent with company practices and one that is readily understood by the construction company's principals. Labor, supplies, permanent materials, subcontracts, and plant and equipment purchases should be listed separately to simplify markup determination.

XIX. ANALYZING CASH FLOW

The construction company's management cannot properly evaluate the markup that should be placed on the estimated cost to arrive at the bid price unless they know the amount of capital that must be invested in the project, the length of time this capital will be required, and the interest applicable to this invested capital. Therefore, after the estimate is completed, a cash forecast should be prepared to establish capital requirements and interest expense.

The contractor must invest capital in a project to provide funds for the purchase of construction plant and equipment; to buy bonds, insurance, etc.; to pay for several months of indirect expenditures; and to finance early expenditures for the project, since the contractor will not receive any large progress payments from the owner at the start of construction.

When the specifications provide for the prepayment of the contractor's expenditures for plant and equipment or include an item in the bid schedule for payment of the contractor's mobilization cost, it is possible for the contractor to receive prompt recovery of some of this expense and thus reduce his capital investment.

The method used for determining the amount of capital that must be invested in a construction project, and the amount of interest on this capital, is to prepare a cash forecast showing the cash expenditures and receipts during the project's construction.

A preliminary cash forecast can be made by totaling the following items and assuming that cash will be required for 50 percent of the contract time.

Preliminary Cash Forecast

Plant and equipment outlay necessary to start the work   _____
Cost of payment and performance bond   _____
First year's insurance premiums   _____
Inventory   _____
Camp construction cost   _____
Cost of any bid item on which payment is to be deferred for any
   reason   _____
Overhead required to start job before any bid item revenue
   is received   _____
Working capital
   Total outlay   _____
Less any plant and equipment prepayment or mobilization pay-
   ment   _____
   Total cash requirement   _____

A more detailed and more accurate cash forecast is prepared by estimating and listing the amount of cash revenue that will be received in each period and the cash expenditures in each period for direct cost, plant and equipment acquisition, camp construction, and indirect cost. Time increments of months, quarters, and halves of a year are used, depending upon the accuracy required and the total length of the construction period. A cash forecast cannot be accurately prepared until the estimated cost and the completed construction schedule are available. To arrive at the total revenue, the estimator must assume a percentage of markup which, when added to the total estimated cost, will give the total revenue. The simplest method of determining the bid revenue per time period is to multiply the direct cost by a factor found by dividing the total direct cost into the estimated total bid price. A more accurate method of determining period revenue would be to spread the plant and equipment to the direct cost items, which often gives a higher factor to use on the excavation items which have early completion, and a lower one to use on the concrete items which are completed later in the job. This type of spreading will reduce the cash requirements. When the contract provides for owner retention, the estimates of earned revenue should also show cash revenue which would be increased by the retained percentage when received after project completion and acceptance.

Receipt of revenue will lag from 30 to 45 days behind the completed construction of pay quantities, because progress payment requests are submitted near the end of each month and because additional time is required by the owner to check the progress payment estimates before making payment.

The Denny Dam example in Chap. 19 shows a typical cash forecast. This cash forecast was prepared using 4-month periods, to simplify its presentation in this book.

When a cash forecast is prepared for a job that is already under construction, it is necessary to vary the form to include additional items such as accounts receivable, accounts payable, deferred charges, and miscellaneous revenue in order to tie into the project's balance sheet and profit and loss statement. Engineers should understand enough accounting to use these statements to establish a starting point for a cash forecast.

XX. MAKING AN IN-HOUSE REVIEW

After the estimate is otherwise complete, and before meeting with the joint venture partners, there will be a comprehensive review of the estimate. Of course, if the work is not to be performed by a joint venture, the in-house review will be the final review prior to determining markup and prior to adjusting plug prices to reflect the quotations actually received for permanent materials and subcontracts.

The in-house review will usually be conducted by the chief estimator with the assistance of the lead estimator. Each estimator will attend, at least during those parts of the review which pertain to his work. The company officers responsible for bidding and for operations will attend, and other people, such as equipment manager and the insurance manager may be called in for consultation.

The meeting will usually start with a briefing by the lead estimator to describe the project for the benefit of those members of management who are not completely familiar with it. General arrangement drawings and other exhibits will be available for this purpose. The lead estimator will then describe in general terms the construction methods and the key items of construction equipment to be employed and the order of work. Plant layout drawings, schedules, and other exhibits will be on hand for this purpose.

The participants in the in-house meeting will have been furnished copies of the spread sheets and copies of the quantity comparisons. They will then review the cost estimate by groups of related items, or item by item when this appears to be necessary. Each estimator will be called upon to explain his work, particularly his assumptions about production rates and equipment selections. There will be careful consideration of indirect costs, plant and equipment costs, and escalation calculations.

During the course of this review the estimators will be informed about the changes which management deems necessary in order to have an adequate but competitive cost estimate. After these changes have been made and entered on a new set of spread sheets the meeting may be reconvened to determine whether further changes are required.

Special Considerations

The lead estimator will have prepared for his principals an abstract of the plans and specifications, listing the principal items of work (and pointing out any major discrepancies in quantities), listing critical dates established by the bid documents, and noting any unusual requirements for performance of the work. General layout drawings will be reduced to a size convenient for inclusion in this package. The synopsis will also state the amounts of liquidated damages which may be assessed, the requirements for minority employment, the requirements for employing minority business enterprises, bonding and insurance requirements, weather and streamflow records, and other information which management will need in order to evaluate the job and to decide upon a proper markup.

The remaining information that the construction company's management should have for markup determinations is the contractor's exposure to contingency items. A *contingency* can be defined as a condition that may or may not occur. If the condition occurs, its extent and duration may be so variable that the cost to the contractor resulting from this condition will vary from a minor to a maximum amount. This cost exposure is often not accurately predictable but the contractor's exposure should be computed, but not included in the summary of estimated cost; instead, it should be listed separately so that the construction company's principals can evaluate its effect on the project markup.

The contractor's exposure to contingency items will be lessened if the specifications contain force-majeure and unforeseen-condition clauses, provided specification restrictions do not restrict their application. The contractor can also reduce his exposure to contingency items by taking out builder's risk insurance and by employing a competent structural geologist.

When the contractor is exposed to contingency items, the risk often arises in one or more of the following areas:

1. Unforeseen Foundation Conditions. The extent of the contractor's exposure to added cost is dependent on the amount of foundation exploration performed by the owner. This exploration is done with drill holes, tunnel adits, open-cut trenches, seismographic investigations, and by geological mapping. Unforeseen conditions may result in deeper excavation, which will result in increasing concrete and excavation quantities, necessitating a longer construction period. When insufficient foundation exploration has been done, the owner may require the contractor to remove the excavation in shallow layers and do exploratory cleanup after each layer is removed, which greatly increases the contractor's excavation cost.

Another cost exposure that cannot accurately be predicted is the amount of confined and dental excavation that may have to be performed. If the final rock surface is very irregular, the cost of foundation preparation is increased because more hand placement and compaction of the initial fill is required. Slides may be encountered, introducing more work quantities and delaying project completion; artesian water may be encountered; large flows of water may come in through the sides of the excavation, etc.

2. Unexpected Floods. This exposure increases with the size of a river's watershed.

3. Unforeseen Inclement Weather. The exposure to extra cost is encountered because the working season may be shortened by many days of rain or freezing weather.

4. Extra Labor Cost. This may be classified as a contingency item when it is indeterminable whether competent labor is available, if there is a possibility that future wage increases will be larger than normal, or if there is a possibility that extra overtime must be worked to attract labor.

5. Concurrent Projects. Equipment or supervision may not be readily available because of similar jobs being bid or under construction.

6. Locale. The job may be in so remote an area that there will be difficulty in securing competent workers, maintaining access, and supplying camp and living accommodations.

7. War, Strikes, Earthquakes, etc. The maximum cost exposure resulting from these occurrences is beyond estimating ability. When the contractor is subject to this exposure, it should be brought to the principals' attention so that they can make their own determination of how it may affect the markup.

8. Escalation. There is a risk on projects of long duration that the cost of labor, fuel, and other supplies, materials, and services may escalate far in excess of reasonable expectations. The bidder may be able to pass these risks on to his suppliers and subcontractors when orders are placed and subcontracts are written, but the quotations from those companies may provide that the prime contractor assume the risks. Some owners may alleviate this situation by incorporating in the contract documents provisions which protect the contractor in whole or in part from unforseeable escalation.

Chapter 18 Example of a Tunnel Estimate

Presented in this chapter is an example of a contractor's estimate of cost for constructing a fictitious hard-rock tunnel by the conventional excavating method of drill, shoot, and muck. If this estimating procedure is thoroughly understood for the conventional tunnel-driving method, then the necessary changes in procedure can be readily made to accommodate any other driving method or any other type of underground construction. Each step in this estimating procedure has the same number as its description given in Chap. 17; frequent reference to that chapter will help in following the estimate. A knowledge of tunneling terminology and of the methods and equipment used in tunnel construction is also necessary; this information is presented in earlier chapters and in the glossary.

In preparing this estimate, labor extensions for each individual tunnel crew were made by using wage rates, working conditions, and fringe benefits in effect during the first half of 1982. This procedure can be changed to eliminate some of the detailed computation. After the total daily crew cost for the first excavation crew and the first concrete crew has been determined, an average cost per manshift can be computed for excavation work and for concrete work. The cost of any additional excavation and concrete crews can then be determined by multiplying the number of men required per day times these average costs per manshift. This change in procedure will not result in decreased estimating accuracy.

The estimate concentrates on underground construction. Therefore, all work except tunnel and shaft work is shown as done by others. Plugged material and subcontract prices were used in the estimate, as shown in Table 18.4. These prices are adjusted to actual quotes at the time of bid submission.

492

I. STUDY OF PLANS AND SPECIFICATIONS

This study was completed and an abstract of the specifications, showing the major contract and specification requirements as well as reduced prints of the job layout and the shaft and tunnel sections, was prepared. (See Figs. 18.1 and 18.2.)

Abstract of Specifications for the Construction of Sierra Tunnel

SCOPE	The work includes the excavation and concrete lining of 23,650 ft of 21-ft-finished-diameter horseshoe-shaped tunnel descending on a 0.1 percent grade from the inlet portal. Also included is the excavation, concrete lining, and first-stage concrete in a 24-ft-finished-diameter gate shaft 214 ft high, which is 200 ft from the inlet portal.
LOCATION	The site of the work is on the South Fork of the China River, 8 miles northeast of Austin in Arizona County, California.
AGENCY	Arizona County Water District, Austin, California.
ENGINEERS	Tunnel Designers, Inc.
BID DATE	May 1, 1982.
BID PROCEDURE	Proposal must be submitted on a serially numbered standard proposal form bound into the specifications specifically issued to the prospective bidder.
	Other than the bid bond, none of the required documents may be detached from the bound book.
	If the proposal is signed by an agent other than an officer of the corporation authorized to sign contracts on its behalf, a power of attorney must be on file with the District prior to the opening of bids or submitted with the proposal.
	Bids shall be submitted in a sealed envelope plainly marked as a bid. The cover shall state the name of the project and the date and time for opening of bids.
DATA TO BE SUBMITTED WITH BID	1. Bid bond, 10% of the contract price.
	2. List of subcontractors performing work in excess of ½ of 1% of the contract price.
	3. Fair Employment Practices certification.
	4. Workers' compensation insurance certification (can be either submitted with bid or executed later by the successful bidder).
	5. Signed copy of each addendum.
BONDS	Bid bond, 10% of bid amount.
	Performance bond, 50% of bid amount.

AWARD

TIME OF COMPLETION

LIQUIDATED DAMAGES

EXTENSIONS OF TIME

PAYMENTS

Labor and materials bond, 50% of bid amount.

Award will be made within 30 days after opening of bids.

The contractor shall begin work within 30 days after receipt of notice to proceed and complete all work in 42 months.

$500 a calendar day for each day after 42 months.

Contractor shall be entitled to an extension of time for completion of the work for delays due to:

1. An order for changes or a suspension of the work for the convenience and benefit of the District.
2. A suspension of the work due to weather conditions such as earthquake, fire, flood, cloudburst, or other cataclysmic phenomena of nature.
3. Failure of the District or its contractors to furnish, within contract time requirements, access to the work, right-of-way, working areas, excavation at portals, drawings, materials, water, or power for which the District is responsible under the contract.
4. Survey error by the District.
5. Changed conditions, i.e.:
   a. Subsurface or latent physical conditions at the site of the work differing materially from those represented in this contract; and
   b. Unknown physical conditions at the site of the work of an unusual nature differing materially from those ordinarily encountered and generally recognized as inherent in work of the character provided for in this contract.
6. Act of the public enemy, governmental act other than an act of the District, epidemic, freight embargo, strike, or labor dispute.

Progress Payments. Monthly payments in an amount equal to 90% of the value of the work performed shall be made to the contractor for the first 50% of the contract amount; thereafter monthly payments shall be made in full. The amount of money retained on the first 50% of the contract shall be paid upon completion of the work.

Force Account Payment. Force account payments for any change shall consist of the actual necessary costs of labor, materials, and equipment and construction equipment used in the portion of work materially affected by the change, plus an allowance on the sum of such costs for superintendence, general expense, and profit determined in accordance with the following schedule:

Sum of Actual Necessary Costs	Allowance
Less than $25,000	16%
$25,000–$100,000	13%
Over $100,000	10%

Changed Conditions. If changed conditions are encountered, an equitable adjustment shall be made in the contract revenue.

LINES AND GRADES

The engineer shall establish centerlines and grades on original ground and cross section of excavation sites prior to excavation and shall provide survey services required for this purpose. The engineer shall also check line and grade in the tunnel.

WAGES

Prevailing wages are contained in the specifications.

ACCESS

Access roads to the inlet and outlet portals, to the adit, and to the top of the gate shaft have been constructed by a previous contractor.

WORK DONE BY OTHERS

Other contractors will do the following:

1. Construct access roads to the inlet and outlet portals, adit location, and the top of the surge tank. Roads will be complete by April 1, 1982.
2. Run power distribution and water distribution lines to these locations. Distribution lines will be complete by April 1, 1982.
3. Complete all opencut excavation at the portals, adit, and surge tank location. This work will be complete by April 1, 1982.
4. Construct inlet and outlet structures after tunnel contract is completed.
5. Place second-stage concrete, gates, hoist, ladders, platforms, control house, and air vent pipe (that is not embedded) for the gate shaft after tunnel contract is completed.

MOBILIZATION PAYMENTS

None mentioned in specifications.

FIRST-AID FACILITIES

First-aid equipment and licensed first-aid attendants are required at each portal used in construction.

SAFETY ENGINEER

The contractor must have a competent and experienced safety engineer.

INSURANCE

The contractor shall be responsible for any liability imposed by law for any damage to any person or property resulting from any cause whatsoever during the progress of the work. Therefore, he must carry insurance on this exposure.

TAXES

The contract price paid for the work shall include full compensation for all taxes which the contractor is required to pay, whether imposed by the federal, state, or local government.

POWER AND WATER

Distribution lines supplied by the owner. Contractor to make own arrangements for cost of power and water.

METHOD OF CONSTRUCTION

The contractor may drive the tunnel from either or both portals or may, at his option, construct a 650 lin ft adit to aid in his tunnel driving. The adit is 10,750 ft from the inlet portal. If contractor elects to construct the adit, he must place a 50 lin ft concrete plug in it upon completion of his work. There will be no separate payment for constructing this adit or placing the concrete plug; so the cost of this work must be absorbed by the contractor in bid-item work.

MUCK DISPOSAL

Adequate space for muck disposal is located at each portal and at the adit, and no special treatment of disposal areas will be required.

INVERT CLEANUP

Cleanup of the tunnel invert does not have to be carried to bare rock. All soft material must be removed from the invert, but sound tunnel muck can be leveled to grade as foundation for invert concrete.

CONCRETE POUR

Use of air placers is permitted.

MAJOR QUANTITIES

Tunnel excavation	431,000 yd³
Steel sets	2,290,000 lb
Timber lagging	400 MBM*
Roof bolts	12,000 lin ft
Shaft excavation	5,300 yd³
Concrete in tunnel lining	111,000 yd³
Concrete in shaft	1,900 yd³
Cement	163,000 bbl
Reinforcement in shaft	285,000 lb
Consolidation grouting	10,000 sacks

II. JOBSITE INSPECTION

An inspection of the jobsite was made. See Chaps. 17 and 19 (estimating step II) for typical information sought and gained.

* MBM is a thousand board feet of measurement.

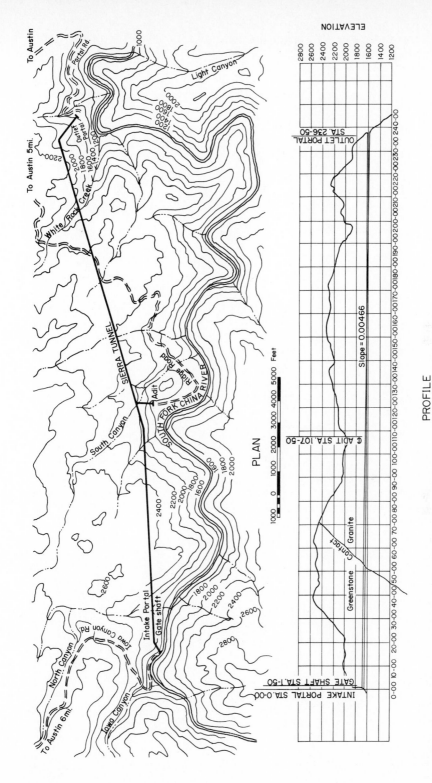

Figure 18.1 Sierra Tunnel general arrangement.

497

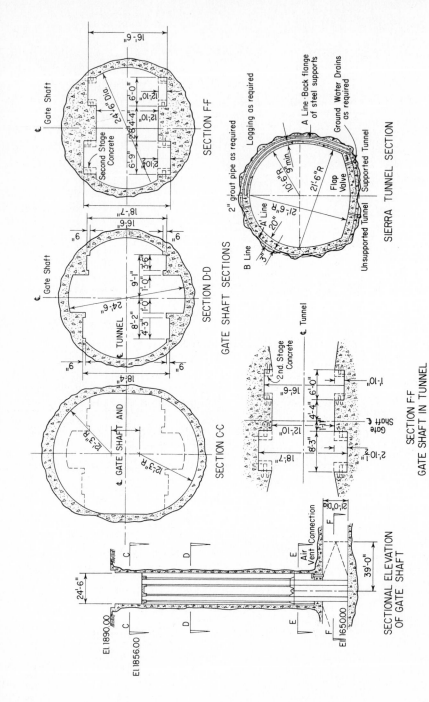

Figure 18.2 Sierra Tunnel gate shaft and tunnel sections.

498

III. GEOLOGICAL REVIEW

A review of the geological report indicated that the first 50 ft of tunnel from each portal must be supported on 4-ft centers.

The gate shaft is in sound granite and will not require supports.

The adit is in sound granite and except for the first 50 ft will not require supports.

The report stated that the outlet portal and adjacent 18,500 lin ft of tunnel will be in granitic formation and that the inlet portal and adjacent 5,150 lin ft of tunnel will be in greenstone. About 10 percent of the granite will require supports on 4-ft centers, and the remainder will be unsupported tunnel except at the contact with the greenstone. For 50 ft on each side of the contact, supports on 2-ft centers will be required, and for 20 ft on each side of the contact, breast boarding will be required. Except for the 50 ft near the contact and the 50 ft from the portal, only 30 percent of the greenstone will require supports and these supports will be on 4-ft centers. An additional 30 percent of the tunnel in the greenstone will require roof bolts in the arch.

This information is summarized in Table 18.1. Figures are rounded out to fit support spacing.

TABLE **18.1   Sierra Tunnel Estimate: Data on Supports**

	Total (lin ft)	Bare Rock (lin ft)	Roof Bolts (lin ft)	Supports on 4-ft Centers (lin ft)	Supports on 2-ft Centers (lin ft)	No. of Sets
Main tunnel:						
Granitic formation	18,500	16,570		1,880	50	495 ea
Greenstone	5,150	2,012	1,520	1,568	50	417 ea
Subtotal	23,650	18,582	1,520	3,448	100	912 ea
Adit	650	600		50		12 ea
Shaft	214	214				

Table 18.2 shows tunnel lengths of the same round spacing and groups them into the two rock classifications. This kind of summary was helpful in computing the excavation crew days after the round calculations were completed.

TABLE **18.2   Sierra Tunnel Estimate: Length of Rounds by Rock Formation**

	Toward Outlet	Toward Inlet	Total
Granite formation:			
Breast boarding		20 lin ft	20 lin ft
2-ft rounds		30 lin ft	30 lin ft

TABLE **18.2   Sierra Tunnel Estimate: Length of Rounds by Rock Formation**   (*Continued*)

	Toward Outlet	Toward Inlet	Total
4-ft rounds supported at portal	50 lin ft		50 lin ft
4-ft rounds supported	1,280 lin ft	550 lin ft	1,830 lin ft
10-ft rounds unsupported	11,570 lin ft	5,000 lin ft	16,570 lin ft
Total granite formation	12,900 lin ft	5,600 lin ft	18,500 lin ft
Greenstone formation:			
Breast boarding		20 lin ft	20 lin ft
2-ft rounds		30 lin ft	30 lin ft
4-ft rounds supported at portal		50 lin ft	50 lin ft
4-ft rounds supported		1,518 lin ft	1,518 lin ft
10-ft rounds with roof bolts		1,520 lin ft	1,520 lin ft
10-ft rounds unsupported		2,012 lin ft	2,012 lin ft
Total greenstone formation		5,150 lin ft	5,150 lin ft
Grand total	12,900 lin ft	10,750 lin ft	23,650 lin ft
Total number of rounds	1,490	1,413	2,903

IV. SUMMARY OF QUANTITY TAKEOFFS

TABLE **18.3   Sierra Tunnel Estimate: Quantity Takeoffs**

		Quantities per Lineal Foot			Total Quantities		
	Lineal Feet	Pay (yd³)	Overbreak (yd³)	Total (yd³)	Pay (yd³)	Overbreak (yd³)	Total (yd³)
Excavation:							
Main tunnel							
Standard section	23,572	18.18	1.85	20.03	428,539	43,608	472,147
Under gate shaft	78	29.52	2.59	32.11	2,303	202	2,505
Total main tunnel	23,650				430,842	43,810	474,652
Shaft					5,299	496	5,795
Adit					4,472	Not used	4,472

TABLE 18.3   Sierra Tunnel Estimate:
Quantity Takeoffs   (*Continued*)

		Quantities per Lineal Foot			Total Quantities		
	Lineal Feet	Pay (yd³)	Over-break (yd³)	Total (yd³)	Pay (yd³)	Over-break (yd³)	Total (yd³)
Concrete:							
Main tunnel							
Standard section	23,572	4.63	1.85	6.48	109,138	43,608	152,746
Under gate shaft	78		2.59	2.59	1,315	202	1,517
Total main tunnel	23,650				110,453	43,810	154,263
Shaft					1,894	496	2,390
Adit					572	28	600

Forms Other than Tunnel and Tunnel Bulkhead

	Straight (ft²)	Curved (ft²)	Transition (ft²)	Slipped (ft²)	Total (ft²)
Tunnel at Y	352				352
Tunnel under gate shaft	1,513		3,091		4,604
Shaft	1,304	658	145	14,218	16,325
Adit	176				176

Weight of Tunnel Sets

	lb per set	Count	Total lb
Main tunnel, 8-in WF 40 lb	2,510	912	2,289,120
Adit, 6-in WF 20 lb	802	12	9,624

V. REQUESTS FOR QUOTATIONS ON PERMANENT MATERIALS AND SUBCONTRACTS

Requests for quotations were sent out for permanent materials, supplies, and the plant and equipment items.

Requests were also made for subcontract prices for furnishing aggregates, for furnishing and installing reinforcing steel, and for drilling and pressure grouting.

A list of plugged prices was prepared as shown in Table 18.4.

TABLE 18.4   Sierra Tunnel Estimate: Plugged Material
and Subcontract Prices Sent Out by Sponsor to
Other Partners*

Bid Item No.	Description	Unit	Permanent-Materials Price	Subcontract Price
2	Steel sets	lb	$ 0.48	
3	Timber	MBM	320.00	

TABLE **18.4   Sierra Tunnel Estimate: Plugged Material and Subcontract Prices Sent Out by Sponsor to Other Partners***   (*Continued*)

Bid Item No.	Description	Unit	Permanent-Materials Price	Subcontract Price
4	Roof bolts	lin ft	1.60	
9	Tunnel lining	Ton of aggregate		$ 6.10
10	Shaft concrete	Ton of aggregate		6.10
11	Cement	bbl	16.00	
12	Steel reinforcing	lb		0.44
13	Embedded anchor bolts	lb	1.60	
14	Embedded vent pipe	lb	1.10	
15	Drilling grout holes	lin ft		10.00
16	Consolidation grouting	ft³		10.00
17	Cement for grouting	Sack		5.00

* Permanent-materials price, FOB job, including sales tax.

VI. COMPUTATION OF LABOR RATES

The labor cost to the contractor is composed of direct wages per hour; overtime payments; workers' compensation insurance; social security and unemployment insurance taxes; fringe benefits which include vacation allowance, health and welfare, apprentice training, subsistence if required by job location; and public liability and property damage insurance.

Amount of overtime payments depends on whether "the job" works 6 days or 5 days a week, whether heading work is carried on continuously, and whether shifts change at the portal or at the face.

Labor rates for the Sierra Tunnel are based on the assumption that the work is in a subsistence area; shifts will change at the tunnel face, so travel time will be paid; work will be carried on continuously, so each shift will run an extra 30 minutes; approximately one-half the crew will be worked through their lunch hour; and the job will work only 5 days a week.

Under these job conditions and assuming an average one-way travel time of 10 minutes for the entire work, the hours to be worked and total hours to be paid per shift for the tunnel crews were tabulated and appear in Table 18.5.

It is seen that the surface crew will have fringe benefits paid for 8.5 hours and actual pay for 8.75 hours per shift. The underground crew will have fringe benefits paid for 9.0 hours and wages paid for 9.625 hours per shift.

Fringe benefits used in this estimate are tabulated in Table 18.6.

Labor burden is calculated on the basis of the wages per shift and includes overtime pay, but insurance is based on a percentage of straight-time wages. Therefore, the nominal rate and the rate used in calculations differ. Furthermore, the social security and unemployment taxes cease after the average employee earns 6 or 7 months' wages, but because of labor turnover, a 40 percent reduction in these rates is reasonable. A token amount of unsched-

TABLE **18.5**   **Sierra Tunnel Estimate:**
Worked, Elapsed, and Straight-Time Hours

	Hours Worked	Elapsed Hours Paid per Shift	Equivalent Straight-Time Hours Paid per Shift
Surface crew:			
Standard shift	7 hr	8 hr	8 hr
Work additional ½ hr	0 hr 30 min	0 hr 30 min	0 hr 45 min
Total 7½-hr shift	7 hr 30 min	8 hr 30 min	8 hr 45 min
Decimal equivalent	7.5	8.5	8.75
Underground crew:			
Average travel time, start of shift		0 hr 10 min	0 hr 15 min
Standard shift	7 hr 0 min	8 hr 0 min	8 hr 0 min
Extra ½ hr of work	0 hr 30 min	0 hr 30 min	0 hr 45 min
Work ½ crew through lunch period	0 hr 15 min	0 hr 15 min	0 hr 22½ min
Travel time, end of shift		0 hr 10 min	0 hr 15 min
Total hours per shift	7 hr 45 min	9 hr 05 min	9 hr 37½ min
Decimal equivalent	7.75	9.08	9.625

uled overtime is to be expected, and it is convenient to include it as a part of burden. Based on these factors, the labor burden was calculated as shown in Table 18.7.

Total cost of hourly workers per shift was computed as shown in Table 18.8.

VII. TABULATION OF EQUIPMENT OPERATING COSTS

The equipment manager was provided with a list of the equipment which would probably be required for the job. He furnished the hourly operating costs, which were used in the estimate and included as supplies. The equipment manager also obtained quotations for the purchase of the equipment.

TABLE **18.6**   **Sierra Tunnel Estimate: Fringe Benefits**

	Laborers	Operating Engineers	Teamsters
Industry fund	$0.02	$0.02	
Health and welfare	0.96	1.10	$0.80
Vacation allowance	1.00	1.40	1.10
Pension	1.60	2.40	1.80
Retired employees' health and welfare			0.40
Drugs, dental and visual care			0.40
Apprenticeship		0.05	
Total per hour	$3.58	$4.97	$4.50

TABLE **18.7**   **Sierra Tunnel Estimate: Labor Burden**

Item	Nominal Rate (%)	Calculated Rate (%)
Workers' compensation insurance	18.67	16.8
Public liability and property damage insurance	1.20	1.1
Social security tax	6.65	4.0
Federal unemployment tax	1.20	0.7
State unemployment tax	3.60	2.2
Unscheduled overtime		5.0
Total labor burden		29.8

These quotations were furnished to the estimator who prepared the estimate of plant and equipment costs. The equipment manager also furnished a list of suitable equipment already owned by the company, together with the value of each item.

VIII. TRANSMITTAL OF ESTIMATING INSTRUCTIONS

It is assumed that the Sierra Tunnel project will be bid as a joint venture and that what is under discussion here is the sponsor's estimate. Therefore, at this point estimating instructions were mailed to all the joint venture partners by the sponsoring company. The following information was supplied:

1. Announcement of a joint venture meeting at the head office of the sponsor, at 9:00 A.M., April 26, 1982.
2. The format for the estimate.
3. Quantity comparison forms.
4. Estimate comparison forms.
5. The geological report.
6. The plugged permanent materials and subcontract prices, as prepared under estimating step V.
7. Wage-rate tabulations, as prepared under estimating step VI.
8. Insurance costs: workers' compensation rates, PL & PD rates, builder's risk insurance rates, and equipment and plant insurance rates. The premium for insuring the contractor's equipment is 1.2 percent of its valuation per year. The premium for insuring the construction buildings and structures is 1.2 percent of their valuation per year.
9. Payroll and property tax rates. The construction plant and equipment has an assessed valuation of 25 percent of market value. The yearly tax is to be 6 percent of the assessed valuation.
10. Bond rates.
11. A list of used equipment owned by the sponsor and available for use by the joint venture, including an offering price for each item.

12. A request that each partner furnish similar lists and prices of used construction equipment available for use by the joint venture.

IX. SELECTION OF CONSTRUCTION METHODS

From inspection of the plans and specifications, four different methods of tunnel driving became apparent. The long distances involved dictated a rail

TABLE **18.8** **Sierra Tunnel Estimate:**
Five-Day Workweek Labor Cost per Shift

	Basic Wage/h	Total Shift, 9.625 h	Taxes and Insurance 29.80%	Fringe Benefits, 9 h	Subsistence/ Shift	Total Cost/ Shift
Underground:						
Shifter, shaft	$13.38	$128.78	$38.38	$32.22	$16.00	$215.38
Miner, shaft	12.18	117.23	34.93	32.22	16.00	200.38
Shifter, tunnel	12.75	122.72	36.57	32.22	16.00	207.51
Miner	11.56	111.26	33.16	32.22	16.00	192.64
Chucktender, nipper	11.18	107.61	32.07	32.22	16.00	187.90
Bull-gang foreman	11.88	114.34	34.07	32.22	16.00	196.63
Bull gang	10.94	105.30	31.38	32.22	16.00	184.90
Nozzleman	11.56	111.26	33.16	32.22	16.00	192.64
Vibrator man	11.18	107.61	32.07	32.22	16.00	187.90
Concrete crew	10.94	105.30	31.38	32.22	16.00	184.90
Brakeman	11.88	114.34	34.07	44.73	16.00	209.14
Conway operator	14.72	141.68	42.22	44.73	16.00	244.63
Motorman	12.68	122.04	36.37	44.73	16.00	219.14
Heavy-duty mechanic	14.72	141.68	42.22	44.73	16.00	244.63
Party chief	15.31	147.36	43.91	44.73	16.00	252.00
Instrument man	14.15	136.19	40.58	44.73	16.00	237.50
Head chainman	12.68	122.04	36.37	44.73	16.00	219.14
Rear chainman	11.18	107.61	32.07	44.73	16.00	200.41
Concrete-belt operator	12.35	118.87	35.42	44.73	16.00	215.02
Pumpcrete or Press Weld	13.75	132.34	39.44	44.73	16.00	232.51
Slip-form jumbo	13.91	133.88	39.90	44.73	16.00	234.51
Front-end loader	15.31	147.36	43.91	44.73	16.00	252.00
	Basic Wage/h	Total Shift, 8.75 h	Taxes and Insurance 29.80%	Fringe Benefits, 8½ h	Subsistence/ Shift	Total Cost/ Shift
Surface crew:						
Powderman	$11.19	$ 97.91	$29.18	$30.43	$16.00	$173.52
Dumpman	10.94	95.72	28.52	30.43	16.00	170.67
Watchman	10.19	89.16	26.57	30.43	16.00	162.16
Compressor operator	13.44	117.60	35.04	42.24	16.00	210.88
Heavy-duty mechanic	14.40	126.00	37.55	42.24	16.00	221.79
Warehouseman	11.26	98.52	29.36	38.25	16.00	182.13
Tractor operator	14.40	126.00	37.55	42.24	16.00	221.79
Crane operator	15.00	131.25	39.11	42.24	16.00	228.60

operation except for a few local situations where rubber-tired equipment was indicated.

Method 1.   A construction adit will not be used. The main tunnel will be excavated by driving a single heading up the slope from the outlet portal. At the inlet portal a small crew will drive approximately 200 lin ft of tunnel with rubber-tired equipment. Then the shaft will be excavated by constructing a small raise from the tunnel. Excavation to the full dimension of the shaft will be done by downdrilling and disposing of the shot muck down the raise. When the shaft excavation is complete, the section of the tunnel under the shaft will be enlarged to its full diameter.

Method 2.   A construction adit will not be used, but tunnel excavation will be done from both portals with two separate single headings. As soon as the tunnel is driven well past the gate shaft at the inlet end, the shaft will be constructed in the same manner as method 1 except that muck disposal will be with rail equipment.

Method 3.   A construction adit will be driven and the main tunnel will then be driven toward the inlet portal by using a single heading. After completion of the excavation of this leg of the tunnel, excavation will be started on the gate shaft and on the section of tunnel from the adit to the outlet portal. The gate shaft will be constructed in the same manner as described in method 1 except that muck disposal will be with rail equipment.

Method 4.   A construction adit will be driven and the main tunnel excavated by single-heading operations until approximately 1,000 lin ft are excavated in each direction. This will be done with rubber-tired equipment. Then this rubber-tired equipment will be used to drive a short section of tunnel at the inlet end and construct the gate shaft. Rail equipment will be installed in the tunnel and, with alternate-heading operations, double-heading operations, and single-heading operations, will finish driving the tunnel. All construction plant will be located near the portal of the adit.

Cost Comparison of the
Four Proposed Tunneling Methods

To determine which of the proposed construction methods would be the most economical, a comparison was made by using method 1 as a base. A preliminary estimate of the deletions or additions in cost for methods 2, 3, and 4, as compared with method 1, are shown in Table 18.9. Based on this preliminary comparison, it was seen that method 4 would result in the least cost and the least construction time; so this method was used in the estimate.

X. SELECTION OF PLANT AND EQUIPMENT

It was decided that the minimum-sized construction adit capable of furnishing clearance for a front-end loader, two vent pipes, and rail equipment would be a 14-ft-wide vertical sidewall horseshoe tunnel. The total length of adit to be excavated will be greater than 650 lin ft because Y branches are required in each heading direction. This Y-branch excavation will make the excavated adit equivalent to a length of 688 lin ft.

The adit and 300 ft of tunnel in each direction from the adit are to be excavated by a rubber-tired front-end loader operating in the load-haul-dump mode, and drilling is to be done with a truck jumbo. The 600 ft of tunnel will provide space for the erection of the equipment needed to drive the main tunnel.

The equipment used at the adit will be moved to the inlet portal, and 200 ft of tunnel will be driven to allow the shaft to be constructed. After excavation of the shaft, the tunnel section under the shaft will be enlarged to full diameter with this equipment.

Each heading of the main tunnel is then to be driven with a Jacobs floor and with a Conway 102 mucker loading into 13-yd^3 cars side-boarded to 15-yd^3 capacity. Drilling is to be done with 4½-in-bore drifters plus one large drill for drilling the burn-cut, all mounted on a gantry jumbo.

Although drilling in granite might very well have justified the purchase of hydraulic drilling equipment, air-operated drills are to be used here to illustrate the selection of drills and the calculation of compressed-air requirements.

To determine the amount of equipment required, it was assumed that 80 holes would be required in the granite and 70 holes in the greenstone. The granite controls the number of drills.

The square feet of face per hole is:

Area of face $= 24.33^2 \times 0.8293 = 491$ ft^2
Granite   491 ft$^2 \div 80 = 6.14$ ft^2 per hole
Greenstone   491 ft$^2 \div 70 = 7.01$ ft^2 per hole

Ten drifter drills plus one burn-cut drill are to be used, that is, one drill for each eight holes, which should result in a balanced cycle.

The compressed air required at 1,600-ft elevation (including one spare unit) was computed as follows:

1 burn-cut drill	600 ft^3/min
10 4½-in-bore drifters at 330 ft^3/min	3,300 ft^3/min
Air required for one heading	3,900 ft^3/min
Air required for two headings	7,800 ft^3/min
Shops and miscellaneous	500 ft^3/min
Subtotal	8,300 ft^3/min
Leakage factor 1.10	9,130 ft^3/min
Altitude correction factor 1.05	9,586 ft^3/min
Therefore, provide nine 1,200 ft^3/min compressors	10,800 ft^3/min

To obtain 4,400 ft^3/min at each heading: A 6-in pipeline will have a pressure drop of 0.94 lb in 1,000 ft, or on the longest heading of 12,700 ft, a total

TABLE 18.9 Sierra Tunnel Estimate: Comparison of Four Methods of Excavation

		Method 1 (Single Heading from Outlet Portal)	Method 2 (Single Headings from Both Portals)	Method 3 (Single Heading from Adit)	Method 4 (Double Heading from Adit)
Geology summary:					
Breast					
boarding	40 lin ft				
2-ft rounds	60 lin ft				
4-ft rounds	3,448 lin ft				
Remainder	20,102 lin ft				
Total	23,650 lin ft				
Estimated heading production:					
Breast					
boarding	5 lin ft/day	8 days			
2-ft rounds	15 lin ft/day	4 days			
4-ft rounds	30 lin ft/day	115 days			
Remainder	60 lin ft/day	335 days			
Start-up		10 days			
Total		472 days			
Average daily production		50 lin ft			
Comparative production:					
Lin ft driven					
Single heading		23,650 lin ft	23,650 lin ft	23,650 lin ft	2,150 lin ft
Alternate heading					4,000 lin ft
Double heading					17,500 lin ft
Total		23,650 lin ft	23,650 lin ft	23,650 lin ft	23,650 lin ft

508

Single-heading crew days at 50 lin ft/day	472 days	472 days	472 days	43 days
Alternate-heading crew days at 80 lin ft/day				50 days
Double-heading days at 100 lin ft/day				175 days
Total crew days	472 days	472 days	472 days	268 days
Concrete days	200 days	200 days	200 days	200 days
Number of headings	1	2	1	2
Calendar months, 20 days/month:				
Start-up and portal-in	3 months	3 months	3 months	3 months
Drive adit			2 months	2 months
Excavation	24 months	12 months	24 months	14 months
Delay before concrete	1 month	1 month	1 month	1 month
Concrete	10 months	10 months	10 months	10 months
Cleanup and move-out	1 month	1 month	1 month	1 month
Total time	39 months	27 months	41 months	31 months
Overhead cost reduction at $30,000/month		$360,000		$240,000
Overhead cost increase			$60,000	
Excavation man-days:				
Excavation crew size/day:				
Single headings, 10 × drills	100	200	100	100
Alternate headings, + 10%				110
Double headings				170
Concrete average crew days	140	140	140	140
Man-days:				
Excavation, single heading	47,200	47,200	47,200	4,300
Excavation, alt. heading				5,500
Excavation, double heading				29,750
Total	47,200	47,200	47,200	39,550
Concrete man-days	28,000	28,000	28,000	28,000
Total man-days	75,200	75,200	75,200	67,550

TABLE 18.9 **Sierra Tunnel Estimate: Comparison of Four Methods of Excavation** (*Continued*)

	Method 1 (Single Heading from Outlet Portal)	Method 2 (Single Headings from Both Portals)	Method 3 Single Heading from Adit	Method 4 (Double Heading from Adit)
Labor cost:				
Total labor cost at $200/man-day	$15,040,000	$15,040,000	$15,040,000	$13,510,000
Labor cost reduction				$ 1,530,000
Supplies:				
Cost reduction based on 25% of labor				$ 382,500
Travel time:				
Estimated travel time, man/day	34 min	20 min	20 min	20 min
Decimal equivalent	0.57 hr	0.33 hr	0.33 hr	0.33 hr
Travel time, hours	42,864 hr	25,067 hr	25,067 hr	22,517 hr
Savings in hours over method 1		17,797 hr	17,797 hr	20,347 hr
Travel payment cost reduction at time and one-half, or $25.00/h		$444,925	$444,925	$508,675
Labor escalation:				
Average labor escalation, 10% per yr*	11.25%	6.25%	12.08%	7.92%
Labor escalation	$1,692,000	$940,000	$1,816,832	$1,069,992
Labor escalation cost reduction		$752,000		$ 622,008
Labor escalation cost increase			$124,832	
Plant and equipment comparison after salvage:				
Construct adit			$+375,000	$+375,000
Rail equipment:				
1 additional Jacobs floor		$+275,000		+275,000
1 additional drill jumbo		+375,000		+375,000
1 additional mucker		+ 90,000		+ 90,000
25 additional muck cars		+125,000		+125,000
additional specialty cars		+ 25,000		+ 25,000
additional compressors		+100,000		+100,000

510

Rubber-tired equipment	−437,500	−437,500	
Savings in track material		−135,000	
Savings in vent line		−115,000	
Savings in fans		−37,500	
Savings in piping		−105,000	
Increased cost, additional portal	+175,000		
Cost reduction in plant and equipment		$ 455,000	
Cost increase in plant and equipment	$ 727,500		$1,365,000

Summary of comparative costs

Cost reductions:			
Overhead	$ 360,000		$ 240,000
Labor			1,530,000
Supplies			382,000
Travel time	445,000	$445,000	509,000
Labor escalation	752,000		622,000
Plant and equipment		455,000	
Subtotal	$1,557,000	$900,000	$3,283,000
Cost increases:			
Overhead		60,000	
Labor escalation		125,000	
Plant and equipment	728,000		1,365,000
NET REDUCTION	$ 829,000	$715,000	$1,918,000

* To midpoint of total time, after deducting initial 12 months (during which there will be no escalation). For simplicity and because of the brief durations, escalation rates have not been compounded.

pressure loss of 12 lb. A 10-in pipeline will have a pressure drop of 0.28 lb in 1,000 ft, or 3.6 lb in 12,700 ft. Therefore, 12-in air pipe to the end of the construction adit and then 10-in pipe in both headings will be used.

A 10-lin-ft round will be pulled which will require the following number of muck cars per round:

Rock in place per lineal foot of tunnel, including overbreak	20 yd³
Rock in place per round	200 yd³
Muck, assuming 50% swell	300 yd³
Capacity of one 15 yd³ car, loaded to 80% of capacity	12 yd³
Required number of cars per round = 300 ÷ 12	25 cars
For two-heading operation with allowance for spares, provide	60 cars

To find the number of locomotives for operating up the 0.1 percent grade:

Weight of car:

Weight of empty car, 15 × 1,150 lb	17,250 lb
Rock, 12 yd³ broken at 2,600 lb	31,200 lb
Total weight of one loaded car	48,450 lb
Or	24.23 tons

Resistance to movement:

Rolling resistance	20 lb/ton
Grade resistance	2 lb/ton
Accleration at 0.2 mph per sec	18 lb/ton
Total	40 lb/ton

To start seven cars, a locomotive with the following drawbar pull is needed:

$$24.23 \text{ ton} \times 40 \times 7 = 6{,}784 \text{ lb}$$

A 15-ton locomotive will start the train, as its starting drawbar pull is 25 percent of its weight, or 7,500 lb.

To find the horsepower of a locomotive required to accelerate seven cars upgrade to 10 mph,

$$\text{hp} = \frac{\text{tractive effort} \times \text{speed in miles per hour}}{375 \times 0.80}$$

Tractive effort is the weight of the train multiplied by the resistance.

Train weight:

7 cars at 24.23 ton	170 tons
Locomotive	25 tons
Total	195 tons

$$\text{hp} = \frac{195 \times 40 \times 10}{375 \times 0.80} = 260 \text{ hp}$$

TABLE 18.10 Sierra Tunnel Estimate
Tunnel-Round Calculations

	With Rubber-tired Equipment (8 yd³ loads)				With Rail Equipment and Jacobs Floor (12 yd³ loads)					
	Adit	Main Tunnel			Main Tunnel					
Type of rock	Granite	Granite	Granite	Granite	Granite	Granite	Granite	Greenstone	Greenstone	Greenstone
Required supports	Unsupported	Supported	Unsupported	Unsupported	Supported	Unsupported	Unsupported	Supported	Unsupported	Unsupported
Length of round	8 ft	4 ft	8 ft	10 ft	4 ft	8 ft	10 ft	4 ft	8 ft	10 ft
Excavation, yd³/lin ft:										
Pay	6.5 yd³	18.18 yd³								
Nonpay	1.0 yd³	1.85 yd³								
Total	7.5 yd³	20.03 yd³								
Excavation, yd³/round:										
Solid	60 yd³	80 yd³	160 yd³	200 yd³	80 yd³	160 yd³	200 yd³	80 yd³	160 yd³	200 yd³
Broken	90 yd³	120 yd³	240 yd³	300 yd³	120 yd³	240 yd³	300 yd³	120 yd³	240 yd³	300 yd³
Number of loads	12 ea	15 ea	30 ea	38 ea	10 ea	20 ea	25 ea	10 ea	20 ea	25 ea
Drilling:										
Drill holes/round	40	80	80	80	80	80	80	70	70	70
Lin ft of drilling	360 lin ft	400 lin ft	720 lin ft	880 lin ft	400 lin ft	720 lin ft	880 lin ft	350 lin ft	630 lin ft	770 lin ft
Number of drills	4	8	8	8	10	10	10	10	10	10
Lin ft/drill	90 lin ft	50 lin ft	90 lin ft	110 lin ft	40 lin ft	72 lin ft	88 lin ft	35 lin ft	63 lin ft	77 lin ft
Drilling speed/hour	60 lin ft	60 lin ft	60 lin ft	60 lin ft	60 lin ft	60 lin ft	60 lin ft	70 lin ft	70 lin ft	70 lin ft
Time required for drilling	1 h 30 min	50 min	1 h 30 min	1 h 50 min	40 min	1 h 12 min	1 h 28 min	30 min	54 min	1 h 6 min
Number of holes/miner	10	10	10	10	8	8	8	7	7	7
Cycle time:										
Move drill jumbo in	5 min	5 min	5 min	5 min	5 min	5 min	5 min	5 min	5 min	5 min
Set steel	0 min	45 min	0 min	0 min	45 min	0 min	0 min	45 min	0 min	0 min
Drill	1 h 30 min	50 min	1 h 30 min	1 h 50 min	40 min	1 h 12 min	1 h 28 min	30 min	54 min	1 h 6 min
Load	40 min	40 min	40 min	40 min	32 min	32 min	32 min	28 min	28 min	28 min
Move jumbo out	5 min	5 min	5 min	5 min	10 min	10 min	10 min	10 min	10 min	10 min
Shoot and ventilate	15 min	15 min	15 min	15 min	15 min	15 min	15 min	15 min	15 min	15 min
Move mucker in	5 min	5 min	5 min	5 min	3 min	3 min	3 min	3 min	3 min	3 min
Muck 3 min/car, 2½ min/truck	30 min	38 min	1 h 15 min	1 h 35 min	30 min	1 h 0 min	1 h 15 min	30 min	1 h 0 min	1 h 15 min
Cleanup and bar down	15 min	15 min	15 min	15 min	15 min	15 min	15 min	15 min	15 min	15 min
Move mucker out	5 min	5 min	5 min	5 min	2 min	2 min	2 min	2 min	2 min	2 min
Delays	10 min	10 min	10 min	10 min	10 min	10 min	10 min	10 min	10 min	10 min
Total cycle	3 h 40 min	3 h 53 min	4 h 25 min	5 h 5 min	3 h 27 min	3 h 44 min	4 h 15 min	3 h 13 min	3 h 22 min	3 h 49 min
Decimal equivalent	3.67 hr	3.88 hr	4.42 hr	5.08 hr	3.45 hr	3.73 hr	4.25 hr	3.22 hr	3.37 hr	3.82 hr
Cycles in 23.25 hours	6.34	5.99	5.26	4.58	6.74	6.23	5.47	7.22	6.90	6.09
Lin ft excavated per day	51 lin ft	24 lin ft	42 lin ft	46 lin ft	27 lin ft	50 lin ft	55 lin ft	29 lin ft	55 lin ft	61 lin ft
USE IN ESTIMATE	45 lin ft*	24 lin ft	42 lin ft	45 lin ft	27 lin ft	50 lin ft	54 lin ft	29 lin ft	55 lin ft	60 lin ft

* Production in adit is reduced because of crew training.

513

Therefore, a 25-ton locomotive with Caterpillar diesel 3406 turbo-charged after-cooled engine which develops 275 hp was selected.

To determine ventilation requirements estimators assumed one locomotive at each heading, one locomotive on main-line haul in each heading, and one locomotive in the adit, or 2½ locomotives per heading, for a total of 687.5 hp per heading, plus a maximum of 30 men underground per heading. California regulations require an air supply of 100 ft^3/min per hp, plus 200 ft^3/min per man. A total requirement of 74,750 ft^3/min was established, and using Table 10.7, a 48-in pipe and 125-hp fans were selected. The first fan is to be set at 320 ft and the others are to be spaced at 960-ft intervals. The total fans per heading:

Toward outlet, 12,900 ft   14 fans
Toward inlet, 10,750 ft plus 650-ft adit   13 fans

In constructing the shaft, a 6- by 6-ft raise will be excavated with an Alimak Raise Climber. The muck will be allowed to fall down the shaft into the tunnel and will be picked up by the Eimco loader and loaded into trucks. The shaft will then be sunk from the top with track-type drills and a small crawler dozer to push the material down the raise.

XI. PREPARATION OF
THE CONSTRUCTION SCHEDULE

To determine tunnel excavation progress, it was necessary to determine the length of time required to excavate rounds in the different types of formations, as shown in Table 18.10. By using this rate of progress, the number of excavation crew days was determined. Concreting crew days were then determined from estimated progress figures. (See Table 18.10.)

Excavation Crew Days

Adit Construction		Length (lin ft)	Time (days)
First 100 lin ft in 15 days at one shift		100	15
Next 500 lin ft at 45 lin ft/day		500	11
Y section, including branch		50	4
		650	30

	Toward Outlet		Toward Inlet	
Tunnel Excavation	Length (lin ft)	Time (days)	Length (lin ft)	Time (days)
Rubber-tired equipment in granite:				
First 160 lin ft from Y	160	8	160	8
4-ft rounds supported at 24 lin ft/day	140	6	140	6
Subtotal	300	14	300	14

Tunnel Excavation	Toward Outlet		Toward Inlet	
	Length (lin ft)	Time (days)	Length (lin ft)	Time (days)
Inlet construction in greenstone:				
First 100 lin ft			100	5
Next 100 lin ft			100	5
Enlarging section				5
Total rubber-tired	300	14	500	29
Rail equipment in granite:				
Alternating heading:				
80% of two-heading production				
4-ft rounds at 22 lin ft/day	200	9	200	9
10-ft rounds at 43 lin ft/day	1,800	42	1,800	42
Total	2,000	51	2,000	51
Rail equipment in granite:				
Double-heading operation:				
4-ft rounds at 27 lin ft/day	837	31	270	10
10-ft rounds at 54 lin ft/day	8,152	151	2,980	55
2-ft rounds at 12 lin ft/day			30	3
Breast boarding			20	3
Subtotal	8,989	182	3,300	71
Rail equipment in greenstone:				
Double-heading operation:				
Breast boarding			20	3
2-ft rounds at 15 lin ft/day			30	2
4-ft rounds at 29 lin ft/day			1,488	51
10-ft rounds at 60 lin ft/day			3,412	57
Subtotal			4,950	113
Total double heading	8,989	182	8,250	184
Rail equipment in granite:				
Single-heading operation:				
4-ft rounds at 27 lin ft/day	213	8		
10-ft rounds at 54 lin ft/day	1,398	26		
Total single heading	1,611	34		
GRAND TOTAL	12,900		10,750	

The main-tunnel driving days used in the estimate and in the preparation of the construction schedule are shown below in summary.

	Toward Outlet (days)	Toward Inlet (days)	Total Days
Single heading, rubber-tired	14	29	43
Alternating heading, rail, excavate in both directions			51
Double heading, rail, excavate in both directions			184
Single heading, rail	34		34

Concreting Crew Days (Work from Adit toward Portals)

	From Adit to Outlet	From Adit to Inlet	Total
Tunnel without sets, lin ft	11,570	8,532	20,102
Tunnel with sets, lin ft	1,330	2,218	3,548
Total lin ft	12,900	10,750	23,650
	In Days	**In Days**	**In Days**
1. Remove vent line at 1,000 lin ft/day	13	11	24
Allowance for start-up	1	1	2
Total	14	12	26
2. Retimber supported ground at 200 lin ft/day	7	12	19
Allowance for start-up	1	1	2
Total	8	13	21
3. Primary cleanup at 500 lin ft/day	26	22	48
Allowance for enlarged station under shaft		1	1
Allowance for start-up	1	1	2
Total	27	24	51
4. Invert concrete at 750 lin ft/day	17	15	32
$(yd^3/day = 750 \times 2 = 1,500)$			
Allowance for enlarged section		1	1
Allowance for equipment setup, moving, and start-up	3	2	5
Total	20	18	38
5. Arch concrete at 300 lin ft/day	43	36	79
$(yd^3/day = 300 \times 4.48 = 1,344)$			
Allowance for section at base of shaft		3	3
Allowance for equipment setup, moving and start-up	4	3	7
Total	47	42	89
6. Low-pressure grout at 300 lin ft/day	43	36	79
Allowance for start-up	1	1	2
Total	44	37	81
7. Final cleanup at 1,000 lin ft/day	13	11	24
Allowance for start-up	1	1	2
Total	14	12	26
Adit plug at 600 yd³ at 5-ft lifts			5

Actual Construction Schedule

Excavation crew days and concreting crew days were thus determined. Crew working days were then converted to calendar days and plotted on the construction schedule. In determining the total excavation time, allowances were made for portaling-in; for assembling the crews and training them; for constructing the adit Y; and for slow progress in the main tunnel near the Y section because of restricted working space. Slow progress must be allowed in the closely supported ground and in the breast-boarded section.

Portaling-in and driving the first 100 lin ft of tunnel will be done one shift per day, and because of the portal construction and crew training, it was estimated that this will take 15 days.

Constructing the Y and driving the first 160 lin ft of main tunnel from the Y in each direction will be slow work, and it was estimated that it will take approximately 4 days to construct the Y and 16 days to drive the 320 lin ft of main tunnel.

When the excavation is in ground supported on 2-ft centers, production should be 20 ft a day. When excavation is in ground requiring breast boarding, it is estimated that an average of only 5 ft a day will be obtained.

In the preparation of the construction schedule, it was assumed that the work will be bid on the first of May and work will be awarded on the first of June. The rubber-tired driving equipment can be delivered in 2 months, so adit driving can start on August 1.

When the main tunnel has been driven 300 ft in the direction of the outlet with the use of rubber-tired equipment, this equipment will be moved to drive 300 ft on the gate shaft leg. While this is being done, the Jacobs floor and rail equipment can be installed in the outlet leg. Upon completion of the 300-ft section in the gate shaft leg, the rubber-tired equipment will be moved to the inlet portal and 200 ft of tunnel will be driven, and then the shaft-raising operations can be started.

Meanwhile the outlet leg will be excavated with rail equipment on a single heading until the Jacobs floor can be erected in the gate shaft (inlet) leg.

After erection of this second Jacobs floor, alternate-heading operations can be started. The remainder of the tunnel will be excavated in this manner as shown on the construction schedule.

Concreting operations have been arranged on the schedules so some of the cleanup can be performed while other work is being done. Since only one invert and one arch concreting setup is planned, this equipment must be transferred from one leg to the other. Concrete will be poured continuously through three shifts, 5 days per week. Bulkheads will be constructed only at the end of each week.

Tunnel construction may be scheduled on a bar chart or on a network (critical path) diagram. Tunnel construction is however, in most respects a lineal operation, and neither the bar chart nor the network diagram shows relationships as graphically as a time-and-location type of schedule. The schedule for Sierra Tunnel appears as Fig. 18.3. This schedule is based on 20 workdays per month, which allows approximately 2 days per month for holidays, bad weather, or other eventualities. The schedule shows that work be-

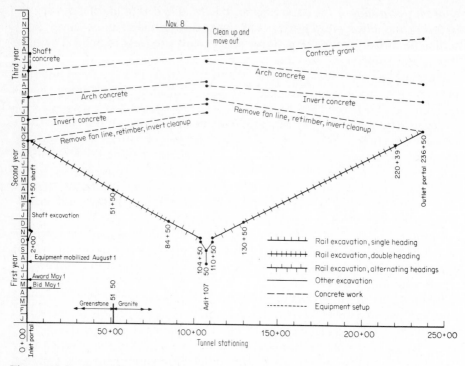

Figure 18.3   Sierra Tunnel construction schedule.

gins at the adit, where the construction plant is located, and proceeds as two simultaneous operations from the adit to the inlet portal and from the adit to the outlet portal. Construction of the gate shaft can, to a large degree, be scheduled independently of the tunnels.

XII. ESTIMATING DIRECT COSTS

Cost of Outside Development

The majority of the outside work, including construction of access roads, construction of power and water distribution lines, construction of the inlet and outlet structures, and the opencut work, will be done by others. The cost of the remaining outside work is readily estimated and is included with the plant and equipment.

Cost of Tunnel Excavation Labor

Bid Item 1

The number of men and the daily cost to the contractor for these men is estimated for each type of excavation crew. This daily crew cost times the crew days computed in step XI is the cost of excavation labor. Part of this labor is charged to steel supports, part to timber, and the remainder to excavation. Tables 18.11 through 18.15 show how crew costs are determined.

TABLE 18.11 Sierra Tunnel Estimate: Crew Cost per Day in Adit—Rubber-Tired Equipment, One-Shift Operation

	No. of Men/ Shift or Day	Rate	Total
Walker	1	$218.44	$ 218.44
Miner	5	192.64	963.20
Front-end loader operator	1	252.00	252.00
Compressor operator	1	210.88	210.88
Mechanic foreman	1	260.12	260.12
Mechanic	1	221.79	221.79
Electrician foreman	1	266.66	266.66
Electrician	1	255.15	255.15
Tractor operator	1	221.79	221.79
Warehouseman	1	182.13	182.13
Crane operator	1	228.60	228.60
Party chief	1	252.00	252.00
Head chainman	1	219.14	219.14
Total	17		$3,751.90

TABLE 18.12 Sierra Tunnel Estimate: Crew Cost per Day in Adit—Rubber-Tired Equipment, Three-Shift Operation

	No. of Men/Shift	No. of Men/Day	Rate	Total
Walker		1	$218.44	$ 218.44
Shifter	1	3	207.51	622.53
Miner	5	15	192.64	2,889.60
Front-end loader operator	1	3	252.00	756.00
Compressor operator	1	3	210.88	632.64
Mechanic foreman		1	260.12	260.12
Mechanic	1	3	221.79	665.37
Electrician foreman		1	266.66	266.66
Electrician	1	3	255.15	765.45
Tractor operator		1	221.79	221.79
Warehouseman		1	182.13	182.13
Crane operator		1	228.60	228.60
Party chief		1	252.00	252.00
Head chainman		1	219.14	219.14
Outside laborer		1	162.16	162.16
Total		39		$8,342.63

TABLE **18.13** **Sierra Tunnel Estimate: Alternating-Heading Crew Cost per Day in Main Tunnel—Rubber-Tired Equipment**

	No. of Men/Shift	No. of Men/Day	Rate	Total
Walker		1	$218.44	$ 218.44
Shifter	1	3	207.51	622.53
Miner	9	27	192.64	5,201.28
Nipper	1	3	187.90	563.70
Front-end loader operator	1	3	252.00	756.00
Compressor operator	1	3	210.88	632.64
Mechanic foreman		1	260.12	260.12
Mechanic	2	6	221.79	1,330.74
Electrician foreman		1	266.66	266.66
Electrician	1	3	255.15	765.45
Tractor operator		1	221.79	221.79
Warehouseman		1	182.13	182.13
Crane operator		1	228.60	228.60
Party chief		1	252.00	252.00
Head chainman		1	219.14	219.14
Outside laborer		1	162.16	162.16
Total		57		$11,883.38

TABLE **18.14** **Sierra Tunnel Estimate: Single-Heading or Alternating-Heading Crew Cost per Day in Main Tunnel—Rail Equipment**

	No. of Men/Shift	No. of Men/Day	Rate	Total
Walker		1	$218.44	$ 218.44
Shifter	1	3	207.51	622.53
Miner	11	33	192.64	6,357.12
Nipper	1	3	187.90	563.70
Mucker operator	1	3	244.63	733.89
Motorman	2	6	219.14	1,314.84
Brakeman	2	6	209.14	1,254.84
Bull-gang foreman		1	196.63	196.63
Bull-gang laborer		5	184.90	924.50
Dumpman	1	3	170.67	512.01
Compressor operator	1	3	210.88	632.64
Mechanic foreman		1	260.12	260.12
Mechanic	2	6	221.79	1,330.74
Electrician foreman		1	266.66	266.66
Electrician	1	3	255.15	765.45
Tractor operator		1	221.79	221.79
Warehouseman		1	182.13	182.13
Crane operator		1	228.60	228.60
Party chief		1	252.00	252.00
Head chainman		1	219.14	219.14
Outside laborer	1	3	162.16	486.48
Total		86		$17,544.25

TABLE 18.15   **Sierra Tunnel Estimate: Double-Heading Crew Cost per Day in Main Tunnel—Rail Equipment**

	No. of Men/Shift	No. of Men/Day	Rate	Total
Walker		2	$218.44	$ 436.88
Shifter	2	6	207.51	1,245.06
Miner	20	60	192.64	11,558.40
Nipper	2	6	187.90	1,127.40
Mucker operator	2	6	244.63	1,467.78
Motorman	5	15	219.14	3,287.10
Brakeman	5	15	209.14	3,137.10
Bull-gang foreman		2	196.63	393.26
Bull-gang laborer		8	184.90	1,479.20
Dumpman	1	3	170.67	512.01
Compressor operator	1	3	210.88	632.64
Mechanic foreman		1	260.12	260.12
Mechanic	3	9	221.79	1,996.11
Electrician foreman		1	266.66	266.66
Electrician	2	6	255.15	1,530.90
Tractor operator		1	221.79	221.79
Warehouseman		1	182.13	182.13
Crane operator		1	228.60	228.60
Party chief		1	252.00	252.00
Instrument man		1	237.50	237.50
Head chainman		1	219.14	219.14
Outside laborer	1	3	162.16	486.48
Total		152		$31,158.26

Summary of Excavation Labor

Excavation Labor in Adit to Be Charged to Plant and Equipment

15 Single-shift days at $3,751.90	$ 56,279
15 Triple-shift days at $8,342.63	125,139
Total excavation labor charged to adit	$181,418

Excavation Labor Charged to Tunnel Excavation, Bid Item 1

	Total	Per yd³ Takeoff, 430,842 yd³
Total excavation labor:		
43 Crew days, rubber-tired, single heading, at $11,883.38	$ 510,985	
51 Crew days, rail, alternate heading, at $17,544.25	894,757	
184 Crew days, rail, double heading, at $31,158.26	5,733,120	
34 Crew days, rail, single heading, at $17,544.25	596,505	

	Total	Per yd³ Takeoff, 430,842 yd³
Weekend maintenance, 71 weekends at $1,675.00	118,925	
	$7,854,292	

Less labor charged to supports:
Cost per lb:

Cost per ½ h/heading on double-heading crew	$335.04	
Lb of steel, one set	2,510 lb	
Labor cost per lb	$0.13	

Total weight:

912 Sets at 2,510 lb = 2,289,120 lb		
Labor for 2,289,120 lb at $0.13	$ 297,586	
Timber 404 MBM at $125.00	50,500	
Labor charge to tunnel excavation	$7,506,206	$17.42

Cost of Tunnel Excavation Supplies

Bid Item 1

a. Powder and Exploders

	Adit and Y	Main Tunnel	
		Granite	Greenstone
Lin ft of tunnel	688	18,500	5,150
Number of rounds	92	2,143	760
Number of holes/round	40	80	70
Number of exploders, std tunnel	3,680	171,440	53,200
Added number under shaft			100
Total number of exploders	3,680	171,440	53,300
Lb of primer powder at ½ lb/hole	1,840	85,720	26,650
Yardage per takeoff	4,472	336,330	94,512
Lb of AN/yd³	6	5	4
Lb of AN total	26,832	1,681,650	378,048
Cost:			
Caps at $0.75 ea	$ 2,760	$128,580	$ 39,975
Powder at $0.55/lb	1,012	47,146	14,658
AN at $0.20/lb	5,336	336,330	75,610
Wire and misc. at $2.00/lin ft	1,376	37,000	10,300
Total cost	$10,484	$549,056	$140,543
Cost per yd³ of takeoff	$2.34	$1.63	$1.49
Cost per lin ft of tunnel	$15.24	$29.68	$27.29
Combining two main tunnel types:			
Total cost			$689,599
Cost per yd³ on 430,842 yd³			$1.60
Cost per lin ft on 23,650 lin ft			$29.16

b. Steel and Bits

	Adit and Y	Main Tunnel
From above:		
Lin ft of tunnel	688	23,650
Number of rounds	92	2,903
Average length round, lin ft	7.5	8.15
Average drilling length of round, lin ft	8.5	9.15
Number of holes	3,680	224,740
Lin ft of standard drilling	31,280	2,056,371
At two holes/round:		
Burn-cut drilling, lin ft	1,564	53,125
Total drilling, lin ft	32,844	2,109,496
With bit life of 200 lin ft:		
Number of standard bits	157	10,282
Number of burn-cut bits	8	266
Estimated cost:		
Steel cost = $60.00 each piece		
One piece of steel has life of 1,000 lin ft		
At 1,000 lin ft = $0.06/lin ft	$ 1,971	$126,570
Standard bits at $47.00 ea	7,379	483,254
Burn-cut bits at $518.00 ea	4,144	137,788
Total cost	$13,494	$747,612
Cost per lin ft of drilling	$0.41	$0.35
Cost per yd^3 of excavation	$3.02	$1.74

c. Power

	Rubber		Rail	
			Single and	Double
			Alternating	Heading
	Adit	Tunnel	Heading	
Power load:				
Horsepower:				
Muckers			165	330
Fans, average	200	400	450	1,000
Compressors	500	1,000	1,250	2,250
Shops	25	50	50	50
Pump		25	50	50
Misc. (shops, office, etc.)	100	100	100	100
Total hp	825	1,575	2,065	3,780
Kilowatts	615	1,175	1,540	2,819
At 50% load factor, kW	308	588	770	1,410
Cost per hour at $0.06/kWh	$18.48	$35.28	$46.20	$84.60
Cost per day	$444	$847	$1,109	$2,030
Equivalent number of days	20	43	85	184
Total cost	$8,880	$36,421	$ 94,265	$373,520
Increase for Saturday and Sunday	888	3,642	9,426	37,352
Grand total cost	$9,768	$40,063	$103,691	$410,872

	Adit	Tunnel
Regroup cost	$9,768	$554,626
Cost per yd^3	$2.18	$1.29

d. Electrical Supplies

Based on past experience, cost without drag-cable replacements on mucker is 9.60 per lin ft of tunnel. Cost including drag-cable replacements is $10.90 per lin ft of tunnel.

Adit cost:	
688 lin ft at $9.60	$6,605
Cost per yd³ on 4,472 yd³	$1.48
Tunnel cost:	
2,000 lin ft at $9.60	$ 19,200
21,650 lin ft at $10.90	235,985
Total cost	$255,185
Cost per yd³ based on 430,842 yd³	$0.59

e. Cost of Fuel, Lube, and Repair Parts

	Unit Cost per Hour	Rubber-Tired Equipment				Rail Equipment in Tunnel					
		Adit		Tunnel		Single Heading		Alternating Heading		Double Heading	
		Oper. hour	Daily Cost	Oper. hour	Daily Cost	Oper. hour	Daily Cost	Oper. hour	Daily Cost	Oper. hour	Daily Cost
Crane, 15 T	$16.00	8	$128.00	8	$128.00	8	$128.00	8	$128.00	8	$128.00
Conway mucker	36.00					12	432.00	24	864.00	24	864.00
Truck for drill jumbo	11.20	3	33.60	6	67.20						
Drill jumbo	1.20	12	14.40	24	28.80	12	14.40	24	28.80	24	28.80
Drills	1.25	60	75.00	108	135.00	132	165.00	264	330.00	264	330.00
Floor	5.00					24	120.00	48	240.00	48	240.00
Locomotives	7.50					72	540.00	72	540.00	120	900.00
Cars	0.15					720	108.00	720	108.00	1,440	216.00
Fans	0.25	36	9.00	72	18.00	312	78.00	144	36.00	312	78.00
Pumps	1.20			24	28.80	48	57.60	24	28.80	76	91.20
Compressors	2.10	48	100.80	96	201.60	120	252.00	120	252.00	216	453.60
Car dumpers	2.50					12	30.00	20	50.00	24	60.00
Tractor	19.00	8	152.00	8	152.00	8	152.00	8	152.00	8	152.00
Front-end loader	21.00	20	420.00	24	504.00	8	168.00	8	168.00	8	168.00
Welding supplies							35.00		65.00		80.00
Miscellaneous							250.00		200.00		250.00
Total			$932.80		$1,263.40		$2,530.00		$3,190.60		$4,039.60
Equivalent number of three-shift days		20		43		34		51		184	
Total cost			$18,656		$54,326		$86,020		$162,721		$743,286
Total cost of tunnel											$1,046,353
Yardage per takeoff			4,472								430,842
Cost per yd³			$4.17								$2.43

525

f. Roadway Gravel, Ballast, and Miscellaneous RR Supplies

	Adit	Tunnel
Length, lin ft	688	23,650
Cost at $5.00/lin ft	$3,440	$118,250
Yd³ of excavation	4,472	430,842
Cost per yd³	$0.77	$0.27

g, h, i. Supply Cost Based on Manpower

	Rubber		Rail		
	Adit	Main Tunnel	Single Heading	Alternating Heading	Double Heading
Crew size	39	57	86	86	152
Cost per day:					
g. Hard hats, raingear, safety at $1.50/man-day	$ 58.50	$ 85.50	$129.00	$129.00	$ 228.00
h. Misc. supplies and sanitation at $7.00/man-day	273.00	399.00	602.00	602.00	1,064.00
i. Small tools and misc. at $5.00/man-day	195.00	285.00	430.00	430.00	760.00
Number of equivalent three-shift days	20	43	34	51	184
Total cost:					
g. Hard hats, etc	$1,170	$ 3,676	$ 4,386	$ 6,579	$ 41,952
h. Misc. supplies, etc	5,460	17,157	20,468	30,702	195,776
i. Small tools and misc	3,900	12,225	14,620	21,930	139,840

	Adit, 4,472 yd³		Tunnel, 430,842 yd³	
Combining:	Total	Per yd³	Total	Per yd³
g. Hard hats, etc	$1,170	$0.26	$ 56,593	$0.13
h. Misc. supplies, etc	5,460	1.22	264,103	0.61
i. Small tools and misc	3,900	0.87	188,615	0.44

Summary of Excavation Supplies

	Adit, 4,472 yd³	Main Tunnel, 430,842 yd³
	Cost per yd³	
a. Powder and exploders	$ 2.34	$1.60
b. Steel and bits	3.02	1.74
c. Power	2.18	1.29
d. Electrical supplies	1.48	0.59
e. Fuel, lube, repair parts	4.17	2.43
f. Gravel, ballast, RR supplies	0.77	0.27

	Adit, 4,472 yd³	Main Tunnel, 430,842 yd³
	Cost per yd³	
g. Hard hats, raingear, safety	0.26	0.13
h. Misc. supplies and sanitation	1.22	0.61
i. Small tools and misc. supplies	0.87	0.44
Total	$16.31	$9.10

**Total Cost of Supplies for Adit
This Cost To Be Tabulated in
Cost of Plant and Equipment**

Supplies from above		
4,472 yd³ at $16.31		$72,938
Supports		
12 Sets at 802 lb at $0.48		4,621
Timber		
4.4 MBM at $320.00		1,408
Portal protection		4,000
Total		$82,967

Cost of Concrete-Lining Labor

Bid Item 9 　　Takeoff quantity　110,453 yd³
　　　　　　　　　　Bid quantity　　　110,000 yd³

This labor is divided into eight different operations. Crew days for each operation were computed in step XI.

a. Remove Vent Line—26 Days

Labor	No. of Men/Shift	No. of Men/Day	Rate/ Manshift	Total
Shifter	1	3	$207.51	$ 622.53
Miner	3	9	192.64	1,733.76
Crane operator	1	3	228.60	685.80
Locomotive operator	2	6	219.14	1,314.84
Brakeman	1	3	209.14	627.42
Mechanic	1	3	221.79	665.37
Total per day		27		$5,649.72
For 26 days				$ 146,893

b. Retimber—21 Days

Labor	No. of Men/Shift	No. of Men/Day	Rate/ Manshift	Total
Walker		1	$218.44	$ 218.44
Shifter	1	3	207.51	622.53

Labor	No. of Men/Shift	No. of Men/Day	Rate/ Manshift	Total
Miner	8	24	192.64	4,623.36
Locomotive operator	1	3	219.14	657.42
Brakeman	1	3	209.14	627.42
Crane operator	1	3	228.60	685.80
Compressor operator	1	3	210.88	632.64
Mechanic		1	221.79	221.79
Electrician		1	255.15	255.15
Party chief		1	252.00	252.00
Instrument man		1	237.50	237.50
Head chainman		1	219.14	219.14
Warehouseman		1	182.13	182.13
Total per day		46		$9,435.32
For 21 days				$ 198,142

c. Primary Cleanup—51 Days

Labor	No. of Men/Shift	No. of Men/Day	Rate/ Manshift	Total
Walker		1	$218.44	$ 218.44
Shifter	1	3	207.51	622.53
Mucker operator	1	3	244.63	733.89
Miner	12	36	192.64	6,935.04
Locomotive operator	2	6	219.14	1,314.84
Brakeman	1	3	209.14	627.42
Electrician foreman		1	266.66	266.66
Electrician		3	255.15	765.45
Mechanic foreman		1	260.12	260.12
Mechanic		3	221.79	665.37
Compressor operator	1	3	210.88	632.64
Crane operator	1	3	228.60	685.80
Outside labor	1	3	162.16	486.48
Warehouseman		1	182.13	182.13
Party chief		1	252.00	252.00
Instrument man		1	237.50	237.50
Head chainman		1	219.14	219.14
Total per day		73		$15,105.45
For 51 days				$ 770,378

d. Invert Concrete—38 Days

Labor	No. of Men/Shift	No. of Men/Day	Rate/ Manshift	Total
Walker		1	$218.44	$ 218.44
Aggregate reclaim	1	3	210.88	632.64
Mix-plant operator	1	3	221.79	665.37
Concrete loader	1	3	162.16	486.48
Mix-plant laborer	1	3	162.16	486.48
Shifter	1	3	207.51	622.53

Labor	No. of Men/Shift	No. of Men/Day	Rate/ Manshift	Total
Locomotive operator	3	9	219.14	1,972.26
Brakeman	1	3	209.14	627.42
Conveyor operator	1	3	215.02	645.06
Bridge operator	1	3	215.02	645.06
Invert screed operator	1	3	215.02	645.06
Vibrator operator	3	9	187.90	1,691.10
Finisher	4	12	192.64	2,311.68
Bridge setup	2	6	192.64	1,155.84
Pipeman and pump	3	9	232.51	2,092.59
Track crew foreman	1	3	196.63	589.89
Track crew	3	9	184.90	1,664.10
Compressor operator	1	3	210.88	632.64
Mechanic foreman		1	260.12	260.12
Mechanic	3	9	221.79	1,996.11
Electrician foreman		1	266.66	266.66
Electrician	1	3	255.15	765.45
Warehouseman		1	182.13	182.13
Party chief		1	252.00	252.00
Instrument man		1	237.50	237.50
Head chainman		1	219.14	219.14
Total per day		106		$21,963.75
For 38 days				$ 834,622
Forms under shaft, 78 ft² at $5.20 for labor				406
Total				$ 835,028

e. Arch Concrete—89 Days

Labor	No. of Men/Shift	No. of Men/Day	Rate/ Manshift	Total
Walker		1	$218.44	$ 218.44
Aggregate reclaim	1	3	210.88	632.64
Mix-plant operator	1	3	221.79	665.37
Concrete loader	1	3	162.16	486.48
Mix-plant laborer	1	3	162.16	486.48
Shifter	1	3	207.51	622.53
Locomotive operator	3	9	219.14	1,972.26
Brakeman	1	3	209.14	627.42
Conveyor operator	1	3	215.02	645.06
Gun operator	1	3	232.51	697.53
Nozzleman	1	3	192.64	577.92
Pipeman	2	6	232.51	1,395.06
Vibrator operator	3	9	187.90	1,691.10
Finish and cure	2	6	192.64	1,155.84
Form foreman	1	3	207.51	622.53
Form mover	4	12	184.90	2,218.80
Pipe mover	2	6	184.90	1,109.40
Compressor operator	1	3	210.88	632.64
Mechanic foreman		1	260.12	260.12

Labor	No. of Men/Shift	No. of Men/Day	Rate/ Manshift	Total
Mechanic	3	9	221.79	1,996.11
Electrician foreman		1	266.66	266.66
Electrician	1	3	255.15	765.45
Warehouseman		1	182.13	182.13
Party chief	1	1	252.00	252.00
Instrument man	1	1	237.50	237.50
Head chainman	1	1	219.14	219.14
Total per day		100		$20,636.61
For 89 days				$1,836,658
Bulkhead forms:				
20 Bulkheads = 1,540 ft^2				
at $6.40 for labor				9,856
Forms under shaft and at adit:				
4,936 ft^2 at $5.20 for labor				25,667
Total				$1,872,181

f. Low-Pressure Grout—80 Days

The estimate is based on 3 ft^3/lin ft of tunnel. If grout progresses 300 lin ft/ day, the same progress as arch, then 900 ft^3/day, working 16 h = 56 ft^3/h. Setup and removal time of one day is included in the 80 days total time.

Labor	No. of Men/Shift	No. of Men/Day	Rate/ Manshift	Total
Shifter	1	2	$207.51	$ 415.02
Driller	2	4	192.64	770.56
Connector	2	4	184.90	739.60
Grout-pump operator	1	2	232.51	465.02
Laborer	2	4	184.90	739.60
Locomotive operator	1	2	219.14	438.28
Outside laborer	1	2	162.16	324.32
Mechanic	1	2	221.79	443.58
Total per day		22		$4,335.98
For 80 days				$ 346,878

g. Final Cleanup—25 Days

Labor	No. of Men/Shift	No. of Men/Day	Rate/ Manshift	Total
Walker		1	$218.44	$ 218.44
Shifter	1	3	207.51	622.53
Track crew foreman	1	3	196.63	589.89
Track crew	6	18	184.90	3,328.20
Pipe crew	4	12	184.90	2,218.80

Labor	No. of Men/Shift	No. of Men/Day	Rate/ Manshift	Total
Finisher	1	3	192.64	577.92
Cleanup labor	4	12	184.90	2,218.80
Locomotive operator	3	9	219.14	1,972.26
Brakeman	1	3	209.14	627.42
Compressor operator	1	3	210.88	632.64
Mechanic foreman		1	260.12	260.12
Mechanic	2	6	221.79	1,330.74
Electrician foreman		1	266.66	266.66
Electrician	1	3	255.15	765.45
Warehouseman		1	182.13	182.13
Crane operator	1	3	228.60	685.80
Outside laborer	1	3	162.16	486.48
Total per day		85		$16,984.28
For 25 days				$424,607

h. Weekend Maintenance Labor—37 Weekends

at $1,675 $61,975

Summary of Labor Cost for Concrete Tunnel Lining, Bid Item 9

a. Remove vent line	$ 146,893	
b. Retimber	198,142	
c. Primary cleanup	770,378	
d. Invert concrete	835,028	
e. Arch concrete	1,872,181	
f. Low-pressure grout	346,878	
g. Final cleanup	424,607	
h. Weekend maintenance	61,975	
Total labor, bid item 9	$4,656,082	
Cost per yd³ based on takeoff quantity of 110,453 yd³		$42.16

Cost of Concrete Lining Supplies

Bid Item 9

Takeoff quantity   110,453
Bid quantity   110,000

The supplies are consumed in seven different operations. Crew days for each operation were computed in step XI.

a. Remove Vent Line—26 Days

	Quantity	Rate	Amount	Cost per Day
Supplies:				
Power and water				
(charged to other items)				
Fuel, oil, lube, repair parts, etc.:				
Locomotives	48 h	$7.50	$360.00	
Crane	24 h	16.00	384.00	
Misc.			100.00	$ 844.00
Other supplies:				
Misc. supplies	27 men	$7.50/day		202.50
Small tools, etc.	27 men	5.00/day		135.00
Rubber, raingear, safety	27 men	1.50/day		40.50
Total per day				$ 1,222.00
For 26 days				$31,772

b. Retimber—21 Days

	Quantity	Rate	Amount	Cost per Day
Supplies:				
Power, 650 hp = 484 kW:				
At 50% load factor	242 kW			
At 24 h/day	5,808 kWh	$0.06		$ 348.48
Electrical supplies				100.00
Fuel, lube, oil, repair parts, etc.:				
Compressors	48 h	$2.10	$100.80	
Locomotives	24 h	7.50	180.00	
Crane	24 h	16.00	384.00	
Misc.			80.00	744.80
Miscellaneous supplies	46 men	$7.50/day		345.00
Small tools and supplies	46 men	5.00/day		230.00
Rubber, raingear, safety	46 men	1.50/day		69.00
Total per day				$ 1,837.28
For 21 days				$ 38,583

c. Primary Cleanup—51 Days

	Quantity	Rate	Amount	Cost per Day
Supplies:				
Power, connected load of 1,300 hp				
970 kW at 50% load and 24 h/day	11,640 kWh	$0.06		$ 698.40
Electrical supplies				100.00
Fuel, oil, lube, repair parts, etc.:				
Front-end loader	24 h	$21.00	$504.00	

	Quantity	Rate	Amount	Cost per Day
Locomotive	48 h	7.50	360.00	
Compressors	96 h	2.10	201.60	
Crane	24 h	16.00	384.00	
Pumps and misc.			120.00	1,569.60
Misc. supplies	73 men	$7.50/day		547.50
Small tools and supplies	73 men	5.00/day		365.00
Rubber, raingear, safety	73 men	1.50/day		109.50
Total per day				$ 3,390.00
For 51 days				$172,890

d. Invert Concrete—38 Days

	Quantity	Rate	Amount	Cost per Day
Supplies, except aggregate, cement and admix:				
Power:				
Connected load—3,100 hp				
Connected load—2,312 kW				
At 25% load factor for 24 h	13,872 kWh	$0.06		$ 832.32
Electrical supplies				125.00
Track supplies	23,560 lin ft	$1.00/lin ft	$23,560	
Cost per day for 38 days (These supplies used for relaying the track on the invert.)				620.00
Fuel, lube, oil, repair parts, etc.:				
Mix plant	24 h	$8.00	$192.00	
Agitator cars	384 h	$1.00	384.00	
Locomotives	72 h	$7.50	540.00	
Belts			100.00	
Bridge			50.00	
Screed			50.00	
Compressors	96 h	$2.10	201.60	
Vibrators	1,500 yd³	$0.30	450.00	
Misc.			100.00	2,067.60
Misc. supplies	106 men	$7.50/day		795.00
Small tools, etc.	106 men	$5.00/day		530.00
Rubber, raingear, safety	106 men	$1.50/day		159.00
Total per day				$ 5,128.92
For 38 days				$194,899
Forms under shaft	78 ft²	$4.50		351
Total supplies				$195,250

e. Arch Concrete—89 Days

	Quantity	Rate	Amount	Cost per Day
Supplies, except aggregate, cement, and admix:				
Power (as above)				$ 832.32
Electrical supplies				125.00
Form hardware	5,124 ft² day	$0.30		1,537.20
Fuel, oil, lube, repair parts, etc.:				
Mix plant	24 h	$8.00	$192.00	
Agitator cars	384 h	$1.00	384.00	
Locomotives	72 h	$7.50	540.00	
Belt			75.00	
Press Weld and pipe	24 h	$2.50	60.00	
Vibrators	1,344 yd³	$0.30	403.20	
Compressors	96 h	$2.10	201.60	
Misc.			100.00	$ 1,955.80
Misc. supplies	100 men	$7.50/day		750.00
Small tools, etc.	100 men	$5.00/day		500.00
Rubber, raingear, safety	100 men	$1.50/day		150.00
Total per day				$ 5,850.32
For 89 days				$520,678
Bulkhead forms:				
20 bulkheads	1,540 ft²	$4.50		6,930
Forms under shaft and at adit	4,936 ft²	$4.50		22,212
Total				$549,820

f. Low-Pressure Grout—80 Days

	Quantity	Rate	Amount	Cost per Day
Supplies:				
Cement by owner				
Sand	450 ft³	$0.30		$ 135.00
Oil, lube, repair parts, etc.:				
Drills	48 h	$1.25	$ 60.00	
Grout pump	24 h	$5.00	120.00	
Locomotive	24 h	$7.50	180.00	
Grout hose and pipe			25.00	385.00
Misc. supplies	22 men	$7.50/day		165.00
Small tools, etc.	22 men	$5.00/day		110.00
Rubber, raingear, safety	22 men	$1.50/day		33.00
Total per day				$ 828.00
For 80 days				$66,240

g. Final Cleanup—25 Days

	Quantity	Rate	Amount	Cost per Day
Supplies:				
Power (as above)				$ 832.32
Electrical supplies				125.00
Fuel, oil, lube, repair parts, etc.:				
Locomotives	72 h	$7.50	$540.00	
Compressor	48 h	2.10	100.80	
Crane	24 h	16.00	384.00	
Other			100.00	1,124.80
Misc. supplies	85 men	$7.50/day		637.50
Small tools, etc.	85 men	5.00/day		425.00
Rubber, raingear, safety	85 men	1.50/day		127.50
Total per day				$ 3,272.12
For 25 days				$81,803

Summary of Cost of Supplies for Concrete Tunnel Lining, Bid Item 9

Cost of supplies as detailed:		
a. Remove vent line	$ 31,772	
b. Retimber	38,583	
c. Primary cleanup	172,890	
d. Invert concrete	195,250	
e. Arch concrete	549,820	
f. Low-pressure grout	66,240	
g. Final cleanup	81,803	
Subtotal	$1,136,358	
Other supplies:		
Cement waste:		
154,263 yd^3 of concrete at $0.25/yd^3	38,566	
Admixture:		
154,263 yd^3 at $0.25/yd^3	38,566	
Total supplies	$1,213,490	
Cost per yd^3, based on takeoff quantity of 110,453 yd^3		$10.99

Cost of Aggregate (Subcontract)

Total yards of concrete, incl. overbreak concrete	154,263 yd^3
Tonnage aggregate at 1.65 tons/yd^3	254,534 tons
Cost of aggregate at $6.10/ton	$1,552,657
Cost per yd^3 of concrete, based on takeoff yardage of 110,453 yd^3	$14.06

Direct Cost of Other Bid Items

Bid Item 2—Steel Sets

Labor/lb (see bid item 1)	$0.13
Permanent materials/lb	0.48
Total per lb	$0.61

Bid Item 3—Timber

Pay timber is within 4 in of outside flange of sets. Estimate assumes that one-half of timber is pay timber.

Labor/MBM from bid item 1	$125.00
Pay timber/MBM	320.00
Nonpay timber supplies/MBM	320.00
Total per MBM	$765.00

Bid Item 4—Roof Bolts

Takeoff = 1,520 lin ft of tunnel has four roof bolts across crown on 4-ft centers, or one roof bolt per lineal foot of tunnel; 1,520 roof bolts of 8-ft length = 12,160 lin ft. Two men on rear-end jumbo drill and set roof bolts. Time to go through roof-bolt section is 1,520 lin ft ÷ 60 = 25.33 days; 25.33 days at 6 men/day = 152 man-days.

		Total	For 12,160 lin ft
Labor for 152 man-days at $192.64/day		$29,281	$2.41
Supplies:			
Drill maintenance:			
2 Drills at 12 h/day = 24 h/day			
24 × 25.33 = 608 h			
608 h at $1.25	$ 760.00		
Misc. cost:			
152 man-days at $18.20	2,766.40		
Bit and steel cost for 12,160 lin ft of drilling:			
Steel at $0.06/lin ft	729.60		
Bits at $47.00/200 = $0.235/lin ft	2,857.60	$ 7,114	0.59
Permanent materials at $1.60/lin ft		19,456	1.60
Total cost		$55,851	$4.60

Bid Item 5—Drill Feeler
Holes ahead of Tunnel Excavation

A length of 24 lin ft is assumed. Therefore, at a 60 lin ft/hr drilling rate, the heading crew will be delayed 24 min, or 0.4 h. On a double heading, crew cost of one-half the crew per hour is $649.12

Labor = $649.12 × 0.4 ÷ 24	$10.82/lin ft
Supplies	5.30/lin ft
Total cost	$16.12/lin ft

Bid Item 6—Drill Grout
Holes as an Aid to Tunnel Driving

This item will be a factor if indications are that grouting-off in front of tunnel driving should be done. Drill 10 holes of 60-ft length with drifters. Because of drill-hole length, cut drilling speed down to 40 lin ft/h.

Number of holes	10	
Number of drills	10	
Lin ft drilled	60	
Time required	$60 ÷ 40 = 1\frac{1}{2}$ hr	
Cost per h for ½ crew of double heading	$649.12	
Cost for 1½ h	$973.68	
Cost per lin ft of labor = $973.68 ÷ (10 × 60)		$1.63
Supplies		5.30
Total cost per lin ft		$6.93

Bid Item 7—Grout as an Aid to Tunnel Driving

This item will be a factor if indications are that grouting-off in front of tunnel driving should be done. Assume grout holes per round will take 100 ft^3 and to place this 100 ft^3 will take 2 h.

Labor for 100 ft^3 = 2 × $649.12 = $1298.24	
Labor cost per ft^3	$12.98
Misc. supplies	2.12
Cement	4.50
Total cost per ft^3	$19.60

Bid Item 8—Shaft Excavation

Takeoff quantities:	
Shaft	214 lin ft
Excavation	5,299 yd^3

Raise

A 6-ft-diameter raise will be excavated from tunnel; muck will fall into bottom of tunnel, and will be picked up with front-end loader and truck. Average progress in raise is 5 ft/shift, or 15 ft/day.

Yd³/lin ft in raise	1.05	yd³
For total raise	225	yd³
Per day	15	yd³
Broken yd/day	22.5	yd³
Number of days = 214 ÷ 15	14.25	days
Allowance for start-up	2.75	days
	17	days
Number of rounds = 214 ÷ 5	43	rounds

Crew Cost per Day for Shaft Raising

	No. of Men/Shift	No. of Men/Day	Rate/ Manshift	Total
Shifter	1	3	$215.38	$ 646.14
Miner	2	6	200.38	1,202.28
Nipper	1	3	193.35	580.05
Front-end loader operator		1	252.00	252.00
Compressor operator	1	3	210.88	632.64
Mechanic		1	221.79	221.79
Electrician		1	255.15	255.15
Total per day		18		$3,790.05

Cost of Shaft Raising

	Total	Cost/lin ft
Labor:		
17 days at $3,790.05	$64,431	$301.08
Supplies:		
Drill steel and bits:		
5-ft round = 7-ft holes at		
18 holes/round = 126		
lin ft round × 43 = 5,418 lin ft		
Steel, 5,418 lin ft at $0.06	325	1.52
Bits, 5,418 lin ft ÷ 200 × $47.00	1,273	5.95
Powder and exploders:		
Exploders, 774 at $0.75	580	2.71
Powder, 225 yd³ at 8 lb at $0.55	990	4.63
Power, $100 day for 17 days	1,700	7.94
Electrical supplies	500	2.34
Fuel, lube, equipment repair parts, etc.:		

	Cost/h	Cost/day
Drills, 24 h	$ 1.25	$ 30.00
Alimak, 24 h	5.00	120.00
Front-end loader, 8 h	21.00	168.00
2 Portable compressors, 48 h	8.00	384.00
Total		$702.00

	Total	Cost/lin ft
For 17 days	11,934	55.77

		Total	Cost/lin ft
Misc. supplies:			
Rubber, raingear, safety	$ 1.50 man-day		
Misc. supplies and sanitation	7.50 man-day		
Small tools, etc.	5.00 man-day		
Total	$ 14.00 man-day		
Daily cost for 18 men	$252.00		
For 17 days		4,284	20.02
Total supplies		21,586	100.87
TOTAL COST		$86,017	$401.95

Downdrill

Total volume	5,299 yd^3	
Less shaft raising	225	
Remainder	5,074	
Average yd^3/lin ft	23.71 yd^3	
Average 80 holes/round at 7 lin ft/hole		
Total drilling	560 lin ft/round	
Use drill jumbo mounting four drifter drills:		
Lin ft drilled/drill each round	140 lin ft	
At 30 lin ft/h	4⅔ h	
Set up jumbo	⅓ h	
Remove jumbo	⅓ h	
Load holes	1⅓ h	
Shoot and ventilate	½ h	
Muck 180 yd^3 of broken material at 30 yd^3/h	6 h	
Hand clean up bottom	1 h	
Total cycle	14.16 h = 2 shifts	
Therefore, base production on 7½ lin ft for three shifts:		
214 ÷ 7½ lin ft	28.5 days	
Allow for start-up	1.5 days	
Total	30 days	

Crew Cost per Day of Downdrilling

	No. of Men/Shift	No. of Men/Day	Rate/ Manshift	Total
Shifter	1	3	$215.38	$ 646.14
Miner	3	9	200.38	1,803.42
Nipper	1	3	193.35	580.05
Front-end loader operator	1	3	252.00	756.00
Compressor operator	1	3	210.88	632.64
Crane operator	1	3	228.60	685.80
Mechanic	1	3	221.79	665.37
Electrician		1	255.15	255.15
Crane oiler	1	3	201.67	605.01
Total per day		31		$6,629.58

	Total	Per yd, 5,074 yd³
Estimated cost:		
Labor, 30 days at $6,629.58	$198,887	$39.20
Supplies:		
Steel:		

		Total	Per yd, 5,074 yd³
Total lin ft in 80 holes of 7 lin ft/hole for 43 rounds = 24,080 lin ft			
24,080 lin ft at $0.06	$ 1,445		
Bits at 24,080 lin ft ÷ 200 × $47.00	5,659	7,104	1.40
Exploders and powder:			
80 × 43 = 3,440 exploders at $0.75	$ 2,580		
Powder, 5,074 yd³ at 4 lb at $0.55	11,163	13,743	2.71
Power, $100/day for 30 days		3,000	0.59
Electrical supplies		1,000	0.20
Fuel, lube, and equipment repair parts:			
Cost per day:			
Crane, 65 T 24 h at $20.00	$ 480		
Drills 48 h at $ 1.25	60		
Compressor 48 h at $ 8.00	384		
Small mucker 12 h at $10.00	120		
Front-end loader 8 h at $21.00	168		
Pumps and fan	25		
Misc.	50		
Total per day	$ 1,287		
For 30 days		38,610	7.61
Misc. supplies:			
Cost per man-day			
Hard hats, raingear, safety	$1.50		
Misc. supplies and sanitary	7.50		
Small tools and misc.	5.00		
Cost per man-day	$ 14.00		
Cost per day for 31 men	$434.00		
For 30 days		13,020	2.56
Total cost of supplies		$76,477	$15.07
TOTAL COST		$275,364	$54.27

Summary of Cost of Shaft Excavation

	Labor	Supplies	Total
Raise	$ 64,431	$21,586	$ 86,017
Downdrill	198,887	76,477	275,364
Total	$263,318	$98,063	$361,381
Cost per yd³ for 5,299 yd³	$49.69	$18.51	$ 68.20

Bid Item 10—Shaft Concrete

Bid quantity	1,900 yd³
Takeoff quantity:	
Pay	1,894 yd³
Overbreak	496 yd³
Total	2,390 yd³

Average cubic yard per lineal foot = 2,390 ÷ 214 = 11 yd³/lin ft. With slip forms, 1 lin ft/h is poured at the same time as the tunnel arch is poured toward the outlet. Two trains are used, one car per train. Total hours of pouring are as follows:

Section	Lin ft	Lifts	Hours
E-E	25	3	36
D-D	155	Slip-form	155
C-C	34	Slip-form	34
Start-up and delays			23
Total			248 hours = 31 shifts

Mixing-plant operation and cost of power and other utilities will not be charged to this operation, as cost is absorbed in tunnel arch concrete.

To clean up shaft and remove tights, allow five shifts.

Labor Cost per Shift

	No. of Men/Shift	Cost of Man/Shift	Total Shift
Shifter	1	$215.38	$ 215.38
Crane operator	1	228.60	228.60
Crane oiler	1	201.67	201.67
Miner	4	200.38	801.52
Front-end-loader operator	1	252.00	252.00
Total per shift	8		$1,699.17
For 5 shifts			$8,496

Supply Cost per Shift

	Unit	Cost/Unit	Cost/Shift
1 Crane, 65 T	8 h	$20.00	$ 160.00
1 Front-end loader	8 h	21.00	168.00
Misc. supplies	8 men	7.50	60.00
Small tools	8 men	5.00	40.00
Rubber and safety	8 men	1.50	12.00
Misc.			50.00
Total per shift			$ 490.00
For 5 shifts			$2,450

Concrete Placing

	No. of Men/Shift	Cost of Man/Shift	Total Shift
Labor for placing:			
Shifter	1	$215.38	$ 215.38
Locomotive operator	2	219.14	438.28
Concrete transfer	1	184.90	184.90
Crane operator	1	228.60	228.60
Crane oiler	1	201.67	201.67
Chute operator	1	200.38	200.38
Vibrator man and laborer	4	200.38	801.52
Finisher	1	200.38	200.38
Mechanic	1	221.79	221.79
Total	13		$ 2,692.90
For 31 shifts			$83,480
Placing supplies:			
Fuel, oil, lube, repair parts:			
Locomotives	16 h at $7.50	$120.00	
Crane, 65 T	8 h at $20.00	160.00	
Misc.		50.00	$ 330.00
Misc. supplies	13 men/shift at $7.50		97.50
Small tools	13 men/shift at $5.00		65.00
Rubber, safety	13 men/shift at $1.50		19.50
Total			$ 512.00
For 31 shifts			$15,872

Slip Forms

		Labor	Supplies	Total
155-ft section	72.33 ft²/lin ft			
34-ft section	89.00 ft²/lin ft			
	161.33 ft²/lin ft			
Purchase 3 lin ft of each size =				
484 ft² at $20.00			$ 9,680	$ 9,680
Wood lining, 8 ft long, 1,291 ft² at $1.00			1,291	1,291
Superstructure		$ 2,500	2,500	5,000
Jack rental			20,000	20,000
Platform rental			5,000	5,000
Labor for raising, 20 shifts for 2 men =				
40 shifts at $207.15		8,286		8,286
Supervising		2,500		2,500
Misc. supplies			2,000	2,000
Total for slip forms		$13,286	$40,471	$53,757

Other Forms

	Labor	Supplies	Total
2,107 ft² at $5.20 and $4.50	$10,956	$ 9,481	$20,437

Contact Grout—4 Shifts

	No. of Men/Shift	Cost of Man/Shift	Total Shift
Labor cost per shift:			
Shifter	1	$215.38	$ 215.38
Crane operator	1	228.60	228.60
Crane oiler	1	201.67	201.67
Grout-pump operator	1	232.51	232.51
Laborer	1	184.90	184.90
Miner	2	200.38	400.76
Locomotive operator	1	219.14	219.14
Mechanic	1	221.79	221.79
Cost per shift	9		$1,904.75
For 4 shifts			$7,619

Supply Cost per Shift

	Unit	Cost/Unit	Cost/Shift
Crane, 65 T	8 h	$20.00	$ 160.00
Locomotive	8 h	7.50	60.00
Grout pump	8 h	5.00	40.00
Misc. supplies	9 men	7.50	67.50
Small tools	9 men	5.00	45.00
Rubber and safety	9 men	1.50	13.50
Misc.			40.00
Total per shift			$ 426.00
For 4 shifts			$1,704
Sand			120
Total supplies			$1,824

Final Cleanup

The same crew used for first cleanup will be used for the final cleanup, for two shifts.

Labor	$1,699 × 2	$3,398
Supplies	490 × 2	$ 980
Total		$4,378

Summary of Cost of Shaft Concrete, Bid Item 10

	Labor	Supplies	Subcontract	Total Direct
Clean shaft, remove tights	$ 8,496	$ 2,450		$ 10,946
Aggregates, 2,390 yd³ at 1.65 tons = 3,944 tons at $6.10			$24,058	24,058
Cement waste, 2,390 yd³ at $0.25		598		598
Admix, 2,390 yd³ at $0.25		597		597
Placing cost	83,480	15,872		99,352
Slip forms	13,286	40,471		53,757
Other forms	10,956	9,481		20,437
Contact grout	7,619	1,824		9,443
Final cleanup	3,398	980		4,378
Total cost	$127,235	$72,273	$24,058	$223,566
Per yd³ based on 1,894 yd³	$ 67.18	$ 38.16	$ 12.70	$ 118.04

Bid Item 11—Cement for Concrete and Low-Pressure Grout

Bid quantity   163,000 bbl

Our takeoff:
Tunnel lining:
| Pay | 110,453 yd³ |
| Nonpay | 43,810 yd³ |

Shaft lining:
| Pay | 1,894 yd³ |
| Nonpay | 496 yd³ |

Total yd³ of concrete   156,653 yd³

Cement at 1 bbl/yd³ for concrete; grouting cement at 1½ sacks/lin ft of tunnel

Total cement   162,565 bbl

Plugged price/bbl delivered to mix plant $16.00

Bid Item 12—Steel Reinforcing

Bid quantity   285,000 lb

Plugged price/lb from subcontractor $0.44
Allow for servicing of the subcontractor   0.03

Total cost per lb $0.47

Bid Item 13—Embedded Anchor Bolts

Plugged price/lb for material	$1.60
Estimated cost of installation:	
Labor	2.50
Supplies	0.20
Total cost per lb	$4.30

Bid Item 14—Embedded Vent Pipe

Plugged price/lb for material	$1.10
Estimated cost of installation:	2.00
Misc. supplies	0.15
Total cost per lb	$3.25

Bid Items 15, 16, and 17—Consolidation Grouting

Plugged subcontract prices:	
Drilling grout holes, per lin ft	$10.00
Consolidation grouting, per ft^3	10.00
Cement for grouting, per sack	5.00

Adit-Plug Concrete

Adit-plug concrete will be charged to plant and equipment. Total concrete is 600 yd.^{3} The pour will be made in 5-ft lifts or 5 days at one shift per day. Crew will be the same as for arch pour, less form movers and pipe movers.

		Labor	Supplies	Total
Arch-pour crew per day	100 men	$20,636.61	$5,850.32	$26,486.93
Less form movers and pipe movers	18 men	3,328.20	252.00	3,580,20
Remainder for three shifts	82 men	$17,308,41	$5,598.32	$22,906.73
Cost for one shift	27 men	$ 5,769.47	$1,866.11	$ 7,635.58
Cost for pouring plug, 5 shifts		$28,847	$9,331	$38,178

Summary Cost of Adit-Plug Concrete

	Labor	Supplies	Total
Aggregates, 990 tons at $6.10		$ 6,039	$ 6,039
Admix, 600 yd^3 at $0.25		150	150
Cement, 600 bbls at $16.96		10,176	10,176
Placing cost	$28,847	9,331	38,178
Forms, 176 ft^2 at $5.20 & $4.50	915	792	1,707
Total cost to plant and equipment	$29,762	$26,488	$56,250

XIII. TABULATION OF DIRECT COST

TABLE 18.16 Sierra Tunnel Estimate: Summary of Direct Cost

Bid Item	Description	Quantity	Labor Unit	Labor Amount	Supplies Unit	Supplies Amount	Permanent Materials Unit	Permanent Materials Amount	Subcontracts Unit	Subcontracts Amount	Total Direct Cost Unit	Total Direct Cost Amount
Tunnel excavation:												
1	Tunnel excavation	431,000 yd³	$ 17.42	$7,508,020	$ 9.10	$3,922,100					$26.52	$11,430,120
2	Steel sets	2,290,000 lb	0.13	297,700			0.48	1,099,200			0.61	1,396,900
3	Timber	400 MBM	125.00	50,000	320.00	128,000	320.00	128,000			765.00	306,000
4	Roof bolts	12,000 lin ft	2.41	28,920	0.59	7,080	1.60	19,200			4.60	55,200
5	Drill exploratory holes	300 lin ft	10.82	3,246	5.30	1,590					16.12	4,836
6	Drill grout holes as aid to tunnel driving	2,400 lin ft	1.63	3,912	5.30	12,720					6.93	16,632
7	Grout as aid to tunnel driving	400 ft³	12.98	5,192	6.62	2,648					19.60	7,840
	Total tunnel excavation			7,896,990		4,074,138		1,246,400				13,217,528
Shaft excavation:												
8	Shaft excavation	5,300 yd³	49.69	263,357	18.51	98,103					68.20	361,460
Concrete:												
9	Concrete tunnel lining	111,000 yd³	42.16	4,679,760	10.99	1,219,890			14.06	1,560,660	67.21	7,460,310
10	Shaft concrete	1,900 yd³	67.18	127,642	38.16	72,504			12.70	24,130	118.04	224,276
11	Cement for concrete and low-pressure grout	163,000 bbl					16.00	2,608,000			16.00	2,608,000
12	Steel reinforcement in shaft	285,000 lb							0.47	133,950	0.47	133,950
13	Embedded anchor bolts	3,200 lb	2.50	8,000	0.20	640	1.60	5,120			4.30	13,760
14	Embedded vent pipe	600 lb	2.00	1,200	0.15	90	1.10	660			3.25	1,950
	Total concrete			4,816,602		1,293,124		2,613,780		1,718,740		10,442,246
Consolidation grouting:												
15	Drill grout holes	4,000 lin ft							10.00	40,000	10.00	40,000
16	Consolidation grouting	10,000 ft³							10.00	100,000	10.00	100,000
17	Cement for grouting	5,000 sacks							5.00	25,000	5.00	25,000
	Total grouting									165,000		165,000
	TOTAL DIRECT COST			12,976,949		5,465,365		3,860,180		1,883,740		24,186,234

546

XIV. ESTIMATING
PLANT AND EQUIPMENT COST

TABLE 18.17 Sierra Tunnel Estimate: Spread Sheet Summarizing Plant and Equipment Cost

Description	Quantity	Cost/ Unit	Freight and Move-in Cost	Labor	Equipment Invoice	Other Invoice	Total Estimated Cost	Salvage	Net Cost P&E*	No. of Months	Rent/ Month	Total Rent
Rubber-tired excavation equipment:												
1 Truck-mounted jumbo	1	$ 50,000	$ 2,500		$ 50,000		$ 52,500	$ 17,000	$ 35,500			
Drifter drills	10	14,500			145,000		145,000	50,000	95,000			
Burn-cut drill	1	27,500			27,500		27,500	12,000	15,500			
Jibs	10	13,200			132,000		132,000	33,000	99,000			
Drill positioners	9	6,900			62,100		62,100	15,000	47,100			
Hydraulic pump & misc.	3	3,050			9,150		9,150		9,150			
Line oilers	9	250			2,250		2,250		2,250			
Portable 1,200-ft³/ min compressors, rent for 3½ mo	3		600				600		600	10½	$ 3,500	$36,750
Total rubber-tired excavation equipment			$ 3,100		$ 428,000		$ 431,100	$ 127,000	$ 304,100			$36,750
Rail-mounted excavation equipment:												
Jacobs floor	2	311,000	30,000	60,000	622,000	20,000	732,000	200,000	532,000			
Gantry jumbos: Structural frame	2	75,000	4,000	20,000	150,000		174,000	24,000	150,000			
Drifter drills	24	14,500			348,000		348,000	87,000	261,000			
Burn-cut drills	2	27,500			55,000		55,000	15,000	40,000			
Jibs	24	13,200			316,800		316,800	75,000	241,800			
Positioners	22	6,900			151,800		151,800	37,000	114,800			
Hydraulic pumps	6	3,050			18,300		18,300		18,300			
Line oilers	22	250			5,500		5,500		5,500			
Conway muckers, new	2	164,000	5000	1,000	328,000		334,000	111,000	223,000			

TABLE 18.17 Sierra Tunnel Estimate: Spread Sheet Summarizing Plant and Equipment Cost (*Continued*)

Description	Quantity	Cost/Unit	Freight and Move-in Cost	Labor	Equipment Invoice	Other Invoice	Total Estimated Cost	Salvage	Net Cost P&E*	No. of Months	Rent/Month	Total Rent
											Rent	
Conway muckers, standby, used	1	82,000	800	500	82,000		83,300	28,000	55,300			
15-yd³ muck cars	60	9,000	12,000	3,000	540,000		555,000	270,000	285,000			
Flat cars	8	2,000	1,600	400	16,000		18,000	6,000	12,000			
Man cars	6	4,000	1,200	300	24,000		25,500	10,000	15,500			
Vent-line cars	2	12,000	400	100	24,000		24,500	10,000	14,500			
Locomotives, 25-ton	8	88,000	16,000	2,400	704,000		722,400	360,000	362,400			
Powder car	2	7,000	400	100	14,000		14,500	4,000	10,500			
Hand-held drills and spades					10,000		10,000	2,000	8,000			
Car passers and misc.					20,000		20,000	3,000	17,000			
Total rail-mounted excavation equipment			$71,400	$87,800	$3,429,400	$ 20,000	$3,608,600	$1,242,000	$2,366,600			
Shaft excavation equipment:												
Alimak Raise Climber	1	$ 58,750	$ 1,500	$ 1,500	58,750		61,750	35,000	26,750			
Stopers	2	4,300		400	8,600	100	9,100	3,000	6,100			
65-ton truck crane, 3 shifts (rental)	1		1,000				1,000		1,000	2	$ 9,000	$18,000
Drill jumbo (drills from truck jumbo)				6,000		2,000	8,000		8,000			
Man & material cages				2,500		2,500	5,000		5,000			
Total shaft excavation equipment			$ 2,500	$10,400	$ 67,350	$ 4,600	$ 84,850	$ 38,000	$ 46,850			$18,000
Tunnel service equipment:												
70-lb rail, 26,000 track-ft	606 ton	$ 475	$15,000			$ 287,850	$ 302,850	$ 90,900	$ 211,950			

548

Item	Quantity	Unit price	(1)	(2)	(3)	(4)	(5)	(6)	(7)	(8)	(9)
Rail hardware	26,000 tf†	1.25				32,500	33,000	10,000	23,000		
Ties	15,000 ea	6.20				93,000	93,000		93,000		
Stringers on concrete under ties	26,000 tf†	1.20	500			31,200	31,200		31,200		
48-in vent pipe	24,000 lin ft	14.00				336,000	336,000	36,000	300,000		
Fans	27 ea	7,500		5,400	202,500	2,700	213,300	50,000	163,300	10,000	50,000
12-in drainage pipe	25,000 lin ft	6.00	2,700			150,000	150,500	37,500	113,000		
12-in air line	1,000 lin ft	6.00	500			6,000	6,050	1,500	4,550		
10-in air line	24,000 lin ft	5.00	50			120,000	120,400	30,000	90,400		
4-in water line	25,000 lin ft	1.60	400			40,000	40,200	10,000	30,200		
Fittings			200			12,000	12,000		12,000		
Pumps					20,000	20,200	20,200		20,200		
Parkway cable	25,000 lin ft	6.50	200			162,500	162,900	50,000	112,900		
Drag cable	4,000 lin ft	10.00	400			40,000	40,100		40,100		
Light cable	25,000 lin ft	1.00	100			25,000	25,100		25,100		
Telephone	25,000 lin ft	.80	100			20,000	20,100		20,100		
Blasting circuit	25,000	0.80	100			20,000	20,100		20,100		
Transformers					80,000		80,500	40,000	40,500		
Outlets			500			4,000	4,000		4,000		
Total tunnel service equipment			$ 20,850	$ 5,400	$ 302,500	$1,382,750	$1,711,500	$ 355,900	$1,355,600	10,000	$50,000
Tunnel concrete equipment:											
Aggregate reclaim conveyor			2,500	12,500	35,000	1,000	51,000	10,000	41,000		
Mix plant			10,000	15,000	400,000	10,000	435,000	200,000	235,000		
Agitator cars, 8-yd³	20 ea	22,000	16,000	4,000	440,000		460,000	220,000	240,000		
Invert bridge and conveyor			1,000	5,000	60,000		66,000	15,000	51,000		
Invert bridge supports			1,000			15,000	15,000		15,000		
Invert screed				1,000		12,000	14,000		14,000		
Arch forms	400 lin ft	900	7,000	20,000	12,000	360,000	387,000	72,000	315,000		
Arch-form traveler	1 ea	50,000	4,000	2,000	6,000	50,000	56,000	12,500	43,500		
Unloading conveyor			400	600			7,000	1,000	6,000		
Press Weld rental			500				500		500		
Pipe and slick line			500				8,000		8,000		
Pipe and jumbo			800	3,200			9,000		9,000		
Vibrators					10,000		10,000		10,000		
Grouting setup	1 ea	20,000	300		15,000	500	15,300	7,500	7,800		
Total tunnel concrete equipment			$ 44,000	$ 63,300	$ 978,000	$ 448,500	$1,533,800	$ 538,000	$ 995,800		

TABLE 18.17 Sierra Tunnel Estimate: Spread Sheet Summarizing Plant and Equipment Cost (Continued)

Description	Quantity	Cost/ Unit	Freight and Move-in Cost	Labor	Equipment Invoice	Other Invoice	Total Estimated Cost	Salvage	Net Cost P&E*	Rent No. of Months	Rent/ Month	Total Rent
Shaft concrete equipment:												
65-ton crane, 3 shifts (rental)	1									2	$ 9,000	$18,000
Concrete buckets			$ 200		$ 6,000		$ 6,200	$ 3,000	$ 3,200			
Swing spout				$ 1,000		3,000	4,000	1,000	3,000			
Air hoist	2	$ 7,150	200		14,300		14,500	7,150	7,350			
Total shaft concrete equipment			400	1,000	20,300	3,000	24,700	11,150	13,550			18,000
Other facilities:												
Office trailers			1,200	5,000	24,000	5,000	35,200	10,000	25,200			
Shop, warehouse, complete				15,000	20,000	20,000	55,000		55,000			
Bins in warehouse				2,000		3,000	5,000		5,000			
First-aid trailer	1		400	1,500	10,000	1,500	13,400	5,000	8,400			
Change house trailers	2		1,600	3,000	45,000	3,000	52,600	15,000	37,600			
Cap and powder house				1,500		4,500	6,000		6,000			
Shop equipment			600	2,400	30,000	2,000	35,000	10,000	25,000			
Front-end loaders	1	70,000	400		70,000	200	70,600	35,000	35,600			
Yard grading and track laying				20,000		10,000	30,000		30,000			
1075-ft³/min compressors, stationary, electric	9	50,300	1,200	9,000	452,700	4,500	467,400	300,000	167,400			
Starters	9	6,400	200		57,600		57,800	20,000	37,800			
Air receivers	2	31,120	400	400	62,240	400	63,440	35,000	28,440			
Pickups and sedans	10	7,000			70,000	1,000	71,000	17,500	53,500			
Ambulance (used)	1	5,000		200	5,000	3,000	8,200	2,000	6,200			
15-ton hydraulic crane	1	126,000	400		126,000	200	126,600	50,000	76,600			
Total, other facilities			$ 6,400	$60,000	$ 972,540	$ 58,300	$1,097,240	$ 499,500	$ 597,740			

* Plant and equipment.
† Track-feet.

550

Item	Quantity	Unit price							
Rail hardware	26,000 t†	1.25	500			32,500	33,000	10,000	23,000
Ties	15,000 ea	6.20				93,000	93,000		93,000
Stringers on concrete under ties	26,000 t†	1.20				31,200	31,200		31,200
48-in vent pipe	24,000 lin ft	14.00	2,700			336,000	336,000	36,000	300,000
Fans	27 ea	7,500		5,400	202,500	5,400	213,300	50,000	163,300
12-in drainage pipe	25,000 lin ft	6.00	500			150,000	150,500	37,500	113,300
12-in air line	1,000 lin ft	6.00	50			6,000	6,050	1,500	4,550
10-in air line	24,000 lin ft	5.00	400			120,000	120,400	30,000	90,400
4-in water line	25,000 lin ft	1.60	200			40,000	40,200	10,000	30,200
Fittings						12,000	12,000		12,000
Pumps			15,000		20,000	200	20,200		20,200
Parkway cable	25,000 lin ft	6.50	200			162,500	162,900	50,000	112,900
Drag cable	4,000 lin ft	10.00	400			40,000	40,100		40,100
Light cable	25,000 lin ft	1.00	100			25,000	25,100		25,100
Telephone	25,000 lin ft	.80	100			20,000	20,100		20,100
Blasting circuit	25,000	0.80	100			20,000	20,100		20,100
Transformers			500		80,000		80,500	40,000	40,500
Outlets			100			4,000	4,000		4,000
Total tunnel service equipment			**$20,850**	**$ 5,400**	**$ 302,500**	**$1,382,750**	**$1,711,500**	**$ 355,900**	**$1,355,600**
Tunnel concrete equipment:									
Aggregate reclaim conveyor			2,500	12,500	35,000	1,000	51,000	10,000	41,000
Mix plant			10,000	15,000	400,000	10,000	435,000	200,000	235,000
Agitator cars, 8-yd³	20 ea	22,000	16,000	4,000	440,000		460,000	220,000	240,000
Invert bridge and conveyor			1,000	5,000	60,000		66,000	15,000	51,000
Invert bridge supports						15,000	15,000		15,000
Invert screed			1,000	1,000		12,000	14,000		14,000
Arch forms	400 lin ft	900	7,000	20,000		360,000	387,000	72,000	315,000
Arch-form traveler	1 ea	50,000	4,000	2,000		50,000	56,000	12,500	43,500
Unloading conveyor			400	600	6,000		7,000	1,000	6,000
Press Weld rental		500	500				500	10,000	50,000
Pipe and slick line			500			7,500	8,000		8,000
Pipe and jumbo			800	3,200		5,000	9,000		9,000
Vibrators						10,000	10,000		10,000
Grouting setup	1 ea	20,000	300	3,200			15,300	7,500	7,800
Total tunnel concrete equipment			**$44,000**	**$63,300**	**$ 978,000**	**$ 448,500**	**$1,533,800**	**$ 538,000**	**$ 995,800**

$50,000

549

TABLE 18.17 Sierra Tunnel Estimate: Spread Sheet Summarizing Plant and Equipment Cost (*Continued*)

Description	Quantity	Cost/Unit	Freight and Move-in Cost	Labor	Equipment Invoice	Other Invoice	Total Estimated Cost	Salvage	Net Cost P&E*	Rent No. of Months	Rent Rent/Month	Rent Total Rent
Shaft concrete equipment:												
65-ton crane, 3 shifts (rental)	1									2	$ 9,000	$18,000
Concrete buckets	1		$ 200	$ 1,000	$ 6,000		$ 6,200	$ 3,000	$ 3,200			
Swing spout						$ 3,000	4,000	1,000	3,000			
Air hoist	2	$ 7,150	200		14,300		14,500	7,150	7,350			
Total shaft concrete equipment			400	1,000	20,300	3,000	24,700	11,150	13,550			18,000
Other facilities:												
Office trailers			1,200	5,000	24,000	5,000	35,200	10,000	25,200			
Shop, warehouse, complete				15,000	20,000	20,000	55,000		55,000			
Bins in warehouse				2,000		3,000	5,000		5,000			
First-aid trailer	1		400	1,500	10,000	1,500	13,400	5,000	8,400			
Change house trailers	2		1,600	3,000	45,000	3,000	52,600	15,000	37,600			
Cap and powder house				1,500		4,500	6,000		6,000			
Shop equipment			600	2,400	30,000	2,000	35,000	10,000	25,000			
Front-end loaders	1	70,000	400		70,000	200	70,600	35,000	35,600			
Yard grading and track laying				20,000		10,000	30,000		30,000			
1075-ft³/min compressors, stationary, electric	9	50,300	1,200	9,000	452,700	4,500	467,400	300,000	167,400			
Starters	9	6,400	200		57,600		57,800	20,000	37,800			
Air receivers	2	31,120	400	400	62,240	400	63,440	35,000	28,440			
Pickups and sedans	10	7,000			70,000	1,000	71,000	17,500	53,500			
Ambulance (used)	1	5,000		200	5,000	3,000	8,200	2,000	6,200			
15-ton hydraulic crane	1	126,000	400		126,000	200	126,600	50,000	76,600			
Total, other facilities			$ 6,400	$60,000	$ 972,540	$ 58,300	$1,097,240	$ 499,500	$ 597,740			

* Plant and equipment.
† Track-feet.

550

TABLE 18.18 Sierra Tunnel Estimate: Summary of Estimated Cost of Plant and Equipment

	Freight and Move-in	Labor	Equipment Invoice	Other Invoice	Total Estimated Cost	Salvage	Net cost P&E*	Rent	Net Cost P&E* plus Rent
Rubber-tired excavation equipment	$ 3,100		$ 428,000		$ 431,100	$ 127,000	$ 304,100	$ 36,750	$ 340,850
Rail-mounted excavation equipment	71,400	$ 87,800	3,429,400	$ 20,000	3,608,600	1,242,000	2,366,600		2,366,600
Shaft excavation equipment	2,500	10,400	67,350	4,600	84,850	38,000	46,850	18,000	64,850
Tunnel service equipment	20,850	5,400	302,500	1,382,750	1,711,500	355,900	1,355,600		1,355,600
Tunnel concrete equipment	44,000	63,300	978,000	448,500	1,533,800	538,000	995,800	50,000	1,045,800
Shaft concrete equipment	400	1,000	20,300	3,000	24,700	11,150	13,550	18,000	31,550
Other facilities	6,400	60,000	972,540	58,300	1,097,240	499,500	597,740		597,740
Adit excavation (from direct cost details)		181,418		82,967	264,385		264,385		264,385
Adit concrete plug (from direct cost details)		29,762		26,488	56,250		56,250		56,250
Total plant and equipment	$148,650	$439,080	$6,198,090	$2,026,605	$8,812,425	$2,811,550	$6,000,875	$122,750	$6,123,625
Sales and use tax at 6%			371,885	121,596	493,481		493,481	7,365	500,846
TOTAL	$148,650	$439,080	$6,569,975	$2,148,201	$9,305,906	$2,811,550	$6,494,356	$130,115	$6,624,471

* Plant and equipment.

551

XV. ESTIMATING INDIRECT COST

TABLE 18.19   **Sierra Tunnel**
Estimate: Details of Indirect Cost
Labor

	Number	Months	Cost/Month	Total
Project manager	1	32	$ 5,000	$ 160,000
Tunnel superintendent	1	31	4,200	130,200
Project engineer	1	32	3,700	118,400
Office engineer	1	32	2,800	89,600
Design engineer	1	6	3,200	19,200
Safety engineer	1	31	2,800	86,800
First-aid attendant*	3	93	2,400	223,200
Accountant	1	32	2,800	89,600
Purchasing	1	31	2,600	80,600
Payroll clerk	1	31	1,800	55,800
Secretary	1	32	1,600	51,200
Equal opportunity recruiter	1	12	3,000	36,000
Total				$1,140,600
Burden, 29.8%				339,899
Total				$1,480,499

* Quartered in trailer at the portal.

Other costs

	Months	Cost/month	Total
Vehicle maintenance	300 VM	$ 350	$105,000
Office and engineering supplies	31	1,200	37,200
Telephone and telegraph	31	1,200	37,200
Computer services	32	1,200	38,400
Entertainment and expense accounts	32	300	9,600
Licenses and fees			4,000
Blueprints, photostats, photographs	31	300	9,300
Office equipment			20,000
Engineering equipment			15,000
First aid and safety			20,000
Outside consultants			5,000
Legal			10,000
Audit			10,000
Move key personnel in and out			15,000
Heat and light for office	31	300	9,300
Signs			2,000
Misc. cost	31	1,000	31,000
Total other cost			$ 378,000

Insurance

		Total
Value of insurable equipment and buildings	$6,100,000	
Insurance per month at $0.10/$100	6,100	
Number of months	30	
Total insurance		$ 183,000

Property tax

		Total
Value of taxable plant and equipment	$6,100,000	
Assessed valuation, 25%	1,525,000	
Yearly tax at $6/$100	91,500	
Number of years	3	
Total tax		$ 274,500

Bond premium

	Amount of Contract	Bond Cost/ $1,000	Amount	Total
First	$ 100,000	$7.50	$ 750	
Next	2,400,000	5.25	12,600	
Next	2,500,000	4.50	11,250	
Remainder	35,000,000	4.00	140,000	
Total	$40,000,000		$ 164,600	
Time premium, 8%			13,168	
Total bond premium				$ 177,768
TOTAL INDIRECT				$2,493,767

NOTE: The bond premium was calculated on the basis of a bid price of $40,000,000. It will be adjusted to reflect changes in estimated costs and markup after the joint venture meeting, and will be further adjusted if actual quotations differ significantly from the plugged prices for permanent materials and subcontracts.

XVI. ESTIMATING CAMP COST

A camp is not required on this project.

XVII. ESTIMATING ESCALATION

TABLE 18.20   Sierra Tunnel Estimate:
Escalation in Labor Cost

	June 1 of 1st Year to June 1 of 2nd Year	June 1 of 2nd Year to June 1 of 3rd Year	June 1 of 3rd Year to June 1 of 4th Year	Total
Labor:				
Direct:				
Tunnel excavation items	$4,200,000	$3,696,990		$ 7,896,990
Shaft excavation	263,357			263,357
Concrete		1,921,000	$2,895,602	4,816,602
Total direct	$4,463,357	$5,617,990	$2,895,602	$12,976,949
Plant and equipment labor	399,080	20,000	20,000	439,080
Indirect labor	500,000	600,000	380,499	1,480,499
Total labor	$5,362,437	$6,237,990	$3,296,101	$14,896,528
Escalation, 10% a year compounded		10%	21%	
Total labor escalation		$ 623,799	$ 692,181	$ 1,315,980
TOTAL LABOR				$16,212,508

Supplies, Permanent Materials, and Subcontracts

Quotations are assumed to be good for the life of the contract so there will be no escalation for these items. (In making an estimate, quotations should be checked, and if they are not good for the life of the contract, escalation amounts should be added.)

XVIII. SUMMARY OF TOTAL COST

TABLE 18.21   Sierra Tunnel Estimate: Total Project Cost

	Total	Total Labor
Direct cost:		
Labor	$ 12,976,949	$12,976,949
Supplies	5,465,365	
Permanent materials	3,860,180	
Subcontracts	1,883,740	
Total direct cost	$24,186,234	
Plant and equipment:		
Freight and move-in	$ 148,650	
Labor	439,080	439,080
Equipment invoice	6,569,975	
Other invoice	2,148,201	
Subtotal	$ 9,305,906	
Salvage	(2,811,550)	
Rent	130,115	
Total plant and equipment	6,624,471	
Indirect:		
Labor	$ 1,480,499	1,480,499
Other cost	378,000	
Insurance	183,000	
Property tax	274,500	
Bond	177,768	
Total indirect	2,493,767	
Labor escalation	1,315,980	1,315,980
TOTAL COST	$34,620,452	$16,212,508

XIX. CASH-FLOW ANALYSIS

TABLE 18.22 Sierra Tunnel Estimate: Cash Forecast
Thousands of Dollars

	Total	1st Year				2nd Year				3rd Year		
		June	Jul-Sep	Oct-Dec	Jan-Mar	Apr-Jun	Jul-Sep	Oct-Dec	Jan-Mar	Apr-Jun	Jul-Sep	Oct-Dec
Revenue:												
Bid item revenue	$40,000		40	3,530	4,970	4,970	4,970	4,700	4,500	4,500	4,400	3,420
Retained percentage			(4)	(353)	(497)	(497)	(497)	(152)				2,000
Subtotal	40,000		36	3,177	4,473	4,473	4,473	4,548	4,500	4,500	4,400	5,420
Salvage revenue	2,812								600	600	700	912
Total cash revenue	42,812		36	3,177	4,473	4,473	4,473	4,548	5,100	5,100	5,100	6,332
Disbursements:												
Direct cost:												
Tunnel excavation	13,218		1,218	2,000	2,500	2,500	2,500	2,500				
Shaft excavation	361			31	330							
Concrete	10,442							492	2,650	2,650	2,650	2,000
Grouting	165										65	100
Labor escalation	1,316					52	156	156	156	219	346	231
Plant and equipment:												
Necessary for excavation	7,682		1,682	3,500	2,000	500						
Necessary for concrete	1,754						200	500	940	30	30	54
Indirect cost:												
Labor and supplies	1,858	58	150	190	190	200	200	200	190	190	190	100
Bond premium	178	178										
Insurance and taxes	458		33			150	75			125	75	
Inventories			250	50	50		150		(50)	(200)	(200)	(50)
Total disbursement	37,432	236	3,333	5,771	5,070	3,402	3,281	3,848	3,886	3,014	3,156	2,435
Revenue less disbursements	5,380	−236	−3,297	−2,594	− 597	+1,071	+1,192	+ 700	+1,214	+2,086	+1,944	+3,897
Accumulated revenue less disbursements		−236	−3,533	−6,127	−6,724	−5,653	−4,461	−3,761	−2,547	− 461	+1,483	+5,380
Working capital required		264	467	873	876	847	889	889	853	839		
Cash required		500	4,000	7,000	7,600	6,500	5,350	4,650	3,400	1,300		
Interest at 18% per annum	1,798	7	180	315	342	293	241	209	153	58		

XX. IN-HOUSE REVIEW

The estimate has been reviewed by the officers of the company. The changes which they considered necessary have been made. The estimate as printed here reflects those changes. Further changes are expected because of the high financing costs developed in the cash-flow analysis. These changes will be discussed at a meeting of the principals after the joint venture participants reach agreement on the basic costs. The changes may be an increase in markup, or may provide for use of equipment presently owned by members of the joint venture. Use of used equipment will probably increase direct costs, but will certainly reduce capital requirements and financing costs.

Chapter 19 Example of a Dam Estimate

This chapter contains a sample cost estimate for the construction of a fictitious concrete dam, located on a fictitious river and near a fictitious town in California. This fictitious dam has been named Denny Dam. Costs are based on the dam being constructed in a subsistence area with a contract award in 1982. The sample estimate is used to illustrate the application of the estimating steps described in Chap. 17.

To simplify the presentation of the estimate and to eliminate unnecessary detail, typical calculations are used to illustrate how many of the estimating steps are performed. To expedite the cross-referencing between the task descriptions in Chap. 17 and their application in this chapter, the same titling and numbering system has been used in both chapters. A concrete dam estimate is presented in preference to an earth- or rock-fill dam estimate, because the construction of a concrete dam is so complex that it requires the use of a more extensive construction system, the application of more types of equipment, and the performance of more construction activities than are required for the construction of either of the other types. If the reader understands the presentation of a concrete-dam construction estimate, he will have little difficulty in using the same basic estimating technique to prepare estimates for other types of dams or for any other large concrete, earth-moving, or rock-fill project. The format that is used for the preparation of the Denny Dam estimate is the same as is used for the preparation of the Sierra Tunnel estimate in Chap. 18.

I. STUDY OF PLANS AND SPECIFICATIONS

This study was completed and the following abstract of the specifications was prepared. (See Figs. 19.1 and 19.2.)

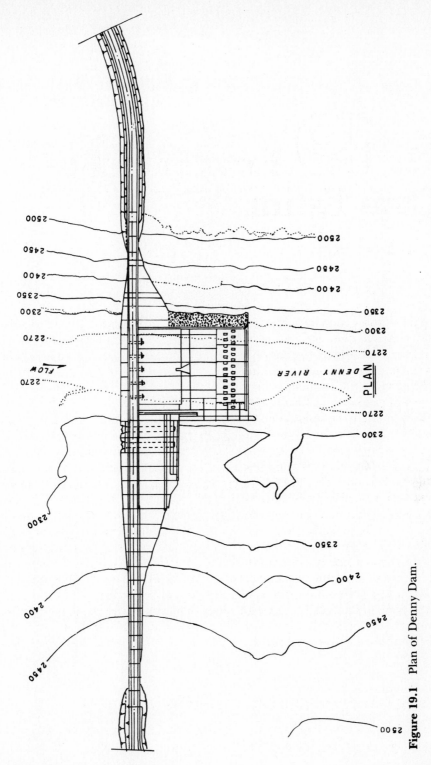

PLAN

DENNY RIVER

FLOW

Figure 19.1 Plan of Denny Dam.

559

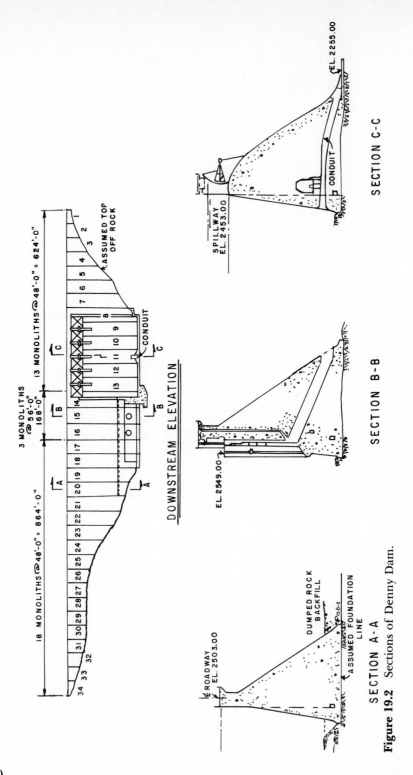

Figure 19.2 Sections of Denny Dam.

560

Abstract of Specifications for the
Construction of Denny Dam

SCOPE

The principal features involved in the construction of Denny Dam are as follows:
1. Construction of a gravity dam with a crest length of 1,656 lin ft and a crest height of 233 ft above streambed
2. Furnishing and installing two penstock gates and hoist, and embedding two penstocks 18 ft 6 inches in diameter by 143 ft long to serve a future powerhouse
3. Furnishing and installing six spillway tainter gates, 40 ft wide by 38 ft high

AGENCY

Scott Water District.

BID DATE

2 P.M. PST, June 10, 1982, at the Scott Water District office, Hilldale, California.

LOCATION

The damsite is located approximately 20 miles from Hilldale, California, on the Denny River in Hill County, California.

ACCESS

A paved double-lane highway from Hilldale passes close to the dam left abutment. The closest rail connection is at Hilldale.

DATA TO BE SUBMITTED

Sealed bid in duplicate, with unit price schedule and acknowledgment of receipt of all addenda attached.
Bid bond.
Bid envelope must be sealed, marked, and addressed as follows: "Bid for Denny Dam to be opened at 2 P.M. PST, June 10, 1982."

BIDDER'S QUALIFICATIONS

Before the bid is considered for award, the district may request the contractor to submit a statement setting forth a detailed account of previous experience in performing comparable work.

FORM OF CONTRACT

The contract form is included in the specifications.

AWARD

A written notice of acceptance of the bid will be forwarded to the successful bidder within 60 calendar days after date of bid opening.

BONDS

Bid bond, 20 percent of bid price.
Payment bond, amount in accordance with the following schedule:

Amount of Contract	Payment Bond
Up to $1 million	50%
$1 million to $5 million	40%
Over $5 million	$2½ million

Performance bond, 50 percent of the contract amount.

TIME OF COMPLETION

The contractor must commence work within 30 days after receipt of notice to proceed, and he will be allowed 1,250 calendar days to complete the project.

LIQUIDATED DAMAGES

The contractor shall pay the owner $1,000 for each calendar day that construction time exceeds 1,250 days.

PAYMENTS

As work progresses, *partial payments* will be made monthly. A *contract retention* of 10 percent of each progress payment will be retained by the owner until all work has been accepted. If work progress is satisfactory, after 50 percent of the work has been performed the agency's engineer has authority to make the remaining payments in full.

Payments will be made for preparatory work and for the acquisition of plant and equipment. These payments will be limited to 75 percent of the contractor's cost. The amount of this prepayment will be deducted from each progress payment in accordance with a schedule presented by the contractor and meeting the approval of the district's engineer. It shall not exceed 10 percent of the total bid price.

OWNER-FURNISHED
MATERIAL

None.

RIVER
DIVERSION

It is the contractor's responsibility to select the method of diversion and to divert and take care of the river. Minimum streamflow shall be maintained below the damsite. The 5 ft 8 in by 10 ft permanent river outlet located in the spillway can be used by the contractor for diversion and for maintaining the minimum flow. The bottom of this permanent outlet is at an elevation of 2,280 ft. The approximate capacities of this outlet at different pool heights are as follows:

Pool Elevation, ft	Capacity, ft³/s
2,300	1,200
2,350	2,600
2,400	3,400
2,450	4,100
2,500	4,700

CONSTRUCTION
CONSTRAINTS
IN
SPECIFICATIONS

Concrete aggregates. Aggregates and sand must be manufactured from rock secured from a rock quarry located 1 mile from the damsite. The quarry is made available to the contractor free of charge.

Aggregate rescreening. This is a requirement. The plant must be located adjacent to the mix plant or on top of it.

Cement storage. Storage for 9,000 bbl of cement must be provided.

Allowable pour height. Contractor has the option of pouring mass concrete either in 7½-ft lifts or in 5-ft lifts.

Concrete transportation. Concrete shall be transported in the same container from the mixing plant to the pour.

Concrete buckets. The maximum amount of concrete that can be dumped in a pour at one time is 8 yd.³

Concrete placing temperature. Concrete shall be placed in the forms at temperatures above 40° and below 45°F.

Concrete cleanup. High-pressure air and water cutting will be acceptable.

Concrete pay lines. The concrete pay lines will be the excavated rock surface.

CLIMATOLOGICAL
DATA

Snowfall. Snowfall is limited to light falls that do not remain on the ground.

Precipitation:

Month	Maximum (in)	Minimum (in)	Mean (in)
January	23.56	0.43	5.99
February	17.02	0.05	5.71
March	18.01	0.11	5.24
April	9.32	0.00	2.38
May	4.27	0.00	1.14
June	2.49	0.00	0.35
July	0.17	0.00	0.01
August	0.16	0.00	0.01
September	3.96	0.00	0.42
October	5.86	0.00	1.58
November	11.08	0.00	2.95
December	13.97	0.35	4.99
Annual	54.58	17.36	30.77

Temperature:

Month	Maximum (°F)	Minimum (°F)	Mean (°F)
January	57	35	46
February	61	37	49
March	66	40	53
April	73	44	58
May	82	48	65
June	90	53	71
July	98	59	80
August	98	57	78
September	92	53	72
October	80	47	64
November	68	40	54
December	58	36	47

RIVERFLOW
DATA

Figure 19.3 shows the hydrographs for the Denny River at the damsite. Recorded elevations of flows at the damsite are:

Flow, ft³/s	Elevation, ft
Minimum flow	2,265
44,600	2,296
80,000	2,305

MAJOR
QUANTITIES

Care and diversion of river	Lump sum
Common excavation	185,000 yd³
Rock excavation	170,000 yd³
Preliminary rock cleanup	7,000 yd²
Close-line drilling	30,000 ft²

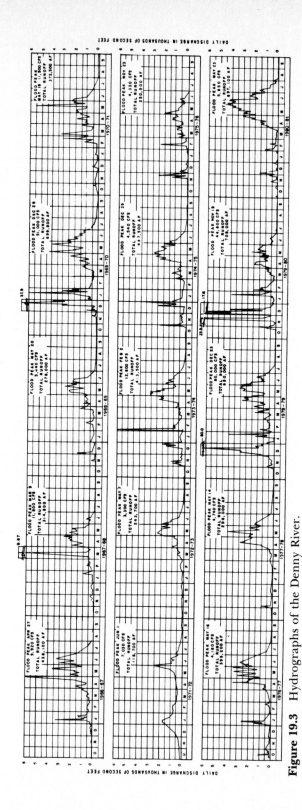

Figure 19.3 Hydrographs of the Denny River.

Final rock cleanup	23,000 yd^2
Drill grout, drain, and exploratory holes	55,500 lin ft
Pressure grouting	8,000 ft^3
Backfill and riprap	23,300 yd^3
Reinforcing steel	3,080,000 lb
Concrete	826,880 yd^3
Cement	63,500 bbl
Gates and guides	2,618,000 lb
Pumps, motors, and piping for gates	Lump sum
Cast-iron pipe and fittings	156,000 lb
Steel pipe and fittings	184,000 lb
Piping, water supply, and plumbing	Lump sum
Structural steel	476,500 lb
Railing and fittings	17,700 lb
Parapet railing	1,152 lin ft
Electrical system	Lump sum

II. JOBSITE INSPECTION

The field investigation was completed and is summarized as follows:

Access to the project is relatively easy since a highway passes close to the left abutment.

A single-status camp will not be required since the surrounding area will provide sufficient living accommodations. Living facilities for key married personnel can be located on a flat area adjacent to the project. Power lines pass this site, and water can be secured from an existing spring. The owner of the area was contacted, and he will rent the area for 4 years for the sum of $20,000.

Power lines pass near the project site. The power company will supply power to the project for $.06 per kWh and will install a suitable substation for a $70,000 "up-and-down" charge.*

Telephone lines pass close to the damsite, and the telephone company will install as many phones as required.

The rock at the damsite is a solid, medium-hard, granitic type overlain with disintegrated granite except at the streambed, where water-washed sand and gravel are encountered.

There is a level flat area on the left abutment which could be easily graded to accommodate a traveling head tower for a cableway.

The streambed is wide and the minimum water flow is so minor that low-

---

* Charge for furnishing, erecting, and removing the substation.

flow diversion can be handled by small dikes and channels. The stream can be forded during the low-water season.

The quarry is composed of exposed, sound diorite not covered with vegetation or overburden and is located in a place where a quarry bench can be rapidly developed.

There is ample space on the left abutment for construction facilities, the project office, and employee parking.

Copies of all applicable labor agreements were secured. Tax rates were obtained from the county assessor.

III. GEOLOGICAL REVIEW

The rocks are granitic and most of the area is devoid of vegetation. It was apparent during the course of the jobsite inspection and from inspection of drill logs and drill cores that there is no occasion for concern about foundations; quality of the rock makes it adequate as a source of concrete aggregate.

IV. SUMMARY OF QUANTITY TAKEOFFS

Following is a summary of the quantity takeoffs for item 24, mass concrete in the dam. (This summary illustrates how quantity takeoffs should be prepared.)

Denny Dam Estimate

Quantity Takeoff Summary—
Bid Item 24, Mass Concrete

1. Concrete quantities:

Bid item quantity		775,000 yd³
Takeoff quantity		
Mass	748,150 yd³	
Blockout concrete	90 yd³	748,240 yd³

2. Number of lifts, pours, and size of pour:

Maximum number of 7½-ft lifts	34
Total number of pours	693
Largest pour	3,033 yd³
Smallest pour	23 yd³
Average pour	1,080 yd³

3. Square feet of form contact:

	Type of Form			
	Built-in-Place (ft²)	**Lift Starter (ft²)**	**Cantilever (ft²)**	**Total (ft²)**
Upstream face	12,600	12,600	219,700	244,900
Downstream face	15,400	15,400	240,700	271,500
Bulkhead	32,000	32,000	378,900	442,900
Subtotal	60,000	60,000	839,300	959,300
Spillway splash wall				9,600
Galleries and other interior forms				83,550
Curved forms				3,250
Blockout concrete forms				3,780
Total				1,059,480
Form ratio based on takeoff yardage				1.416

4. Special forms:
   8-in-diameter formed hole · · · · · · · · · 4,410 lin ft
5. Surface finish:
   Float · · · · · · · · · 72,410 ft²
   Steel trowel · · · · · · · · · 11,152 ft²
6. Concrete cleanup · · · · · · · · · 2,455,900 ft²
7. Rock cleanup · · · · · · · · · 23,450 yd²

V. REQUESTS FOR QUOTATIONS ON PERMANENT MATERIALS AND SUBCONTRACTS

Requests for quotations were sent out for permanent materials, supplies, subcontracts, and the plant and equipment items. A list of plugged prices was prepared as shown in Table 19.1.

VI. COMPUTATION OF LABOR

The basic labor agreements in effect at Denny Dam are the same as those in effect at Sierra Tunnel (Chap. 18), and the methods of calculating effective labor costs per shift for each classification of labor are the same. The calculations will not be repeated. The different costs per shift for each classification of labor are due to the following factors:

1. Workers' compensation insurance rates are lower for work on a surface project than for work on an underground project.
2. The basic scale for tunnel laborers is higher than the basic scale for other laborers.

TABLE 19.1 **Denny Dam Estimate: Plugged Material and Subcontract Prices Sent Out by Sponsor to Other Partners***

Bid Item No.	Work Description	Bid Quantity	Permanent Material Price*	Subcontract Price	Painting Subcontract Unit	Painting Subcontract Amount
7	Drill 1½-in-diameter grout holes	25,000 lin ft		$ 11.00		
8	Drill 3-in-diameter drain holes	17,000 lin ft		13.00		
9	Core drill 3-in exploratory holes	13,500 lin ft		15.00		
10	Grout connections	800 each		25.00		
11	Pressure testing	800 test		25.00		
12	Grouting cement	8,000 ft³		5.00		
13	Pressure grouting	8,000 ft³		15.00		
18	Reinforcing steel	3,080,000 lb		0.40		
25	Cement	635,000 bbl	$ 16.00			
26	12-in-diameter RCCP	100 lin ft	6.50			
27	18-in-diameter RCCP	100 lin ft	12.00			
28	Copper water stop	12,300 lin ft	7.50			
29	6-in-diameter porous concrete pipe	28,000 lin ft	2.75			
30	12-in-diameter half-round porous pipe	1,950 lin fit	5.00			
31	Miscellaneous doors	400 ft²	25.00		$1.50	$ 600

No.	Item	Quantity	Unit price		Amount
32	Miscellaneous metal castings	Lump sum	15,000.00		1,200
33	Slide and bulkhead gates	286,000 lb	2.00	0.03	8,580
34	Pumps, motors, piping for gates	Lump sum	60,000.00		750
35	Penstock gates, frames, and guides	288,000 lb	2.20	0.04	11,520
36	Penstock-gate hoist and supports	Lump sum	120,000.00		3,000
37	Penstock lining and frames	535,000 lb	4.00	0.20	107,000
38	Tainter gates and accessories	670,000 lb	1.80	0.25	167,500
39	Tainter-gate trunnion anchors	320,000 lb	2.20	0.10	32,000
40	Tainter-gate hoist and accessories	150,000 lb	4.00	0.05	7,500
41	Trashrack guides and beam	260,000 lb	1.25	0.04	10,400
42	Stop logs, protection plates, and beams	168,000 lb	1.50	0.04	6,720
43	Cast-iron pipe and fittings	156,000 lb	4.50	0.02	3,120
44	Steel pipe and fittings	184,000 lb	4.25	0.04	7,360
45	Piezometer piping and fittings	Lump sum	5,000.00		500
46	Sump pump and piping	Lump sum	32,000.00		1,000
47	Water supply and plumbing	Lump sum	45,000.00		600
48	Structural steel	66,500 lb	1.40	0.05	3,325
49	Railing and fittings	17,700 lb	1.50	0.05	885
50	Parapet railing	1,152 lin ft	30.00		
51	Steel in spillway bridge	410,000 lb	1.40	0.05	20,500
52	Electrical system	Lump sum	600,000.00		3,000
	Total				$396,460

*Permanent-materials price, FOB job, including sales tax.

3. Travel time underground is eliminated. Travel time aboveground is reflected in production rates.
4. Annual earnings for dam employees will be lower than for underground employees, and hence the calculated rate for burden on labor is nearer the nominal rate than is the case for Sierra Tunnel.
5. The net effect of the above factors is a substantial decrease in labor costs per shift in every classification of labor.

VII. TABULATION OF EQUIPMENT OPERATING COSTS

Equipment operating costs for Denny Dam were obtained in exactly the same manner as they were for Sierra Tunnel (Chap. 18).

VIII. TRANSMITTAL OF ESTIMATING INSTRUCTIONS

This was done in exactly the same way as is described in Chap. 18 for the Sierra Tunnel project. The sponsor included a tabulation of available used equipment, as shown in Table 19.2.

IX. SELECTION OF CONSTRUCTION METHODS

The decision between crane or cableway service for placing concrete and for handling other loads was dependent upon the scheme for river diversion, and will be discussed under estimating step X. Only the decisions affecting the diversion scheme will be discussed here.

The combination of a wide streambed at the damsite, minimum water flows during 6 months of each year, absence of restrictions on maximum heights between adjacent monoliths, and absence of restrictions on height differentials between the lowest and highest monoliths made diversion through the dam seem the most economical method. The periods of minimum water flow, which control when succeeding diversion stages can be started, will restrict the speed of concrete placement.

An examination of the hydrographs and a review of the history of floods in northern California indicated that flood flows that exceed 13,000 ft³/s will only occur during the months of November or December; that the maximum recorded flood flow of 80,000 ft³/s occurred during the year that maximum flood flow was recorded for this area in northern California, which makes the possibility of having a flood flow exceeding 80,000 ft³/s extremely remote; and that the possibility of having a water flow of over 13,000 ft³/s in any calendar year is 1 chance in 8.

The recorded water elevation at the damsite for a flow of 80,000 ft³/s was elevation 2,305 ft. The cross section of the river channel at the damsite shows approximately 11,700 ft² of channel area below this elevation.

TABLE 19.2   Denny Dam Estimate:
Available Used Equipment*

Units	Weight (tons)	Price
1 4½ yd³ diesel shovel, including buckets	125	$250,000
10 35-ton rear-dump trucks, price and weight for each unit	29	80,000
1 3,000-gal water truck		15,000
1 Kenworth Lo-boy truck		37,500
1 lube truck		10,000
1 48- × 72-in jaw crusher, motor, and starter	130	100,000
1 No. 12 Caterpillar grader	13	25,000
1 30-in × 450-ft mixing-plant charging conveyor, including supports and motors	35	42,500
1 aggregate rescreening plant	80	100,000
1 mixing plant, containing 1,030 yd³ of aggregate storage and 3 4-yd³ mixers; equipped for cold-air circulation through aggregate storage bins	240	300,000
1 2,300-bbl cement-storage silo	28	10,750
1 cement screw conveyor	2	3,750
1 cement pump	9	12,000
1 100-ton per day ice machine, including screw conveyor for mixing-plant charging	5	25,000
550 tons of refrigeration equipment for producing cold air to circulate through the mix plant, and cold water to use in the mix and in cooling coils	120	375,000
2 25-ton diesel locomotives, price and weight for each unit	25	37,500
2 12½-ton hydraulic cranes, price and weight for each unit	17	30,000
1 lot of concrete forms for pouring 7.5-ft lifts	145	250,000
This lot includes:		
650 lin ft of upstream face forms		
650 lin ft of downstream face forms		
2,000 lin ft of bulkhead forms		
100 lin ft of upstream overhang forms		
100 lin ft of downstream overhang forms		
1,000 ft² of wall forms		
1 ogee crest form		

* Equipment owned by the sponsor and available for use on the project for the listed prices, FOB haul trucks at the sponsor's equipment yard.

A review of the preliminary construction schedule indicated that it would take approximately 11 months to obtain and erect the concrete plant and equipment. This established June 1 of the second calendar year as the earliest date that concrete placement could commence. Therefore, it was decided to handle river diversion as shown by Figs. 19.4 and 19.5 and by the construction schedule prepared under estimating step XI.

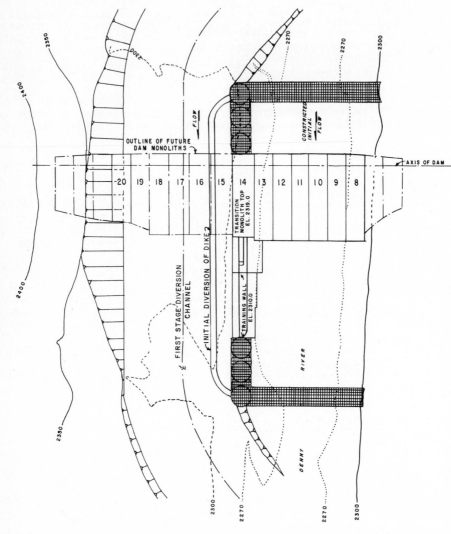

Figure 19.4 First-stage diversion for Denny Dam.

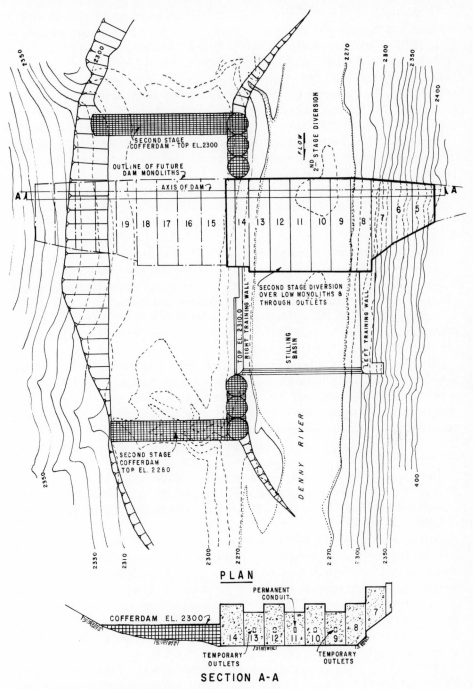

Figure 19.5 Second-stage diversion for Denny Dam.

To prepare for river diversion in the fall of the first year, a diversion channel is to be excavated on the right abutment passing through monoliths 15 to 19. This channel is to be excavated to provide a minimum area of 10,000 ft^2 below elevation 2,305 ft. Since this will be a slightly smaller area than that below this elevation in the original channel, when first-stage diversion is accomplished, the upstream elevation for a riverflow of 80,000 ft^3/s will exceed elevation 2,305 ft but should not exceed elevation 2,307 ft. Prior to construction of the diversion facilities, this plan is to be checked by model studies.

The material excavated for the diversion channel from beneath future dam monoliths 15 to 19 is to be paid excavation, and the cost of removing the remainder will be charged to diversion.

Diversion preparation includes the construction of three cofferdam cells upstream of monolith 14, to an elevation of 2,312 ft, and the construction of three cofferdam cells downstream of the right training wall, to an elevation of 2,310 ft.

Initial diversion of the low-water flows is to be made in May of the second contract year. This initial diversion will consist of diverting the minimum flow of water through the diversion channel. To force the minimum flow into the channel, it is necessary to construct the upstream and downstream cofferdams to elevation 2,276 ft. To contain the flow in the channel, it is necessary to construct a 6-ft high longitudinal dike along the edge of the channel connecting the upper and lower cofferdams.

During this low-flow diversion, excavation will be completed for dam monoliths 9 through 14, for the training wall, and for the spillway. Concrete placement in these structures will be started, and excavation and concrete placement will be expedited for monolith 14 and the right training wall. The concrete in the right training wall and monolith 14 will be raised to elevation 2,319 ft, prior to the start of the high-water flows, to take the place of the dirt dike as a longitudinal cofferdam.

Also, prior to November 1 of the second calendar year, the upstream cofferdam is to be raised to elevation 2,312 ft, and the downstream cofferdam raised to elevation 2,310 ft. This will give a minimum of 5 ft of freeboard for a flood flow of 80,000 ft^3/s.

During the high-water season, concrete placement will be completed in the left training wall and in the spillway apron and will continue in blocks 9 to 14 and in abutment blocks. When blocks 9, 11, and 13 reach elevation 2,326.5 ft, concrete placement in these blocks will be stopped. These blocks will remain at this elevation to pass the flood flows during the high-water season starting at the end of the third contract year. Four diversion conduits of the same size and with the same invert elevation as the permanent outlet will be formed in monoliths 9, 10, 12, and 13.

In May of the third year, the upstream and downstream cofferdams are to be removed and the second-stage cofferdams will be placed upstream and downstream of blocks 15 to 19. The upstream cofferdam will be constructed to elevation 2,290 ft to pond the water so that low-water flows will pass through the diversion conduits and the permanent outlet. The downstream cofferdam will be constructed to elevation 2,270 ft to protect monoliths 15 to 19 from the tailwater.

During low-flow diversion, excavation underneath monoliths 15 to 19 will be completed, and concrete placement will be expedited. By November 1 of the third year, the lowest monolith in this group will be poured to elevation 2,341.5 ft. Prior to this time, the three upstream cofferdam sheet-pile cells will be removed and the upstream cofferdam leveled. The three downstream cofferdam cells and the downstream cofferdam can be removed at any time prior to project completion.

From November 1 of the third year until all concrete is placed, low-flow diversion will be through the diversion conduits and through the permanent outlet. Monoliths 9, 11, and 13 will be kept 15 ft lower than other monoliths to handle flood flows until all floods can be handled by the combination of reservoir storage and conduit capacity. Diversion conduit plus permanent-outlet capacity will be five times the capacity of the permanent outlet. This capacity for different dam elevations is:

Elevation of Dam, ft	Capacity, ft³/s
2,300	6,000
2,350	13,000
2,400	17,000
2,450	20,500
2,500	23,500

Each of three low blocks will be able to handle flows of 24,500 ft³/s, or the three can handle 73,500 ft³/s before the upstream water level reaches a high enough elevation to pass flood flows over other low blocks. These flows were computed using the broad-crested weir formula found in all hydraulics handbooks.

Finally, at the start of the low-water season in the fourth year, all the water can be carried by the permanent river outlet. Flow through the diversion conduits will be stopped by dropping concrete stop logs over their entrances, and they will be filled with concrete. This will complete all the work required for river diversion.

X. SELECTION OF PLANT AND EQUIPMENT

The decision made in the performance of estimating step IX, to divert the river through the dam, made a cableway the preferable equipment to use for placing concrete in Denny Dam. Since the cableway towers are located on the dam abutments, their erection is not dependent on completing the dam-foundation excavation. Also, a cableway will provide hook coverage of all the monoliths in the dam as soon as it is erected, and it will not interfere with the flow of the water in the natural streambed or diversion channel. Cableway coverage for the construction of Denny Dam can be secured using a radial cableway as shown in Fig. 6.1. A radial cableway is preferred to a parallel cableway, since its use eliminates one traveling tower and its runway, while giving adequate coverage.

A cableway was preferred to a trestle and cranes for placing concrete in Denny Dam because the erection of the trestle would create scheduling problems. As shown by Fig. 6.9, a trestle cannot be erected across the streambed until the dam foundation is excavated. Also, it cannot be completed until second-stage diversion is started. Its use would cause considerable delay in work scheduling and would not permit the early concrete placement in the dam monoliths located on the dam abutment.

One 8-yd^3 cableway will have adequate placing capacity, as shown by the construction schedule. Using the rule of thumb for mixing-plant capacity (which is 1½ times the cableway bucket capacity), three 4-yd^3 mixers are required, or a total mixer capacity of 12-yd^3. The used mixing plant owned by the sponsor fits this requirement.

The head tower of the cableway, the concrete track, the mixing plant, and the aggregate plant are to be located on the left abutment, since ample flat space is available. This location for the aggregate plant will also permit the storage piles to serve as surge piles for the mixing plant charging conveyor.

The following calculations were made to check the capacity of the ice-making machinery and the refrigeration equipment that is obtainable from the sponsor. These calculations showed they are adequate.

Denny Dam Estimate

Computations Made to Check the Selected Cooling Method and to Determine the Required Tonnage of Refrigeration Equipment

Assumed Summer Conditions

Temperature rise from mix plant to final placement will be 1°, so concrete should leave the mix plant at 39°F.

	°F
Mean dry-bulb temperature at time of pour	80
Mean wet-bulb temperature at time of pour	65
Temperature of riverwater	65
Temperature of cement	90
Coarse aggregate piles will be equipped with sprinkler facilities so aggregates will be wet-bulb temperature of	65
Sand storage will be shaded; temperature of	80

Table 19.3 gives the estimate for the assumed mass-concrete mix, showing the percentage and weight of surface moisture and the specific heat.

TABLE 19.3   Denny Dam Estimate: Mass-Concrete Mix

	Lb of Materials yd³ of Concrete	Moisture, %	Lb of Surface Water	Temperature (°F)	Specific Heat
Cement	292	0.0	0.0	90	0.23
Large aggregate:					
6 by 3 in	355	0.5	1.8	65	0.23
3 by 1½ in	778	0.7	5.4	65	0.23
1½ by ¾ in	848	1.0	8.5	65	0.23
Total large aggregate	1,981		15.7		
¾-in by ¼-in aggregate	810	1.5	12.2	65	0.23
Sand	740	4.0	29.6	80	0.23
Subtotal	3,823		57.5		
Water:					
Surface	57.5				1.00
Free	114.5				1.00
Total water	172				
Total mix	3,995				

Table 19.4, estimating heat balance, is based on the requirement of producing 39°F concrete from the mixing plant, which entails air-cooling of large aggregates to 35°F and the use of 35°F water plus ice in the mix.

TABLE 19.4   Denny Dam Estimate: Heat Balance

Ingredient	Concrete (lb/yd³)	Temperature Change (°F)	Specific Heat	BTU
Cement	292	90–39	0.23	+ 3,425
Sand	740	80–39	0.23	+ 6,978
¾- by ¼-in aggregate	810	65–39	0.23	+ 4,844
Other coarse aggregates	1,981	35–39	0.23	− 1,823
Surface water on sand	29.6	80–39	1.00	+ 1,214
Surface water on ¾- by ¼-in aggregate	12.2	65–39	1.00	+ 317
Surface water on other sizes	15.7	35–39	1.00	− 63
Chilled water	9.5	35–39	1.00	− 38
Ice, heat of fusion*	105	32–39	144	−15,120
Ice, as chilled water*	105	32–39	1.00	− 735
Mechanical heat of mixing				+ 1,000
Total	3,995			−1

* Identical ingredient; only one included in total weight.

The above balance shows that concrete cooling can be accomplished using the assumed system. The high proportion of ice will be required under only the most severe conditions. Table 19.5 establishes refrigeration needs.

TABLE **19.5   Denny Dam Estimate: Refrigeration Required for Each Cubic Yard of Concrete**

Ingredient	Concrete (lb/yd³)	Temperature Change (°F)	Specific Heat	BTU
Cement	292	90–39	0.23	3,425
Sand	740	80–39	0.23	6,978
Coarse aggregates	2,971	65–39	0.23	16,690
Surface water on sand	29.6	80–39	1.00	1,214
Surface water on coarse aggregates	27.9	65–39	1.00	725
Free water	114.5	65–39	1.00	2,977
Mechanical heat of mixing				1,000
Total	3,995			33,009

Since 1 ton of refrigeration is 12,000 Btu per hour, 2.75 tons of refrigeration will be required to cool 1 yd³ of concrete per hour. With a maximum mixing-plant capacity of 240 yd³ per hour, 660 tons of refrigeration will be required.

The aggregate plant should be of a size adequate to produce as many tons of aggregate in 1 hour as the mixing plant consumes when it is mixing concrete at its maximum rate. This capacity is computed by multiplying the maximum production of the mix plant in cubic yards per hour by 1.85. Three 4-yd³ mixers with a 3-minute cycle can produce 240 yd³ of mixed concrete an hour; this multiplied by 1.85 gives a maximum demand for aggregates of 444 tons per hour. This demand was rounded to 500 tons per hour to establish aggregate plant capacity. The basic components selected for the aggregate plant are shown in Fig. 7.4.

The quarry must be equipped to supply 500 tons per hour of broken rock to the aggregate plant. Since a bank cubic yard of diorite weighs approximately 2.5 tons, this establishes the required production in the quarry as 200 bank yd³ per hour.

In order to secure good rock breakage, it was established that the explosive requirements will be 1.25 lb per bank cubic yard and that relatively small-diameter holes on a close drill pattern will be necessary. Five-in-diameter holes were planned with a 10-ft burden and on a 12.5-ft drill-hole spacing. The bench height is to be 30 ft, the subdrilling depth will be 3 ft, and the holes will be stemmed for 7 ft. The hole pattern requires that 0.24 lin ft of hole be drilled to break 1 bank yd³ of rock. Allowing for subdrilling and stemming, 0.19 lin ft of hole can be loaded with explosive for 1 bank yd³ of rock. The capacity of a 5-in-diameter hole is 7.66 lb of pneumatically loaded AN/FO per lineal foot of hole. Therefore, the planned blast-hole diameter permits a maximum loading of 1.46 lb of AN/FO per bank cubic yard of rock. For a shift production of 200 blank yd^3 an hour, 48 lin ft of hole must be drilled per hour. Since the estimated production of one drill is 30 lin ft per hour, two drills will be required. Air consumption per drill is 900 ft^3 per minute, so two 1,100-ft^3/min stationary, electrically driven compressors, will be installed.

The 4½-yd^3 shovel owned by the sponsor is to be purchased for loading the rock. Since the aggregate plant will be located adjacent to the mixing plant, the quarried rock must be hauled 1 mile, requiring the use of four of the used 35-ton rear-dump trucks owned by the sponsor. To take care of equipment availability, five will be purchased. One new D-8 bulldozer will be purchased for cleanup around the shovel and for constructing the quarry road.

Since quarry operations will not start until the last of May of the second contract year, the quarry equipment can be used to excavate the dam foundation. Additional excavation equipment will be provided by purchasing four more of the used 35-ton rear-dump trucks available from the sponsor. New excavation equipment purchases will be two light track-mounted percussion drills for drilling the foundation rock, two D-8 tractors, one 10-yd^3 articulated front-end loader, and a ¾-yd^3 backhoe.

Miscellaneous equipment to be purchased includes a 90-ton truck crane which will be used to construct the cofferdam cells, to erect the construction plant and equipment, and to have for general use. Approximately 2,200 ft^3 of stationary air and 1,800 ft^3 of portable air will be purchased for use during the excavation of the dam foundation and during the concrete placement. Other equipment purchases will consist of a grader, a water wagon, service trucks, shops, a warehouse, an office, and parking facilities.

Quotations were requested on all required new equipment. A plant and equipment list was started, and this list was completed and priced under estimating step XIV.

XI. PREPARATION OF THE CONSTRUCTION SCHEDULE

The construction schedule prepared for Denny Dam is shown by a bar chart in Fig. 19.6. A precedence diagram is not shown, since it is too large to be included in this book.

Work scheduling is controlled by the date of the receipt of the notice to proceed, the periods of minimum water flow, the date that all work must be completed, the time required for plant and equipment mobilization, the dates that each stage of diversion can be started, the time required for dam foundation excavation, and the dates that concrete placement can be started in the dam center monoliths. Of these controls, the two crucial ones are the time required for plant and equipment mobilization and the dates that each stage of diversion can be started, since these control when the dam's foundation can be excavated and when concrete placement can start.

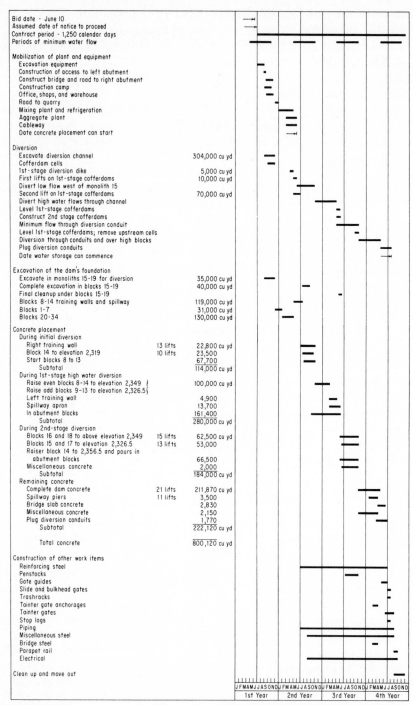

Figure 19.6 Construction schedule for Denny Dam.

The excavation plant and equipment can be mobilized rapidly, since they consist basically of used equipment owned by the sponsor. This permits excavation to be started on September 1 of the first contract year. Excavation of the diversion channel can be started prior to completion of the bridge to the right abutment, since the river is fordable during the low-water season. The placement of concrete cannot start until June 1 of the second contract year, since receipt of the cableway will require 9 months from the time of ordering, and 2 months will be required for its erection.

The diversion plan controls the sequence of placing concrete in the dam monoliths. This concrete-placing sequence is discussed in the previous section under selection of the diversion method.

Concrete placement during the first stage of diversion will be limited to the volume required to bring the low alternate monoliths of the spillway section up to elevation 2,326.5 ft and the volume of concrete that can be placed in the abutment monoliths. Concrete placement during second-stage diversion will be limited to the volume that can be placed in monoliths 15 to 19 and the concrete that can be placed in the abutment monoliths.

The concrete-placement equipment cannot be worked to full capacity when first- and second-stage diversion are being accomplished. However, the placement of concrete in the abutment monoliths during these two periods permits the concrete system to operate at a reasonable capacity and also reduces the amount of face forms that must be purchased. The total concrete placed in the dam by June 1 of the third year and by November 1 of the third year is shown on the pour diagram, Fig. 19.7

After November 1 of the third year, concrete will be placed across the entire center section of the dam, leaving the alternate spillway monoliths lower than the remaining dam monoliths to provide for flood protection. The rate of concrete placement will be limited by the requirement that only one lift per week can be poured and not by the capacity of the placement equipment.

When concrete placement is started, the volume that can be placed per month will be low, since only a few pouring areas will be available. As additional monoliths are started, production will increase. When the dam is being topped out, the placement rate will again decrease since all pours will be small. After the high monoliths have been topped out, it will take 3 weeks to top out the low monoliths. Mass-concrete placement will be completed by the end of March of the fourth contract year, or 21 months after its start. Mass concrete will be placed during the night and swing shifts, 5 days a week, resulting in a maximum hourly placement rate of 150 yd^3, which is well within the capacity of an 8-yd^3 cableway. During the day shift, the cableway will be used for yarding. The main concrete-cleanup crew will be used on the day shift, which will make it available for completing large concrete pours whose placement extends into the day shift.

From April through July of the fourth year, the monthly concrete placement rates are so minor that the concrete production facilities will only be operated when required.

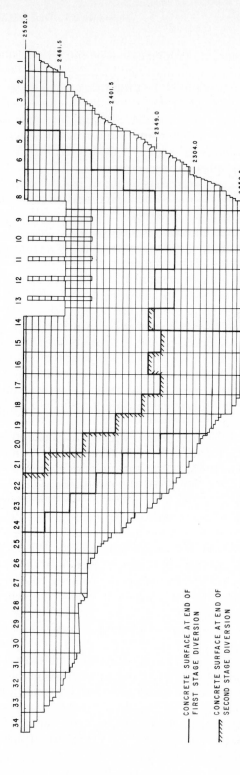

Figure 19.7 Concrete pour diagram for Denny Dam.

582

The following is a monthly summary of the concrete planning schedule:

Concrete Placement Schedule (Including Extra Volume at Diversion Channel Excavation)

	Yd³ of Concrete
Second contract year:	
July	11,000
August	23,000
September through December,	
4 months at 40,000 yd³/mo	160,000
Total for second year	194,000
Third contract year:	
January through May, 5 months at 40,000 yd³/mo	200,000
June through October, 5 months at 36,800 yd³/mo	184,000
November through December, 2 months at 44,000 yd³/mo	88,000
Total for third year	472,000
Fourth contract year:	
January and February, 2 months at 44,000 yd³/mo	88,000
March	25,000
April	14,120
May	3,000
June	3,000
July	1,000
Total for fourth year	134,120
Total concrete	800,120

The construction schedule (Fig. 19.6) and the pour diagram (Fig. 19.7) indicate that approximately 528 lin ft of both upstream and downstream face forms will be required for six 48-ft and three 56-ft dam monoliths. The maximum requirement for bulkhead forms will be 1,640 lin ft. These will be required when monolith 14 is at elevation 2,319 ft, requiring 320 lin ft of bulkhead forms. At this time monoliths 8, 10, and 12 will be at elevation 2,259 ft, requiring 1,100 lin ft of bulkhead forms, and an additional 130 lin ft of bulkhead forms will be required in the right training wall.

The scheduling of the construction of the remaining bid items is controlled by the concrete placement schedule, since these items are either embedded in or supported by the concrete.

XII. ESTIMATING DIRECT COST

The direct cost section of a dam estimate is more voluminous than the remainder of the estimate since it consists of the detailing and pricing of all the work required to construct the project. To present here the direct cost estimating details for all the bid items for the construction of Denny Dam would be impractical. For the sake of brevity, only the direct cost estimating details for bid item 24, mass concrete in the dam, are presented. These details will amply illustrate how direct cost estimating is performed, since they are representative of all the major activities required for the construction of a concrete dam.

After these details are given, a summary of the estimated direct cost for all the bid items is presented. Calculation of the costs for these remaining bid items, although not presented in detail, was based upon production rates from other jobs. Labor unit costs are summations, derived from hourly labor costs divided by hourly production rates for each of the various operations necessary for each bid item. Supply costs were derived in the same manner. Permanent materials costs and subcontract costs are extensions of the plugged prices tabulated in estimating step V.

Table 19.6 is a summary of the components of the direct cost for bid item 24, mass concrete in the dam, and estimating details for each of the component costs is thereafter discussed.

Direct Cost of Stockpiled Aggregates

It was necessary to establish the number of shifts that the quarry and aggregate plant must be operated and the average production that will be achieved per shift before the direct cost of aggregates could be estimated. Since the aggregate plant will have 70,000 tons of aggregate storage in its finished

TABLE 19.6 Denny Dam Estimate: Summary of Direct Cost Bid Item 24

	Cost/yd³ of Concrete		
	Labor	Supplies	Total
Mixed concrete:			
Cost of aggregates in stockpiles	$ 3.44	$2.59	$ 6.03
Aggregate reclaiming, rescreening, and concrete mixing	1.99	0.67	2.66
Cement loss		0.19	0.19
Admixture		0.20	0.20
Aggregate and concrete cooling	0.66	0.42	1.08
Subtotal mixed concrete	$ 6.09	$4.07	$10.16
Place, cleanup, cure, finish:			
Concrete haul	0.52	0.09	0.61
Cableway operating cost	1.20	0.87	2.07
Cableway dead time on day shift	0.10	0.08	0.18
Placement crew	1.28	0.40	1.68
Cleanup and cure	2.10	0.22	2.32
Patch and finish	0.69	0.08	0.77
Utility service	0.28	0.14	0.42
Subtotal placement	$12.26	$5.95	$18.21
Form cost	6.12	1.92	8.04
Total direct cost	$18.38	$7.87	$26.25

stockpiles (enough aggregate for 37,840 yd³ of concrete), the quarry and aggregate plant can operate independently of the maximum hourly demand of the mixing plant. This amount of storage permits the quarry and aggregate plant to be operated to meet the scheduled maximum monthly rate of concrete placement. It also permits stopping the operation of the quarry and aggregate plant when the aggregate storage exceeds that required by the concrete remaining to be placed.

If the quarry and aggregate plants are worked one shift per day, 5 days per week, they will produce 4,000 tons of aggregate per shift, or 84,000 tons per month. This is sufficient aggregate for 45,400 yd³ of concrete. This monthly production slightly exceeds the scheduled concrete placement rates, so that this shift schedule can be used to estimate aggregate cost. If for any reason actual concrete placement exceeds the amount scheduled, additional shifts can be worked. Aggregate cost was estimated based on suspending the operation of the quarry and aggregate plant on March 15 of the fourth year, when only 33,620 yd³ of concrete will be left to pour. The total number of shifts required is one shift per day from July 1 of the second year to March 15 of the fourth year, or 431 shifts. The concrete takeoff showed a total concrete yardage of 800,120. Using 1.85 tons of aggregate per cubic yard of concrete requires the production of 1,480,222 tons of aggregate. This tonnage, divided by 431 shifts, results in an average shift production of 3,435 tons, or 1,374 bank yd³. With the average shift production figures and the total aggregate requirements, the direct cost of stockpiled aggregates was determined by estimating the direct cost of four operations.

1. Quarry Development

	Labor	Supplies	Total
Allowance for quarry development	$60,000	$60,000	$120,000
Cost/ton (1,480,222 tons)	$0.04	$0.04	$0.08

2. Drill and Shoot (3,435 tons/shift)

	Labor	Supplies	Total
Shift cost:			
2 drillers	$ 335.10		$ 335.10
2 chuck tenders	314.36		314.36
Drill parts and supplies		$ 75.00	75.00
1 compressor operator	194.56		194.56
Compressor parts, supplies,			
and maintenance	25.00	75.00	100.00
1 drill doctor and bit grinder	204.58		204.58
1 powder truck	220.40	35.00	255.40
1 powderman	167.55		167.55
Drill steel:			
Cost/set $1,200, Life 2,000 lin ft			
Cost/lin ft of hole $0.06			

	Labor	Supplies	Total
Hole drilled/shift 330 lin ft			
Cost/shift		198.00	198.00
Bits:			
Cost/bit $525, Life 250 lin ft			
Cost/lin ft of hole $2.10			
Cost/shift		693.00	693.00
Powder and exploders:			
Cap and powder cost/hole $3			
No. of holes/shift 30			
Cost/shift			
AN/FO cost/lb $0.20			
Cost/yd³ $0.25			
Cost/shift		858.75	858.75
Small tools and supplies		100.00	100.00
Total/shift	$1,461.55	$2,034.75	$3,496.30
Cost/ton	$ 0.43	$ 0.59	$ 1.02

3. Load and Haul to Aggregate Plant

Truck cycle:	
Load	3.5 min
Haul 1 mile at 15 mph	4.0 min
Dump in hopper	2.0 min
Return	4.0 min
Total	13.5 min
Trips per 50-min h	3.7
Truck capacity/h	129.5 tons
Truck capacity/shift	1,036.0 tons
Trucks required for 3,435-ton/shift	3.3 or 4

Cost/Shift (3,435 tons/shift):	Labor	Supplies	Total
1 quarry foreman	$ 231.16		$ 231.16
1 truck spotter	157.18		157.18
1 4½-yd³ shovel	672.15	$ 188.20	860.35
1 bulldozer	255.73	109.42	365.15
4 35-ton rear dumps	908.70	600.80	1,509.50
Road maintenance	100.00	75.00	175.00
Small tools and miscellaneous		100.00	100.00
Total/shift	$2,324.92	$1,073.42	$3,398.34
Cost/ton of aggregates	$ 0.68	$ 0.31	$ 0.99

4. Operation of Aggregate Plant

Cost/Shift (3,435 tons/shift):	Labor	Supplies	Total
1 foreman	$ 231.16		$ 231.16
1 primary crusher operator	196.50		196.50
1 operator for other crushers	196.50		196.50
1 operator for screens and classifiers	196.50		196.50

Cost/Shift (3,435 tons/shift):	Labor	Supplies	Total
4 oilers	699.80		699.80
2 laborers	314.36		314.36
1 welder	204.58		204.58
2 mechanics	409.16		409.16
Power 700 kW		$ 336.00	336.00
Repair parts		1,000.00	1,000.00
Oil, grease, and misc. supplies		250.00	250.00
Total cost/shift	$2,448.56	$1,586.00	$4,034.56
Cost/ton	$ 0.71	$ 0.46	$ 1.17

Summary of the Direct Cost of Stockpiled Aggregates

	Labor	Supplies	Total
Quarry development	$0.04	$0.04	$0.08
Drill and shoot	0.43	0.59	1.02
Load and haul	0.68	0.31	0.99
Aggregate plant	0.71	0.46	1.17
Total cost/ton	$1.86	$1.40	$3.26
Cost/yd^3 of concrete	$3.44	$2.59	$6.03

Aggregate Reclaiming, Mix-Plant Charging, Rescreening, Cement Handling, and Concrete Mixing

In accordance with the construction schedule, mass-concrete placement will be completed 21 months after it is started. Mixing and placing will be done two shifts per day, totaling 441 days or 882 shifts. During this period 779,000 yd^3 of mass concrete will be poured. Thus the average production per shift will be 883 yd^3.

Cost/Shift

	Labor	Supplies	Total
1 foreman	$ 231.16		$ 231.16
1 belt operator	194.56		194.56
1 rescreen plant operator	194.56		194.56
1 mix-plant operator	196.50		196.50
1 oiler	174.95		174.95
1 concrete dispatcher	196.50		196.50
1 laborer	157.18		157.18
Maintenance labor	409.16		409.16
Power		$ 42.80	42.80
Repair parts and supplies		450.00	450.00
Miscellaneous supplies		100.00	100.00
Total cost/shift	$1,754.57	$592.80	$2,347.37
Total cost/yd^3	$ 1.99	$ 0.67	$ 2.66

Cement Loss and Admixture Cost

The cost of cement per cubic yard of concrete is $12.43. Cement loss will be 1½ percent of this, or $0.19/yd³.

Admixture cost will be approximately $0.20/yd³.

Aggregate and Concrete Cooling

Cost/Shift

	Labor	Supplies	Total
Operator of ice equipment	$179.94		$179.94
Compressor operator	194.56		194.56
Maintenance labor	204.58		204.58
Power and water		$200.00	200.00
Repair parts and supplies		125.00	125.00
Miscellaneous		50.00	50.00
Cost/shift	$579.08	$375.00	$954.08
Cost/yd³	$ 0.66	$ 0.42	$ 1.08

Concrete Haul

Two locomotive and car units will be used.

	Labor	Supplies	Total
Cost/shift	$454.72	$80.00	$534.72
Cost/yd³	$ 0.52	$ 0.09	$ 0.61

Cableway Operation

The cableway will handle approximately 800,000 yd³ of concrete, or 1,600,000 tons. The track cable should be good for the life of the project and will not require a replacement. However, approximately three travel-line replacements and approximately 14 hoisting-line replacements will be required. The direct cost of each aspect of cableway operation is tabulated below in cost per cubic yard of concrete.

Cost/yd³

	Cable replacements		
	Labor	Supplies	Total
3 travel lines	$ 9,000	$ 39,000	$ 48,000
14 hoisting lines	21,000	266,000	287,000
Total	$30,000	$305,000	$335,000
Cost/yd³	$0.04	$0.38	$0.42

Cableway operation—pouring concrete (883 yd³/shift)			
	Labor	**Supplies**	**Total**
Cost/shift:			
1 operator	$ 214.83		$ 214.83
1 oiler	174.95		174.95
1 rigger	179.73		179.73
1 bellboy	179.73		179.73
Maintenance labor	280.00		280.00
Power		$180.00	180.00
Repair parts and supplies		250.00	250.00
Total/shift	$1,029.24	$430.00	$1,459.24
Cost/yd³	$ 1.16	$ 0.49	$ 1.65
Cable replacements	0.04	0.38	0.42
Total/yd³	$ 1.20	$ 0.87	$ 2.07
Operation yarding, 1 shift/day, 442 shifts			
Cost/shift	$1,029.24	$430.00	$1,459.24
Cost 442 shifts	$454,924	$190,060	$644,984
Cost distribution of day-shift operation:			
Reinforcing steel	55,000	15,000	70,000
Other material items	60,000	22,000	82,000
Concrete cleanup and cure	175,000	63,000	238,000
Forms	85,000	30,000	115,000
Dead time charged to concrete	$ 79,924	$ 60,060	$139,984
Cableway dead time cost/yd³,			
based on 779,000 yd³	$0.10	$0.08	$0.18
Placement crew (883 yd³/shift)			
Cost/shift:			
1 foreman	$ 171.91		$ 171.91
5 vibrator men	796.80		796.80
1 bucket dumper	159.36		159.36
Miscellaneous supplies		$ 88.30	88.30
1 carpenter (charged to forms)			
Total	$1,128.07	$ 88.30	$1,216.37
Cost/yd³	$ 1.28	$ 0.10	$ 1.38
Furnishing and maintenance			
of vibrators		0.30	0.30
Total/yd³	$ 1.28	$ 0.40	$ 1.68

Cleanup and Cure

This work was estimated on a total time basis, since crews will be required on each of the three daily shifts and part of the cableway time on the day shift will be charged to this operation. Total days are 442.

	Labor	Supplies	Total
Cost/day			
Day shift:			
1 foreman	$ 171.91		$ 171.91
4 air-tool operators	650.53		650.53
4 laborers	637.44		637.44
Swing shift:			
1 foreman	171.91		171.91
2 air-tool operators	325.26		325.26
2 laborers	318.72		318.72
Night shift:			
1 foreman	171.91		171.91
2 air-tool operators	325.26		325.26
2 laborers	318.72		318.72
Maintenance and weekend curing labor	210.00		210.00
Parts and supplies		$ 250.00	250.00
Total/day	$ 3,301.66	$ 250.00	$ 3,551.66
For 442 days	$1,459,334.00	$110,500.00	$1,569,834.00
Cableway charge	175,000.00	63,000.00	238,000.00
Total	$1,634,334.00	$173,500.00	$1,807,834.00
Cost/yd^3, based on 779,000 yd^3	$ 2.10	$ 0.22	$ 2.32

Patch and Finish

	Labor	Supplies	Total
Daily crew (1,766 yd^3/day):			
1 foreman	$ 187.04		$ 187.04
3 finishers (1/shift)	544.76		544.76
3 laborers (1/shift)	478.08		478.08
Supplies		$150.00	150.00
Cost/day	$1,209.88	$150.00	$1,359.88
Cost/yd^3	$ 0.69	$ 0.08	$ 0.77

Utility Service

	Labor	Supplies	Total
Daily cost (1,766 yd^3)	$500	$250	$750
Cost/yd^3	$0.28	$0.14	$0.42

Form Cost

As previously stated, most of the charges for forms were estimated by adjusting unit cost experienced on past jobs for changes in labor rates.

The cantilever forms required for the project are to be purchased from the sponsor and the following calculations were made to determine their write-off:

Purchased forms	$250,000
Freight and move-in	7,500
Reconditioning	50,000
Total cost	$307,500
Assumed net salvage value	67,500
Amount to be written off to project	$240,000
Write-off on:	
Mass concrete	748,240 yd³
Spillway piers	3,500 yd³
Training walls	27,700 yd³
	779,440 yd³
Write-off/yd³	$0.31
Amount to be written off to mass concrete	$231,954

For the area of forms for bid item 24, refer to the quantity takeoff summary in this chapter.

Cost/Ft² of Forms

	Labor	Supplies	Total
Purchased form write-off		$ 231,954	$ 231,954
Cableway charges	$ 85,000	30,000	115,000
8-in-diameter formed hole, 4,410 lin ft at $7.00	22,050	8,820	30,870
Built-in-place forms, 60,000 ft² at $13.50	585,000	225,000	810,000
Blockout forms, 3,780 ft² at $14.50	39,690	15,120	54,810
Curved forms, 3,250 ft² at $12.80	26,975	14,625	41,600
Gallery and other interior forms, 83,550 ft² at $11.10	509,655	417,750	927,405
Shop-built lift-starter forms, 12,000 ft² at $6.60	43,200	36,000	79,200
Set, strip, and raise panel forms, 908,900 ft² at $4.10	3,272,040	454,450	3,726,490
Total form cost	$4,583,610	$1,433,719	$6,017,329
Cost/ft², based on 1,059,480 ft²	$4.33	$1.35	$5.68
Cost/yd³ of mass concrete (748,240 yd³)	$6.12	$1.92	$8.04
Man-hours/ft² of formed contact, based on $22.50/h labor cost	0.1924		

XIII. TABULATION OF DIRECT COST

TABLE 19.7 Denny Dam Estimate: Spread Sheet Summarizing Direct Cost

Bid Item	Description	Bid Quantity	Labor Unit	Labor Amount	Supplies Unit	Supplies Amount	Permanent Materials Unit	Permanent Materials Amount	Subcontract Unit	Subcontract Amount	Total Direct Cost Unit	Total Direct Cost Amount
			$	$	$	$	$	$	$	$	$	$
1	Care and diversion of river	Lump sum		754,000		664,750						1,418,750
	Excavation:											
2	Common	185,000 yd³	2.02	373,700	1.40	259,000					3.42	632,700
3	Rock	170,000 yd³	4.66	792,200	2.80	476,000					7.46	1,268,200
4	Preliminary rock cleanup	7,000 yd²	3.90	27,300	0.54	3,780					4.44	31,080
5	Close-line drilling	30,000 ft²	1.07	32,100	1.15	34,500					2.22	66,600
6	Final rock cleanup	23,000 yd²	17.80	409,400	2.02	46,460					19.82	455,860
	Total excavation			1,634,700		819,740						2,454,440
	Drill and grout:											
7	1½-in-diameter grout holes	25,000 lin ft							11.00	275,000	11.00	275,000
8	3-in-diameter drain holes	17,000 lin ft							13.00	221,000	13.00	221,000
9	3-in-diameter cored exploratory holes	13,500 lin ft							15.00	202,500	15.00	202,500
10	Grout connections	800 ea							25.00	20,000	25.00	20,000
11	Pressure testing	800 test							25.00	20,000	25.00	20,000
12	Grouting cement	8,000 ft³							5.00	40,000	5.00	40,000
13	Pressure grouting	8,000 ft³							15.00	120,000	15.00	120,000
	Total drill and grout									898,500		898,500
	Backfill:											
14	Uncompacted backfill	4,300 yd³	4.90	21,070	4.35	18,705					9.25	39,775
15	Dumped rock backfill	18,000 yd³	4.25	76,500	3.14	56,520					7.39	133,020
16	Backfill material	560 yd³	5.40	3,024	4.80	2,688					10.20	5,712
17	Dumped riprap	1,000 yd³	5.13	5,130	3.32	3,320					8.45	8,450
	Total backfill			105,724		81,233						186,957
18	Reinforcing steel	3,080,000 lb	0.018	55,400	0.005	15,400			0.40	1,232,000	0.423	1,302,340
	Concrete:											
19	Stilling basin	13,700 yd³	23.62	323,594	8.18	112,066					31.80	435,660
20	Roadways, sidewalks, bridge	2,850 yd³	136.40	386,012	35.40	100,182					171.80	486,194

No.	Item	Quantity	$	$	$	$	$	$	$	$	$	$
21	Slabs, beams, and columns	4,150 yd³	69.40	288,010	20.80	86,320					90.20	374,330
22	Spillway piers	3,500 yd³	77.77	272,195	24.35	85,225					102.12	357,420
23	Training walls	27,700 yd³	26.47	733,219	7.94	219,938					34.41	953,157
24	Mass in dam	775,000 yd³	18.38	14,244,500	7.87	6,099,250					26.25	20,343,750
	Total concrete	826,880 yd³		16,247,530		6,702,981						22,950,511
25	Cement	635,000 bbl					16.00	10,160,000			16.00	10,160,000
	Miscellaneous:											
26	12-in-diameter pipe	100 lin ft	11.83	1,183	5.47	547	6.50	650			23.80	2,380
27	18-in-diameter pipe	100 lin ft	17.63	1,763	6.02	602	12.00	1,200			35.65	3,565
28	Copper water stops	12,300 lin ft	5.80	71,340	1.04	12,792	7.50	92,250			14.34	176,382
29	6-in-diameter porous concrete pipe	28,000 lin ft	6.08	170,240	1.11	31,080	2.75	77,000			9.94	278,320
30	12-in-diameter half-round porous pipe	1,950 lin ft	7.90	15,405	1.36	2,652	5.00	9,750			14.26	27,807
31	Miscellaneous doors	400 ft²	5.11	2,044	0.55	220	25.00	10,000	1.50	600	32.16	12,864
	Total miscellaneous			261,975		47,893		190,850		600		501,318
	Gates and guides:											
32	Miscellaneous metal castings	Lump sum		18,120		2,056		15,000		1,200		36,376
33	Slide and bulkhead gates	286,000 lb	0.17	48,620	0.05	14,300	2.00	572,000	0.03	8,580	2.25	643,500
34	Pumps, motors, piping for gates	Lump sum		1,000		320		60,000		750		62,070
35	Penstock gates, frames and guides	288,000 lb	0.39	112,320	0.08	23,040	2.20	633,600	0.04	11,520	2.71	780,480
36	Penstock-gate hoist and supports	Lump sum		46,562		2,460		120,000		3,000		172,022
37	Penstock lining and frames	535,000 lb	0.03	16,050	0.01	5,350			4.20	2,247,000	4.24	2,268,400
38	Tainter gates and accessories	670,000 lb	0.23	154,100	0.01	6,700	1.80	1,206,000	0.25	167,500	2.29	1,534,300
39	Tainter-gate trunnion anchors	320,000 lb	0.19	60,800	0.01	3,200	2.20	704,000	0.10	32,000	2.50	800,000
40	Tainter-gate hoist and accessories	150,000 lb	0.16	24,000	0.01	1,500	4.00	600,000	0.05	7,500	4.22	633,000
41	Trashrack guides and beam	260,000 lb	0.23	59,800	0.01	2,600	1.25	325,000	0.04	10,400	1.53	397,800
42	Stop logs, protection plates, and beams	168,000 lb	0.32	53,760	0.02	3,360	1.50	252,000	0.04	6,720	1.88	315,840
	Total gates and guides			595,132		64,886		4,487,600		2,496,170		7,643,788

TABLE 19.7 Denny Dam Estimate: Spread Sheet Summarizing Direct Cost (Continued)

Bid Item	Description	Bid Quantity	Labor Unit	Labor Amount	Supplies Unit	Supplies Amount	Permanent Materials Unit	Permanent Materials Amount	Subcontract Unit	Subcontract Amount	Total Direct Cost Unit	Total Direct Cost Amount
	Mechanical and miscellaneous metal:											
43	Cast-iron pipe and fittings	156,000 lb	0.03	4,680	0.01	1,560			4.52	705,120	4.56	711,360
44	Steel pipe and fittings	184,000 lb	0.03	5,520	0.01	1,840			4.29	789,360	4.33	796,720
45	Piezometer piping and fittings	Lump sum								5,000		5,000
46	Sump pump and piping	Lump sum								32,500		32,500
47	Water supply and plumbing	Lump sum								46,000		46,000
48	Structural steel	66,500 lb	0.42	27,930	0.05	3,325	1.40	93,100	0.05	3,325	1.92	127,680
49	Railing and fittings	17,700 lb	1.39	24,603	0.18	3,186	1.50	26,550	0.05	885	3.12	55,224
50	Parapet railings	1,152 lin ft	5.50	6,336	2.50	2,880	30.00	34,560			38.00	43,776
51	Steel in spillway bridge	410,000 lb	0.19	77,900	0.05	20,500	1.40	574,000	0.05	20,500	1.69	692,900
	Total mechanical and miscellaneous metal			146,969		33,291		728,210		1,602,690		2,511,160
52	Electrical system	Lump sum								603,000		603,000
	TOTAL DIRECT COST			$19,801,470		$8,430,174		$15,566,660		$6,832,960		$50,631,264

594

XIV. ESTIMATING
PLANT AND EQUIPMENT COST

Only the summary of this part of the estimate will be presented here (Table 19.8). The estimate was prepared item by item and group by group, using the same methods used in Chap. 18 in calculating the plant and equipment costs for Sierra Tunnel.

The used equipment offered by the sponsor is included. The additional equipment which will be required is to be purchased new. The major pieces of new equipment include a 90-ton truck crane, a 10-yd^3 front-end loader, three 300-hp bulldozers, two hydraulic track drills for excavation, two track drills for the quarry, additional crushing equipment, new conveyors for the aggregate plant, a 25-ton cableway, and concrete buckets. Major installation costs occur for the aggregate plant and for the cableway because of the great volume of rock excavation required. The only rental equipment consists of a diesel pile hammer and a diesel pile extractor.

TABLE 19.8 Denny Dam Estimate: Summary of Plant and Equipment Cost

	Freight and Move-in	Labor	Plant Invoice	Used Equipment	New Equipment	Total Cost	Net Salvage	Net Cost	Rental Cost	Total
General:										
Access		$ 104,400	$ 137,500			$ 241,900		$ 241,900		$ 241,900
Buildings and shops	$ 1,800	87,700	179,200	$ 10,000	$ 88,750	367,450	$ 32,920	334,530		334,530
Service eqpt. and facilities	7,400	132,200	298,100	92,500	1,087,550	1,617,750	393,350	1,224,400		1,224,400
Total general	9,200	324,300	614,800	102,500	1,176,300	2,227,100	426,270	1,800,830		1,800,830
Dewatering	2,200		12,000		62,500	76,700	20,000	56,700	$14,950	71,650
Excavation	18,500	2,300	10,000	320,000	1,237,500	1,588,300	778,750	809,550		809,550
Concrete:										
Quarry and haul to aggregate plant	19,100	55,800	63,620	656,000	607,910	1,402,430	421,300	981,130		981,130
Aggregate plant	46,600	275,500	413,610	100,000	1,577,700	2,413,410	559,230	1,854,180		1,854,180
Mixing-plant charging	6,000	40,600	36,250	142,500		225,350	50,000	175,350		175,350
Mixing plant	12,000	87,100	93,750	300,000		492,850	150,000	342,850		342,850
Cement handling	1,900	25,550	31,250	26,500		85,200	10,000	75,200		75,200
Ice manufacturer	650	19,720	46,200	25,000		91,570	10,000	81,570		81,570
Cold air and cold water	9,150	151,980	193,750	375,000		729,880	125,000	604,880		604,880
Concrete delivery	7,500	17,850	51,000	75,000		151,350	37,500	113,850		113,850
Cableway	61,000	203,100	424,660		2,373,000	3,061,760	1,186,500	1,875,260		1,875,260
Concrete placing	2,950			60,000	239,600	302,550	189,730	112,820		112,820
Total concrete	166,850	877,200	1,354,090	1,760,000	4,798,210	8,956,350	2,739,260	6,217,090		6,217,090
TOTAL COST	$196,750	$1,203,800	$1,990,890	$2,182,500	$7,274,510	$12,848,450	$3,964,280	$8,884,170	$14,950	$8,899,120

596

XV. ESTIMATING INDIRECT COST

TABLE **19.9** **Denny Dam Estimate: Details of Indirect Cost**
        Labor

	No. of Employees	No. of Man-months	Monthly Salary	Total Cost
Job management:				
Project manager	1	44	$5,600	$ 246,400
Secretary	1	44	1,400	61,600
Total	2			$ 308,000
Job supervision:				
General supt.	1	42	4,100	$ 172,200
Swing-shift supt.	1	22	3,600	79,200
Graveyard-shift supt.	1	21	3,600	75,600
Excavation supt.	1	15	3,600	54,000
Carpenter supt.	1	40	3,600	144,000
Mechanical supt.	1	40	4,000	160,000
Electrical supt.	1	40	4,000	160,000
Rigging supt.	1	40	3,600	144,000
Secretary/clerk	2	60	1,250	75,000
Total	10			$1,064,000
Engineering:				
Project engineer	1	44	3,600	$ 158,400
Office engineer	1	42	2,750	115,500
Cost & progress engineer	1	40	2,400	96,000
Design engineer	2	24	2,750	66,000
Draftsman	2	52	2,100	109,200
Concrete engineer	3	67	2,200	147,400
Chief surveyor*	1	38	3,250	123,500
Surveying crew*	2	60	3,000	180,000
Secretary	1	44	1,250	55,000
Total	14			$1,051,000
Office:				
Accountant	1	44	3,200	$ 140,800
Cost accountant	1	41	2,800	114,800
Personnel manager	1	40	2,800	112,000
Key puncher	1	36	1,200	43,200
Timekeeper and clerk	3	72	1,400	100,800
Secretary	1	44	1,250	55,000
Total	8			$ 566,600
Purchasing & warehouse:				
Purchasing agent	1	40	2,750	$ 110,000
Warehouseman*	3	83	3,300	273,900
Yard crane operator*	1	40	4,040	161,600
Laborer*	1	40	3,380	135,200
Truck driver*	1	40	3,760	150,400
Total	7			$ 831,100

TABLE **19.9   Denny Dam Estimate:
Details of Indirect Cost**   (*Continued*)

	No. of Employees	No. of Man-months	Monthly Salary	Total Cost
Safety and first aid:				
Safety engineer	1	40	2,250	$ 90,000
Nurse	3	110	1,650	181,500
Total	4			$ 271,500
Janitorial & security				
Watchman & janitor*	2	80	3,150	$ 252,000
Total labor	47			$4,344,200
Burden on nonunion labor† 19%				582,844
Total labor and burden				$4,927,044

Other Cost

	Months	Cost/ Month	Cost
Operation of pickups and sedans,			
520 vehicle months	520	$ 350	$182,000
Office supplies	40	1,200	48,000
Computer cost	40	1,750	70,000
Telephone, telegraph, and cable	40	1,500	60,000
Office rent (temporary)	4	1,200	4,800
Heat, light, and power for office	40	500	20,000
Engineering supplies	40	1,200	48,000
Office equipment			30,000
Engineering equipment			25,000
First-aid and safety supplies	40	600	24,000
Entertainment and travel	40	400	16,000
Blueprints, photostats, and photographs	40	400	16,000
Outside consultants			7,500
Legal cost			12,500
Audit cost			15,000
Testing laboratory expense			5,000
Job fencing			15,000
Mobilization of personnel			40,000
Licenses and fees			5,000
Sign cost			2,400
Drinking water	40	550	22,000
Miscellaneous cost—supplies and			
maintenance for warehouse truck and crane	40	1,600	64,000
Total other cost			$732,200

Insurance Premiums

	Value	Cost/Year	Total Cost
Acquisition cost of construction			
equipment	$ 9,457,010		

Insurance Premiums

	Value	Cost/Year	Total Cost
Yearly insurance premium at 1.8% per year		$170,226	
Value of buildings	$ 268,700		
Yearly insurance premium at 1.2% per year		3,224	
Total yearly premiums		$173,450	
For three years			$520,350
Builder's risk insurance			186,000
Total insurance			$706,350

Taxes

Acquisition cost of equipment	$ 9,457,010		
Buildings	268,700		
	$ 9,725,710		
Value at completion of project	3,964,280		
	$13,689,990		
Average value	$ 6,844,995		
Assessed valuation—25%	$ 1,711,249		
Yearly tax—5%		$ 85,562	
For three years			$256,686

Bond Premiums‡

	Contract Value	Rate/ $1,000	Total
First	$ 100,000	$7.50	$ 750
Next	2,400,000	5.25	12,600
Next	2,500,000	4.50	11,250
Remainder	80,000,000	4.00	320,000
Total	$85,000,000		$344,600
Time premium, 18%			62,028
			$406,628

* Union labor.

† Payroll burden for union labor has already been added to wage rate.

‡ Bond premiums are based on nontowner rates and on an assumed bid total of $85,000,000.

Summary of Indirect Cost

Labor:

Job management	$ 308,000
Job supervision	1,064,000
Engineering	1,051,000
Office	566,600
Purchasing and warehouse	831,100

Summary of Indirect Cost (*Continued*)

Safety and first aid	271,500
Janitorial and security	252,000
Subtotal	$4,344,200
Burden for nonunion personnel	582,844
Total labor	$4,927,044
Other cost	732,200
Insurance premiums	706,350
Taxes	256,686
Bond premium	406,628
Total indirect cost	$7,028,908

XVI. ESTIMATION OF CAMP COSTS

A single-status camp will not be required. In order to have key supervisory personnel live adjacent to the project, a married-quarters camp will be built. This camp will consist of 16 three-bedroom house trailers and a separate laundry building. Breakdown of camp costs is given in Table 19.10.

TABLE 19.10   Denny Dam Estimate: Camp Cost
Camp Construction

	Labor	Invoice	Total
Grading and surfacing roads	$ 6,000	$ 11,000	$ 17,000
Water supply	7,000	5,000	12,000
Sewage disposal	10,000	15,000	25,000
Electrical distribution	6,000	12,000	18,000
16 3-bedroom, 60-ft long house trailers, at $21,250		340,000	340,000
Support, close in bottoms, build entrances and porches	16,000	16,000	32,000
Laundry building	10,000	10,000	20,000
Washers and dryers	2,000	10,000	12,000
Total construction cost	$57,000	$419,000	$476,000
Rent property		20,000	20,000
	$57,000	$439,000	$496,000
Salvage on trailers		100,000	100,000
Net camp cost	$57,000	$339,000	$396,000

Camp Operation

Rent per unit, $300/month
Rent for 16 units, $4,800/month

34 months' rental (revenue)	$163,200
Maintenance, 34 months at $1,200	40,800

XVII. ESTIMATION OF ESCALATION

TABLE 19.11 **Denny Dam Estimate: Escalation in Labor Cost**

		Labor Expenditures by Years			
	Total Labor	June 1 of 1st Year to May 31 of 2nd Year	June 1 of 2nd Year to May 31 of 3rd Year	June 1 of 3rd Year to May 31 of 4th Year	June 1 of 4th Year to May 31 of 5th Year
Direct labor:					
Diversion and care of river	$ 754,000	$ 474,000	$ 100,000	$ 100,000	$ 80,000
Excavation and backfill	1,740,424	1,000,000	634,700	65,724	40,000
Concrete and reinforcing	16,302,530		7,880,000	7,975,000	447,530
Remainder	1,004,516	123,516	325,000	352,000	204,000
Total direct labor	$19,801,470	$1,597,516	$ 8,939,700	$ 8,492,724	$ 771,530
Plant and equipment labor	1,203,800	1,103,800			100,000
Indirect labor	4,927,044	1,080,000	1,719,000	1,463,544	664,500
Camp construction labor	57,000	50,000			7,000
Camp operation and maintenance labor	27,200	5,000	10,000	10,000	2,200
Labor cost before escalation	$26,016,514	$3,836,316	$10,668,700	$ 9,966,268	$1,545,230
Estimated labor cost increase (10% per annum)			10%	21%	33.1%
Labor escalation	3,671,257		1,066,870	2,092,916	511,471
Total labor cost	$29,687,771	$3,836,316	$11,735,570	$12,059,184	$2,056,701

601

XVIII. SUMMARY OF TOTAL COST

TABLE 19.12   **Denny Dam Estimate: Total Project Cost**

	Labor Cost	Total Cost	
Direct cost:			
Labor	$19,801,470	$19,801,470	
Supplies		8,430,174	
Permanent materials		15,566,660	
Subcontracts		6,832,960	
Total direct cost			$50,631,264
Plant and equipment cost:			
Freight and move-in		$ 196,750	
Labor	1,203,800	1,203,800	
Plant invoice		1,990,890	
Used equipment		2,182,500	
New equipment		7,274,510	
Subtotal		$12,848,450	
Net salvage		3,964,280	
Net cost		8,884,170	
Outside equipment rental		14,950	
Total P&E cost			$ 8,899,120
Indirect cost:			
Labor	4,927,044	$4,927,044	
Other cost		732,200	
Insurance premiums		706,350	
Taxes		256,686	
Bond premium		406,628	
Total indirect cost			$ 7,028,908
Camp cost:			
Construction cost	57,000	$ 496,000	
Operation and maintenance	27,200	40,800	
Total cost		$ 536,800	
Camp revenue		163,200	
Salvage revenue		100,000	
Total revenue		$ 263,200	
Net camp cost			$ 273,600
Cost before escalation	$26,016,514		$66,832,892
Escalation:			
Labor	$ 3,671,257	$ 3,671,257	
Materials			
Total escalation	$ 3,671,257		$ 3,671,257
TOTAL ESTIMATED COST	$29,687,771		$70,504,149

XIX. CASH-FLOW ANALYSIS

TABLE 19.13 Denny Dam Estimate: Cash Forecast

Thousands of Dollars

	Total	1st Year*		2nd Year*			3rd Year*			4th Year*		
		May–Aug*	Sept–Dec	Jan–Apr	May–Aug	Sept–Dec	Jan–Apr	May–Aug	Sept–Dec	Jan–Apr	May–Aug	Sept–Dec
Cash revenue:												
Plant and equipment prepayment	$ 8,500	610	2,530	2,550	2,550	260						
Bid item net revenue	76,500		1,620	2,350	4,290	11,570	13,080	12,680	12,910	11,110	4,970	1,920
Total bid	$85,000	610	4,150	4,900	6,840	11,830	13,080	12,680	12,910	11,110	4,970	1,920
Retained percentage	163	−61	−415	−490	−684	−1,183	−1,308	−109				4,250
Camp revenue			11	19	19	19	19	19	19	19	19	
Salvage of plant and equipment	3,964						779				1,285	1,900
Salvage of camp	100											100
Total revenue	89,227	549	3,746	4,429	6,175	10,666	12,570	12,590	12,929	11,129	6,274	8,170
Cash disbursements:												
Direct cost	1,419		518	336	144	48	48	72	72	36	145	
Excavation and backfill	2,641		354	700	700	700						
Concrete, cement and reinforcing	34,413				1,444	6,776	6,776	6,513	6,854	5,391	659	
Other items	12,158		450	450	500	258	1,000	2,000	3,000	3,000	1,300	200
Plant and equipment cost:												
Excavation and diversion	1,680	1,665	7								8	
Other	11,183	153	3,600	3,600	3,600							230
Indirect cost	5,659	263	353	477	597	656	656	656	656	656	468	216
Bond premium	407	407										
Insurance and taxes	963	110		86	250		85	250	60	60	96	
Camp cost:												
Construction	496	296	200									
Operating cost	41		4	4	5	5	5	5	5	4	4	
Labor escalation	3,671				267	356	356	611	698	698	434	251
Inventories		100	100	100	200	200			−200	−200	−200	−100
Total disbursements	$74,731	2,994	5,591	5,753	7,707	8,999	8,926	10,114	11,145	9,731	2,974	797
Revenue less disbursements	14,496	−2,445	−1,845	−1,324	−1,532	1,667	3,644	2,476	1,784	1,398	3,300	7,373
Accumulative		−2,445	−4,290	−5,614	−7,146	−5,479	−1,835	641	2,425	3,823	7,123	14,496
Working capital		555	710	786	854	821	865	641				
Capital from partners		3,000	5,000	6,400	8,000	6,300	2,700					
Interest at 18%	1,884	180	300	384	480	378	162					

* Four-month periods were used to conserve space. The use of monthly periods produces a more accurate cash forecast.

XX. IN-HOUSE REVEIW

The estimate has been reviewed by the officers of the company, and the changes which they considered necessary have been made. The estimate as presented here reflects those changes.

As a result of a general discussion of the exposure to risks, there is consensus that contingency allowances are not required for the construction of Denny Dam since unusual expense should not be incurred. Unknown foundation conditions should not be encountered, since the foundation rock is sound granitic material. Work access will not be a problem since there is an existing highway on the left bank of the river. A labor contingency will not be required since there is an adequate supply of construction labor in northern California. Finally, unusual diversion expenses should not be experienced, since the diversion plan provides for a flood flow of 80,000 ft³/s and the chance of floods of greater magnitude is very remote.

Because of mobilization payments by the owner, capital requirements and financing costs are less than had been anticipated, and the $85,000,000 bid price which has been assumed carries a markup of 17.9 percent. It is believed that, considering the low risk, no markup in excess of 15 percent will be competitive. It was decided, however, that no changes will be made until the joint venture meeting. Changes will then be made on the basis of agreed costs and on the basis of agreement by the joint venture partners on the optimum markup.

Chapter 20 Joint Venturing and Bid Preparation

This chapter discusses the reasons for forming joint ventures in heavy construction and how these joint ventures operate, and describes the prebid meetings held by the joint venture partners to arrive at a mutually agreed upon estimate. The steps necessary to prepare a bid are described and illustrated, using the Sierra Tunnel project as an example.

If a bid is submitted by a single contractor, there will be no prebid meeting, but the steps necessary to work up the bid will be the same as those for a joint venture. In some cases the contractor employs an outside consultant to make a separate estimate with which he compares his own. The procedure in comparing the two estimates is similar to that of a joint venture prebid meeting.

JOINT VENTURES

In order to have the capital, bonding capacity, and experienced personnel to bid on large construction jobs, it is common practice for a contractor to bid in a joint venture with other contractors. A joint venture is a partnership of two or more contractors who join together to bid a particular construction job. Joint ventures are formed for only one job and are disbanded after this job is bid or constructed. The advantages of joint venturing are:

1. It is easier for each individual partner to secure a bond for his percentage of the contract than to bond the entire contract.

2. It is easier for each partner to raise a share of the capital necessary to finance the work than to raise the entire sum.

605

3. Since each major partner generally prepares an estimate and since an estimate-comparison meeting is held before submitting a bid, a check on the various estimates is provided.

4. Special talents of any of the participating companies can be utilized.

5. The contracting risk is spread, and one company is able to have an interest in many other projects, which decreases the financial loss should one project be unprofitable.

6. Joint venturing works to the benefit of the owner as well as the contractor, as it assures the owner of more competition on the large projects. If a large contract is put out for bids, the number of contractors with sufficient bonding capacity not committed to other construction work may be very few. By use of the joint venture practice, contractors without free bonding capacity to bid the complete work by themselves are able to join together in groups to submit bids in competition with other joint ventures similarly formed. Many large contracts receive 10 or more bids from 10 or more joint ventures. On the assumption that each joint venture is composed of 4 contractors, this means that 40 or more contractors are interested in any one project. This keeps the heavy construction industry a very competitive field.

The formation of a joint venture is usually initiated when one contractor decides to bid on a certain project and contacts other contractors with whom he has been associated in the past or with whom associations are desired, to determine if they are interested in participating in the joint venture. The initiating contractor is the sponsor; the other participating companies are partners in the joint venture. The sponsor also inquires what percentage each contractor desires in the joint venture. When enough contractors indicate interest to subscribe for 100 percent of the work, the joint venture is formed.

The sponsor takes the largest percentage of participation in the joint venture. In the majority of cases, the sponsor is in charge of the work and receives only advice on major decisions and financial help from the other partners. In some cases a committee composed of a member from each of the major partners may be formed to meet and make the major decisions. In other joint ventures the work is divided among the partners, with each partner responsible for certain phases of construction, but each shares in the common profit. Others are formed where one partner is responsible for one phase of the work and takes the profit on this phase, and other partners are responsible for other phases and take the corresponding profits from their phases. It is the practice with some organizations that the sponsoring partner gets an additional percentage of profits for sponsoring and administering the work, but this is not commonly done. Usually, each partner puts up working capital in accordance with his subscribed percentage and shares in the profits accordingly; the sponsor does not receive any additional compensation for manning and directing the work.

JOINT VENTURE AGREEMENT

A prebid joint venture agreement is signed by all the partners prior to the bid submission. If the bid is successful, then before the work is started on the project, a joint venture agreement and a power of attorney are signed by all the partners. The joint venture agreement differs from the prebid joint venture agreement only in that:

1. It is dated after the award of the contract.
2. It establishes the joint venture and designates a name for it.
3. If part of the work is to be performed by one of the partners, that work is defined, together with the terms of payment.
4. It designates the sponsoring partner's personnel who have been nominated as attorneys-in-fact for the joint venture, and states the authority that has been delegated to them. This includes authority to make purchases, award subcontracts, open bank accounts, etc.
5. It provides for audits.

The text of the prebid joint venture agreement for the Sierra Tunnel project is given below.

JOINT VENTURE AGREEMENT

THIS AGREEMENT, made and entered into as of April 5, 1982, by and between

ABC Contracting Company,
KLM Contracting Company, and
XYZ Construction Company
WITNESSETH:

WHEREAS, the parties hereto have agreed to enter into a Joint Venture for the purpose of submitting a joint bid to the Arizona County Water District, Austin, California, for the construction of the Sierra Tunnel, the bid date being on or about May 1, 1982, and

WHEREAS, the parties desire to enter into this Joint Venture Agreement for the performance of the aforesaid contract in the event that the Joint Venture is the successful bidder for said work and is awarded a contract for the performance thereof, and desire to fix and define between themselves their respective interests and liabilities in the performance of the work under said construction contract in the event of an award.

NOW, THEREFORE, in consideration of the mutual covenants hereinafter contained, it is hereby agreed as follows:

1. The parties hereto hereby associate themselves as joint venturers for the purpose of performing and completing the work contemplated by the aforesaid construction contract in the event it is awarded to them. All such work shall be performed under the names of the parties or such fictitious name as may be agreed upon by the parties, and all money, equipment, materials, supplies, and other property acquired by the Joint Venture shall be held jointly in such name.

2. The parties hereto hereby designate ABC Contracting Company as the sponsoring joint venturer, and the construction contract shall be carried out and performed on behalf of the Joint Venture under the direction of the sponsoring joint venturer, acting through such of its officers, employees, or agents as it may hereafter at any time or from time to time designate, and the parties hereto hereby authorize the performance of the construction contract under the direction of the officers, employees, and agents of the sponsoring joint venturer so designated.

3. The interests of the parties hereto in and to any profits and assets derived from the performance of the construction contract, and in and to any property acquired by this Joint Venture in connection with the work to be performed thereunder, and in and to all contributions required, all moneys received and losses incurred in the performance of the construction contract shall be those percentages set opposite their respective names as follows:

ABC Contracting Company	55%
KLM Contracting Company	25%
XYZ Construction Company	20%

4. Each of the parties agrees to execute all applications and indemnity agreements required by the sureties upon any bond or bonds required in connection with the said bid and contract. All financial obligations assumed by the Joint Venture in connection with the performance of the construction contract, all liabilities assumed by or charged to the Joint Venture as contractor, guarantor, or indemnitor in connection with any surety bond or other bonds which may be given or executed in connection with the construction contract, and all other obligations and liabilities of any kind or character which are assumed or undertaken by the Joint Venture in connection with and for the benefit of the performance of the construction contract shall be shared by the parties hereto proportionately and in accordance with their respective interests as set forth in paragraph 3 hereof.

5. All necessary working capital, when and as required for the performance and prosecution of the construction contract, shall be furnished by the parties hereto proportionately in accordance with their respective interests as set forth in paragraph 3 hereof. If any party borrows funds to meet its obligation hereunder, such borrowing shall be the sole and separate obligation of the party and shall not be the debt or obligation of the Joint Venture.

If any party fails or is unable to provide its proportionate share of the funds required by the Joint Venture, the interest of said party in the return of investment and profits of this Joint Venture shall be decreased to the proportion that the amount actually provided by it bears to the total amount of the funds provided by all parties, and the interest of any party which may have contributed more than its proportionate share of such funds shall be increased in the same proportion. Nothing contained herein shall increase or decrease the proportionate liability of the parties hereto for losses suffered or sustained by the Joint Venture.

6. All funds advanced by the parties or borrowed by the Joint Venture for its account and all progress and final payments or other revenue received as a result of the performance of the construction contract shall be deposited to the account of the Joint Venture in an account to be established at such bank or banks as the sponsoring joint venturer may designate. Checks may be drawn on said account or accounts by signature or signatures of such persons as may be agreed upon from time to time by the joint venturers, and unless and until otherwise agreed, the sponsoring joint venturer is authorized to designate the persons to be authorized to draw checks on said bank accounts. The Joint Venture may also maintain payroll or other accounts at such bank or at such branch or at such other bank as the sponsoring joint venturer may designate, and checks may be drawn on such accounts by signature or signatures of such persons as may be agreed upon from time to time, or, unless and until otherwise agreed, by the signature or signatures of such persons as may be designated by the sponsoring joint venturer.

7. Adequate books of account shall be maintained by the Joint Venture and such books of account may be examined by any of the joint venturers at all reasonable times. Reports of the financial condition of the Joint Venture and the progress of the work shall be made to each joint venturer periodically.

8. It is specifically understood and agreed between the parties hereto that this Joint Venture Agreement extends only to the performance of the construction contract, together with any changes or additions thereto or extra work thereunder, but not to other or different work. The term *construction contract,* as used herein, is intended to and shall include the changes, additions, or extra work hereinabove mentioned.

9. In the event of the insolvency, bankruptcy, or dissolution of any of the parties hereto, this Joint Venture shall immediately upon the occurrence thereof cease and terminate. Thereafter, the successors, receivers, trustees, or other legal representatives, hereinafter called *the representatives* of any party so affected shall cease to have any interest in and to the Joint Venture or the assets thereof. In any such case, the remaining parties shall have the right to wind up the affairs of the Joint Venture and in that connection to carry out and complete the performance of the construction contract. Upon such completion or sooner termination and receipt of payment of all amounts due under the construction contract, the remaining joint venturers shall account to the representatives of the party or parties so affected, and such representatives shall then be entitled to receive from the remaining joint venturers an amount equal to the sums advanced by the party represented, plus such party's proportionate share of the profits, or less such party's proportionate share of the losses resulting from the performance of the construction contract to the date of such termination of the Joint Venture; provided, however, that the profit or loss computed as of the date of such termination shall be in the same proportion to the whole profit or loss resulting from the performance of the construction contract as the amount of work done thereunder at such time bears to all of the work which is done thereunder. In the event the share of the losses chargeable to the party so represented exceeds the sums advanced by such party, the representatives shall promptly pay unto the remaining joint venturers any such excess. The books of the Joint Venture shall be conclusive in establishing whether a profit has been realized or a loss sustained, the amount thereof, and the proportionate amount of work done as of any given time or date.

10. None of the parties shall make any charges against the Joint Venture for any ordinary overhead expenses or for time which may be expended in connection with the performance of the aforesaid construction contract by any of such parties, their officers or employees, except such officers or employees as may be employed by the Joint Venture in actually carrying on the performance of the aforesaid construction contract.

11. No payment shall be made by the Joint Venture to any party hereto in reimbursement of expenses incurred by such party in connection with the preparation of the bid for, and securing the award of, the construction contract.

12. It is the intent of the parties hereto that the joint bid contemplated and provided for herein shall be satisfactory and acceptable to each of the parties. If all the parties are unable to agree upon a joint bid, this Joint Venture shall terminate.

13. No party hereto shall sell, assign, or in any manner transfer its interest in the Joint Venture without first obtaining the consent of the other parties hereto.

14. Upon completion of the performance of the construction contract, the Joint Venture shall render a true and correct account to the parties of all expenses incurred on account of such performance and all moneys received as a result thereof, and shall settle and adjust all accounts in connection with the performance of the construction contract, and shall pay to the parties such sums as will result in each of the parties receiving its proportion of all profits arising from the performance of the construction contract, or bearing its proportion of all losses arising therefrom, in accordance with paragraphs 3, 5, and 9 hereof.

15. Subject to the foregoing provisions herein contained, this agreement shall inure to the benefit of, and be binding upon the parties hereto, their successors, trustees, assigns, receivers, and legal representatives, but shall not inure to the benefit of any other person, firm, or corporation.

IN WITNESS WHEREOF, the parties hereto have executed this agreement as of the date first hereinabove written.

ATTEST:

ABC CONTRACTING COMPANY

/s/   Peter Finch
_____
Secretary

/s/   Ralph M. Peters
_____
President

ATTEST:

KLM CONTRACTING COMPANY

/s/   John L. Smith
_____
Secretary

/s/   Arthur K. Martin
_____
Vice President

ATTEST:

XYZ   CONSTRUCTION   COMPANY

/s/   Robert L. Jones
_____
Secretary

/s/   James P. Arthur
_____
President

These instruments are executed by all partners.

A joint venture has the advantages and disadvantages of any other partnership and is entered into by contractors who have mutual trust in each other's construction and financial abilities and honesty.

Bonds

Four types of bonds are usually required by the owner, and four types of bonds are frequently required of subcontractors or suppliers by prime contractors.

Bid bonds are usually required only by owners. This bond, commonly in the range of 20 percent of the amount bid, ensures that the bidder will execute the contract, even if he has second thoughts about his bid after seeing that all other bidders have submitted substantially higher bids. This bond lapses when the contract is executed.

Performance bonds and payment bonds are usually required by owners and frequently required by contractors. These bonds are sometimes combined, and are most often in the amount of the bid. These bonds ensure that the contract is performed and that all wages and accompanying liabilities, and all bills for supplies, materials, and services are paid by the party which incurs these costs, thus protecting the owner or the prime contractor, as the case may be.

Supply bonds are most often required of suppliers by contractors for the same reasons performance and payment bonds are required.

Warranties are usually required by owners, and in turn by prime contractors. Warranties are to ensure the proper performance of equipment, roofs, etc., after the completion of the contract. Warranties are effective for stated periods of time, as specified in the contract documents.

Although a bond is ultimately a responsibility of the bonding company, the terms by which a contractor secures a bond states that all of the contractor's assets are to be exhausted before the bonding company suffers a loss. Bonding companies therefore look carefully at a contractor's financial position before making a commitment to supply bonds for a project. A bond is invariably written as a joint and several agreement by the bonding company and the contractor, whereby each signatory is bound to make good on the whole of the bond, even if the other party becomes bankrupt. In the case of a joint venture between a financially strong contractor and a weaker partner, the stronger partner may insist upon cross-bonding for protection against failure by the weaker partner, or may be able to obtain from the bonding company a limitation on indemnities to the percentage of participation in the joint venture.

The principals of a joint venture will agree in advance upon the bonds which will be required of potential subcontractors and suppliers, and will sign the bid bond at the same time as they sign the bid documents. The other bonds are not obtained until and unless a contract is awarded.

Insurance

There are a number of types of insurance policies which may be required by law or by the specifications, or which may be optional with the contractor.

Workers' compensation insurance is required by law in all jurisdictions, and this requirement is usually recited in the specifications. This insurance provides medical attention, compensation to the injured employee who is sick or disabled because of his employment, and death benefits for surviving family. No proof of negligence by the employer is necessary for an award, and no proof of negligence by an employee will interfere with an award. Workers' compensation insurance is not applicable if the injury is caused by another contractor or subcontractor on the same job. Injuries which occur on floating plant may not fall under this form of insurance.

Public liability and property damage insurance is usually required and the minimum limits are fixed by specification. It covers all liability for injury or property damage to people and property not employed by the insured.

Fire insurance and equipment floaters protect the contractor against fire losses on buildings or against accidental damage to equipment. This protection generally extends only to the contractor's property. There are usually substantial deductibles on equipment floaters.

Care and custody insurance protects the insured against damage or loss of property, such as owner-furnished equipment, which is in his care and custody.

Fidelity bonds are a form of insurance against embezzlement of funds by employees.

Builder's risk insurance protects the insured against damage to permanent work or temporary facilities, such as cofferdams, caused by fire, flood, earthquake, landslide, etc. There is usually a large deductible. Some specifications require this insurance and establish minimum coverage.

Special forms of insurance may be required for operations adjacent to railroads and may be required by specification to protect the owner and engineer against liabilities which may arise from the contract. A few contractors have insured themselves against the costs of quantity overruns on fixed price contracts, and against the possible failure of the owner to obtain necessary funds to complete the work, but policies of this type are very seldom used. Some contractors place *completed contract* insurance to protect them against claims after other insurance policies are terminated.

The principals of a joint venture will agree in advance upon the insurance which should be placed if their bid is successful. Certificates of insurance for required insurance need not be furnished until after a contract award.

Some owners elect to provide *wrap-up* insurance, covering workmen's compensation, perhaps public liability and property damage, and perhaps builder's risk insurance.

ESTIMATORS' MEETING

Separate estimates are usually made by all or several of the various partners. To arrive at a mutually agreed upon estimate on which to base the bid, an estimators' meeting is held prior to bid submittal. At this meeting estimates are compared, construction methods discussed, and adjustments made in each participating partner's estimate for anything left out of his estimate, any mistakes, or any cost changes resulting from improved methods of construction that are brought out at the meeting by other estimators. The estimators then arrive at a mutually satisfactory estimate that is approved by each estimator present. This estimate is then presented to the principals, who have a meeting to set the markup. Many principals attend the estimators' meeting since this is one of the best methods for them to become acquainted with the job.

The chairman of the estimators' meeting is a representative of the sponsoring company, and as well as presenting his company's estimate, he moderates the opinions of the representatives of the other companies in order to arrive at an agreement.

In some cases the estimators do not come to an agreement. When this happens, a tabulation of the various estimates is presented to the principals, who then must reach an agreement on bid price without having the advantage of an agreed estimate.

At the estimators' meeting there is a limited time available to reach an agreement; so in order to facilitate the comparison of estimates and take advantage of each estimator's approach, the sponsoring company sends out to each partner, as soon as the information is available, some or all of the following information:

1. Forms for comparison of estimates
2. The format for preparation of the estimate
3. Wage rates applicable to the job (See Table 18.8.)
4. Payroll burdens
5. Insurance rates
6. Bond rates
7. Cost of power and "up-and-down" charges for the power supply
8. Plugged, or bogie, prices for materials and subcontracts (See Table 18.4.)
9. List and cost of used construction equipment, if available
10. Any geological or materials engineering reports on the project site or material sources

The partners' estimators also often circulate each firm's quantity takeoffs so that agreement on quantities is reached on these before the estimators' meeting. This leaves more time to resolve other differences.

The use by all estimators of the same plugged, or bogie, prices for permanent materials and subcontract items results in a prearranged agreed estimated cost for these items and saves time and effort at the prebid estimators' meeting. These plugged prices do not have any effect on the final bid estimate as the cost ultimately used for these items is obtained from quotations submitted by suppliers and subcontractors just prior to bid submittal. Upon receipt of these quotations the sponsoring company substitutes the quotation prices for the plugged prices and adjusts the total estimate and the total bid accordingly. Therefore, the use of these plugged prices allows each estimator to arrive at a total estimate of cost without spending time on items that will change upon the receipt of final quotes. This practice also saves time in the prebid estimators' meeting as each estimate and the agreed estimate will contain identical amounts for these items.

The first order of business at the meeting is for each partner or estimator to read his figures as tabulated on the estimate comparison sheets, while the other estimators write them down (see Table 20.1). After the figures have all been read, discussions on cost, progress, methods, and equipment are held, and each partner or estimator makes such adjustments to his estimate as he feels are proper after due considerations of the factors presented in the discussions. After adjustments have been made to each estimate, further adjustments are made to arrive at a mutually agreed upon estimate to be used as a basis for bid preparation (See Table 20.2).

Estimators should prepare their estimates in accordance with the format sent out by the sponsoring company and should completely fill out all forms transmitted to them for comparison of estimates. Estimates should be arranged so that unit cost and total cost of each operation are readily available for comparison. The estimators should know all the details of their estimate thoroughly and should have the construction schedule and construction procedure in an orderly form both in their estimates and in their memories. It happens too often that meetings are delayed while one estimator tries to find in his estimate the answer to some simple question asked him by another estimator. The estimate form used in Chap. 18 arranges all information so that it is readily available. If the sponsoring company sends out a different estimating format than the one the partner's estimator used, the estimator should have enough flexibility to adjust his estimate to suit the format.

TABLE 20.1 **Sierra Tunnel Bid Preparation: Estimate Comparison Sheets**

	ABC Contracting Co.		KLM Contracting Co.		XYZ Construction Co.		Agreed cost
	Original	Revised	Original	Revised	Original	Revised	
Direct cost:							
Labor	$12,976,949	$13,469,429	$14,961,000	$12,967,000	$14,555,000	$13,987,000	
Supplies	5,465,365	5,465,365	6,092,000	5,717,000	5,501,000	5,501,000	
Permanent materials	3,860,180	3,860,180	3,745,000	3,745,000	3,795,000	3,795,000	
Subcontracts	1,883,740	1,828,002	1,884,000	2,050,000	1,884,000	1,938,000	
Total direct cost	$24,186,234	$24,622,976	$26,682,000	$24,479,000	$25,735,000	$25,221,000	$24,622,976
Plant and equipment cost:							
Freight and move-in	$ 148,650	$	$ 124,000	$ 124,000	$ 102,000	$ 102,000	
Labor	439,080		196,000	196,000	477,000	477,000	
Equipment cost	6,569,975		5,527,000	7,879,000	6,310,000	6,060,000	
Plant invoice	2,148,201		2,056,000	2,056,000	2,034,000	2,034,000	
Subtotal	$ 9,305,906		$ 7,903,000	$10,255,000	$ 8,923,000	$ 8,673,000	
Salvage	(2,811,550)		(2,273,000)	(3,260,000)	(2,076,000)	(2,431,000)	
Rent	130,115						
Total P&E cost	$ 6,624,471	$ 6,624,471	$ 5,630,000	$ 6,995,000	$ 6,847,000	$ 6,242,000	$ 6,624,471
Indirect cost:							
Labor	$ 1,480,499	$	$ 2,146,000	$ 1,666,000	$ 2,075,000		$ 1,750,000
Other cost	378,000		484,000	484,000	665,000		450,000
Insurance	183,000		137,000	137,000	137,000		183,000
Property tax	274,500		142,000	142,000	179,000		275,000
Bond	177,768		199,000	199,000	174,000		180,000
Total indirect cost	$ 2,493,767	$ 2,493,767	$ 3,108,000	$ 2,628,000	$ 3,230,000	$ 3,230,000	$ 2,838,000
Labor escalation	1,315,980	1,118,583	1,170,000	1,243,000	1,132,000	1,203,000	1,138,172
Total Cost	$34,620,452	$34,859,797	$36,590,000	$35,345,000	$36,944,000	$35,896,000	$35,223,619

615

TABLE 20.1 Sierra Tunnel Bid Preparation: Estimate Comparison Sheets (*Continued*)

	ABC Contracting Co.		KLM Contracting Co.		XYZ Construction Co.		Agreed cost
	Original	Revised	Original	Revised	Original	Revised	
Summary of labor:							
Direct	$12,976,949	$13,469,429	$14,961,000	$12,967,000	$14,555,000	$13,987,000	$13,469,429
Plant and equipment	439,080	439,080	196,000	196,000	477,000	477,000	439,080
Indirect	1,480,499	1,480,499	2,146,000	1,666,000	2,075,000	2,075,000	1,750,000
Subtotal	$14,896,528	$15,389,008	$17,303,000	$14,829,000	$17,107,000	$16,539,000	$15,658,509
Labor escalation	1,315,980	1,118,583	1,170,000	1,243,000	1,132,000	1,203,000	1,138,172
Total	$16,212,508	$16,507,591	$18,473,000	$16,072,000	$18,239,000	$17,742,000	$16,796,681
Direct cost comparison:							
Items 1–7, tunnel excavation	$13,217,528	$13,710,008	$16,538,000	$14,040,000	$14,605,000	$14,067,000	
Item 8, shaft excavation	361,460	361,460	362,000	362,000	457,000	457,000	
Items 9–14, concrete	10,442,246	10,386,508	9,615,000	9,910,000	10,506,000	10,530,000	
Items 15–17, consolidation grouting	165,000	165,000	167,000	167,000	167,000	167,000	
Total direct cost	$24,186,234	$24,622,976	$26,682,000	$24,479,000	$25,735,000	$25,221,000	$24,622,976
Breakdown of item 9:							
Aggregate	$ 1,560,660	$ 1,504,922	$ 1,356,000	$ 1,522,000	$ 1,426,000	$ 1,480,000 (1)	
Cement loss and admix	77,700	77,700	66,000	66,000	83,000	83,000	
Remove vent line and retimber	417,360	417,360	345,000	345,000	451,000	451,000	
Primary cleanup	947,940	947,940	988,000	988,000	1,193,000	1,193,000	

Invert concrete	1,035,630	1,035,630	955,000	1,000,000	1,078,000	1,048,000 (4)	
Arch concrete	2,434,230	2,434,230	2,829,000	2,913,000	2,508,000	2,508,000	
Low-pressure grout	415,140	415,140	386,000	386,000	463,000	463,000	
Final cleanup	509,490	509,490	464,000	464,000	504,000	504,000	
Weekend maintenance	62,160	62,160	63,000	63,000	75,000	75,000	
Total	$ 7,460,310	$ 7,404,572	$ 7,452,000	$ 7,747,000	$ 7,781,000	$ 7,805,000	$ 7,404,572
Statistics:							
Excavation, main tunnel:							
Portals used	Adit		Outlet		Adit		Adit
Number of single-heading days, rubber	43						71
Number of single-heading days, rail	34		480		90		31
Number of alternate-heading days, rail	51						51
Number of double-heading days, rail	184				210		169
Number of men, single heading, rubber	57						57
Number of men, single heading, rail	86		100		104		97
Number of men, alternate heading, rail	86						97
Number of men, double heading, rail	152		168		168		160

TABLE 20.1 Sierra Tunnel Bid Preparation: Estimate Comparison Sheets (*Continued*)

	ABC Contracting Co.		KLM Contracting Co.		XYZ Construction Co.		Agreed cost
	Original	Revised	Original	Revised	Original	Revised	
Average wage rate/shift	$204		$203		$203		$204
Concrete, main tunnel:							
Progress:							
Number of days, invert pour	38		40		42		40
Number of days, arch pour	89		95		90		90
Size crew:							
Men per day, invert pour	106		100		110		106
Men per day, arch pour	100		90		100		100
% of overbreak	40		20		30		35
Total no. of mo. to complete all construction	31		39		33		31
Yearly increase in labor cost	10.0%		8.0%		8.0%		8.5%
% Escalation on total labor	8.8%		6.8%		6.6%		7.3%
Maximum capital required to finance job	$7,600,000						
Date last capital will be returned	25 months						
Interest on capital invested	$1,798,000						

TABLE **20.2**   **Sierra Tunnel Bid Preparation Adjustments to Sponsor's Estimate to Conform to Agreed Estimate**

Original total estimate	$34,620,452	
Increase crew size, tunnel excavation	+492,480	(direct labor)
Reduce volume of overbreak concrete	−55,738	(subcontract)
Increase indirect labor	+269,501	(indirect cost)
Increase indirect, other cost	+72,000	(indirect cost)
Increase property tax	+500	(indirect cost)
Increase bond	+2,232	(indirect cost)
Increase labor escalation per above increases; decrease escalation rate	−177,808	(labor escalation)
Revised total estimate	$35,223,619	

The agreed estimate is somewhat of a compromise, but advantage should be taken at the prebid meeting of the best construction methods presented and the best estimates of cost. Individual estimators should relinquish any pride of authorship and be willing to adjust their estimates to take into account any method or cost which is better than they originally had in their estimates. In the majority of cases the final agreed estimate will reflect the best method of construction and the closest approximation to the actual cost of doing the work. It should always be remembered that the final agreed estimate is only an estimate and that the final cost of doing the work will vary to some degree from this estimate. (Contracting is more hazardous than most businesses, since the contractor is the only manufacturer who prices his product before it is manufactured.)

After the estimators have arrived at an agreed cost, they should discuss the amount of capital that will be required from the joint venture partners to finance the construction, and the amount of interest charges on this invested capital. These two amounts should be presented to the principals so that they can be included in the markup figure. For an illustration of the method of determining the amount of capital required and the interest expense on this capital, refer to Table 18.21 for the Sierra Tunnel estimate.

The estimators should discuss the overall concept of the job to determine whether it is a good job to bid or one with a high percentge of risk. Jobs may be classified risky for the following reasons:

1. If payment is on a lump sum or lineal foot basis, all the costs resulting from encountering bad ground, water, support requirements, etc., are the contractor's responsibility.

2. The job may be in a remote area and may therefore present problems in securing competent workers, maintaining access, and supplying camps and living accommodations.

3. The job may extend for a considerable period of time with a large exposure to raises in wages, changes in fringe benefits, subsistence increases, changes in working hours and working conditions, and strikes.

4. Relief from unforeseen conditions may not be present in the specifications, and the job may be in ground of such a geological nature that it is difficult to determine the type of rock that will be encountered.

5. Equipment or supervision may not be readily available because many similar jobs are being bid or are under construction.

All conclusions reached at the estimators' meeting should then be presented by each estimator to his principal. Table 20.2 shows the log of changes prepared by the sponsor's chief estimator.

PRINCIPALS' MEETING

After the estimators have concluded their meeting, the results of this meeting are submitted by the estimators to their respective principals. The principals then hold a meeting to determine the amount of markup to be added to the agreed estimate to arrive at the bid price.

Principals should be well informed about the various aspects of the job. This job information is secured by discussing the work with their estimators, reviewing the results of the estimators' meeting, and studying the bidding documents and job description. Many principals attend the estimators' meetings where exposure to different construction methods and their estimated costs will assist them in evaluating the proposed project.

The markup determined by the principals is composed of the following elements:

1. A proportional share of each participating company's home office expense.
2. Interest cost of the capital required to finance the work.
3. Contingency. In setting the contingency, the principals start with the job risk information that is presented to them by their estimators and with knowledge they have formulated by job review or attendance at the estimators' meeting. The amount reflects their evaluation of whether the markup should contain allowance for job contingency or estimating contingency or both.
   a. Job contingency, discussed in the preceding section, includes a measure of the reliability of the forecast of ground conditions on which the estimate is based. The contingency for variations in ground conditions should be taken into account more on tunnel construction than on other types of construction, since the progress and cost of tunnel excavation is largely dependent on the kind of material that will be encountered. Even if the tunnel line has been completely drilled, it is impossible for any geologist or engineer to determine accurately how ground

conditions will affect tunnel driving. Compared with other types of construction, a large percentage of markup is used in tunnel work to offset the risk of unforeseen conditions.

b. Estimating contingency. On most divisions of estimated cost, such as materials, supplies, and subcontracts, the estimator has previous job records or quotes that he can use to check his estimate. With labor, he must rely to a large extent on his own judgment, since he will not receive any quotes on labor, and labor cost varies greatly from job to job. For this reason, and also because labor in tunnel construction represents a larger percentage of the total cost than in any other type of construction, profitable tunnel work is more reliant on estimating judgment and should therefore carry a larger estimating contingency than other types of construction.

4. Profit. In setting the job profit, the principal is influenced by the following job requirements: the amount of the contractor's capital required to finance the job; how many of the contractor's key personnel will be required to supervise the job, and for how long; and the estimated salvage value of the plant and equipment. This salvage value is a matter of judgment, since it must be predetermined for a future date when the equipment may or may not be marketable, depending on whether it will fit any tunnel being advertised. Sometimes tunnel equipment does not sell for several years after it has been placed on the market and therefore requires storage and care, and property taxes must be paid on the equipment until it is sold or junked.

If the bid documents have not been circulated and signed prior to the prebid meeting, they are signed at the principals' meeting. At the conclusion of this meeting, the markup and other bidding instructions are given to the chief estimator of the sponsoring company, who has the responsibility for the bid preparation.

For the Sierra Tunnel bid, it is assumed that the principals agreed that interest rates would decline to an average rate of 15 percent; they further agreed on a 20 percent markup on total costs, as follows:

Total before financing costs		$35,223,619
Financing costs (at 15 percent)		1,524,438
Total cost		$36,748,057
Home office expense* at 1½ percent	$ 551,221	
Profit at 18½ percent	6,798,390	
Total markup		7,349,611
Bid price before adjustments for quotations		$44,097,668

---

* The home office expense accrues to the sponsor, but since it is a part of profit, no home office expense will be earned if the job fails to show a profit.

It was decided that money would be shifted from later work to earlier work to provide early revenue. This is regarded as insurance in case interest rates fail to decline as anticipated.

EFFECT ON THE BID PRICE
IF USED EQUIPMENT IS AVAILABLE

To illustrate the effect that used plant and equipment bears on the bid price, it is assumed that one of the partners of the Sierra Tunnel joint venture has used plant and equipment available. This equipment could be furnished to the project for 50 percent of its replacement value. Use of this equipment would prompt the following changes, which could result in a revised bid price:

1. Plant and equipment net charge to the job would be reduced. If job salvage of the used plant and equipment is valued at 25 percent of replacement cost, the net charge to the job for plant and equipment is 25 percent where it is utilized. Table 20.3 illustrates the effect this may have on the bid price.

2. Job maintenance of equipment would increase, as it costs more to maintain used equipment than new equipment. The total cost of maintaining the equipment as tabulated in the Sierra Tunnel estimate is approximately $2,170,000. If it is assumed that the maintenance cost will increase 15 percent, additional maintenance labor and supplies will amount to $325,500.

TABLE **20.3   Sierra Tunnel Bid Preparation:**
Reduction in P&E Net Charge

	Sierra Estimate Agreed Cost for New Equip.	Revised for Used Plant and Equip.
Freight and move-in	$148,650	$148,650
Labor	439,080	439,080
Plant and equipment:		
Equipment that can be replaced with used	$5,000,000	$2,500,000
Remainder of equipment purchased new	1,569,975	1,569,975
Plant invoice that can be replaced with used	1,000,000	500,000
Remainder of plant invoice purchased new	1,148,201	1,148,201
Subtotal, equipment and plant invoice	$8,718,176	$5,718,176
Total	$9,305,906	$6,305,906
Salvage	(2,811,550)	(1,311,550)
Rent	130,115	130,115
Total	$6,624,471	$5,124,471
Net job savings on plant and equipment		$1,500,000

3. Estimated production rates would decrease on this hard-rock, fast-production job because the older equipment will not have all the improvements that have been incorporated in the newer equipment. After tunnel equipment has driven several jobs, obsolescence generally makes it more economical to buy new equipment. The equipment being considered here has been used on only one job and is relatively modern, so the loss in production is estimated to be approximately 5 percent. Hence, increased cost of tunnel excavation due to decreased production is estimated to be:

Cost of tunnel excavation labor and supplies for item 1, plus labor for items 2 and 3	$11,777,820
Cost of loss of production, 5%	$ 589,000
Cost of additional overhead:	
18 months @ 0.05 = 0.9 month at $90,000	81,000
Total cost of loss of production	$ 670,000

4. Interest costs would decrease. The decrease can only be approximated because there is not time to rework the cash-flow analysis for this purpose before bid time. The cash-flow analysis will be reworked after a decision is reached about whether or not to purchase used equipment. The reworked cash-flow analysis will then reflect all the agreed charges in costs, job duration, interest rates, and markup.

For the present purpose, it is found that the first cost of plant and equipment is reduced by one-third, from $8,718,176 to $5,718,176. By examination of the cash-flow analysis, it is seen that the maximum capital required is at the time when $7,182,000 has been spent for plant and equipment. If this plant and equipment valued at $7,182,000 were reduced in cost 33 percent because of used equipment to $4,788,000, then cash requirements would be reduced a corresponding amount, or $2,394,000. Similarly, interest would be reduced by shortening the time period during which the capital would be required and decreasing the money required per month. Again, by examining the cash forecast, it is seen that plant and equipment expenditures amount to 50 percent of the cash required. If these expenditures were reduced by one-third, the total interest could be reduced by one-third of 50 percent, or by 17 percent. The reduction in interest costs will approximate 17 percent of $1,798,000 or $306,000, if the interest rate is 18 percent, or $255,000, if the interest rate is 15 percent.

Table 20.4 summarizes the effect the use of the used equipment would have on the Sierra Tunnel bid.

In theory, with all other parts of the estimate being the same and with a consistency of markup, the contractor who has modern used equipment partly amortized can always submit a lower bid than one who has to purchase new equipment. In practice this does not occur, since tunnel construction methods may cause much larger variations in cost than the savings from used

TABLE 20.4   **Sierra Tunnel Bid Preparation:
Effect of Used Equipment on Bid Price**

	Sierra Tunnel Agreed Bid	Theoretical Bid with Used Equipment
Agreed total cost	$35,223,619	$35,223,619
Changes in cost due to used equipment:		
Reduction in plant and equipment		(1,500,000)
Increased cost of maintenance		325,500
Increased cost due to slower progress		670,000
Total before financing costs	$35,223,619	$34,719,119
Financing costs (at 15 percent)	1,524,438	1,269,438
Total cost	$36,748,057	$35,988,557
Markup at 20 percent	7,349,611	7,197,711
Bid price	$44,097,668	$43,186,268

equipment. For example, on the Sierra Tunnel, which is a relatively simple job, it was seen that the use of different methods would have resulted in a variation of $2,022,000. (See Table 18.9). If one adds 20 percent markup to this, the total variation is $2,426,400. Most tunnel jobs are more complicated than the Sierra Tunnel and could be constructed by many more methods, with resulting greater cost differences. Furthermore, progress and crew sizes used vary so much among estimators that these factors also cause large differences in estimated cost, and there is no consistency in markup among contractors. Therefore, although a contractor with used equipment has an advantage in bidding, he may not be the low bidder because of other factors.

In the present case it was decided that the joint venture will bid the Sierra Tunnel job on the basis of purchasing new equipment, but that the joint venture may elect to purchase some items of used equipment if awarded the contract. This later decision will depend in part upon the outcome of another tunnel bid which the same joint venture will be submitting one week after the Sierra Tunnel bid, since there is not sufficient used equipment in a good state of repair and readily available to supply both jobs.

BID PREPARATION

General

It is the responsibility of the sponsoring company's estimator to prepare the final bid. The period immediately before bid submittal is a very hectic period. In the short time available, the estimator must spread all cost and markup to individual items, check all quotations and make quotation adjustments as necessary, prepare the bidding papers in final form, check all extensions and additions, and transmit the bid to the bid opening agency.

Spreading Cost

There often are discussions by owners and engineers about "balanced" and "unbalanced" bids without a good understanding of what a balanced bid is. In preparing a tunnel estimate, the direct cost is estimated for each bid item, but the other divisions of cost—namely, plant and equipment cost, camp cost, indirect cost, and escalation—are estimated for the total job and must then be divided or spread among the individual bid items.

A *balanced bid* is one on which these other divisions of cost and profit are spread to the individual bid items which cause this cost and profit to occur. To arrive at a truly balanced bid, a very detailed study is necessary, and this study would require much more time than there is available between the prebid meeting and bid submission time. Because of this time limitation, it is impossible to submit such a bid on heavy construction work. Cost spreading is therefore done by various approximations, with reliance on the experience of the estimator. The methods of spreading vary among companies, and since the human element is involved, no two estimators with similar cost to spread would spread the cost in the same manner.

The easiest method of cost spreading (but the one which takes the least recognition of where cost occurs) is to divide the total direct costs into the total bid price to secure a factor by which each direct unit cost is multiplied to get the unit bid price. By this method plant and equipment cost, indirect cost, labor escalation, etc., are spread against many items which would not incur this cost, which results in an *unbalanced bid*. If time becomes critical, every contractor must resort to this simplified method in order to submit the bid by the deadline. Some contractors, in order to simplify bid submission, use this method on every bid, but it is not recommended except in critical bid submission conditions.

The preferred cost-spreading method is to use experience and judgment in arbitrarily placing cost against the proper items so that as good a unit cost is produced as is possible in the time available. This will not be a truly balanced bid but only an approximation. True unit cost is very important if the final pay items overrun or vary in quantity from the estimated quantities shown on the bid comparison sheets, since the final pay amounts will then reflect the true cost of the work. For instance, if a tunnel is bid where it is hard to determine how much rock will be supported and how much of the tunnel will be driven without supports, spreading can be done so that the cost of driving an unsupported tunnel is placed against tunnel excavation and the additional cost of driving the tunnel in supported ground is placed against supports. Then, if the contractor drives a tunnel that is mostly unsupported, the final pay quantities charged to the owner will reflect the lesser cost of that type of tunnel work. On the other hand, if the tunnel is supported through the majority of the ground, the contractor will receive payment comparable with the cost he incurred.

By using this cost-spreading method:

1. Plant and equipment cost after salvage is spread against the major bid items that cover the type of work which caused its purchase.

2. Indirect cost is spread primarily against prime contract work, with only a small percentage placed against purchased materials and subcontracts.

3. Camp construction and operation is spread against prime contract work, as a subcontractor must reimburse the prime contractor if he uses the camp.

4. Contingencies are spread against high-risk items.

5. Labor escalation is spread against those items that involve labor and are scheduled in the years that the labor escalation occurs.

6. Markup is spread primarily against prime contract work with a small percentage against purchased materials and subcontracts. Markup should also be spread to round out bid unit prices so that the extensions are even dollar amounts to facilitate their use. In spreading markup, the unit bid prices can be rounded out so that the extensions will always be in even dollar amounts and cents will not have to be tabulated.

7. After bid unit prices are established, they should be reviewed by the estimator and adjusted to be sure that the contractor is protected against high-risk items. For instance, if the roof bolts overrun, work may be delayed, which will cause extra cost to the contractor. The estimator may therefore decide to increase the bid price on this item and lower the bid price on another item. At this point in bid preparation, the estimator has a fixed bid sum to adhere to, and any increase in one item necessarily means a corresponding decrease in other items or item.

These seven steps represent the *preferred spread.* There is seldom time to complete all these steps, so approximations are made and reliance placed on the experience of the man directing the spread.

Table 20.5, the spread sheets on Sierra Tunnel, is given as an example of one method of spreading cost. The spread sheets have been set up so that they have an arithmetical check. If all totals posted are computed by multiplying the quantity by the unit price and then the columns are all added down and checked by adding across, arithmetic errors should be eliminated.

In the Sierra Tunnel spread of plant and equipment, cost for the special shaft excavation equipment was placed against the shaft excavation item, concrete equipment was spread against concrete, and all other plant and equipment cost was distributed to these same bid items.

Indirect cost was spread only to prime contract items, on the basis of direct costs.

Labor escalation was spread against concrete and excavation in accordance with the labor escalation computation.

Markup was spread at the rate of 5 percent to materials and subcontracts and the remainder at the rate of 27 percent to prime contract work. This combination of rates works out to the agreed 20 percent overall markup, plus financing costs, included here for convenience.

Bid prices were not adjusted for high-risk items, as none occur in this tunnel. If the tunnel were in soft ground, more cost might have been placed against tunnel supports in order to reimburse the contractor for delays in tunnel driving. Bid prices for the tunnel and shaft excavation group of bid items were increased to provide early revenue. Bid prices for the concrete group were decreased to offset the increases in the tunnel driving group. The amount shifted was the whole of the markup on the concrete group. The money shift was purposely accomplished by shifting markup to avoid the necessity for adjustments to costs, since cost-accounting systems compare actual costs with budgeted costs, and are not concerned with markups.

To be competitive in the bidding and to give the owner a lower price for the work, some contractors spread cost heavily to items done early in the contract. This gives the contractor early money, reduces his financing costs, which in turn reduces his interest expense, and permits a lower markup. This type of spreading may work to the advantage of both the owner and the contractor.

The owner often recognizes the large sum of money that is required by the contractor to finance the work and to relieve the contractor of some of this financing includes a pay item for mobilization which partly reimburses the contractor for this expense. Such payment clauses often specify reimbursement for 75 percent of the equipment cost when this equipment arrives at the jobsite and is paid for by the contractor. This mobilization item helps the contractor to finance the work, but is to his disadvantage if quantities overrun. If the contractor places his equipment cost in the mobilization item, the bid is "unbalanced" in that the bid prices for the various items do not include any charges for equipment write-off.

If there is a major overrun in quantities, the job will be extended in time, and the equipment will be older and have more hours of use, with a resulting decrease in salvage. This causes the net cost of equipment to the job to increase. But since equipment is a fixed sum under mobilization, the contractor is not reimbursed for this added equipment cost. If equipment write-off is spread to the direct cost items, then as the work increases the reimbursement for its additional use is included in the payment for the additional quantities. To express this in another manner, placing equipment cost in mobilization fixes the equipment write-off, irrespective of the amount of work required.

A contractor makes his own quantity takeoffs to check the accuracy of bid quantities furnished by the owner's engineers. If the bid-item quantities are incorrectly stated so that an underrun will occur when final pay quantities are determined, the contractor must increase the unit prices in order that he may properly recover his cost, since a large percentage of the cost spread over these items is fixed and varies only slightly with the amount of work performed. Such fixed costs will not be recoverable on the quantities of work bid but not done or paid for unless the contractor increases the spread in propor-

TABLE 20.5 Sierra Tunnel Bid Preparation
Spreading Plant and Equipment, Indirect, and Escalation Costs to Arrive at Total Cost

Bid Item	Description	Pay Quantity	Agreed Direct Cost		Agreed Plant and Equipment Cost		Agreed Indirect Cost		Agreed Escalation Cost		Agreed Total Cost	
			Unit	Amount	Unit	Amount	Unit	Amount	Unit	Amount	Unit	Amount
Tunnel excavation:												
1	Tunnel excavation	431,000 yd³	$ 27.66	$11,921,460	$ 10.21	$4,440,510	3.61	$ 1,555,910	$0.82	$ 353,420	$ 42.30	$18,231,300
2	Steel sets	2,290,000 lb	0.61	1,396,900			0.08	183,200			.69	1,580,100
3	Timber	400 MBM	765.00	306,000			99.98	39,992			864.98	345,992
4	Roof bolts	12,000 lin ft	4.60	55,200			0.60	7,200			5.20	62,400
5	Drill exploratory holes	300 lin ft	16.12	4,836			2.11	633			18.23	5,469
6	Drill grout holes as aid to tunnel driving	2,400 lin ft	6.93	16,632			0.91	2,184			7.84	18,816
7	Grout as aid to tunnel driving	400 ft³	19.60	7,840			2.56	1,024			22.16	8,864
	Total			$13,708,868		$4,400,510		$ 1,790,143		$ 353,420		$20,252,941
Shaft excavation:												
8	Shaft excavation	5,300 yd³	68.20	361,460	19.13	101,389	8.91	47,223			96.24	510,072
Concrete:												
9	Tunnel lining	111,000 yd³	66.71	7,404,810	18.50	2,053,500	8.73	969,030	7.06	783,660	101.00	11,211,000
10	Shaft concrete	1,900 yd³	118.04	224,276	36.96	70,224	15.43	29,317			170.43	323,817
11	Cement for concrete and low-pressure grout	163,000 bbl	16.00	2,608,000							16.00	2,608,000
12	Steel reinforcing in shaft	285,000 lb	0.47	133,950							0.47	133,950
13	Embedded anchor bolts	3,200 lb	4.30	13,760			0.56	1,792			4.86	15,552
14	Embedded vent pipe	600 lb	3.25	1,950			0.42	252			3.67	2,202
	Total			$10,386,746		$2,123,724		$ 1,000,391		$ 783,660		$14,294,521
Consolidation grouting:												
15	Drilling grout holes	4,000 lin ft	10.00	40,000							10.00	40,000
16	Consolidation grouting	10,000 ft³	10.00	100,000							10.00	100,000
17	Cement for grouting	5,000 sacks	5.00	25,000							5.00	25,000
	Total			$ 165,000								$ 165,000
	Total from extensions*			$24,622,074		$6,625,623		$ 2,837,757		$1,137,080		$35,222,534
	Total from agreed estimate and instructions from principals†			$24,622,976		$6,624,471		$ 2,838,000		$1,138,172		$35,223,619

* Totals are adjusted from agreed direct cost to give even unit prices.

† It is not necessary to bid exactly the agreed amount as bid extensions can seldom be made to total the agreed bid.

TABLE 20.5 Sierra Tunnel Bid Preparation
Spreading Markup and Making Quotation Adjustments to Bid Items

Bid Item	Description	Pay Quantity	Total Cost Unit	Total Cost Amount	Markup* Unit	Markup* Amount	Bid Price Unit	Bid Price Amount	Quote Adjustments Unit	Quote Adjustments Amount	Adjusted Bid Price Unit	Adjusted Bid Price Amount
Tunnel excavation:												
1	Tunnel excavation	431,000 yd³	$ 42.30	$18,231,300	$ 11.42	$4,922,020	$ 60.27	$25,976,370	$+0.54	+232,740	$ 60.81	$26,209,110
2	Steel sets	2,290,000 lb	0.69	1,580,100	0.19	435,100	1.00	2,290,000			1.00	2,290,000
3	Timber	400 MBM	864.98	345,992	233.66	93,464	1,230.00	492,000			1,230.00	492,000
4	Roof bolts	12,000 lin ft	5.20	62,400	1.40	16,800	7.50	90,000			7.50	90,000
5	Drill exploratory holes	300 lin ft	18.23	5,469	4.92	1,476	26.00	7,800			26.00	7,800
6	Drill grout holes as aid to tunnel driving	2,400 lin ft	7.84	18,816	2.12	5,088	11.00	26,400			11.00	26,400
7	Grout as aid to tunnel driving	400 ft³	22.16	8,864	5.99	2,396	32.00	12,800			32.00	12,800
	Total			$20,252,941		$5,476,344		$28,895,370		$+232,740		$29,128,110
Shaft excavation:												
8	Shaft excavation	5,300 yd³	96.24	510,072	26.00	137,800	136.65	724,245			136.65	724,245
Concrete:												
9	Tunnel lining	111,000 yd³	101.00	11,211,000	27.25	3,024,750	101.00	11,211,000	-1.29	-143,190	99.71	11,067,810
10	Shaft concrete	1,900 yd³	170.43	323,817	46.04	87,476	171.00	324,900			171.00	324,900
11	Cement for concrete and low pressure grout	163,000 bbl	16.00	2,608,000	0.80	130,400	16.00	2,608,000			16.00	2,608,000
12	Steel reinforcing in shaft	285,000 lb	0.47	133,950	0.02	5,700	0.50	142,500			0.50	142,500
13	Embedded anchor bolts	3,200 lb	4.86	15,552	1.31	4,192	5.00	16,000			5.00	16,000
14	Embedded vent pipes	600 lb	3.67	2,202	0.99	594	4.00	2,400			4.00	2,400
	Total			$14,294,521		$3,253,112		$14,304,800		$-143,190		$14,161,610
Consolidation grouting:												
15	Drilling grout holes	4,000 lin ft	10.00	40,000	0.50	2,000	10.50	42,000			10.50	42,000
16	Consolidation grouting	10,000 ft³	10.00	100,000	0.50	5,000	10.50	105,000			10.50	105,000
17	Cement for grouting	5,000 sacks	5.00	25,000	0.25	1,250	5.25	26,250			5.25	26,250
	Total			$ 165,000		$ 8,250		$ 173,250				$ 173,250
	Total from extension			35,222,534		8,875,506		44,097,665		89,550		44,187,215
	Total from agreement			$35,223,619		$8,874,049		$44,097,668		$ 88,952		$44,186,620

* Includes 20 percent markup and financing costs.

tion to the amount each item will underrun. This increase in unit prices will increase the bid total and, if the bid-quantity error is not caught by every contractor, there will be an error in bid preparation, with the low bidder receiving the work without sufficient compensation. Owner's engineers too often do not understand this principle of bidding and do not pay sufficient attention to the accuracy of the bid quantities. In one case, the engineers threw out all the original bids submitted, then completely redesigned the job with smaller quantities but readvertised the work with the original bid quantities. This made it impossible for the contractors to submit a bid that would add up to a proper total and still allow the contractor proper reimbursement. When such flagrant errors in bid quantities occur, it forces the contractor to place as much cost as possible against lump sum items or items for which quantities are correctly stated.

Spreading cost is an important item in bids, worthy of more attention than is generally given it by estimators, owners, and engineers.

BIDDING DOCUMENT PREPARATION AND ADJUSTMENTS FOR LOW QUOTES

As soon as the spread sheets are finished and bid prices computed, these units and extensions should be transferred to the bidding schedule. Because the Sierra Tunnel project contains only 17 bid items, this posting procedure is comparatively short and simple, but many tunnel jobs are quite complicated, with hundreds of bid items. Posting of the bid may take hours and, in some cases, days. All permanent-materials prices and estimated subcontract prices are still subject to last-minute quotation adjustments, so one, two, or three large bid items are left open, with prices not written in the bidding papers; in this way, all final quotation adjustments can be made in these bid items. When this is done, all extensions written in should be added, so that this total will be available when the total adjusted bid price is determined at the bid closing time.

It is very important that the bid prices be checked, that the extensions be checked, and that the additions be checked on the bidding papers. The unit bid price controls; if the extensions are wrong, they are corrected by the owner's representative and a new total developed. Some contracts are lost because sloppy extensions or erroneous unit prices are written in the bid. The need for accuracy in this portion of the bid cannot be overstressed.

Keeping track of the latest quotes on materials and subcontracts is very important, as the receipt of a final low quote may make the difference between having the low bid or being just another bidder. There are many ways of keeping track of these quotations so that adjustments can be readily made to the bid. The estimator should decide on the method that suits him best and be familiar with its procedure. Table 20.6 shows an "over-and-under" sheet

TABLE 20.6 Sierra Tunnel Bid Preparation: Adjustments to Bid Price for Final Permanent-Materials and Subcontract Quotes

Bid Item	Description	Bid Quantity	In Estimate Incl. Sales Tax		Lowest Quotation Incl. Sales Tax		Plus	Minus	Name of Supplier or Subcontractor
			Unit	Amount	Unit	Amount			
2	Steel sets	2,290,000 lb	$ 0.48	$1,099,200	$ 0.56	$1,282,400	$ 183,200		Steel Fabrication Co.
3	Timber	800 MBM*	320.00	256,000	314.50	251,600		$ 4,400	North Woods Lumber Co.
4	Roof bolts	12,000 lin ft*	1.60	19,200	1.57	18,840		360	True Steel Co.
9	Aggregate in tunnel lining	247,143 tons*	6.10	1,507,572	6.44	1,591,601	84,029		Crusher Aggregate Producers
10	Aggregate in shaft concrete	3,944 tons*	6.10	24,058	6.44	25,399	1,341		Crusher Aggregate Producers
11	Cement for concrete and low-pressure grout	163,000 bbl	16.00	2,608,000	14.77	2,407,510		200,490	Kiln Cement Co.
12	Steel reinforcement in shaft	285,000 lb	0.44	125,400	0.425	121,125		4,275	Sutter Steel Co.
13	Embedded anchor bolts	3,200 lb	1.60	5,120	1.55	4,960		160	Sutter Steel Co.
14	Embedded vent pipe	600 lb	1.10	660	1.55	930	270		Sutter Steel Co.
15	Drilling grout holes	4,000 lin ft	10.00	40,000	9.56	38,240		1,760	Hardrock Drilling Co.
16	Consolidation grouting	10,000 ft³	10.00	100,000	11.11	111,100	11,100		Hardrock Drilling Co.
17	Cement for grouting	5,000 sacks	5.00	25,000	6.00	30,000	5,000		Kiln Cement Co.
	Aggregate in tunnel plug	990 tons*	6.10	6,039	6.44	6,376	337		Crusher Aggregate Producers
	Total			$5,816,249		$5,890,081	$ 285,277	$211,445	
	Net change due to quotations						73,832		
	Markup, 20%						14,766		
	Subtotal						$ 88,598		
	Adjust bond premium						354		
	Additions to agreed bid price						$ 88,952		
	Agreed bid price						44,097,668		
	Adjusted bid price						$44,186,620		

* Takeoff quantity.

NOTE: All bid prices and extensions had been written in the bid except for item 1, tunnel excavation, and item 9, tunnel concrete, so the bid adjustment was made in these two items. If the bid is successful and a budget estimate is prepared, each adjustment should be worked back into the estimate.

which is a very simple and effective method by which a running total of adjustments may be kept and the total adjustment to the agreed bid may be determined at any time. In using this method, the lowest quote for any item is posted and compared with the amount in the agreed cost, and the adjustment is tabulated. Any new quotation received is compared with the one posted; if it is higher, it is put aside; if lower, the one on the over-and-under sheet is erased and the new one posted, extensions are made, and the new adjustment is made. This is a foolproof method, as each quote needs to be handled only once.

The receipt of quotes must be halted sufficiently ahead of bid closing time to allow time for entering the final adjusted bid figures on the bidding documents and to allow time for the transportation of the bidding documents to the point of submission. In order to reduce this bid closing time to a minimum, a temporary office can be used near the location where the bid must be submitted. Hotel rooms are commonly used for this purpose. To close the bids, the adjustments to the total bid price must be totaled on the over-and-under sheet. The bid items left open on the bidding papers have to be adjusted to reflect this adjusted total, the adjusted bid total has to be written in, these computations should be checked, the bid envelope containing all the bidding documents must be sealed, and the bid must be carried and deposited at the location stated in the bidding instructions. After bid closing time, bids are opened by the owner's representative, read, and the apparent low bidder announced. The owner's representative then checks all bids for any errors in extensions or additions and determines the low bidder. As previously stated, the unit prices control and the apparent low bidder may not be low when the extensions are verified and the additions are checked on all the bids.

BUDGET ESTIMATE

If the bid is successful, the sponsoring company's estimator should then prepare a budget estimate. This budget estimate is used by the job management to compare the production and cost achieved during construction with the production and cost that was estimated for the work. The estimator should also make these comparisons so that he can reflect this work experience in his next estimate.

To prepare the budget estimate, the bid estimate must be adjusted to the agreed cost estimate. The detailed sheets in the estimate must also be changed to reflect the final quotations for materials and subcontracts.

The cash forecast should also be adjusted for major changes made at the prebid meeting by actual spread of plant and equipment, final markup, and final quotations so that it can be used for job control.

Glossary of Heavy Construction Terminology

A Line Dimensioned line in a tunnel beyond which rock projections are not permitted.

Adit A short length of tunnel driven from the surface and connecting to the main tunnel. Often driven to enable more headings to be driven in the main tunnel; sometimes used to provide permanent access to the tunnel.

Air Gun A steel cylindrical-shaped vessel with gates for receiving concrete and with air connections and discharge pipe connections. After the gun is charged, the concrete is forced through the discharge pipe by air pressure.

Air Legs Air-activated, pipe jacks used for positioning light drills and maintaining pressure on the drills during the drilling operation. Also known as *feed legs*.

Air Locks Pressure chambers located in the air bulkhead of a tunnel driven under air pressure. Men, material, and equipment pass in and out of the tunnel through these locks. While they are in the locks, the air pressure is slowly adjusted to that which will next be encountered in the direction of travel. Separate locks are provided for personnel.

Alternating Crew A crew of miners who are switched from one heading to another in a double-heading location; use of an alternating crew has the advantage of keeping drilling and mucking operations continuous with one crew of men for each operation.

AN/FO An explosive consisting of a mixture of ammonium nitrate prills and a sensitizing agent. Originally the agent was fuel oil, hence the initials.

Arch Pour All the tunnel lining except the bottom section; also, the process by which this lining is poured. See Invert pour.

Arrow Diagraming A method of making a construction schedule using arrows to indicate the sequence of operations or events.

B Line Pay line for excavation in a tunnel.

Back Top of the arch of a tunnel.

Backfill Grout   See Low-pressure Grout on p. 640.

Baloney Cable   Heavy insulated electric drag cable often used to supply power to movable cableway towers, cranes, shovels, and tunnel equipment.

Bank Yard   A cubic yard of material in its natural state. See Loose Yard, Compacted Yard.

Barrage   A diversion dam.

Barring and Wedging   Procedure used to remove the last portion of unsuitable rock from the foundation surface of a dam.

Bathtub   The concrete container on a dumpcrete truck.

Bellboy   Man who transmits voice and signal directions to the operator of a cableway or crane to control the spotting of a concrete bucket.

Bentonite Slurry trench   A deep trench excavated through common material and filled with bentonite slurry to maintain the sides of the excavation.

Bid Quantities   Quantities published on the bid schedule for unit-price items. Bid quantities are extended by unit prices to arrive at the total bid price for each unit-price bid item.

Bifurcation   The dividing of one penstock into two or more smaller penstocks, each of which is then connected to a turbine scroll case.

Block Holing   Secondary breakage of blasted rock accomplished by drilling holes in the rock and then using explosives in these holes.

Blocking   (1) Wood blocks installed between the lagging (or steel sets or a collar bracing) and rock surfaces of a tunnel to transfer stress to the supports. (2) Any timbers with a thickness greater than 4 in spanning the space between sets.

Block-out Concrete   The concrete surrounding gate guides and other similar members that is not placed when the main structure is poured. After completion of the main structure, the steel gate guides are placed and aligned, and the block-out concrete is then placed.

Blowout   Quick release of air from a heading driven under air pressure.

Bogie Prices   See Plugged Prices.

Bootleg   Amount of drilled rock in the tunnel face not broken by the explosion.

Bottom Heading   The excavation of the bottom half of a tunnel after the top half has been excavated for the full tunnel length.

Break   Fragmentation of solid rock as a result of the explosion of an explosive mixture in drill holes at the heading.

Breast Boarding   Partial or complete, braced support across the tunnel face which holds back soft ground during tunnel driving. Breast boards are braced to steel sets or to a shield or are supported with hydraulic jacks mounted on a breast-board jumbo.

Bucket Car   A railcar fabricated to carry as many as four 8-yd^3 buckets, usually equipped with a catwalk at the proper elevation to expedite hooking and unhooking the concrete buckets from a cableway. If an automatic hook is used on the cableway, no catwalk is needed.

Bucket Dock   A landing platform for concrete buckets with a length that permits the cableway to land a bucket when its traveling tower or towers are at any travel position. The bucket dock is located adjacent to and slightly below the transfer track, permitting the transfer car to discharge concrete directly into the bucket.

Budget Estimate   A bid estimate adjusted for late quotes and other bid adjustments, used as a reference for control of the cost of constructing a project.

Built-in-place Forms   Forms with such irregular shape that they must be built in place and hand-tailored to fit the opening. An example is the first lift forms for dam concrete which are placed against the irregular rock surface forming the dam foundation.

Bulkhead Forms (Dam) Forms which separate the concrete in a dam into a series of individual blocks or monoliths.

Bulkhead Forms (Tunnel) Forms placed in the top and sides of the pour between the forms and the tunnel surface. These forms are placed in a vertical plane to form a vertical construction joint.

Bulkhead Pour Concrete placed in a short section of tunnel between two bulkhead forms or between another bulkhead pour and a bulkhead form.

Bull Gang Crew of laborers usually used on the day shift to lay rail track, install pipes, clean up the tunnel, and perform any other necessary work that is located in the tunnel but not at the face.

Burn-cut Drilling pattern of large relief holes drilled horizontally in the center of the face to provide space for the expansion of rock broken by a blasting agent.

Burn-cut Drill Large drill used for the large-diameter holes in the center of a tunnel face.

Button Line A line used on older-type cableways to control the spacing along the track cable of the slackline carriers.

Buy-out Prices Prices actually paid for equipment and materials required during the construction of a project.

Cable Sag Amount of track-cable sag at the center of the span of a cableway. Six percent of the span is a rough approximation of its extent.

Cable Tender Man who takes care of the power cable running to the mucker.

Cage An enclosed platform, similar to an elevator car, used to transport men and miscellaneous material up and down a shaft.

California Switch A portable platform that rides on a track in a tunnel and is used for passing cars and trains. It has space for two or more tracks, crossovers, and switches, and has sliding and tapered end rails that ride on main-line rails.

Candy Wagon Service truck.

Car Passer Portable tracks used for switching cars. The portable tracks are of sufficient length to accommodate one car when the portable tracks are resting on top of the main-line track. The rails are knife-edged on each end so a car can be placed on them from the main track. These tracks and car can be rolled on other rails to the side a sufficient distance to clear the main tracks.

Carriage The frame and wheel sheaves that are pulled along the track cable of a cableway by the conveying line. Framed beneath the wheels are sheaves over which pass the hoisting line which suspends and controls the elevation of the cableway hook.

Change House A building containing toilets, showers, lockers, and clothes-drying facilities for the tunnel crews.

Cherry Picker Any frame capable of lifting a car vertically so a train may pass beneath it; used for passing cars. Also a light crane.

Chuck Tender Assistant to the driller. (Before the use of modern jibs, the driller's assistant had to help position the drill steel while starting the hole in the face; hence the term *chuck tender*.)

Cleanup The operation of removing loose muck, water, and other material from a tunnel prior to concreting.

Collar Bracing Struts installed between sets, capable of taking compressive forces.

Compact Yards Measure, in cubic yards, of the space occupied by compacted backfill or a compacted embankment.

Conveying Line Endless line used for

transporting concrete buckets to pour location.

Crown Bar   Timber or other member installed in tunnel roof above sets.

Curb   Small pour made on the bottom and at each side of a tunnel and generally located outside the minimum concrete line; used as a reference point and support for invert and arch pour in the tunnel.

Curing Compound   Liquid sprayed on a freshly poured concrete surface which prevents water from evaporating from the surface, thus retaining sufficent water in the concrete to assure complete hydration of the cement.

Cut   A drilling pattern in the tunnel face which provides relief for an explosive charge. See also Burn-cut.

Day Shift   Shift that extends from 8:00 A.M. to 4:00 P.M. See Swing Shift, Graveyard Shift.

Delay Cap   Exploder used in the tunnel face; cap has delay provision so that charges are fired in rotation. Delay caps are used so that the rock will be exploded into an area into which it can expand.

Dental Excavation   Excavation performed in pockets, shafts, or drifts in the foundation of a dam to remove unsuitable foundation rock.

Dewatering   Removal of water from within soil or rock, generally via a well-point system. Compare Unwatering.

Diamond Cut   Inclined, short drill holes near the center of the tunnel face, so arranged that when the first shots are exploded in the round, a diamond-shaped wedge of rock is removed, which allows relief for the remaining rock when the delay exploders set off the charges in the other drill holes.

Disposal Area   Area where excavated materials can be wasted.

Doorknob Estimate   A rough approximation of the construction cost of a construction contract prepared by a few man-hours of work.

Down-the-hole Drill   Percussion drill that follows the bit down the drilled hole.

Double Heading   Two headings in a tunnel that can be driven from one plant location. Double headings are possible, for example, when a tunnel passes through and daylights in a depression or canyon or when two headings are driven from one adit.

Double-heading Crew   Crew of men, located at one entrance to the tunnel, who drive two headings simultaneously from this entrance.

Dredger Tailings   Piles of gravel left by gold dredges as they dredged river channels or old river channels.

Drift   A short section of tunnel not connecting to another tunnel nor daylighting at one end.

Drifter   A heavy drill for drilling nearly horizontal holes in the tunnel face. Drifters require support from jibs.

Drifter-type Drill   Percussion drill with bore diameter of from 3½ to 5 in.

Drill   Machine that transmits striking or rotating force (or both) to the drill steel and bits; used to bore holes which can be loaded with powder in rock tunnels.

Drill Jumbo   Movable frame on which drill positioners, jibs, and drills are mounted.

Drill Positioner   Mechanical control for moving, rotating, and controlling jibs.

Dry Packing   Pea gravel or other material blown in between the lagging and the excavated surface to furnish support when solid lagging is used in a supported tunnel.

Dummy Activities   Nonevent arrows required by PERT or CPM scheduling to enable the schedule to show proper sequence of activities.

Dumpcrete   Concrete hauled from the mixer to the pour in open-top-tank-type truck bodies. These tanks may or may not be equipped with paddles for agitating the concrete.

Elastic Fractionation   Method of removing lightweight materials from concrete aggregates, based on the principle that the denser aggregates will rebound a greater distance when they are dropped on a steel plate.

Elephant Trunk Steel pipe or rubber hose used to drop concrete vertically into a pour.

Embankment Yard Volume of fill material in an earth- or rock-fill embankment in its compacted state.

Endless Line Another name for the conveying line of a cableway. See Conveying Line.

Equipment Availability Percentage of time that construction equipment will be available for use. The remainder of the time it will be out of service for repair or servicing.

Equipment Leveling Scheduling the construction of noncritical work items for periods when construction equipment will be available.

Erector Arm Swing arm on boring machine or shield, used for picking up supports or liner segments and setting them in position.

Escalation The amount that labor, material, and equipment costs will increase during the life of a construction project.

Exploratory Drilling Drilling performed to secure information about dam foundations, borrow pits, quarries, tunnels, etc.

Face Vertical wall at the end of the excavation in a tunnel.

Feed Legs See Air Legs.

Feeler Hole Hole driven ahead of the excavation for exploratory purposes.

Fineness Modulus Measure of the fineness of sand computed from its screen analysis.

Finish Jumbo Traveling support for the concrete finishers when repairing concrete and applying curing compound to arch concrete.

First-stage Concrete The principal structural concrete. The term is used only when second-stage concrete is to be poured at a later stage of construction.

Floor Self-propelled platform used in tunnel excavation.

Force Majeure Clause in the specifications that relieves the contractor of financial responsibility for certain events not under his control. Examples include acts of God, war, earthquakes, floods, and strikes.

Forepoling Members of timber or steel underneath the second rib from the face and above the first rib from the face to provide support around the excavated area. See Spiling.

Form Contact Square feet of concrete surface area that must be formed.

Form Hardware Form ties, she-bolts, snap ties, etc., used to hold forms in position.

Form Oil Oil painted on the contact surface of concrete forms to expedite form stripping and protect the formed surface.

Form Ratio Square feet of formed surface required per cubic yard of concrete.

Form Traveler Traveling frame used to strip, collapse, transport, and erect full-circle or arch forms.

Foundation Cleanup The removal of all loose material and water from a rock surface before it is covered with concrete.

Foundation Slope Correction Excavation that must be performed to comply with specifications that require that the rock surface under the impervious core of an earth- or rock-fill dam have an upstream slope in relation to the dam axis when looking downstream. This excavation must be done in areas where the rock slopes downstream.

Full-circle Pour Process by which the complete concrete lining in a tunnel is poured in one operation. See Arch Pour, Invert Pour.

Full Face Tunnel blasted out to full bore size with each round.

Gabion A group of medium-sized (8 to 24 in) rock particles contained in a wire netting.

Gantry Jumbo Drill jumbo which has an open space in the center large enough for muckers, cars, and locomotives to pass through; supported by separate rails, tracks, or tires.

Getting off the Rock The completion of the first concrete lift on a dam foundation.

Gob Hopper Hopper located under the mixers in a concrete-mixing plant into which the mixers dump the concrete. Concrete buckets, cars, or trucks can then be filled with concrete from the gob hopper, freeing the mixer for charging and mixing.

Gouge Finely ground up material found in fault areas.

Grasshopper Traveling frame with hinged ends, riding on separate wide-gauge rail track and used for car passing. Cars are run up on this frame to permit trains to pass beneath.

Graveyard Shift Shift that extends from midnight to 8:00 A.M. See Day Shift, Swing Shift.

Green-sheet Estimate Another name for a doorknob estimate. See Doorknob Estimate.

Grizzly A heavy screen with large openings used to scalp off large rocks from quarried rock or from concrete aggregates. Often grizzlies consist of spaced rails.

Ground The material through which a tunnel is driven, whether it is solid rock or running mud.

Grout A mixture of water and cement or water, cement, and sand. See Low-pressure Grout, High-pressure Grouting. Chemical grouts are available which do not use portland cement.

Grout Curtain A continuous zone of grouted rock, formed by pumping grout into a regular pattern of drill holes.

Gunite Pneumatically placed mortar, used to prevent air slacking of a tunnel's excavated surface; also applied as a support either directly on the excavated surface or over other supports in order to strengthen them.

Gut A section of unsuitable foundation material that passes through the foundation of a dam. See also Main Gut.

Headframe Tower built over a shaft to support the ropes for raising and lowering men, equipment, and material in the shaft.

Heading Space adjacent to the face of a tunnel where the excavation crew works.

Head Tower The tower of a cableway that contains or is near the drums and hoist that operate the cableway.

Heavy Ground Ground requiring strong supports.

Heavy-media Separation A method of eliminating lightweight particles in concrete aggregates by flotation in a mixture of water, magnetite, and ferro-silicon.

High Line Slang terminology for a cableway.

High-pressure Grouting Consolidation grouting used to strengthen rock or to cut off water inflows.

Hog Rod A steel turnbuckle rod installed horizontally across a shield-driven tunnel and fastened on each side to the liner plate; it remains in position until the shield tail void is filled with pea gravel and grouted; it prevents deflection of the liner plate.

Hoist Line A line that controls the elevation of the cableway or crane hook.

Hole Burden The horizontal distance from the rock face of an excavated area or quarry to the blast holes.

Hole Pattern Horizontal spacing of drill holes.

Hole Spacing The horizontal distance

between drill holes measured parallel to the rock face.

Holing Through Reaching that point in excavation when a tunnel face daylights at a portal or meets another face, giving a continuous tunnel.

Hydraulic Monitor A nozzle that discharges a jet of high-pressure water used as a means of removing common materials from a dam foundation and as a means of improving rock consolidation in a rock-fill dam.

Impervious Zone The clay or silt zone of an earth- or rock-fill dam which provides a water barrier.

Inclined Drilling Drilling so that the holes are parallel to the vertical face of a quarry. Also used to describe nonvertical exploratory, grouting, or drainage holes.

Incompetent Ground Ground that requires support when a tunnel is excavated through it.

Invert On a circular tunnel, the invert is approximately the bottom 90° of the arc of the tunnel. On a square-bottom tunnel, it is the bottom of the tunnel.

Invert Pour Process by which the invert of a tunnel is poured separately. See Arch Pour.

Invert Struts Compression struts installed across the invert to resist inward movement of tunnel ribs.

Jackhammer Light, hand-held percussion drill.

Jib Horizontal support for drifter drills bolted to a jumbo. These arms allow mechanical positioning of the drill at the tunnel face. Some jibs allow rotation of the drills so that side holes and lifters can be drilled closer to the final grade. Also an articulated extension on the tip of a crane boom.

Jim Crow Manual or hydraulic rail bender.

Joe Magee Cableway A construction cableway used during the start of a construction project to provide temporary access to one dam abutment. Often it consists of a cable rigged across the canyon from a logging winch on a tractor.

Jumbo Any traveling frame used to support other items, such as drills, men, pipe for conveying concrete, conveyors, etc.

Jump Set Tunnel rib (support) installed between two previously placed ribs.

Lagging Wood or other structural members spanning the area between the ribs.

Laitance Dehydrated cement mortar that forms on the top of a concrete lift.

Left on the Table The difference between the low bid and the second bid.

Lift The vertical thickness of concrete placed during one pouring operation.

Lift Differential The vertical distance between the top of the lowest and highest blocks (monoliths) of a concrete dam or between adjacent monoliths.

Lifter Holes Holes drilled in the bottom of the invert to fragment the rock in the invert. Because drills cannot be positioned to drill on the invert line, they have to be pitched slightly down in order to prevent tights.

Lift-starter Forms Noncantilever forms used after the first pour has been made on a dam foundation, and until sufficient height has been poured in each monolith to provide clearance for cantilever forms.

Liner Plate Pressed-steel plate with turned-back edges on each side containing bolt holes so that the plates can be bolted together to support the arch sides, and in some cases the invert of a tunnel. Liner-plate sections can also be fabricated from steel plates and structural members, or they can be of cast iron, cast steel, or precast concrete.

Long-hole Drilling Procedure of drilling long holes to contain powder for breaking the rock in underground excavation. As an illustration, a tunnel may be excavated by first driving a pilot drift near the center of the tunnel. Then approximately 10-ft lengths of this drift are enlarged to full tunnel dimensions at intervals of 100 to 200 ft. These excavated areas are used as working chambers to drill holes in the remaining rock to the next enlarged section. Holes are then loaded and exploded, with relief for the explosion provided by the pilot drift.

Loose Yards Volume of space occupied by material that has been excavated and deposited in a haulage vehicle. Properly loose cubic yards.

Low-pressure Grout Grout pumped through the concrete lining under low pressure to fill any voids between the concrete lining and the tunnel surface or between steel tunnel liners and backfill concrete.

Luffing Cableway A cableway with two fixed towers which provide transverse hook coverage by drifting the tops of the towers as the lengths of the side guys are adjusted.

Main Gut Track cable of a cableway.

Main-line Jumbo Drill jumbo that travels on the main rail line; has man platforms on the sides that can be folded back when not in use. After drilling out the round, the jumbo must be moved back from the face to a passing track before the mucker can be moved in.

Mass Concrete Concrete in massive structures which has a low form ratio.

Miscellaneous Concrete Concrete in slender structures that has a high form ratio.

Mixed Face The juncture at which the tunnel passes through both rock and unconsolidated material and both are exposed simultaneously at the face.

Mole A tunnel-boring machine.

Monkey Slide Inclined trackage equipped with a winch-controlled car.

Monolith A block of concrete in a dam separated from the others by bulkhead forms.

Muck Broken rock or other material produced at the face of a tunnel by the excavation process.

Mucker Machine for loading muck into haulage units.

Mud A slang term for concrete.

Mudcapping A method of performing secondary rock breakage wherein an explosive is placed on the surface of large rocks and covered with mud.

Mudsill Continuous horizontal member installed along the side of a tunnel at the lowest excavation line to support sets or take the load of the supports from the wall plate.

Multiple Drift Excavation of a tunnel by driving two small drifts along each side of the tunnel, which allow the side support to be placed. A top drift is then driven and widened out slowly to take the roof support.

Multiple-sheave Drive A double set of sheaves controlling the travel of a conveying line of a cableway.

Multiple-shift Schedule The operation of a construction project for two or more shifts per day.

Nipper Laborer on the drilling crew who handles drill steel replacements and bits and does general housekeeping.

Nontelescopic Forms Full-circle or arch forms that, when stripped and collapsed, will not pass through other forms erected in place.

Off-highways Trucks with heavy bodies and with such a payload capacity that the weight on each tire will exceed that permitted for highway usage.

Old-man Slackline Carriers The two slackline carriers at each end of a cableway that are very seldom moved except when the carriage is brought into a tower for servicing.

On-highways Trucks constructed with light bodies and with payloads limited to weights that will permit the load on each tire to conform to highway vehicle codes.

On-site Cost   Cost generated or money expended at the site of a construction project.

Optimum Bank Height   The height that a bank of rock or common material must be to secure maximum production from a power shovel.

Over-and-under Sheet   Method of tabulating prebid quotes for materials and subcontracts.

Overbreak   Excavation that extends past pay lines.

Overburden   The mantle of earth overlying rock formations.

Panning   The channeling of water down the sides of a tunnel behind metal strips before concreting. After concrete is placed, these channels can be grouted off.

Parallel Cableway   A cableway that has movable head and tail towers which maintain the same relative position on parallel or concentric runways.

Payload   The weight of the material that can be carried by each haulage unit.

Pay Quantities   Quantities of work performed for each unit bid item. Pay quantities times unit bid price will equal the final payment for each unit-price bid item.

Penetration Rate   The length of blast hole produced by a drill in a unit time interval.

Pickup Point   The position under a cableway where the cableway transfers concrete buckets or where a concrete bucket is filled by a transfer car.

Pipe Jumbo   Traveling support for the discharge line from a concrete placer.

Plenum Method   Excavating a tunnel under air pressure.

Plugged Prices   Approximate prices used in an estimate for permanent materials and subcontracts until quotes are received from suppliers and subcontractors.

Portal   Point where tunnel enters the earth's surface.

Portal-in   To start the tunnel excavation at the portal face.

Pot   A steel pressure vessel used for transporting concrete. At the point of placement, the concrete is ejected from the pot by use of compressed air.

Powder Factor   The pounds of explosives required to break one bank cubic yard of rock.

Pozzolan   Finely ground shale, pumice, volcanic material, or reclaimed flue dust produced by the combustion of coal (fly ash), used to disperse cement in concrete and in some instances to react with the free alkali in the cement.

Precedence Diagramming   One method of preparing a logic diagram for a construction schedule, using rectangles or other shapes to represent work items, all set forth in a logical sequence of events.

Prepack Concrete   Concrete formed by first placing the coarse aggregates and then pumping sand, cement, grout, and a dispersing agent into the pre-placed aggregates.

Press Weld   Brand name of an air-activated concrete placer.

Proportional Slackline Carriers   Wheeled carriers traveling on the track cable of a cableway; the carriers support the operating ropes. Proportional carriers space themselves across the span by differences in travel-speed gearing. They are actuated by the movement of the endless line.

Pull   Length of rock broken when a round is shot. In estimating, it is assumed to be at least 1 ft shorter than the length of the drill holes.

Pumped Storage Power Projects   Power-generating installations that can function as hydrogenerating units and as motor-driven pumps. Such plants are located between two reservoirs that are at different elevations. During periods when excess power is available from thermal plants, they pump water from the lower reservoir to the higher. During periods of maximum power demands, the water flow is reversed and they act as generators.

Punch List   A list of deficiencies in an almost completed construction unit or project.

Pusher   A power unit used to assist in the loading of a scraper.

Quarry Face   The nearly vertical surface of unbroken rock in a rock quarry.

Radial Cableway   A cableway with a fixed tower and a movable tail tower. The movable tower travels on a track which is a constant distance from the fixed tower.

Raise   To excavate a shaft from a tunnel below instead of sinking it from the top down.

Random Fill   Fill placed in an embankment which may consist of a variety of materials.

Reactive Aggregate   Concrete aggregate that contains siliceous ingredients that combine with the free alkali in the cement to form an alkali silica gel. Over a period of time, the concrete may be broken by the osmotic swelling of this gel.

Reclaim Tunnel   A tunnel under stockpiled material, usually containing a conveyor belt. The purpose of the tunnel is to facilitate loading out of the stockpiled material.

Refraction Seismograph   An instrument for determining the characteristics of rock by measuring the velocity at which seismic waves travel through the material.

Rescreening Plant   An aggregate-screening plant located adjacent to or on top of the concrete-mixing plant. The rescreening plant is used to correct any size degradation caused by rehandling the various sizes of aggregates.

Resteel   A slang term for reinforcing steel; also called rebar.

Rib   Section of tunnel between the spring line and back. Also an H- or I-beam steel support. See Set.

Rock Cleanup   The cleanup of a rock surface on which concrete is to be placed.

Rock Ladder   Fabricated steel framework placed under the discharge of aggregate stockpiling conveyor belts to restrict the free fall of the aggregate to between 2 and 3 ft.

Rock Necklaces   Large rocks that have been drilled and strung on wire ropes so that when they are used to close a water-diversion opening, they can better resist the erosive force of the water.

Rock Throw   The distance broken rock will be thrown when the round is exploded.

Rod Busters   Men who install reinforcing steel.

Roof Bolt   Bolt equipped with an expandable anchor at one end and a nut and washer at the other. Installed in drilled holes to tie rock together. Also known as *rock bolt*. May also be grouted in place.

Round   The length of tunnel that is drilled and shot in one operation.

Run-in   Flow of material into the tunnel from the tunnel face or from the tunnel circumference.

Runner   Operator of a shovel or crane.

Running Ground   Material of fine particles which flows into the excavated area while the tunnel is being excavated. Water is usually present in the material.

Sand Hog   Miner who works in a tunnel driven under air pressure.

Scalping   The removal of oversized rock or boulder particles from the remainder of the material.

Screed   A slip form used on an invert pour; anything used to strike off a concrete pour.

Scrubber   The equipment installed on diesel engines for the purpose of dissolving or neutralizing exhaust gases.

Second-stage Concrete   Concrete placed within a partially completed structure to embed turbine scroll cases or similar items.

Secondary Breakage The breaking of oversized rocks into smaller particles in quarry operations.

Self-raising Forms Forms equipped with powered jacks. After concrete has been poured and set within the forms, the jacks raise the forms to the proper position to form the next concrete lift.

Set One structural support for the sides and roof of a tunnel; used more when tunnel supports were built up with wood members (post and crown). See Rib.

She-bolt Form hardware used to anchor vertical forms. A she-bolt is threaded externally to accept clamps which bear upon the forms and internally to accept the threads of spacer rods. All parts are salvageable after a pour is completed except the spacer rods.

Shell The zones of an earth- or rock-fill dam which support the impervious zone.

Shield Steel tube shaped to fit excavation line of the tunnel and used to provide support for the tunnel, provide space within its tail for erecting supports, protect the men excavating and erecting supports, and if breast boards are required, provide supports for them.

Shifter Foreman of heading crew in a tunnel.

Shotcrete Method for applying concrete lining on tunnel arch immediately after excavation. This lining would replace tunnel supports and also serve as the permanent lining. Shotcrete concrete contains coarse aggregate as distinguished from earlier methods utilizing fine aggregate only.

Shrink The reduction in volume of material when earth or rock is compacted. It may be measured from bank yards or from loose yards.

Silicosis A lung disease caused by breathing dust from rock drills over a long period of time. Rocks with high silica content are the most harmful.

Single-heading Crew Crew of men located at one entrance to the tunnel who only excavate one heading at any one time from this entrance.

Sink-float A method for removing lightweight portions from concrete aggregates. The lightweight particles are floated off using a mixture of water, magnetite, and ferrosilicon. See Heavy-media Separation.

Sinking The conventional method of excavating a shaft from the top down.

Skeleton Bay Future generating bay of a hydro powerhouse in which only the first-stage concrete has been poured.

Skinner Operator of a tractor or similar type of equipment.

Skip Metal boxlike container used to hold debris cleaned from the rock surface of a dam foundation or from concrete surface areas. Also used in hoisting muck from a shaft.

Slackline Carrier Wheeled frame riding on the track cable of a cableway and supporting the operating cables.

Slashing The operation of enlarging a pilot shaft to full diameter by conventional methods.

Slick Line Section of the discharge line from a concrete placer that is embedded in the fresh concrete during arch or full-circle pour.

Sliding Crown Bar Crown bar that is slid forward over sets as mucking advances, in order to protect the workmen and support the back.

Slip Form Form that is raised as a concrete pour progresses so that there is a continuing formed surface within which concrete can be poured.

Slurry An explosive consisting of a mixture of water, ammonium nitrate prills, a sensitizer, and other ingredients. Also a bentonite slurry to control groundwater as an aid to excavation.

Slusher Train A muck train composed

of a locomotive and articulated muck cars. During the mucking cycle, the locomotive furnishes power to operate a scraper which runs along the top of the train; as the lead car is loaded by the mucking machine, the scraper drags the material back from the lead car and loads the others.

Snap Ties Form hardware consisting of small-diameter rods spanning between two wall forms to hold them at a constant distance from each other. When the concrete has set they can be broken at two deformed points located back from the surface of the concrete.

Soffit Form Form supporting the bottom of a concrete beam or other suspended concrete member.

Sonotube Form Circular concrete form made from treated paper.

Spider Temporary support which maintains a penstock in circular shape while it is being transported and erected.

Spiling Supporting wood or steel members driven from under the second set and over the first set from the face; supports the roof of the tunnel. See Forepoling.

Spotter The individual who instructs a truck driver where to spot his truck for loading or for dumping.

Spread Sheet Estimate form used to distribute the various general construction costs to individual bid items.

Spring Line Point on the side of a tunnel where tunnel or set starts curving into the arch.

Squeezing Ground Material that exerts heavy pressure on the circumference of the tunnel after excavation has passed through that area; may cause inward movement of sets, rising of invert, breaking of blocking, etc., and may necessitate the installation of invert struts and jump sets and further excavation to relieve pressure.

Stemming Inert material packed in a

drilled hole on top of the explosive charge to contain the force of the explosion in the rock.

Step Excavation Excavation done on the abutments of a dam to provide level foundation surfaces that form steps up the abutments.

Stoper Drill designed for drilling overhead holes.

Strut Compression support placed between tunnel sets.

Subdrilling The distance drilled beneath the desired surface of an excavated area or quarry floor to ensure that all rock will be broken to the required elevation.

Subsistence The daily living allowance paid to workers in remote areas.

Subsistence Area Area defined by labor agreement in which the employer must pay living allowances to his employees.

Surface Grouting Grouting the rock below the surface of the dam foundation to improve its characteristics. Also known as consolidation grouting.

Surface Moisture The amount of water on the surface of concrete aggregates.

Swell Increase in rock volume when it passes from the solid to the broken state.

Swing Shift Shift that extends from 4:00 P.M. to midnight. See Day Shift, Graveyard Shift.

Tail Tower The tower of a cableway that does not contain the operating machinery. See Head Tower.

Takeoff Quantities Quantities of work calculated from the contract drawings by the contractor.

Telescopic Form Full-circle or arch form that, when stripped and collapsed, will pass through other forms that are erected in place in a tunnel.

Tetrahedron A concrete block with four sides used to close off a diversion opening. The shape of the block allows block interlocking which permits it to resist the transporting effect of flowing water.

Tetrapod   A concrete block with four legs that interlock when used to close a diversion opening. See Tetrahedron.

Tie Rods   Tension members between sets to maintain spacing. These pull the sets against the struts.

Tight   Projection of rock in a tunnel past the A line.

Top Heading   Upper section of the tunnel. Also, a tunnel excavation method where the complete top one-half of the tunnel is excavated before the bottom section is started.

Top Heading and Bench   Method of tunnel driving where the top heading is carried about 1½ times the length of one round ahead of the lower heading, or bench.

Track Cable   The main cable of a cableway on which the cableway carriage travels. It is made with a track lay that presents flat surfaces of the wire strands of the cable to resist the wear of the wheels of the carriage. The general range in cable diameter is from 1½ to 4 in, depending on the capacity and span of the cableway.

Transfer Car   A railroad car which may be self-propelled or may be pulled by a locomotive, containing concrete compartments of the same size as the concrete buckets used on a cableway. Used to transfer concrete from the mix plant to the bucket dock of a cableway. At the bucket dock the car dumps each compartment into a concrete bucket hooked to the cableway.

Transfer Track   The railroad track running between the mixing plant and a cableway. See Transfer Car.

Travel Line   Another name for conveying line. See Conveying Line.

Tunnel-boring Machine   Machine that excavates a tunnel by drilling out the heading to full size in one operation.

Unclassified Excavation   Bid item terminology used when both rock and common excavation are undifferentiated for payment purposes and are paid for under one bid item.

Unwatering   Removal of surface water, as from excavations, from cofferdams, etc. See Dewatering.

Up-and-down Charge   Charge made by a utility company to furnish, erect, and dismantle a power substation.

Vacuum Process   The process used to cool concrete aggregates by causing surface water to evaporate from the aggregates in a partial vacuum.

Vibroflotation   A patented process of compacting sandy material.

V Cut   Inclined short drill holes near the center of the tunnel face, drilled so that when the first shots are exploded in the round, a wedge of rock is removed, which allows relief for the remaining rock when the delay exploders set off the charges in the other drill holes.

Walker   Shift superintendent in a tunnel.

Wall Plates   Continuous horizontal members installed along the sides of the tunnel at or near the spring line to support arch sets when a top heading or multiple heading is driven. Often two I beams are used for wall plates. After excavating the bottom heading, supports are installed to support the wall plate from the mudsill.

Wellpoints   A series of pipes installed in shallow wells around an area to be dewatered. The pipes are connected to headers and pumps to remove the water.

Work Constraints   Specification or labor contract requirements that specify how construction work must be performed, or the type of labor and working rules that must be used during construction.

Metric Equivalents for U.S. Customary Units

U.S. Unit of Measure	Metric Equivalent	U.S. Unit of Measure	Metric Equivalent
in	25.40 mm	gal	3.785 L
ft	0.3048 m	acre-foot	1234 m³
ft²	0.0929 m²	mi/h	1.609 km/hr
ft³	28.32 L	lb/in²	0.0703 kg/cm²
yd	0.9144 m	lb/ft	1.488 kg/m
yd²	0.8361 m²	lb/ft²	4.882 kg/m²
yd³	0.7646 m³	lb/yd³	0.5933 kg/m³
mile	1.609 km	ft-lb	0.1383 kg-m
acre	0.4047 ha	ft³/min	28.32 L/min
lb	0.4536 kg	yd³/h	0.7646 m³/h
ton	907.2 kg	°F	0.5556 °C

Index

ABOUT THE AUTHORS

The late ALBERT D. PARKER, a registered Civil Engineer in California and a Fellow of the American Society of Civil Engineers, had over 30 years experience in the supervision and estimating of heavy construction projects for Kaiser Companies in the United States and abroad. He also had six years of experience with construction of pipelines, refineries, and harbor facilities in California and Arabia for Standard Oil Company. He was author of the two McGraw-Hill titles now expanded upon in this volume: *Planning and Estimating Underground Construction* and *Planning and Estimating Dam Construction*.

In DONALD S. BARRIE's more than 30 years of experience as a Project Manager and construction executive he has been responsible for 25 miles of tunnel, 7 dams, 6 powerhouses, and airport and highway projects throughout the United States and Canada. The author of two highly regarded books on construction, *Professional Construction Management* and *Directions in Managing Construction*, Mr. Barrie has lectured at Stanford, the University of Wisconsin, and the University of California. A graduate of California Institute of Technology, he is presently Vice President at Raymond Kaiser Engineers.

ROBERT M. SNYDER is a registered civil engineer and holds a Class A contractor's license in California. He has worked for the U.S. Navy, the Corps of Engineers, the Perini Corporation, and most recently for Kaiser Engineers where his title was Chief Heavy Construction Engineer. Today, as an independent consultant, he estimates and bids all types of heavy construction for major U.S. contractors.